DIRTY
WARS

DIRTY WARS

A CENTURY OF COUNTERINSURGENCY

SIMON INNES-ROBBINS

To Carol

First published 2016

The History Press
The Mill, Brimscombe Port
Stroud, Gloucestershire, GL5 2QG
www.thehistorypress.co.uk

© Simon Innes-Robbins, 2016

British Library Cataloguing in Publication Data.
A catalogue record for this book is available from the British Library.

ISBN 978 0 7524 6411 4

Typesetting and origination by The History Press
Printed and bound in Great Britain by TJ International

CONTENTS

ACKNOWLEDGEMENTS

I am most grateful for and very appreciative of the forbearance and support of my colleagues past and present at the Imperial War Museum during the long gestation of this volume, notably in the Department of Research: Suzanne Bardgett, Emily Fuggle, Toby Haggith, Emily Peirson-Webber, Hilary Roberts, Roger Smither and the late Rod Suddaby; and in Publishing: Elizabeth Bower, Abigail Ratcliffe, Madeleine James and Peter Taylor.

The number of people who have contributed to my researches over many years are too numerous to name but I am thankful in particular for the extensive knowledge of Professor Matthew Hughes, Professor Peter Lieven, Professor Martin Thomas, Dr Huw Bennett, Dr Jan-Georg Deutsch, Dr Norrie MacQueen, Terry Charman, William Gee, Guy Robbins and Roger Smither, who have contributed to, read and made detailed and very useful comments on the manuscript.

Above all, I am grateful to my wife, Carol, and to my two sons, Jasper and Toby, for their fortitude, good humour and wonderful support during another long campaign.

THE GUERRILLA THREAT

The commitment of troops to Iraq and Afghanistan ended with ignominious withdrawal between 2011 and 2014 and saw the emergence of the cyber battle-field and of groups such as Boko Haram in Nigeria and Islamic State in Syria and Iraq. This has shown that there is a pressing need for a better understanding by the public, politicians and the military of guerrilla warfare and of the meth-ods that can be employed to defeat it. Guerrilla warfare is arguably the future of conflict and in response armies need to function within the increased tempo and complexity of today's multi-media information age. Since involvement in Iraq and Afghanistan the American and British militaries have been looking at ways to avoid getting enmeshed in any further counterinsurgency campaigns, although these operations have generated a renaissance in doctrine and writing on counterinsurgency.[1]

In an attempt to respond to the modern information environment, the British Army formed 77 Brigade in early 2015 as part of a major restructuring of the military organisation and development of a new doctrine of 'Integrated Action' which would enable the United Kingdom to fight in the information age. The brigade took its name, insignia and inspiration from the Chindits, an unconventional force that fought behind Japanese lines in Burma during the Second World War. The new unit was created specifically to draw together existing and developing capabilities essential to meet the challenges of modern conflict and to learn the lessons of Iraq and Afghanistan. It would comprise soldiers who are skilled not just in the use of conventional weapons but also in 'soft power', notably social media, such as Facebook, Instagram and Twitter, new technology and the dark arts of psychological operations (psyops), in order to win hearts and minds through unconventional 'non-lethal' means.

As long ago as 1971, General Sir Frank Kitson commented on 'an analysis of world trends showing that subversion and insurgency are current forms of warfare which the army must be ready to fight'.[2] Indeed, Colonel Roger Trinquier, the French counterinsurgency expert, argued somewhat prematurely as early as 1961 that traditional (or conventional) warfare no longer existed and had been replaced by a new clandestine, subversive and revolutionary warfare, which differed 'fundamentally from the wars of the past'.[3] General Sir Rupert Smith has also stated more recently (2005) that conventional war on the battlefield 'no longer exists'.[4] Smith also believes that future wars will not be state against state on well-defined battlefields but intra-state where the battlefield is less well defined.[5] This is also a theme of David Kilcullen's latest book.[6]

Sir Robert Thompson, like many others before and later, saw the war against subversion and revolutionary warfare as one of the most important struggles for domination of the world as a whole.[7] Just as the insurgencies in Malaya, Vietnam and elsewhere became intertwined during the Cold War with a perceived communist threat based on Mao's concept of revolutionary war, so recent struggles in Afghanistan and Iraq were linked to a Muslim *jihad* as part of the 'Global War against Terror'. As Kitson noted, 'the main characteristic which distinguishes campaigns of insurgency from other forms of war is that they are primarily concerned with the struggle for men's minds'.[8] *Indian Army Doctrine* observed in 2004 that 'the likelihood' of such wars 'becoming the form of warfare tomorrow is being discussed quite widely' and noted that involvement in unconventional, irregular wars, increasingly risks becoming mired in complex 'internal unrest and disturbances' and operating 'in an environment of ambiguity'.[9]

Guerrilla warfare is irregular or unconventional warfare in which small groups of combatants (usually civilians) frustrate and eventually defeat much larger, organised but less mobile conventional security forces (police and regular army). Usually, the guerrillas are attempting to overthrow an established government or rebelling against a foreign invader. To achieve their goals guerrillas adopt the strategy of fighting as irregulars and employing fast-moving, mobile, small-scale tactics to draw the orthodox enemy forces on to terrain unsuited to them and they employ their greater mobility and surprise to attack any vulnerable targets. As they are usually fighting against larger but less mobile conventional forces, guerrillas move quickly and keep their battles short.

The classic guerrillas' tactic is 'hit and run', attacking only when they can do so with overwhelming superiority of numbers and quickly disbanding to hide among a sympathetic population in order to avoid being annihilated by the subsequent reaction of the security forces. By surprising their enemy and then retreating almost immediately, they prevent their foes from either defending themselves adequately or staging a counter-attack. Che Guevara emphasised that the guerrilla tactic was to 'hit and run, wait, lie in ambush, again hit and run, and thus repeatedly, without giving any rest to the enemy'.[10] Successful guerrillas are reluctant to meet regular armies on the open battlefield where they are likely to suffer heavy defeat. The insurgents use their superior mobility and knowledge of often difficult and remote terrain to ambush any occupying regular forces who have been lulled into a false sense of security. As Guevara noted, 'the fundamental characteristic of a guerrilla band is mobility'.[11] Mao summed up guerrilla tactics:

> When guerrillas engage a stronger enemy, they withdraw when he advances; harass him when he stops; strike him when he is weary; pursue him when he withdraws. In guerrilla strategy, the enemy's rear, flanks, and other vulnerable spots are his vital points, and there he must be harassed, attacked, dispersed, exhausted and annihilated.[12]

The strategy is one of 'avoiding the enemy when he is stronger and attacking him when he is the weaker, now scattering, now regrouping one's forces, now wearing out, now exterminating the enemy, determined to fight him every-where, so that wherever the enemy goes he would be submerged in a sea of armed people who hit back at him, thus undermining his spirit and exhausting his forces'.[13] Typically, a small guerrilla force seeks to concentrate its strength against the weaknesses of the enemy's forces, such as garrisons and outposts or lines of communication and logistics, to strike suddenly, and then to disappear into the surrounding countryside or local population. The guerrilla relies on using his greater flexibility, mobility and speed to retain the initiative and keep his more powerful enemy off-balance. Colonel C.E. (later Major General Sir Charles) Callwell noted the tactics employed by guerrillas:

> They revel in stratagems and artifice. They prowl about waiting for their opportunity to pounce down upon small parties moving without due precau-tion. The straggler and camp follower are their natural prey. They hover on the flanks of the column, fearing to strike but ready to cut off detachments which may go astray.[14]

While Mao emphasised that 'in order to mislead, decoy and confuse the enemy, there should be constant use of stratagems, such as making a feint to the east but attacking in the west, appearing now in the south and now in the north, hit-and-run attacks, and night actions'.[15] Thus, the guerrilla is able 'to extend guerrilla warfare all over this vast enemy-occupied area, make a front out of the enemy's rear, and force him to fight ceaselessly throughout the territory he occupies'.[16] Mao noted that 'a guerrilla unit ... can strike sudden blows and then vanish into hiding without a trace, thus reducing the enemy to a level where he does not feel secure whether he is withdrawing or advancing, attack-ing or defending, moving or remaining still, sitting or lying down'.[17]

The psychological aspect of sudden attack and ambushes by guerrillas on the security forces and their administration and communications has always been a very important component of the guerrilla strategy. This wears down the resistance of an enemy who is conventionally superior in terms of firepower, numbers and technology by using irregular (or unconventional) rather than conventional warfare. Robert Taber notes that while the guerrilla wages 'the war of the flea', his military enemy 'suffers the dog's disadvantages: too much to defend; too small, ubiquitous and agile an enemy to come to grips with'.[18] Thus, over the centuries frustrated military commanders have consistently damned guerrillas as bandits, barbarians, brigands, outlaws, savages, and terrorists.

A Brief History of Guerrilla Warfare

The practitioners of guerrilla warfare have been called guerrillas, insurgents, irregulars, partisans and rebels. In Spanish, the word *guerrilla* (the diminutive of *guerra*, 'war') means 'little war', hence the frequent use of the term 'Small Wars' as well as insurgency to describe guerrilla wars. Use of the term 'guerrilla' originates from the Napoleonic Wars and the Duke of Wellington's campaigns during the Peninsular War (1807–14), in which small bands of Spanish and Portuguese irregulars, or *guerrilleros*, helped the British to drive the French from the Iberian Peninsula. But, in fact, guerrilla warfare and guerrilla tactics have a long history dating back to ancient times, notably to the ideas of Sun Tzu (Sunzi), the Chinese military strategist who lived more than 2,000 years ago and argued in *The Art of War* that all warfare involves the employment of one's strength to exploit the weakness of the enemy, especially one who is more numerous and better equipped.

In 512 BC the Persian warrior-king Darius I ('the Great', 550–486 BC), who ruled the largest empire and commanded the best army in the world, was frustrated by the hit-and-run tactics of the nomadic Scythians. The Macedonian king Alexander the Great (356–323 BC) also faced serious guerrilla opposition from the Persian general Beseus in the Hindu Kush and the Bactrians. The Carthaginian general Hannibal (247–c.183 BC) faced considerable guerrilla opposition in crossing the Alps into Italy. Later, following victories such as Cannae, he was frustrated by the delaying strategy of the Roman general Quintus Fabius Maximus, from whom the term 'Fabian tactics' is derived. The Romans themselves fought against guerrillas in their 200-year conquest of Spain.

In the twelfth century the Crusader invasion of Syria was obstructed by the guerrilla tactics of the Seljuk (Seljuq) Turks, while the Normans were similarly thwarted in their conquest of Ireland (1169–75). Two invasions of Vietnam by Kublai Khan's Mongols during the 1280s were defeated by Tran Hung Dao, who used guerrilla tactics. Edward I (1213–1307) and Edward II (1284–1327) of England struggled through long, hard, and expensive campaigns to subdue the Welsh under Llywelyn ap Gruffyd and to conquer Scotland against the brilliant guerrilla operations of William Wallace and Robert the Bruce (Robert I). Bertrand du Guesclin (*c.*1320–80), a Breton guerrilla leader in the early part of the Hundred Years' War (1337–1453), reversed the English run of success under Edward III and his son, the Black Prince, by using Fabian tactics of avoiding battle, harassment, surprise and ambush. Shivaji Bhonsle (1630–80), founder of the Maratha Empire, used guerrilla tactics as part of his brilliant military strategy to seize strongholds from the declining Mughal Empire during the 1660s. In America, colonists adopted American Indian tactics in the wars against the French and their Indian allies (1754–63). The founder of the guerrilla tradition in America is considered to be Major Robert Rogers of Connecticut, who organised and trained Rogers's Royal American Rangers in 1756 to carry the war deep into enemy territory and whose publication, *Rogers' Rules for Ranging* (1757), remains a classic.

Guerrilla warfare became a useful adjunct to and complemented orthodox military operations both inside enemy territory and in areas seized and occupied by an enemy. For example, during the first two Silesian Wars (1740–45) and the Seven Years' War (1756–63), when Hungarian, Croatian and Serbian irregulars supporting the Austrian army several times forced Frederick II ('the Great') of Prussia to retreat from Bohemia and Moravia with heavy losses. During the American War of Independence (1775–83), bands of irregulars under leaders such as Francis Marion, Andrew Pickens and Thomas Sumter supported the conventional operations of George Washington and Nathanael Greene, helping to drive the British under Lord Cornwallis from the Carolinas (1780–81) and to defeat at Yorktown, Virginia. Similarly, Wellington in Spain was supported by guerrillas, who attacked the French communications. In Italy the peasants of Calabria fought against the French invaders (1806–09). In 1812, in the long retreat from Moscow, the army of Napoleon I suffered thousands of casualties inflicted by Russian partisans and Cossacks. Following the Prussian defeats at Jena and Auerstädt in 1806, there was a mass levy of the *Landsturm* (People's Army) during the War of Liberation against Napoleon (1813–14). Thus, during the early nineteenth century, guerrilla warfare was linked to a savage People's War, epitomised by the wars of liberation in Germany and Spain against Napoleon's occupation.

Guerrilla wars flourished in the nineteenth century as indigenous irregulars in Africa, Asia and New Zealand tried, usually in vain, to resist colonisation by the great powers. Resistance against the Russians in the Caucasus in the mountains of Dagestan, Chechnya and Avaria between 1832 and 1859 was led by the charismatic Shamil. The T'ai P'ing Rebellion (1850–64) in China, a peasant uprising against the Qing (Manchu) dynasty, killed an estimated 20 million Chinese before it was suppressed. During the American Civil War, the outnumbered Confederate forces employed guerrillas, whose leaders included Colonel John Singleton Mosby and General Nathan Bedford Forrest, to attack Union communications. A particularly fierce guerrilla war was waged in the border states of Kansas and Missouri, where William Quantrill became notorious for his daylight raid and destruction of the city of Lawrence (1863). Quantrill's followers included Frank and Jesse James and the Younger brothers, destined to become prominent outlaws in the post-war years.

Indian tribes in North America stubbornly fought the opening up of the West, notably the Seminole in Florida who fought three wars of resistance (1818–58). Considered one of the premier practitioners of guerrilla warfare after the Civil War, the American Indian proved a formidable and elusive foe. Before being ultimately defeated, the Indians occasionally inflicted stunning reverses on the regular army, notably in the Fetterman fight (1866) and the defeat of George Custer's 7th Cavalry at the Battle of the Little Bighorn (1876). Filipino guerrillas fought the Spanish and then the Americans (1898–1902). In the South African War, Boer commandos led by a group of able leaders who included Louis Botha, Koos de la Rey and Christian de Wet held off a large British army for two years (1900–02) before agreeing to a settlement. Emiliano Zapata and Pancho Villa employed guerrilla warfare to achieve specific political goals in the Mexican Revolution (1910–20). Assisted by T.E. Lawrence, Britain's Arab allies supported their campaign in Palestine against the Ottoman Empire during the First World War. The Easter Rising in 1916 was the forerunner for a ferocious guerrilla war fought by the Irish Republican Army (IRA) that ended in the partition of Ireland in 1921 and an uneasy peace.

The People's War of the nineteenth century developed during the twentieth century, notably during the Cold War era (1945–91), into revolutionary guerrilla warfare. This evolved out of Marxist–Leninist ideology and the wars of national liberation against colonialism and 'neo-colonialism' of countries seeking independence from European political rule. Not all guerrilla campaigns have, however, been ideologically Marxist–Leninist.[19] The role of guerrilla warfare had expanded considerably during the Second World War, when guerrilla groups, both communist and non-communist, fought against the German, Italian and Japanese occupiers and political ideology became a more pronounced factor in the numerous guerrilla campaigns thereafter. While consolidating their hold on the country, some of these groups spent as much time eliminating indigenous opposition as they did fighting the enemy, but most of them contributed substantially to the Allied war effort. For example, Tito's communist partisans in Yugoslavia, and by the French and Belgian Maquis. Many subsequently challenged existing governments after the war.

In particular, the victory of Mao Tse-tung and the communists in China confirmed not only the importance of revolutionary warfare but also transferred the emphasis within Marxism–Leninism from reliance on the support of an urban elite to mobilisation of the peasants in the countryside. Mao was profoundly influenced by Chinese culture, traditions and previous peasant wars. He was inspired not only by Sun Tzu's classical text, *The Art of War*, but also by heroic tales of popular bandits who challenged the effete and corrupt court in historical novels. These included *Romance of the Three Kingdoms*, which records the turbulent final years of the Han dynasty, and *The Water Margin*, which chronicles the exploits of the outlaws of Liangshan, who led a peasant rising during the Northern Sung (Song) Dynasty. Mao had also studied the great T'ai P'ing rebellion, the mid-nineteenth-century peasant revolt that almost brought down the Manchu dynasty. From these ideas Mao gradually developed the concept of guerrilla war as a revolutionary struggle during the bitter and protracted contest with both Chiang Kai-shek's Nationalists (the Kuomintang) and the Japanese in the 1930s and 1940s. Mao rejected the dogmatic doctrine of Marxism-Leninism-Stalinism advocated by the Chinese Communist Party's leadership, who slavishly followed the dogma preached by Stalin in Moscow. This stated that the revolution should be led by the urban proletariat. Mao realised that, however suitable this ideology may have been for industrialised countries of the West, it was not appropriate for China. Instead, Mao proposed that the revolution should rely on the peasant masses to encircle the cities from gradually expanding territorial bases in the rural areas. The establishment of a communist government in China in 1949 was an inspiration to all revolutionaries. This was especially true of China's neighbour, Vietnam, in the struggle for liberation against the French led by Ho Chi Minh and Vo Nguyen Giap, who adopted the techniques used by Mao in China.[20]

Thus, guerrilla warfare became 'a formula of the revolutionary masses for carrying out insurrection stage by stage' and for seizing power.[21] Organising and mobilising 'the forces of all the people and of the entire country' guerrillas waged not only 'an armed struggle' but also a 'fierce political struggle' for the support of the population as part of a 'national liberation war to regain power and overthrow the imperialists' yokes'.[22] As Robert Taber noted in 1970, guerrilla warfare 'has become the political phenomenon of the twentieth century, the visible wind of revolution, stirring hope and fear on three continents'. Insurgency had become '*revolutionary* war, engaging a civilian population, or a significant part of such a population, against the military forces of established or usurpative governmental authority'. Guerrilla warfare had thus become 'the agency of radical or political change'.[23] For Che Guevara, the overthrow of the Batista dictatorship in Cuba 'showed plainly the capacity of the people to free themselves by means of guerrilla war from a government that oppresses them'.[24]

Mao's political goal was the communist takeover of China. Guerrilla warfare alone, he realised, could not achieve this. But it was an indispensable weapon during a protracted revolutionary war, allowing the insurgents to delay and wear down the enemy until orthodox armies could take to the field. Mao saw guerrilla warfare as part of a series of three merging phases for political control of the state and to destroy and replace the existing society and its institutions. Phase One was the organisation, consolidation and preservation of support in the countryside for the rebels, who develop their bases and wage a guerrilla war against the enemy and their supporters. Phase Two was one of expansion of the rebel support and bases, while simultaneously mounting relentless insurgent operations to wear down the enemy forces. Phase Three was the destruction of the enemy as a significant proportion of the irregular force completed its transformation into conventional, regular forces that were capable of engaging the enemy in decisive battles, supported by continued guerrilla operations that were closely co-ordinated with the conventional campaign.[25]

Mao noted that 'leaders in guerrilla war must exert their utmost to build up one or several guerrilla units and, in the course of the struggle, develop them gradually into guerrilla corps and eventually into regular units and regular corps'.[26] The Vietnamese provided a good example of this transition from a guerrilla force to a regular army. As Giap emphasised:

> In twenty years, it has gradually developed from guerrilla units and masses' self-defence units into independent armed groups; from guerrilla cells into increasingly concentrated units, including main-force, regional, and militia units; and from poorly equipped infantry units into armed forces with numerous branches and services operating with modern equipment.[27]

A rash of new insurgencies, both communist and non-communist, followed Mao's example to end the rule of the Dutch in Indonesia, the French in Indochina and Algeria and the British in Palestine, Malaya, Kenya, Cyprus and South Arabia. The overthrow by Fidel Castro and Che Guevara of the corrupt regime of Fulgencio Batista in Cuba in 1959 provoked other rural insurgencies throughout Latin America, Asia, the Middle East and Africa. Old and new insurgencies flourished during the Cold War in Peru, Colombia, El Salvador, Nicaragua, the Philippines, Thailand, Sri Lanka, India, Kashmir, Lebanon, Syria, Morocco, Angola, Mozambique, Northern Ireland and Spain. In the following decades during the prolonged Cold War period, the Soviet Union and United States supported a series of widespread guerrilla insurgencies in a series of costly proxy wars. These were fuelled by ethnic and religious rivalries in which numerous guerrilla forces of various political beliefs were showered with money, modern weapons and equipment. In Afghanistan (1978–92) a coalition of Muslim guerrillas known as the mujahideen, commanded by regional warlords who were heavily subsidised by the United States, fought against the government who were supported by Soviet forces during 1979–89. The Soviets withdrew in 1989, leaving the various Afghan factions to fight on in a civil war.

In the post-Cold War period, the collapse of the Soviet Union in 1991 did little to alter the emphasis on guerrilla warfare, which had become the primary form of conflict. But, while variants of communist ideology, Marxist or Maoist, continued to fuel insurgencies in Mexico, Turkey, Nepal and East Timor, other guerrilla groups, notably in Colombia, Peru, Northern Ireland and Spain, have turned to criminal terrorism on behalf of drug barons and other Mafia-style overlords. Increasingly, however, many insurgencies, notably in Palestine, Kashmir, Afghanistan and Chechnya, were fuelled by Islamic fundamentalism. The growth of this religious factor was seen in the emergence of renegade terrorist organisations. Al-Qaeda, for example, had been patched together by Osama bin Laden, a wealthy Saudi Arabian expatriate and religious zealot, who recruited fellow religious fanatics from various countries to form a worldwide network of followers that had carried out terrorist attacks since the 1990s. The most famous of these was the 11 September attacks on the United States in 2001, which led to the American invasions of Afghanistan (to eliminate bin Laden's headquarters there) and Iraq and the subsequent 'war on terror'. Nevertheless, al-Qaeda and other such groups have continued to launch terrorist attacks and insurgencies in Africa, the Middle East and Asia.

How to Win a Guerrilla War

Traditionally, guerrilla warfare has been employed to nullify the conventional superiority of the forces deployed either by a ruling government or by an invader. As Mao appreciated:

> Guerrilla warfare … is a weapon that a nation inferior in arms and military equipment may employ against a more powerful aggressor nation. When the invader pierces deep into the heart of the weaker country and occupies her territory in a cruel and oppressive manner, there is no doubt that conditions of terrain, climate, and society in general offer obstacles to his progress and may be used to advantage by those who oppose him. In guerrilla warfare, we turn these advantages to the purpose of resisting and defeating the enemy.[28]

Thus, guerrilla warfare is 'the war of the broad masses of an economically backward country standing up against a powerfully equipped and well trained army of aggression'.[29] In essence, guerrilla war allows 'a brave, intelligent, stalwart, and resourceful people' to use 'a small, weak army to fight and defeat the huge, strong, aggressive army of an imperial power whose vast and populous country has a great economic and military potential and modern technical equipment'.[30]

The overall strategy of guerrilla warfare is to enmesh the much larger and stronger enemy's forces in a long-drawn-out war and to gradually wear them down. The broad strategy underlying successful guerrilla warfare is that of continuous harassment that is designed to break the enemy's will to continue. The object is to gain time. This allows the rebels either to develop sufficient military strength to defeat the enemy forces in orthodox battle (as did Mao in China) or to subject the enemy to sufficient internal and external military and political pressures to force him to accept a peace favourable to the guerrillas (as in Algeria, Angola, Mozambique and Vietnam). As Giap noted in Vietnam:

> Only a long-term [guerrilla] war could enable us to utilise to the maximum our political trump cards, to overcome our material handicap and to transform our weakness into strength.[31]

Owing to 'not only an imbalance of numerical strength and population but also a great imbalance of technical equipment', guerrillas are forced to fight a long, protracted war because 'we must have time gradually to weaken and exterminate enemy forces, to restrict his strength and aggravate his weaknesses, gradually to strengthen and develop our forces, and to overcome our deficiencies'.[32] Guerrillas thrive on the political miscalculations made by their enemies. The Japanese in invading China in 1937 made all the mistakes that drove the population into the arms of the resistance. This produced the environment that allowed the guerrillas to flourish and assert credibly that they were the politico-military manifestation of an all-inclusive nationalism against a foreign invader:

> They advanced far into the interior, only holding key cities and lines of communication. They sought to destroy the Chinese field armies ... but neglected to search out and scotch the beginnings of guerrilla Resistance. They continued to cherish the hope that the capture of this or that city would bring the capitulation of the Chinese Government. They permitted their troops to treat the population with great brutality and disgraced their army by the ferocious sack of Nanking, the wanton attacks on universities and other cultural institutions, and the slaughter of prisoners ... They could not have adopted policies more calculated to rouse the Chinese people to enduring opposition.[33]

As Karl Marx had noted, 'insurrection is an art quite as much as war or any other, and subject to certain rules of proceeding, which, when neglected, will produce the ruin of the party neglecting them'.[34] Leaders who do not respect the principles of guerrilla warfare soon find themselves in trouble, particularly against effective counterinsurgency forces. Greek communist guerrillas were defeated (1946–49) because they lost both the sanctuaries and supplies provided by Tito's Yugoslavia and popular support in northern Greece as a result of their cruel treatment of the local population. During the 1940s and 1950s Filipino, Malayan, and Indonesian guerrillas failed because of weak organisation, poor leadership, and lack of external support. Uruguayan and Guatemalan insurgents lost support because of their use of indiscriminate terrorist tactics. Similarly, Basque guerrillas became unpopular in Spain because of their brutal assassinations and the Provisional IRA suffered loss of financial support from previously sympathetic Irish–Americans. Angolan and Mozambican guerrillas split into several factions and became pawns in the Cold War conflict between Cuba (and by extension the Soviet Union), South Africa and the United States.

Fought largely by independent, irregular bands, sometimes working in support of regular forces, the guerrilla employs ambushes, raids, sabotage, subversion and terrorism to wear down the enemy in a protracted political–military struggle. The aim of the guerrilla leader is 'to exhaust little by little by small victories the enemy forces and at the same time to maintain and increase' his own.[35] The tactics employed have been described by Robert Taber, who observed Fidel Castro's guerrillas in Cuba, as 'The War of the Flea', in which:

> The flea bites, hops, and bites again, nimbly avoiding the foot that would crush him. He does not seek to kill his enemy at a blow, but to bleed him and feed on him, to plague and bedevil him, to keep him from resting and to destroy his nerve and morale. All this requires time. Still more time is required to breed more fleas. What starts as a local infestation must become an epidemic, as one by one the areas of resistance link up, like spreading ink spots on a blotter.[36]

The guerrilla is often on the defensive and forced to undertake a strategic retreat, which 'is a planned step taken by an inferior force for the purpose of conserving its strength and biding its time to defeat the enemy, when it finds itself confronted with a superior force whose offensive it is unable to smash quickly'.[37] Guerrillas should be reluctant to meet regular armies in the open field, where they would be liable to heavy defeat. Giap stressed that in Vietnam:

> To maintain and increase our forces, was the principle to which we adhered, contenting ourselves with attacking when success was certain, refusing to give battle likely to incur losses to us or to engage in hazardous actions.[38]

Mao summed up the attitude required by the guerrilla in avoiding battle on unfavourable terms:

If we do not have a 100 per cent guarantee of victory, we should not fight a battle, for it is not worthwhile to kill 1,000 of the enemy and lose 800 killed ourselves. Especially guerrilla warfare such as we are waging, it is difficult to replace men, horses, and ammunition; if we fight a battle and ourselves lose many men, and horses, and much ammunition, this must be considered a defeat for us.[39]

The Nagas who fought the British on India's North-East Frontier during the nineteenth century were natural guerrillas and employed model guerrilla tactics out of necessity, given their numerical inferiority:

On approaching the enemy's territory, they collect their troops and advance with great caution. Even in their hottest and most active wars, they proceed wholly by stratagem and ambuscade. They place not their glory in attacking their enemies with open force. To surprise and destroy is their greatest merit of a commander, and the greatest pride of his followers… Such a mode of warfare may be supposed to flow from a feeble and dastardly spirit, incapable of any generous or manly exertion. But … these tribes … not only defend themselves with obstinate resolution, but attack their enemies with the most daring courage … The number of men in each tribe is so small, the danger of rearing new members amidst the hardships and dangers of savage life so great, that the life of a citizen is extremely precious, and the preservation of it becomes a capital object in their policy.[40]

The guerrillas employ a wide variety of weapons, some home-made, some captured, and some supplied from outside sources. In the early stages of an insurgency, weapons have often been primitive before they can be replaced by arms and ammunition stolen from army and police depots. The Mau Mau in Kenya initially relied on knives and clubs while the Viet Cong in Vietnam frequently employed home-made rifles, hand grenades, bombs, mines and booby traps. Nearly every guerrilla campaign has relied on improvisation, both from necessity and to avoid a cumbersome logistic tail. Molotov cocktails and plastic explosives are cheap but extremely effective. Mao noted that 'a kitchen knife, a wooden cudgel, an axe, a hoe, a wooden stool, or a stone can all be used to kill people'.[41] The worldwide proliferation of weapons during the decades of the Cold War added a new dimension to guerrilla capabilities, as the superpowers and other states provided modern assault rifles, machine guns, mortars, and more sophisticated weapons such as rocket-propelled grenades and anti-tank and anti-aircraft missiles. The collapse of the Soviet Union has also produced a proliferation of weapons.

Sympathetic neighbouring countries provide not only material support but also refuge for the guerrillas. Vietnamese guerrillas, in their struggle against

France, relied on China for sanctuary, training, and supply of arms and equipment. Later, in the war against the United States, they used Laos and Cambodia for refuge. Similarly, Nicaraguan guerrillas found sanctuary in Honduras. Palestinians have benefitted from refuge in Arab states bordering Israel, and a wide variety of militant groups were given sanctuary in Afghanistan during the 1990s. Basque ETA terrorists took shelter in France, while the IRA had hideouts in southern Ireland and Chechen guerrillas went into hiding in adjacent Ingushetia and Georgia.

Successful, large-scale guerrilla campaigns have a number of other important features in common. One is favourable terrain, which is rugged and difficult to operate in and often remote from the main population centres. A second is plenty of space within which the guerrillas can manoeuvre and evade the security forces. A third is popular support from the local population, which provides the guerrillas with superior intelligence and hence superior mobility while denying them to the enemy's more conventional forces. A fourth is foreign support and aid, which often proves crucial in maintaining and expanding the fledgling guerrilla movement. This strategy allows the guerrillas to have a cumulative effect while building outwards from base areas that are safe from the enemy's main conventional forces. These essentials of guerrilla warfare were grasped long before the twentieth century but it was not until then that these techniques were linked to revolutionary warfare, mobilising the peasantry for political ends.[42]

Guerrillas operate outside the formal constraints of the military and, therefore, outside the laws of war, often taking up arms in response to an invasion and using terrorist tactics directed at civilians to intimidate perceived 'collaborators' with the enemy or government within the local population. This often results in a brutal struggle for control and support of the population in which terror is one of the most basic and widely used techniques. Tactically, its purpose is to intimidate the security forces by killing their soldiers and attacking their garrisons. At a strategic level it is used to eliminate political and military leaders and officials in order to destabilise the government; to persuade the general populace to offer sanctuary, money and recruits; and to maintain discipline and prevent defections within the organisation. It is also employed to gain international recognition of and support for the rebel cause, including financial and military assistance, while simultaneously maintaining internal morale and attracting recruits.

The lack of a viable political cause or goal to galvanise support from the population has often been the key factor in an insurgency's failure. As Mao warned, 'if guerrilla warfare is without a political objective, it must fail; but if it maintains a political objective which is incompatible with the political objectives of the people, failing to receive their support, participation, assistance, and active co-operation, then this too must fail'.[43] He also observed that 'because guerrilla warfare basically derives from the masses and is supported by them, it can neither exist nor flourish if it separates itself from their sympathies and co-operation'.[44] Mao made a famous analogy to compare the relationship between the local population and the guerrillas in which 'the former may be likened to water and the latter to the fish who inhabit it' and emphasised that 'it is only undisciplined troops who make the people their enemies and who, like the fish out of its native element, cannot live'.[45] When outlining the flexibility required from guerrilla commanders in adjusting their tactics 'to the enemy situation, to the terrain, and to the prevailing conditions', Mao developed this analogy:

> The leader must be like the fisherman, who, with his nets, is able both to cast them and to pull them out in awareness of the depth of the water, the strength of the current, or the presence of any obstructions that may foul them. As the fisherman controls his nets through the lead ropes, so the guerrilla leader maintains contact with and control over his units. As the fisherman must change his position, so must the guerrilla commander.[46]

At first the insurgents operate in remote base areas where they can mobilise support and are undisturbed by the security forces before expanding outwards into the massed population. The guerrillas are 'waging a war of attrition, slowly nibbling away the rural areas, gradually expanding the free territories and building a military force with captured arms while strangling the army in its barracks'.[47] Once sufficient base and operational areas are established, guerrilla operations can be extended to include cities and vulnerable lines of com-munication. If a guerrilla force is to survive and prosper it must control safe areas where its guerrillas can recuperate, resupply and repair arms and equip-ment, and where recruits can be indoctrinated, trained and equipped. Mao emphasised that 'guerrilla warfare cannot exist and develop over a long period without bases'.[48] Such areas are traditionally located in remote, rugged terrain, usually mountains, forests and jungles. In Vietnam, the guerrillas built 'secret armed bases' in 'forests, mountains, and, occasionally, in the swampy plains' from which they could 'step up political struggle in combination with armed struggle'. They sought 'continually to enlarge the guerrilla-infested areas and guerrilla bases that had been established throughout the enemy rear area'[49] in order 'to arouse and organise the masses and to train guerrilla units and armed forces'.[50] It is only when all the rural areas are under their control and they are convinced that they outnumber the opposition, that the guerrillas come out into the open and take part in conventional warfare.

Thus, an insurgency attempts to gain control of a country through the use of irregular military forces and an illegal political organisation which builds up support amongst the population, usually in conjunction with a larger political–military strategy. The rebels rely on the support of the local population to provide logistic support and hide guerrillas from retribution and searches by the security forces, providing not only recruits for the insurgent bands but also money, food, shelter, refuge, transport, medical aid and intelligence. A successful guerrilla movement is very careful to cultivate the support of the population and to be disciplined in its behaviour, avoiding alienating their allegiance by brutal or arrogant treatment. This support and intelligence from a population, which is sympathetic to the political aims of the insurgency, also allows the insurgents to terrorise the supporters of the government while simultaneously denying the security forces the ability to locate and track down the guerrillas. Giap observes that:

> Submerged in the great ocean of people's war, the enemy finds that his eyes and ears are covered. He fights without seeing his opponent, he strikes without hitting, and he is unable to make effective use of his strong combat methods. For this reason, even though the enemy has many troops and much equipment, his forces are scattered, weakened, and unable to develop their efficiency as he wants.[51]

Regular armies are at a serious disadvantage when unable to gather intelligence from a hostile population and their communications are being continually harassed by a seemingly invisible foe. For example, Callwell remarked that in Cuba at the end of the nineteenth century:

> The Spanish troops were obstructed by the intense hostility of the inhabitants. They could get no information of the rebel movements, while the rebels were never in doubt about theirs. An insurgent was distinguished from a peaceful cultivator only by his badge which could be speedily removed and by his rifle which was easily hidden. Hence the Government forces, whether in garrison or operating in the country, were closely surrounded by an implacable circle of fierce enemies who murdered stragglers, intercepted messages, burned stores, and maintained a continual observation.[52]

Callwell also comments that 'guerrilla warfare is what the regular armies always have most to dread, and when this is directed by a leader with a genius for war, an effective [counterinsurgency] campaign becomes nigh impossible' because 'no amount of energy and strategic skills will at times draw the enemy into risking engagements, or induce him to depart from the form of warfare in which most irregular warriors excel and in which regular troops are almost invariably seen at their worst'.[53] The search for solutions to this threat over the last century are now examined.

INTRODUCTION

THE CHANGING NATURE OF COUNTERINSURGENCY SINCE 1899

Political Nature of Counterinsurgency

Over the centuries, regular armies have been faced with defeating guerrillas in numerous counterinsurgency or counter-guerrilla campaigns. Given the difficulty of locating and defeating the elusive insurgent or rebel, the emphasis in most successful campaigns has been on finding a political solution rather than in gaining a victory by purely military means. Thus, 'winning hearts and minds', a phrase that is usually attributed to Field Marshal Sir Gerald Templer in Malaya but which was in fact first used by Colonel Robert Sandeman in the Punjab in 1866 and again by Colonel Charles E. Bruce on the North-West Frontier of India in the 1930s,[1] is often used to describe the emphasis placed on gaining the support of the local population rather than on killing the rebels. In short, as David Kilcullen (the Australian counterinsurgency expert) points out:

> Hearts means persuading people their best interests are served by your success; minds means convincing them that you can protect them, and that resisting you is pointless. Note that neither concept has anything to do with whether people like you.[2]

The key and crucial factor in any counterinsurgency is the inherently politicised nature of the struggle between the insurgents and the state that seeks to assert its authority.[3] Major General Charles Gwynn concluded that the insurgents' aim 'is to show defiance of Government, to make its machinery unworkable and to prove its impotence; hoping by a process of attrition to wear down its determination'. To do this, the rebels employ political agitation, rioting, sabotage, subversion and guerrilla warfare. Also 'by terrorising the loyal or neutral elements of the population, they seek to prove the powerlessness of the government to give protection, and thus provide for their own security, depriving the Government of sources of information and securing information themselves'.[4]

In insurgencies, as Kitson explained, 'the mixture of harassing the government and mobilising international opinion is a theme that constantly recurs'.[5] In what 'becomes a battle of wits' with the rebels,[6] the government must have a clear strategy 'designed to achieve its aim of regaining and retaining the allegiance of the population'. It must reassure the people that the security forces can control the country and protect them against intimidation by the insurgents.[7] 'Controlling the level of violence is a key aspect of the struggle' because 'a more benign security environment' allows the civilian agencies and the security forces to operate, 'regaining the population's active and continued support' for the government, which 'is essential to deprive the insurgency of its power and appeal'.[8] In other words, a 'competition in government' in which the insurgents and the security forces strive to show through good governance that they deserve the support of the population is at the centre of any counterinsurgency campaign.[9]

The War Office concluded in 1949 that 'civilian morale is most important': If it cracks, the whole situation may be lost. Great attention must therefore be paid to maintaining civilian morale and to leading civilian thought in the right direction.[10] Thus, counterinsurgency 'is usually a project dealing with the social, economic, and political development of the people'.[11] The political nature of counterinsurgency also requires close civil–military co-operation, ensuring that the political objectives are built into the counterinsurgency campaign and that humanitarian aid and reconstruction are part of that campaign.[12]

Kitson commented that:

> For this to happen, security-force commanders from the top to the bottom must work closely together with national and local politicians and officials to implement the programme.[13]

To ensure this co-operation there must be some recognition that a political rather than a purely military solution is required. As the French counterinsurgency expert, David Galula, pronounced:

> Essential though it is, the military action is secondary to the political one, its primary purpose being to afford the political power enough freedom to work safely with the population.[14]

Moreover, 'the government must give priority to defeating the political subversion, not the guerrillas'. In other words, 'if the guerrillas can be isolated from the population, i.e. the "little fishes" removed from "the water", then their eventual destruction becomes automatic'. Also, an effective organisational framework, such as the committee system employed by the British in Malaya, must be set up at all levels of the government structure to co-ordinate administrative, military and political activities.[15] To use another metaphor used frequently by the British, the government sought to clear the swamp rather than trying to kill the elusive mosquitoes that emerged continuously from it. It focused on 'the political dimension of the campaign' – the population and their support for the insurgency – rather than on 'the strictly military problem of killing terrorists'.[16]

Yet, as one commentator notes, 'a regular army must make fundamental changes to its conventional doctrine, tactics, and procedures to be successful in counterinsurgency'.[17] The emphasis placed by regular armies on a capability to wage conventional warfare and their reluctance to abandon such conventional methods in order to be able to defeat an insurgency are constant and universal themes. The negative impact of fighting guerrillas using their ability to fight a conventional war reinforces their reluctance to adapt. Thus, in a sense, armies often defeat themselves by refusing to make the radical changes required to fight counterinsurgencies because they threaten their most basic institutional and hierarchical structures.[18]

In such circumstances, 'dynamic conduct' of the counterinsurgency campaign is very important, as operations need to 'break free of set patterns, stereotyped plans and rigid responses' to the insurgency.[19] Thus, a major factor in a successful campaign is the character and competency of counterinsurgency leaders. The role of individuals such as Field Marshal Sir Gerald Templer in Malaya and General Sir George Erskine in Kenya can be overemphasised but nevertheless is significant in providing charismatic, energetic and flexible leadership.[20] Just as crucial is dynamic political leadership, notably by men of the calibre of Tunku Abdul Rahman in Malaya and Ramon Magsaysay in the Philippines.[21]

Equally, the actions of the troops on the ground could have great repercussions in undermining the winning of hearts and minds. It is of great importance that each individual soldier is aware of the political dimension and behaves accordingly. Soldiers need to accept that military demands should always be subordinate to political considerations and show 'cultural awareness', respecting the population and understanding that brutality or retaliation are counterproductive, aiding the insurgents.[22] General Peter Chiarelli (1st US Cavalry Division in Baghdad, 2006–07) stressed that 'cultural awareness and an empathetic understanding of the impact of Western actions on a Middle East society were constantly at the forefront of all operational considerations', notably understanding their effect 'through the lens of the Iraqi culture and psyche'.[23] Similarly, the 1st US Infantry Division (Major General John Batiste) in Samarra during 2004 operated on 'the basic premise' that 'no one platoon could win the campaign but any platoon could lose it' through 'the way soldiers conduct operations and treat people'.[24] The population is unlikely to support the government and its security forces and provide information about the rebels if they have been mistreated, beaten up or tortured.

This requires good training both prior to entering and during service in the theatre. Training includes inducting the troops not only in the peculiarities of an area of operations, for example jungle warfare in Malaya or the Philippines, but also in the requirements of counterinsurgency, notably in fighting an elusive enemy and treating the population well. Major General Anthony Deane-Drummond observed that:

> The change in role from conventional military operations to internal security and para-military duties is neither rapid nor easy. Intense – and time-consuming – periods of training are required to prepare troops tactically and psychologically for a role which although less lethal in terms of overall casualties than conventional war is equally demanding and stressful.[25]

British doctrine emphasised training in ambushes, field-craft, patrolling and shooting.[26] Moreover, the counterinsurgency force must have two skills, which are not needed in conventional warfare:

> First, it must be able to see issues and actions from the perspective of the domestic population; second, it must understand the relative value of force and how easily excessive force, even when apparently justified, can under-mine popular support.[27]

As one officer, who served in Oman, comments 'low-tech war' needs greater preparation and training than 'high intensity war', as it requires additional skills:

> The patience and tolerance to live harmoniously in an unfamiliar culture; the fortitude to be content with less than comfortable circumstances for prolonged periods; an understanding of and sympathy for a foreign history and religion; a willingness to learn a new language; the flexibility, imagination and humility necessary to climb into the head of people who live by a very different set of assumptions; none of these are found automatically in modern developed Euro-Atlantic culture.[28]

Typical of those who lacked this empathy were American advisers in Vietnam, who were self-assured and driven by the need for results,[29] but lacked the necessary language and cultural skills. They were 'too inflexible, too mechanical, and not realistically adapted to the Vietnam battlefield' so that 'the language barrier and cultural difference ... formed a wide and seemingly unbridgeable gap'.[30] Similarly, in Afghanistan and Iraq, the shortage of personnel who could speak the local languages frustrated not only effective communication with the local population but also the collection and analysis of intelligence that is key to a successful counterinsurgency campaign.[31] One observer of the US Army in Iraq during 2003–04 noted that 'many personnel seemed to struggle to understand the nuances' of the environment in Iraq and that 'at times their cultural insensitivity, almost certainly inadvertent, arguably amounted to institutional racism'.[32]

The co-ordination of the various civilian and military elements of the security forces under a unified command is also vital to overcome inertia and bureaucracy, ensuring a speedy, efficient and flexible response to the threat posed by an insurgency.[33] But it 'is one of the most sensitive and difficult-to-resolve issues' given the 'political sensitivities' and often 'inherently problematic' relationships between civilian and military agencies.[34] The security forces must offer a political solution, which will win and retain the voluntary political support of a good majority of the population.[35] The 'centre of gravity' for the traditional counterinsurgency campaign is the population,[36] protecting the civilian inhabitants against the insurgents, because as Trinquier noted:

> We know that the *sine qua non* of victory in *modern warfare* is the unconditional support of a population.[37]

As a British Army field manual noted, 'the success or failure of an insurgency is largely dependent on the attitude of the population' and thus 'counterinsurgency is about gaining and securing the support of the people in the theatre of operations and at home'.[38] Galula explained that:

> If the insurgent manages to dissociate the population from the counterinsurgent, to control it physically, to get its active support, he will win the war, because, in the final analysis, the exercise of political power depends on the tacit or explicit agreement of the population or, at worst, on its submissiveness.[39]

'Success' in counterinsurgency is based on the ability to recognise the reality of the situation and to acknowledge what is possible given those circumstances. Given the intractability of insurgencies, which may be able to gain significant support from the population, counterinsurgency campaigns have lasted for decades rather than years. Moreover, the methods of measuring progress, which are used in conventional warfare, such as statistics of prisoners and weapons captured, are of little use in assessing the impact of the counterinsurgency campaign on the insurgency, as the Americans discovered in Vietnam. In the 1950s and 1960s the acceptance of the inevitability of decolonisation by the British political elite allowed the undermining of the insurgency in Malaya by granting independence. This was in stark contrast to the inflexible resistance to decolonisation by the British in Kenya, Cyprus and Aden and by the French in Indochina and Algeria until withdrawal became inevitable. Such opposition committed the security forces to achieving the military defeat of the insurgents and did not permit a withdrawal based on a political compromise.[40] Similarly, the Portuguese opposed decolonisation in Africa.[41] Great powers seek to obtain 'victory' against insurgencies but usually have to settle for something far less in order to achieve a lasting settlement. It is essential to have a realistic 'endstate' or objective, which can be achieved by the government. In successfully combating an insurgency this often requires the negotiation of a political settlement based on some concessions to the insurgents or the population which provides them with support.[42]

In 1936 Gwynn recorded that the suppression of insurgencies 'unless nipped in the bud, is a slow business, generally necessitating the employment of numbers out of all proportion to the actual fighting value of the rebels'.[43] Politicians in particular must be aware that although so-called 'low-intensity conflicts' involve fighting with companies, platoons and sections rather than battalions, brigades or divisions, counterinsurgency campaigns are often protracted and tie down large numbers of troops. These represent a significant commitment in peacetime. For example, 45,000 Commonwealth troops were deployed in Malaya during the Emergency that lasted some twelve years. These large numbers were necessary in order to keep the violence from escalating beyond levels that were politically acceptable. However, this means the cost of such campaigns is high.[44] As Sir Robert Thompson, a member of the Malayan Civil Service who had served as Staff Officer (Civil) to the Director of Operations, Lieutenant General Sir Harold Briggs, and played a key role in the implementation of the Briggs Plan during the Emergency,[45] admonished:

> It is a persistently methodical approach and steady pressure which will gradually wear the insurgent down. The government must not allow itself to be diverted either by countermoves on the part of the insurgent or by the critics on its own side who will be seeking a simpler and quicker solution. There are no short-cuts and no gimmicks.[46]

In the long term, central to the strategic planning of a counterinsurgency is winning the hearts and minds of the civilian population by building up ideological support for the government and improving their lives. Attempts to win hearts and minds are supported by 'civic action', providing education, medical care and amenities to improve the standard of living of the population. In some cases special military units are raised to provide 'civic action'. The concept of 'winning hearts and minds' is often misunderstood and maligned, conjuring up images of soldiers building playgrounds and handing out sweets to smiling children. In fact, it consists of analysing what has caused the population to rebel and devising a campaign to redress the deep-seated causes of discontent and unrest, such as unemployment and political dissatisfaction. Long-term political reform is often necessary in order to prevent a resurgence of the insurgency. Above all, this can only be achieved by providing security for the population, protecting and separating them from intimidation by the insurgents.[47]

Consequently, since the insurgents hide among the population, 'like fish in water', in the words of Mao Tse-tung (Mao Zedong), and their support was 'indispensable' to the guerrillas, one crucial tactic was to cut the rebels off from their support from within the local community.[48] The British manual, *The Conduct of Anti-Terrorist Operations in Malaya*, concluded that 'the most important factor' in destroying the insurgent 'is to complete his isolation from the rest of the community', allowing him 'no money, no food or clothing; no help of any sort'.[49] As Galula explained, success for the security forces is not made possible by the destruction of 'the insurgent's forces and his political organisation' but by 'the permanent isolation of the insurgent from the population' from which all else follows.[50] By concentrating their efforts on the population, the security forces make the most of their assets, notably control of administrative, economic and media resources. This avoids the need to catch the elusive insurgent, who becomes isolated from and loses the support of the populace.[51] This phenomenon is reinforced by the fact that the population tends to move away from the insecurity of contested areas in the countryside towards the relative safety of government-controlled urban areas, as shown in Vietnam and during the Soviet occupation of Afghanistan.[52]

There must be a realisation that there is a link between an appropriate use of force and a successful counterinsurgency campaign. For example, British theorists advocated the use of 'minimum force' when dealing with insurgencies. They realised that the Empire relied on the support of the population and imperial policing could not rely on force alone.[53] The concept of using restraint or 'minimum force' is of great importance in conducting a successful counterinsurgency. One historian of the US Army's experience of counterinsurgency commented that:

> Atrocities have taken place in virtually all wars, but the frustrations of guerrilla warfare, in which the enemy's acts of terror and brutality often add to the anger generated by the difficulty of campaigning, create an environment particularly conducive to the commission of war crimes.[54]

As Major General Sir Charles Gwynn noted, 'ruthlessness in military methods' often leads to controversy.[55] The prospect of alienation is increased if the troops are unsympathetic or have been riled by insurgent attacks, leading to accusations of brutality against the population and of damage to and looting of property. Such behaviour often occurs in areas associated with the insurgents when the troops have become frustrated by the lack of clear-cut progress or enraged by the deaths of colleagues.[56] Having served in Palestine during the Arab Revolt and as the last Governor of Cyprus, Sir Hugh Foot summed up that:

> ... action against a subversive or terrorist movement ... must be selective and not indiscriminate. That sounds obvious, but in Palestine and then in Cyprus there was often a tendency to attempt to make up for a lack of Intelligence by using the sledgehammer – mass arrests, mass detentions, big cordons and searches and collective punishments. Such operations can do more harm than good and usually play into the hands of the terrorists by alienating general opinion from the forces of authority. It is not by making the life of ordinary people intolerable that a nationalist movement is destroyed – it is by a selective drive against the terrorist leadership undertaken by small numbers of skilled forces acting intelligently on good information.[57]

Kitson stressed 'the importance of handling the population correctly in wars of counterinsurgency'.[58] Gwynn commented that officers 'must rely on their own judgment to reconcile military action with the political conditions', deciding 'what is the minimum force they must employ rather than how they can develop the maximum power at their disposal'. He then noted that:

> Mistakes of judgment may have far-reaching results. Military failure can be retrieved, but where a population is antagonised or the authority of government seriously upset, a long period may elapse before confidence is restored and normal stable conditions are re-established.[59]

Following Operation Banner in Northern Ireland, the British Army emphasised that the 'desired end-state' was one 'which allowed a political process to be established without unacceptable levels of intimidation'. This could only be achieved by the 'recognition of the need for a sensitive approach to the use of military force, and avoiding overreaction'.[60] The US Marines *Small Wars Manual* elaborated:

> The aim is not to develop a belligerent spirit in our men but rather one of caution and steadiness. Instead of employing force, one strives to accomplish the purpose by diplomacy. A Force Commander who gains his objective in a

small war without firing a shot has attained far greater success than one who resorted to the use of arms.[61]

This principle has been reinforced in the post-colonial era on the grounds that excessive use of force, such as torture, massacres of the local populace, or 'shoot-to-kill' policies, merely provides more recruits for the insurgency by generating 'the three "Rs": resentment, resistance and revenge'.[62] Recently, the concept of 'minimum force' has been interpreted as requiring that the actions of the security forces 'be projected in a transparent, honest and positive manner for maximum psychological gains' and 'aim at respecting and protecting human rights, reducing the threat to the people and inspire a sense of security'.[63] For example, the modern Indian Army seeks 'conflict management rather than conflict resolution', aiming to achieve 'low profile and people-friendly operations rather than high intensity operations related only to body and weapon counts'.[64] But, as Gwynn notes:

> Excessive severity may antagonise the neutral or loyal element, add to the number of rebels, and leave a lasting feeling of resentment and bitterness. On the other hand, the power and resolution of the government forces must be displayed. Anything which can be interpreted as weakness encourages those who are sitting on the fence to keep on good terms with the rebels.[65]

A balance is needed in keeping the support of the local populace while destroying the insurgents but this equilibrium is difficult to detect or sustain. The implication is that while coercion plays a role in counterinsurgency it must underpin rather than undermine the strategic objectives, a technique of combining the 'stick' (punitive measures) with the 'carrot' (winning hearts and minds).[66]

Thus, as early as 1940 the US Marine Corps manual recommended 'a serious study of the people, their racial, political, religious and mental development' in order to deduce 'the reasons for the existing emergency' and 'to apply the correct psychological doctrine' in response.[67] Recently there has been much more emphasis on studying 'the human terrain', notably the cultures, languages, ideologies, religions, economics, politics, group interests and social structures, of the area which is being operated in. This is to understand and neutralise the insurgents and gain and maintain popular support, which is 'an essential objective for successful counterinsurgency'.[68] Believing that 'a policy of reprisals is always dangerous', Gwynn stated that:

> Punishments of a nature humiliating to a community, or which outrage religious susceptibilities, are contrary to the principle that no lasting feeling of bitterness should be caused.[69]

The US Marine doctrine also recommended 'minimum interference' with the sovereignty of the 'host nation', avoiding 'unnecessary' humiliation of the local population, learning 'a working knowledge' of the local language. Indeed, 'all ranks' should be indoctrinated to display 'the proper attitude toward the civilian population' in order to avoid antagonising them.[70] The policy of 'minimum interference' has proved difficult with rulers such as Diem in South Vietnam or Karzai in Afghanistan who presided over deeply corrupt regimes that were dominated by a small, unrepresentative elite, prone to cronyism and hostile towards reform.

What is often forgotten, however, is that winning hearts and minds is often part of a carrot-and-stick strategy in which one element is also explicitly coercive. Thus, very often the key to winning over the population, so important to a successful counterinsurgency campaign, is not economic or political reforms or 'civic action' but establishing control over the population. Indeed, extensive repression, such as employed by the British in Kenya and Malaya and by the French in Algeria, can be a very effective counterinsurgency strategy.[71] In the past the principle of 'minimum force' has not necessarily precluded harsh treatment of the insurgents, particularly if they are a religious or ethnic minority lacking widespread support. It is possible to retain the wider support of the population for the security forces if the use of force is well targeted as a result of good intelligence. But it has excluded brutality towards or mistreatment, such as looting and rape, of civilians in general. When badly targeted as a result of poor intelligence or a growing frustration at being unable to locate the insurgents the use of force is likely to lose the security forces support from the wider population. Sir Robert Thompson believed that 'the government must function in accordance with the law' not only to maintain the principle of law and order but also to avoid the very real risk of alienating the support of the population.[72] Thus, in some cases, the security forces seek to control the population and separate it from the insurgents while using the winning of hearts and minds as a fig leaf to hide its true intention.

The French in Algeria advocated a 'tight organisation' to control the local community in order 'to cut off, or at least reduce significantly, the contacts between the population and the guerrillas'. They conducted a census of the entire population to find out where they lived, restricting and controlling movement and the distribution of food supplies, and using checkpoints, curfews, informants and identity cards to maintain surveillance of their movement and activities.[73] Galula emphasised that:

> Control of the population begins obviously with a thorough census. Every inhabitant must be registered and given a foolproof identity card.[74]

The British employed similar restrictions in the Strategic Hamlets of Malaya, requiring the population to carry an identity card with a photograph and thumbprint.[75] Such systems are easier to implement under colonial or totalitarian regimes and the clear implications for the civil liberties of inhabitants may preclude their use in a democracy.[76]

In the independent Malaya and Kenya the new regimes retained authoritarian powers, such as detention without trial, which had been introduced during the Emergency and employed the security apparatus left by the British to keep political rivals under surveillance.[77] As a legacy of its colonial past, the head of state of Kenya still 'controls a formidable security and intelligence service'.[78] Implicit in the control of the population is the concept of divide and rule. This was often employed to re-establish control, notably in Tsarist Russia and the Soviet Union, when economic, ethnic, political, religious or tribal divisions could be exploited either to split the insurgents or separate them from the population.[79] Security forces, notably the British in Malaya, Kenya and Cyprus and the French in Algeria, often created 'prohibited areas' or 'forbidden zones' that were closed to the population and anyone found there was assumed to be an insurgent and shot on sight.[80]

One widely pursued policy was the forced resettlement of the population, whose 'fundamental aim' is 'to isolate the insurgent both physically and politically from the population'.[81] This was part of the effort to break the hold of the insurgents over the local inhabitants and to prevent them from supporting the insurgency.[82] In Kenya large numbers of suspected Mau Mau were detained by the British, removing the guerrillas from contact with the population, while more than a million people were controlled in villages by the local home guard and under constant state surveillance.[83] Relocating the population away from the guerrillas meant that they could be better protected from attack and intimidation while being coerced, restrained and watched over by the security forces. This technique also destroyed the underground infrastructure, which supported the insurgents. It is important that the resettlement should be accompanied by 'civic action' providing welfare and amenities to ensure that the resettled population is persuaded by 'the carrot' of a better standard of living to accept 'the stick' of being forced to relocate. However, as Galula warned, resettlement is 'a radical measure' which is both 'complicated' as large numbers of people need to be moved, housed and given modern facilities and 'dangerous' because 'nobody likes to be uprooted and the operation is bound to antagonise the population seriously'.[84] For example, the resettlement of the Boers in 'concentration camps' by the British in South Africa (1901–02) was an example of the disastrous consequences of poor conditions and a failure to accompany 'population concentration' with any attempt to win hearts and minds.[85] Other unsuccessful resettlement projects were those employed by the

Japanese in China in the 1940s, the French in Algeria in the 1950s, and the South Vietnamese in the 1960s.[86]

The Key Importance of Intelligence

Intelligence is the key to success in defeating an insurgency. As early as the 1890s, Charles Callwell believed that 'in no class of warfare is a well organised and well served intelligence department more essential than in that against guerrillas'.[87] General Lord Bourne noted that good intelligence is 'the key to success in dealing with bandits or with a full-scale rebellion',[88] while Julian Paget believed that it 'is undoubtedly one of the greatest battle-winning factors in counterinsurgency warfare'.[89] As experts such as Trinquier, Thompson and Kitson have concluded, the main difficulty in suppressing the insurgents, whose nature is largely anonymous and clandestine, is locating them.[90] The War Office noted in 1949 that:

> Good military intelligence is essential. Nothing is so wasteful of troops or of any kind of security forces as using them on internal security tasks without sufficient information for them to know what they are trying to do and where they are likely to find their opponents.[91]

Operations in Iraq after 2003 taught that:

> Without good intelligence, a counterinsurgent is like a blind boxer wasting energy flailing at an unseen opponent.[92]

Good intelligence is the difference between victory and defeat and, ideally, one single director or organisation should be responsible for obtaining and disseminating information. Intelligence provides an ability to understand the insurgent organisation, intentions and capabilities. It additionally enables the security forces to launch counterinsurgency operations that are targeted against clearly identified targets, such as insurgent bases, cells and supporters. Poor intelligence results in ill-considered and poorly targeted reactions to insurgent attacks, such as overly aggressive and intrusive patrols and stop-and-search operations, which alienate the population. Good intelligence also supports the political objectives of the counterinsurgency campaign by, for example, identifying divisions within the insurgent leadership and its support among the population. Furthermore, it allows assessment of likely acquiescence by the local inhabitants towards government policies and operations.[93] But, as Galula noted:

> Intelligence is the principal source of information on guerrillas, and intelligence has to come from the population, but the population will not talk unless its feels safe, and it does not feel safe until the insurgents' power has been broken.[94]

The police are an important factor in any counterinsurgency campaign as 'the primary front line COIN force'.[95] As 'a static organization reaching into every corner of the country', it possesses 'long experience of close contact with the population'.[96] According to Gwynn, 'in all internal trouble the basis of the intelligence system must depend on police information',[97] because of 'their minute knowledge of the area to which they belong'.[98] The police are one of the primary intelligence gathering agencies, employing their local expertise and contacts in a particular area to identify and detain the political and supply networks supporting the insurgency, but also to provide security for the population. In contrast to paramilitary or militarised police, who tend to alienate the population, community policing provides the intelligence necessary for effective operations and the means for 'population control' and to win hearts and minds. These are vital if the security forces are to undermine and defeat the insurgents. It also helps to build the bridges with the local inhabitants that foster an environment conducive to collecting intelligence and providing a long-term settlement.[99] Protected by the army, the police control the population, identify and destroy the insurgent infrastructure, search, interrogate and detain suspected rebels, and gather information.[100] However, under the pressure of constant attack by insurgents there is always the danger of paramilitarism, in which the police become heavily armed for self-protection and 'bunkered down' in fortified police stations, alienating the local population that ideally they are meant to serve.

The intelligence provided by the police as the unit of the security forces serving among the community is the key factor in the maintenance of internal order and control. As such, the police are one of the main targets for insurgent terrorism in order to break down public security and destroy the organisation's effectiveness. This deprives the security forces of intelligence and thus their ability to function and defeat the insurgents while also demonstrating the incompetence of the incumbent government. The campaigns in Ireland, Palestine, Cyprus, Aden and Northern Ireland show the effectiveness of systematic tactical assassination and targeted terrorism as a political weapon when employed by insurgents to break the resolve of the police.[101] The British experience highlights the difficulties for the security forces when a substantial proportion of the population is at best neutral and at worst actively hostile towards the government's rule.[102] In such cases, 'the normal civil control does not exist, or has broken down to such an extent that the Army becomes the main agent for the maintenance of or for the restoration of order'.[103] The question for the security forces is how to implement civilian control and to protect the police so they can work effectively within the community.

However, politicians and soldiers also need to be aware that repression of insurgents can also degenerate into 'state terror', in which intelligence can be exploited by unscrupulous officials not just to contain opposition by the rebels but to eliminate other groups that have been identified as opponents of the state. An insurgency provides a political environment where arbitrary and even extreme brutality can be condoned in the name of security.[104] For example, the British saw their counterinsurgency campaigns very much in the light of the Cold War. In Malaya the detentions under Emergency regulations were employed to curb left-wing opinions and Malay nationalism generally,[105] while in Cyprus, having declared a State of Emergency in November 1955, they promptly banned AKEL (the Communist Party) and detained 128 of its leading members, even though it opposed the insurgency.[106]

Similarly, methods employed against urban terrorists in Algiers and Constantine during the Algerian War could be transferred for use against Algerian protestors in Paris during 1961.[107] Operations in Northern Ireland drove a demand for population surveillance in which a census of the population would include the taking of photographs and fingerprints to allow the issue of identity cards and the setting up of computer databases.[108] It has also been noted, as regards Afghanistan and Iraq, that:

> Intelligence in a counterinsurgency needs a national computerised database that can be readily shared by the police and coalition military.[109]

The government and its security forces also needs to win the propaganda war, as counterinsurgency 'is a battle of ideas'.[110] The US Marine manual noted the importance of maintaining 'cordial relations with the local, American, and foreign press'.[111] As Jamie Shea, NATO's chief spokesperson in Kosovo, warned, 'a media strategy won't win you the war, but if it is bad it will lose you the war'.[112] Above all, the information services need to be co-ordinated and working in close co-operation with its intelligence organisation. The security forces must gain the initiative by developing their own themes and conveying a unified 'message'. This seeks to boost the government's image and rally the population to it while sowing dissension amongst the insurgents and their supporters and wherever possible inducing them to surrender or support the government.[113] As Tony Jeapes commented:

> Persuading a man to join is far cheaper than killing him. Words are far, far less expensive than bullets, let alone shells and bombs. Then, too, by killing him

you merely deprive the enemy of one soldier. If he is persuaded to join the Government forces the enemy again become less one but the Government forces become one more, a gain of plus two. So the use of information services is not only a humane weapon of modern warfare but a singularly cost-effective one.[114]

The use of amnesties, pardons and rewards to get insurgents to surrender was a major feature of many counterinsurgency campaigns, notably in Malaya, the Philippines, and Northern Ireland. It is therefore interesting that the more modern terminology, 'amnesty, reconciliation and reintegration' (AR2), is not mentioned in either the American (FM 3-24) or British doctrines (British Army Field Manual, Volume 1, Part 10) that were developed as a response to operations in Iraq and Afghanistan.[115]

The central importance of the media in the vital propaganda and psychological battle to win the support of the population and the exploitation of imagery by insurgents as the primary instrument of their campaign has increased with the development of global communications since the end of the Cold War. 'The propaganda of the deed', in which highly visible and sensational terrorist attacks by organisations such as the Irish Republican Army (IRA) and Palestine Liberation Organization (PLO) were employed to catch the attention of the world and its media, took on a new significance as a result of the growing power of communications and the dispersal of populations across the globe. The news value of an act of violence now outweighs its tactical value and an insurgent campaign now measures its success in terms of recognition, notoriety and activism rather than the amount of territory it holds or the number of governments it has overthrown.[116]

The revolution in digital communications and the torrent of news and imagery that followed has given insurgency a global dimension that campaigns in isolated colonies during the 1950s and 1960s simply did not have.[117] However, a significant number of campaigns, notably Ireland, Palestine, Algeria and Vietnam, had an international dimension that foreshadowed that of the early twenty-first century. This undermines the credibility of the claim that colonial campaigns could somehow be fought in 'splendid isolation'. Moreover, the population affected by a modern insurgency is no longer just the local population in one well-defined area but multiple, global populations spread across the world. Modern technology and social media erode the cohesion and increase the complexity of the counterinsurgency campaign, making it difficult to define its centre of gravity. Suppressing an insurgency is no longer a local, internal matter but one carried out by multiple, multinational agencies and forces against a global jihad under the glare of international media scrutiny.[118] Intervention to 'stabilise' Afghanistan and Iraq has done little to develop a strategy as a solution to this challenging situation; rather it has exacerbated the problem.

Securing Territory

One important characteristic of counterinsurgency throughout the twentieth century is the principle of securing 'the more highly developed areas of the country' containing 'the greatest number of the population' which 'are more vital to the government'. This allows the government to limit the expansion of the insurgency and secure the support of the population, creating a secure base from which the security forces can 'work methodically outwards' into insurgent-controlled areas.[119] Linked to this concept is deployment of units to specific areas for an extended duration so that they could get to know the terrain and the local population.[120] Apart from using a 'grid' system to partition the countryside and thus restrict the movement of the guerrillas, the most important task for such units was that 'of constantly maintaining pressure' on the insurgents 'to deprive them of any chance of resting or of reorganizing and preparing new operations'.[121] But there was also 'the temptation to dissipate military forces far and wide in order to provide protection for all vulnerable points', dispersing the security forces in 'scattered detachments'.[122]

The types of operation employed against the insurgents are also crucial. Large-scale sweep operations by troops when directed by good intelligence disrupt potential guerrilla operations and allow the security forces to maintain a tempo of active operations that break up and harry any large insurgent bands. However, without intelligence such 'drives' tend to be ineffective and counterproductive. As well as requiring large numbers of men to carry them out, cordon-and-search operations to gather intelligence and to locate arms, documents, supplies, the insurgents and their supporters, too often expose civilians to a greater risk of violence. The benefits, which are gained from cordon-and-search operations, are often limited by the poor search techniques of the troops and the alienation of the community by their brutality and misconduct when conducting those searches. They are thus likely to alienate and antagonise the local population, especially if it is already sympathetic to the insurgency. Given the mobility of the insurgents who employ unconventional tactics, such as ambushes or hit-and-run raids, the security forces need to abandon large-scale sweep operations, which usually fail, unless directed by good intelligence. Moreover, aggressive and arbitrary 'cordon-and-search' operations do not achieve the security forces' main objectives of killing insurgents or more crucially destroying their infrastructure.[123]

Thus, a clear-and-hold strategy is key to a successful counterinsurgency campaign. Clearance of the guerrillas from an area by large-scale operations must be immediately followed up by holding operations to prevent the insurgents from re-infiltrating.[124] Local forces, such as militias, auxiliary police, special constabularies and home guards, recruited from indigenous peoples are an important element in the clear-and-hold operations of a counterinsurgency campaign. As Kitson concluded, the key factor 'is the ability of the police and locally raised forces to hold the pacified area for the government when the soldiers move elsewhere', preventing it from slipping back under insurgent control.[125] Indigenous forces provide not only extra manpower, releasing troops from static defence duties, which secure the population for the government, but also more crucially the ability to gather intelligence from the local inhabitants, knowledge of the indigenous culture, such as complex tribal relationships, and language skills. They also provide the ability to 'turn' captured or surrendered insurgents, who could provide intelligence about the operations, leadership and organisation of the insurgency and help to undermine their morale and induce further surrenders. Such troops, however, raise questions of loyalty, on the one hand often colluding with or 'turning a blind eye' to the activities of the insurgents, and on the other alienating the population by being diehard 'loyalists', committing vigilante atrocities against suspected guerrillas or their supporters.[126] Another danger is that 'such units tend to become private armies, owing allegiance less to the government than to some territorial local figure'.[127] There is also the risk of counter-infiltration of indigenous units by rebels who, as in Afghanistan recently, mount 'green on blue' attacks on the occupying troops. A recent development has been an emphasis on 'clear-hold-build' in which the next stage after first clearing and then holding and securing an area is to engage with the population, gaining its confidence and support by 'civic action', such as the provision of facilities and the repair of roads.[128]

The break up of rebels' units in an area by large-scale operations should be exploited by small units, which can pursue and harry any survivors. The US Marines training manual stated that the platoon of 'efficient, mobile, light infantry, composed of individuals of high morale and personal courage' is 'the basic unit best suited' to combat the tactics of guerrilla warfare that was 'conducted by small hostile groups in wooded, mountainous terrain'.[129] Small 'hunter-killer' units of specialist troops can match the mobility of the elusive insurgents, often employing indigenous troops such as trackers and 'turned' guerrillas and operating from bases in the areas affected by the insurgency. They provide maximum flexibility and the ability to surprise the rebels in their own domain while maintaining the initiative in order to locate and hunt down the guerrillas in rugged terrain or in inaccessible or remote areas. 'Small Team operations' acting on good intelligence increase the chances of contact with and success against guerrillas, especially if the security forces are overstretched across a large area. 'Pseudo-gangs' can also infiltrate the insurgents, collecting invaluable intelligence and destroying guerrilla bands. A garrison should hold the area for a lengthy period to ensure that the insurgent infrastructure does not return before being handed over to the Civil Power.[130]

One obvious measure for a counterinsurgency campaign is to attempt to cut off the insurgents from external support. This external support can come in various forms. The powerful Irish–American lobby in the United States provided moral support for the republicans in Northern Ireland during the 'Troubles'. Greece gave political support for Cypriot self-determination by raising the issue in the United Nations General Assembly. Egypt furnished material support (mainly military, financial and logistical) for the National Liberation Front (NLF) in Aden and South Arabia during the 1960s. Zambia gave sanctuary support (cross-border bases, hiding places and training facilities) to insurgents fighting in Mozambique and Rhodesia. Although not ensuring success, external support is often crucial for the success of insurgents and preventing such support frequently becomes a major objective for the security forces, which frequently attempt to close borders and to shut off support from outside. In some cases they may launch raids or use surrogate forces across the border to destroy such bases. When an insurgency has political support and a ready supply of weaponry from external sources as the British faced in South Arabia, the Americans in Vietnam, and the Soviets in Afghanistan, then it is very difficult to suppress it.[131]

Conclusion

This book looks at counterinsurgency during the twentieth century by examining a number of the campaigns and the experiences of six nations (Germany, Portugal, Russia, France, the United States of America and the United Kingdom) whose failures and successes highlight the complex theory and practice that has grown up as a response to various types of insurgency. As will be seen, practice and reality was often very different from theory and indeed the 'classic principles' of counterinsurgency were either unknown to or ignored by the practitioners who faced insurgencies on the ground. The book will then end by examining the recent conflicts in Afghanistan and Iraq and their impact upon the more recent development of counterinsurgency. Counterinsurgency is no longer a local, internal matter but one carried out by multiple, multinational agencies and forces against a global jihad, and conducted under the glare of international media scrutiny owing to the revolution in digital communications. Interventions to 'stabilise' Afghanistan and Iraq have not provided a long-term solution to this challenging situation. As a result, as after Algeria and Vietnam, the Americans, and their allies, are now showing a distinct disinclination to become involved in such costly and divisive 'small wars'. These are perceived as being not only very difficult to wage successfully but also often counterproductive both domestically and internationally in their consequences. The main principles of counterinsurgency remain the same, but new challenges such as new information technology and Islamic fundamentalism now have to be faced. Using modern technology and social media, organisations such as al-Qaeda and Islamic State have reacted to circumstances more quickly and effectively, and have exploited security voids in fragile states such as Iraq, Mali, Somalia, Syria and Yemen.[132] The activities of al-Shabab militants in Kenya, Boko Haram in Nigeria, Naxalite-Maoists in India, Uighurs in China, and the Taliban in Afghanistan all indicate that the problem will not go away and cannot be ignored.

1

TERRORISM AND BRUTALITY

THE GERMAN COUNTERINSURGENCY EXPERIENCE, 1900–45

Introduction

The German Army has the reputation of being probably the most professional, disciplined and efficient organisation to take the field of battle during
the late nineteenth and early twentieth centuries. It has often been admired
for its military competence, effectiveness, and obduracy, which underpinned
many astonishing victories against numerically superior but less skilful adversaries during the two world wars. This assessment tends to neglect the German
approach to counterinsurgency – the other side of the coin to the conventional
warfare that dominated the country's military culture and thinking during
the twentieth century. Germany's response to the problem of defeating insurgents during various campaigns reveal a specific 'German way of war'. German
military culture responded to civilian or colonial resistance with much greater
brutality than any other European power, displaying a phobia towards guerrilla warfare and an obsession with annihilating, rather than merely defeating,
insurgencies. Germans instituted the use of terror on the grounds of 'military
necessity'. This institutional antipathy to guerrilla warfare originated with the
Prussian officer corps, a conservative elite, who viewed with repugnance the
'un-Prussian' concept of citizens in arms, recognising that it posed a revolutionary challenge to the established order. This aversion manifested itself as early as
1806 following the Prussian defeats at Jena and Auerstädt and during Prussia's
nationalist War of Liberation against Napoleon of 1813–14 when the mass levy
of the *Landsturm* (People's Army) was supervised closely by the Prussian officer
corps and disbanded as soon as national liberation had been achieved.[1]

While the abhorrence towards irregular warfare, which formed the long-standing institutional mindset of the Prussian Army (and later the German Army), was not unique, it was certainly more pronounced within the German officer corps when combined with the right-wing ideology of the totalitarian regimes ruling Germany. This fear of an unseen enemy who avoided battle, was indistinguishable from the civilian population and employed indirect and unconventional tactics against the occupying forces, brought out the worst in the German Army as an institution that historically regarded terror rather than the use of 'minimal force' and the winning of hearts and minds as the most effective way of pacifying a defeated population. This *modus operandi* would coincide with Nazi ideological doctrine during the Second World War, sanctioning a security policy based on terror and genocide. Irregular warfare, which was perceived to be unclean, indecent and illegal, challenged German mastery of conventional warfare in which superior tactics and technology employed by well-co-ordinated, conventional armies overwhelmed inferior, inefficient and poorly co-ordinated opponents in battles that were viewed as clean, decent and fought according to well-defined rules. Of course, Germany's opponents, as those of the United States later, soon learned to ignore the rules and to fight in an unconventional manner as the only means of overturning German conventional superiority. The German officer corps regarded guerrilla warfare 'as the appalling antithesis of its own doctrine'.[2]

France 1870–71

When the Prussians defeated the armies of Emperor Napoleon III in the summer of 1870, and occupied most of France, the French Republican government refused to admit defeat. Instead it opted for guerrilla warfare, urging its citizens to form irregular forces to 'cut off convoys, harass the enemy and hang from trees all the enemies they can take'.[3] German troops were ambushed and sniped at by the guerrillas, known as *francs-tireurs,* and telegraph wires cut. The destruction of the railway bridge at Fontenay was one of the most dramatic achievements of the guerrillas, who employed small units instead of conventional forces to harass the invading forces besieging Paris. This unexpected French tactic came as a nasty surprise to the meticulous planners of the Prussian General Staff. The German armies in France, consisting of some 500,000 men, were operating in a hostile country populated by 36 million people, and were at the end of lengthy lines of communication, supplied by railways that were particularly vulnerable. Around 120,000 men, nearly a quarter of the German force, were pinned down guarding lines of communication, and the Germans suffered some 1,000 casualties to the guerrillas, whose significance tended to be exaggerated.[4]

Rather naively, the Germans hoped that the French population would support them against the 'illegitimate' insurgents, issuing proclamations that ordered the surrender of all arms and proscribed demonstrations. Seeing the *francs-tireurs* as criminals rather than soldiers, the Prussians reacted harshly. Crown Prince Frederick reported that 'single shots are fired, generally in a cunning, cowardly fashion, on patrols, so that nothing is left for us to do but to adopt retaliatory measures by burning down the house from which the shots came or else by the help of the lash and forced contributions'. Bismarck noted that the *francs-tireurs* were 'not soldiers' and were being treated as 'murderers'.[5] Collective punishments such as fines were the response to guerrilla attacks and many towns and villages suspected of hiding guerrillas were burnt. *Francs-tireurs* themselves (and inevitably some innocent peasants) were summarily executed, prisoners mistreated and hostages taken. In October 1870 in retaliation for an attack by *francs-tireurs*, the town of Ablis was razed and its male inhabitants killed, while in January 1871 the Prussian 57th Regiment burned the village of Fontenoy-sur-Moselle and bayoneted the villagers in reprisal for an attack on an outpost by *francs-tireurs*.[6] Apart from such horrific and exceptional brutality, the Prussians more typically employed fines and imprisonment rather than fire and sword in tackling the guerrillas. Moreover, despite the revolutionary traditions of the National Guards and the precedent of resistance to the First Republic by monarchists in the Vendée, the guerrillas never represented a serious threat, being poorly equipped and led and lacking mass support – only 57,000 men joined them – and had waned by early 1871.[7]

Nevertheless, although the impact of the *francs-tireurs* upon their communications, supplies, and manpower was not decisive, the psychological shock for the Prussians was immense, notably on the sense of security among their troops serving in France. Attitudes, which formed the core of the army's institutional mindset, were reinforced and, following the foundation of the German Empire in 1871, the officer corps remained opposed to the concept of a nation in arms, which they maintained was 'illegal' under international law. In future wars the German Army would again deal harshly and brutally with any civilian interference with their plans, resorting to terror to subdue an occupied country that continued to resist a 'lawful' invader. The German use of reprisals was condemned by the military commentators of other nations, who believed that they were carrying out a new 'horrible' form of warfare. Edward Hamley, Commandant of the British Staff College between 1870 and 1877, observing the Prussian experience of 1870–71, noted prophetically that the 'grand mistake of the Germans is that, while ascribing great influence to fear, they ignore the counter influence of desperation'.[8]

The Boxer Rebellion

The employment of fear and terror was exemplified by the methods employed by German soldiers who volunteered to fight in the Boxer Rebellion (a traditional peasant uprising that was fuelled by resentment of foreigners and Christians) in China (1900–01), and campaigns against the Herero and Nama in German South-West Africa (now Namibia, 1904–08), and in German East Africa (now Tanzania, 1905–07). Indeed, General Lothar von Trotha, who was chosen personally by Kaiser Wilhelm II to lead German troops in South-West Africa, and the German soldiers deployed there had already been 'thoroughly brutalised' by their previous service in German East Africa and China.[9] The Kaiser's infamous *Hunnenrede* (Hun Speech), made in July 1900 to the 10,000 German soldiers who were about to sail from Bremerhaven to suppress the Boxer Rebellion and avenge the death of the German minister Klemens von Ketteler set the tone for the rest of the twentieth century. The Kaiser demanded that no mercy should be shown and no prisoners taken.[10]

On arrival in China in September 1900, after peace negotiations had already started and a month after Peking had been taken, German forces participated in no fewer than fifty punitive expeditions to 'mop up' the Boxers. While British and French troops were also involved in the destruction and massacre of villages, the German troops were known for their brutality even in friendly settlements. In these 'cleansing' or pacification operations, Peking and other cities were ransacked and plundered, women raped and suspected Boxers executed. As it was difficult to identify the Boxers among the civilian population, the German troops acted on mere suspicion. Suspected Boxers and local officials accused of aiding the insurgents were executed. 'Boxer villages' were burnt down and all inhabitants killed. Huge numbers of civilians were made homeless. Others were merely robbed or forced to pay fines. The brutality of 'punitive actions' led by Field Marshal Alfred Graf von Waldersee, commander-in-chief of the eight nations international force, increased as the Boxers continued to launch guerrilla attacks on the foreign troops. The respected German weekly newspaper *Die Zeit* detailed the mass murder of the Chinese civilian population by soldiers, some of whom wrote letters home graphically recording their atrocities. Large numbers of these letters were published in German newspapers, where they became known as *Hunnenbriefe* (Hun Letters).[11] One of these letters noted that 'everything that came across our way, be it man, woman or child, everything was slaughtered'.[12]

Encouraged by the Kaiser, the commander of the German troops in China, Trotha, behaved harshly. This left behind a legacy of hatred for Germany and provided a precedent for German colonial campaigns that was adversely commented upon by other nations. This brutalisation of warfare during the Boxer expedition foreshadowed the atrocities committed by German forces later in the twentieth century. The German Army displayed a ferocity in its campaigns against indigenous uprisings in Africa and Asia that surpassed that of its imperialist peers. It is clear that terror, which was also an integral part of American, British, French and Portuguese colonial military thought, was employed much more ruthlessly by the Germans. They crossed the boundaries of what was deemed to be correct behaviour in China, and again when dealing with the major counterinsurgencies that developed within their African colonies prior to 1914.[13]

German East Africa

In East Africa the Hehe, a highly organised military power in the south, offered the longest resistance to the Germans between 1891 and 1898, routing a column under Emil von Zelewski in August 1891. This was the commencement of a long German campaign to destroy the Hehe and its leader, Mkwawa. In October 1894 forces under the command of Colonel von Schele attacked, overran, and sacked Kulenga, Mkwawa's main fortress, capturing 1,500 women and children, 30,000lb of gunpowder, hundreds of rifles and guns, and livestock. But Mkwawa escaped and for four years between 1894 and 1898, German forces pursued him and his followers who, supplied and supported by the local population, proved very elusive. The Hehe were finally beaten by the *Schutztruppe* (Defence Force or 'protection troops') under Tom von Prince and Lothar von Trotha. Harried by the Germans and without food or supporters, Mkwawa eventually committed suicide in July 1898 and his skull was displayed as a trophy before being sent to Germany.[14]

Between 1891, when Zelewski was killed, and 1898, when Mkwawa died, German counterinsurgency strategies and tactics were transformed to provide success in African conditions. Unable to match the guerrilla tactics and staying power of their indigenous foe, the Germans resorted to a 'scorched earth' strategy to destroy the population's means of subsistence by burning villages and seizing cattle and supplies. The *Schutztruppen* operated by sector, moving systematically and relentlessly from valley to valley in turn, destroying all food and water sources, killing the elders and men, and taking prisoner the young women and children, who would be forced to work. This policy resulted in the depopulation of the land, which was then unable to support the enemy forces. The local population was expected to give unconditional allegiance to and support for the German forces. Failure to do so resulted in death. Zelewski, commander of the *Schutztruppen*, prior to his death and Tom von Prince, commander of the Iringa District, were particularly brutal. The insurrection was eventually suppressed by systematic starvation, which was directed at the civilian population rather than against the rebels.[15]

This methodical use of scorched earth and famine was also applied in German East Africa against the Maji-Maji insurrection during 1905–07, which was one of the greatest anti-colonial uprisings in Africa prior to the First World War, with similar results for the indigenous population. The enemy was worn down by starvation. The main cause of the revolt was the coercive recruitment of labour to work on the plantations, which grew cash crops, such as cotton, to ensure colonial profitability. This aroused even stronger African opposition than the brutal collection of hut taxes since it threatened the very foundation of African societies. It resulted in July 1905 in an African rebellion much greater than any previously experienced by a modern colonial power. The rebellion spread quickly, with killings of Europeans throughout the south. The *Schutztruppe*, supported by white volunteers, were hard pressed, and the colonial government was compelled to request army and marine infantry reinforcements from as far away as New Guinea. In September 1905 a Catholic bishop was killed and a number of missions, one near to Dar-es-Salaam, were burnt down. By late autumn in 1905 the rebels, joined by the Ngoni Tribe, were in control of the southern fifth of the country.[16]

The arrival of German reinforcements, however, turned the tide against the rebels, who lacked modern weapons and co-ordination. The rebels regrouped and, avoiding large-scale confrontations, reverted to classic guerrilla tactics, such as ambushes and hit-and-run attacks by small, highly mobile groups of warriors. By April 1906 the war was effectively over, although the Ngoni were not defeated until 1907. Having a unified command, highly disciplined troops, a good logistics system, the use of African allies and mercenaries, and superior weapons, the Germans did not attempt to gain the support of the population but instead laid waste to vast areas of the south in a step-by-step reconquest employing a brutal famine strategy of destroying settlements, crops and stores, which provided the infrastructure for the rebels. Implementing orders from Gustav von Götzen, the governor, to execute a 'scorched earth' strategy, which gave no clemency to the population, Major Kurt Johannes' ruthless counter-insurgency campaign in Ungoni was an extreme example of colonial violence. He eliminated the entire Ngoni elite by executing some 100 elders in the spring of 1906. He also pursued their warriors and supporters without mercy, killed their leaders, employed captured warriors as forced labour, and took women and children as hostages to prevent supplies reaching the warriors. Between 75,000 and 100,000 Africans and several hundred Germans had been killed, causing considerable damage to European investments. The colonial financial deficit increased phenomenally during the rising and, combined with the war in German South-West Africa, created a major political crisis in Germany.[17]

German South-West Africa

Similarly, the war against the Herero and Nama of 1904–07 was a full-scale attempt to destroy the indigenous resistance and exterminate those portions of the African population opposed to German rule of the colony. When the Herero uprising began on 21 January 1904, it surprised the authorities, and more than 100 Europeans, mostly farmers, were killed. Although taken by surprise, many German colonials took the opportunity presented by the war to destroy once and for all the indigenous social order that obstructed unfettered development of the colony. At the outbreak of the revolt there were only 766 *Schutztruppe* in the colony who were spread over some 600 miles (960km) on outpost duty in strong and well-supplied forts or garrisoning the five towns. The important towns of Windhoek and Okhandja remained besieged until the end of January 1904. Although outnumbered, Major Theodor Leutwein, the commander of German forces in the colony, could assemble ten artillery pieces and five Maxim guns against the Hereros, of whom only a quarter had firearms. Reinforcements arrived prior to the defeat of the main Herero force during the Battle of Watersberg in August 1904.[18]

At first, superior technology was negated by the barren and harsh terrain, which presented formidable logistical difficulties. A narrow-gauge railway supplied only 30 tons per day along the 250 miles (400km) of coast between Swakopmund and Windhoek. From the railheads, the troops relied on wagon trains that took twenty-five days to travel the 136 miles (250km) of the 'Bauweg Trail'. The Namib Desert formed a strip 75 miles (120km) inland from the coast while further inland there was the Kalahari Desert, areas that German maps showed to be devoid of waterholes. Thus, the main problem was water, which had to be carried, increasing hugely the supplies requiring transportation and ensuring that the colonial troops suffered constantly from thirst while the indigenous guerrillas thrived. With some justification, German commanders complained of the impossibility of fighting in the parched and rugged 'bush', which resulted in the failure of the German spring offensive of 1904, when the three columns under Leutwein were forced to withdraw.[19]

Von Trotha assumed command of the *Schutztruppe* in June 1904, bringing 2,000 reinforcements. He was well versed in colonial wars, having served in a series of small, brutal campaigns in East Africa and to suppress the Boxer Rebellion in China. He had gained a reputation as a successful colonial soldier in Germany and on arriving in South-West Africa set about adding to it. Described by fellow soldiers as a 'human shark' and as 'a bad leader, a bad African, and a bad comrade',[20] Trotha believed that 'the tribes of Africa' responded only to the use of 'terrorism and even brutality'.[21]

The German tradition of military independence carried over into colonial affairs, creating a tendency to make the military completely independent of civilian control and to subordinate the civil government to them. This militarisation reached its greatest heights in South-West Africa in 1905, when the Colonial Department lost a bureaucratic fight with the Army over the conduct of the Herero War. Governor Leutwein, a professional soldier with much experience of colonial warfare, was sidelined while Trotha was given command of military operations. A protégé of General Alfred von Schlieffen, Trotha answered directly to the Chief of the General Staff. Having been granted *de facto* independence from civilian authorities in both Africa and Berlin by Wilhelm II, Trotha was given *carte blanche* to win a decisive victory over the Herero.[22]

After the initial success of the Herero, the colonists retaliated with a campaign of extermination that Trotha and his specially recruited troops continued following their arrival and after the decisive defeat of the Herero at Waterberg on 11 August 1904. Faced with guerrilla warfare, the Germans began 'an orgy of killing', implementing a systematic policy of scorched earth and famine, which was also used in German East Africa during the Maji-Maji rising with similar results for the victims. On 2 October Trotha announced his policy of *Schrecklichkeit* (terror or frightfulness) and attempted to exterminate the Herero by driving the survivors of Waterberg to die in the Omaheke Desert, a waterless wilderness of 2,000 square miles. He refused to negotiate with them, placing prices on the heads of their leaders, authorising his troops to kill Herero on sight and issuing an infamous 'extermination order' to kill or starve all Herero men, women and children. Water supplies were poisoned and the survivors, once hunted down, were transported to concentration and labour camps. The Herero population was devastated. Of the estimated 80,000 Herero living in South-West Africa in 1904 before the war, fewer than 20,000 remained by 1906, and many survivors were held in concentration camps on the coast, working and dying in droves as slave labour. By the time of the 1911 census, only 15,130 Hereros were left alive. The smaller Hottentot population was reduced from 20,000 to 9,781. Those inhabitants who survived were reduced to servitude – deprived of property and cattle, subject to pass laws and brutal whippings. The official report recorded that the Herero had 'ceased to exist'. Trotha was recalled in 1905 after criticism in the *Reichstag* (Parliament) and the extermination order cancelled, as his methods proved too ruthless even for the colonial lobby as it resulted in a level of slaughter that menaced the future labour supply of the colony. Nevertheless, between May and November 1905, it still proved remarkably difficult for the Germans to pin down Herero guerrilla leaders.[23]

The Herero were not alone. The Nama, the previously loyal Hottentots in the south, saw that it would soon be their turn and led by Hendrik Witbooi went to war later in 1904. It was not until 1907 that they were 'pacified' because their military operations were more effective than those of the Herero. They avoided large set piece battles, instead employing classic hit-and-run tactics, and their leadership, particularly Hendrik Witbooi, Cornelius Frederiks, Jacob Marenga and Abraham Morris, was outstanding. Jacob Marenga, a chief who had risen on his own merits and was unusually conscious of the political situation in which the Nama were operating, was a particularly dangerous foe in the Karas Mountains. Cornelius Frederiks also emerged as a master of guerrilla tactics. The Nama were also joined by the Herero, notably Andreas, after their own uprising had been crushed. Less numerous than the northern tribes, the Nama nevertheless proved dangerous opponents as the conflict, in the words of one German staff officer, 'degenerated into guerrilla warfare', small bands hiding in the inhospitable terrain of Namaland and conducting destructive raids and stinging reprisals against colonial forces. The Nama used their mobility, bush skills and horsemanship to good effect.[24]

Trotha's successor, Deimling, introduced new methods, notably the division of rebel territory into zones that each had a flying column for pursuing the insurgents. He also ordered the removal of all livestock, which provided food for the guerrillas and potential targets for rebel attacks on military installations. They also guarded wells and waterholes. The Hottentot bands were also pursued relentlessly by the German columns, who in the end won by superior organisation and firepower. They also relied heavily on indigenous auxiliaries to act as scouts and trackers. Between 6 and 30 August 1906, four German columns marched 500 miles (800km) and fought three 'battles' to destroy the Namas led by Johannes Christian. These counterinsurgency practices of denying resources and relentless pursuit that constantly harassed the enemy proved effective and the last of the rebels surrendered just before Christmas 1906. The Nama were officially declared defeated in March 1907, although even then one leader, Simon Kopper, continued to raid into the German colony from British Bechuanaland, until he was bought off with an annual pension. But the suppression of the Nama – a 'small war' of attrition that lasted almost two years – proved expensive, costing the Germans nearly 500,000 *Reichsmark* and casualties of 2,500 officers and men, of whom 900 were wounded and 700 lost to disease.[25] The official German war history noted that 'our enemies were equal in skill and marksmanship to the Boers, but exceeded them in military efficiency and resolute action'.[26] The defeat and genocide of the Herero in South-West Africa and the extreme brutality of similar anti-guerrilla campaigns against 'inferior' races in China and East Africa loosened further the German Army's inhibitions about the use of terror against an occupied population, foreshadowing the brutality of the Second World War.[27]

Belgium 1914

The German Army, with its loathing and contempt of guerrilla warfare that had been nurtured since the Napoleonic and Franco-Prussian Wars, was acutely aware of the dangers posed by a popular uprising or *Volkskrieg* (People's War) as it advanced through Belgium and northern France at the beginning of the First World War. In the eyes of the Germans, who refused to accept the legitimacy of their military resistance, the Belgians were guilty of refusing to allow them safe passage. Haunted by the spectre of resistance by the civilian population, the Germans responded with brutality. As a result, the advance towards Liège in early August 1914 was accompanied by mass executions of some 850 civilians and around 1,500 buildings were burnt down. Other measures included the taking and execution of hostages and the use of civilians as 'human shields'. Belgian civilians were condemned by the German high command for conducting *Volkskrieg* and illegally participating in the fighting. Similarly, when a few pot-shots were fired at an officer in Louvain in mid-August 1914, the Germans overreacted with an orgy of killing, pillage and pyromania with the intention of destroying the most important parts of the city. One notable casualty of this destruction was the priceless collection held in the university library. Men were dragged out of the houses in front of their terrified families. For instance, Hubert David-Fischbach, 83 years old, having been tied up and forced to watch his house being burnt down, was beaten with bayonets and then shot. Others were killed while fleeing from their burning houses.[28]

This was only one among many incidents that resulted in a death toll of some 6,500 from the civilian populations of Belgium and northern France who were killed by the German armies in response to an illusory popular uprising. With the *francs-tireurs* of the Franco-Prussian War still fresh in its institutional memory, the German Army took drastic countermeasures in an overreaction against mostly imaginary ambushes. But such incidents were short-lived and ended once the front line stabilised in the autumn of 1914. Some of the viciousness displayed by the troops may have resulted from frustration and exhaustion – the result of marching enormous distance in an attempt to meet the utterly unrealistic schedule. This was set by the Schlieffen Plan, which sought to defeat France in a lightning campaign within six weeks before the Russians could mobilise on Germany's Eastern Front. This increased their propensity to 'lash out' at the slightest provocation. It also reflected the belief of the officer corps, which sought to achieve a conventional victory through superior firepower and manoeuvrability, that resistance by *francs-tireurs* was unwarranted. This mindset was also fed by a paranoid world view that was shared by many officers with the same pan-German movement that would eventually spawn the post-war Nazi Party. It produced directives permitting harsh and brutal treatment of the population. The officer corps saw French and Belgian guerrillas as having the same characteristics as the Reich's internal enemies (Catholics, workers, and Francophiles, especially the inhabitants of Alsace-Lorraine, the provinces lost by France to the German Empire in 1871). Furthermore, if an enemy violated its standards of proper conduct by, for example, resorting to guerrilla warfare, then the German Army felt free to ignore international law and retaliate with reprisals against enemy civilians, especially those considered racially inferior. In China and East Africa the shooting of enemy warriors was widespread and justified by their status as being outside the law as 'rebels', although the complete annihilation of the male population did not become an objective, as it had been in South-West Africa. In both East Africa and South-West Africa the destruction of villages and food supplies resulting in depopulation of the rebellious areas became the norm as the Germans sought total victory.[29]

In particular, occupied Belgium and northern France suffered from harsh military control, including large-scale economic exploitation, forced labour and deportation, which reduced their populations to extreme poverty and mass starvation. At least 10,000 French and 13,000 Belgian civilians were deported to Germany in 1914 and more men and women were rounded up during the subsequent occupation of northern France and Belgium. They were held in poor conditions across Germany, partly as punishment, partly as a security or deterrence measure to prevent a popular uprising, and partly for forced labour, from which many did not return. In one parish alone in France out of 500 rounded up 179 returned while 321 died or disappeared. As late as 1924 80,000 Belgians and 250,000 French workers were claiming for wages not paid by Germany. In a precursor of future practice that was to be driven by ideology, it was not just colonial races that were marked for scorn. Clearly, the German Army had few inhibitions about punishing an occupied population, foreshadowing the mass use of slave labour and the brutal methods of the Second World War.[30]

Ukraine 1918

In 1918 the German Army was confronted by a real rather than an imaginary insurgency in the Ukraine, which was sponsored by the Bolsheviks in Moscow and supported by small local landowners. The Germans and Austro-Hungarians occupied the Ukraine in February 1918 in support of 'The Ukrainian People's Republic', which had been declared by a national council, the so-called 'Central Rada'. The Rada contested control of the countryside with the Bolsheviks. Although by mid-May the German occupation had been completed, the countryside was far from being under military control, as only around twenty divisions were available to undertake the challenging task of pacifying the vast territory of the Ukraine. In April 1918 the Rada was overthrown by a former Russian general, Pavlo Skoropadskiy, who established himself as 'Hetman' (leader) of an authoritarian regime that was welcomed by the Germans, the landowners and former Tsarist elite but which lacked popular support. The Bolsheviks continued their underground opposition with considerable support from Moscow. Neither the socialist government of the Rada, nor its national-ist–authoritarian successor, received substantial approval from the Ukrainian population. The Chief of Staff of the Army Group in the Ukraine General (Karl Eduard) Wilhelm Groener commented in March 1918, 'the power of the [Ukrainian government] reaches only as far as our bayonets' and, following the German withdrawal in late autumn 1918, the Hetman regime quickly fell as the Ukraine descended into civil war.[31]

As one of the main reasons for the German intervention in the Ukraine was to obtain large-scale grain supplies, the majority of the civilian population did not welcome the occupation and the peasants resisted handing over their grain. Quickly realising that they could not match German firepower and tactical skills in open battle, the Bolsheviks adopted guerrilla tactics, confronting them with an invisible enemy who wore civilian clothes and did not obey the laws of war. This was especially disconcerting for the German Army, which for more than three years had faced only conventional forces, and it retaliated with drastic measures. This response was identical to that of the summer of 1914 on the Western Front as then Germans tried to suppress the large-scale partisan war solely by brute force. Severe orders were issued to execute all captured 'bandits', destroy houses, impose collective punishments on villages and to take hostages. Rebellious villages were punished mainly by the levy of 'natural contributions' and fines, but sometimes entire settlements were also burnt down. By mid-April 1918, German troops were sacking and burning villages in a vicious, never-ending circle of repression and the situation in the Ukraine was beginning to slip out of their control.[32]

While German behaviour towards insurgents in the Ukraine was very similar in many ways to that of other European colonial powers it was excep-tional because of its ideological basis. The most striking similarity in German behaviour on the Eastern Front during both world wars was the treatment of Bolshevik and, later, Soviet partisans, who could expect no mercy if captured, being shot on the spot. Orders relating to the treatment of captured Bolsheviks stated explicitly that they 'are our enemies and thus are to be treated according to martial law'. The German Army displayed a ferocious anti-Bolshevism from the first day of the occupation. The Bavarian Cavalry Division was particularly ruthless as its officers were recruited almost entirely from the nobility, who were only too aware of what had happened in Bolshevik Russia to members of their social class. Reporting on an engagement against Bolshevik forces, the Bavarian Cavalry Division noted that, 'The enemy had high casualties. No prisoners were taken and the number of fatal casualties is estimated to be several hundred.' A subordinate unit, the 1st Bavarian Cavalry Brigade, supported the ruthless policy of killing prisoners of war on the grounds that it terrorised the Bolshevik forces. In dealing with captured Bolshevik partisans and their civilian supporters, some units issued orders that gave effectively a *carte blanche* for atrocities. Anyone carrying a weapon, offering resistance or hiding weapons after the deadline for their surrender to the Germans was to be shot. One brigade order noted that 'detentions are of limited assistance, they are to be avoided if possible'. Such draconian methods caused 'great bitterness', according to one order issued in late April 1918, although the punishments, such as the expropriation of cattle or grain, levied in the Ukraine were mild compared with the mass executions behind and on the Eastern Front during the Second World War.[33]

By the late summer of 1918, demonstrating a considerable grasp of what made a counterinsurgency campaign successful, the 4th Bavarian Cavalry Brigade had identified in one of its monthly reports the techniques that had made pacifying the territory possible. These included a well-developed intelligence service, close co-operation with the Ukrainian authorities and an extensive propaganda campaign. Thus, despite limited political support, the Germans were able to suppress the Bolshevik insurgency by early summer 1918, and relative peace reigned in the Ukraine until German troops withdrew in November 1918. During June and July 1918 the Germans subdued large Bolshevik uprisings in the area south of Kiev, having rebuffed a substantial Bolshevik amphibious landing near Taganrog on the coast of the Black Sea earlier in June 1918.[34]

This resulted in the worst German atrocity in the Ukraine. Colonel Bopp, commanding the 52nd Württemberg *Landwehr* Infantry Brigade, ordered the execution of more than 2,000 captured Bolsheviks and civilians, including women and children, from surrounding villages. This soon became common knowledge from letters written by German soldiers involved in the massacre and was even discussed in the *Reichstag*. When asked for an explanation General Gustav von Arnim (Bopp's commander), who disliked taking prisoners in battle, stated that the incident was 'not only humanly and legally justified, but also militarily necessary'. This was a result of earlier Bolshevik atrocities that had caused bitterness among German troops. No further action was taken. Several other lapses of discipline occurred in the Ukraine, although they were no longer officially condoned. Continual demands for food, which continued to be resisted by the peasants and frequently required the use of force, further alienated the rural population. Such methods meant that the long-term success of the German counterinsurgency campaign remained dubious because too few Ukrainians supported their invasion. Furthermore, ill discipline, such as looting, by the German troops continued as a result of low morale within the Army on the eve of defeat in November 1918. This was accompanied by a persistent anti-Slav sentiment, which as part of Nazi ideology would play a major role during the Second World War.[35]

In the aftermath of the First World War the German officer corps underwent a process of radicalisation, with devastating effects for both the planning and exercise of German anti-partisan operations during the Second World War. The experience of war on the Eastern Front had shown the officer corps the 'primitive' living conditions and the 'bestial' conduct of Russian troops during their invasion of East Prussia in August 1914, and, following the 1917 Russian revolution, conflict with the Bolsheviks. Together with experiences of extremely brutal fighting with the *Freikorps* (Free Corps) against insurgencies by the communists at home, the Poles in Upper Silesia and the Bolsheviks in the Baltic States, this reinforced the institutional paranoid fear of *francs-tireurs*. It also reinforced the traditional barbaric response to such resistance. In the 1920s *Reichswehr*, and in the *Wehrmacht* from 1935, a conventional war between two mass armies continued to dominate German military thinking. Apart from a short-lived flirtation in the mid-1920s with fighting a people's war to defend Germany, irregular warfare played no role in the German concept of a future war. The failure to develop any doctrine for irregular warfare was to haunt the *Wehrmacht*. The inability to counter insurgencies was the result of strict adherence to conventional warfare, which under the leadership of General Hans von Seeckt provided the basis for modern German military thought and early successes during the Second World War. Service in the East and ideological radicalisation influenced and brutalised the *Wehrmacht*'s conduct of anti-partisan warfare during the Second World War, building on previous thinking that had emphasised the ruthless repression of any opposition by irregulars.[36]

Second World War

During the Second World War, these 'most ruthless measures,' notably on the Eastern Front and in the Balkans, 1941–45, reached a magnitude and dimension that went far beyond that of the First World War and extended to a new dimension the brutal methods that had been employed earlier in the Franco-Prussian War, Africa and China. Mass executions of 'bandits', the deportation of civilians as forced labour and the wiping out of complete villages were systematically employed by the Army, although methods varied considerably from unit to unit. German soldiers performed well in conventional warfare, as a result of their training and experience, but not in the role of counterinsurgency, for which they were neither specifically trained nor equipped. The need to win hearts and minds, between 1941 and 1943, was precluded by the Army's long-standing institutional mindset, which had been reinforced by Nazi ideology. The Germans employed a scale of killing in its counterinsurgency operations that went far beyond military necessity. During 1942 and 1943 large-scale 'cauldron' (search-and-destroy) operations to surround and trap guerrillas were employed by German troops despite an explicit caution from OKH (*Oberkommando des Heeres*, the Army High Command) in autumn 1941 against their use. The 'cauldron' normally consisted of three stages. In the first, troops from different starting points encircled a suspected partisan area. The huge operational areas often meant that the partisans usually escaped. In the second phase, the Germans tightened the cauldron by advancing from all sides and looking for partisans and their supporters. In the third and final stage, the area was searched for a few days to mop up the surviving partisans. This manoeuvre was in use until late 1943, when the Germans developed a new tactic in the Balkans of advancing methodically along a front in order to force the partisans back on to a cordon line.[37]

These types of large-scale pacification operations were always brutal and ruthless affairs in which the Germans attempted to separate the partisans from their bases and support by turning whole regions into 'desert zones'. The Germans were well aware that they lacked sufficient manpower either to catch the quickly dispersing partisans during the operation or to garrison permanently any areas that had been captured. Thus, villages were burned down, the local population forcibly evacuated and all supplies seized or looted. Many units, notably the police battalions, did not waste time on evacuating civilians but instead just shot them on the spot. When the Germans withdrew, the partisans were still left in control of the area alongside the surviving population. This German strategy did little to win the hearts and minds of the local population as many officers and men showed little respect for the occupied Slavs, relying instead on violence and terror to coerce.[38]

The ultimate explanation for the failure of German occupation forces to suppress the partisan problem in the Balkans and on the Eastern Front was the fact that they were trained and deployed *strictly* as combat units, and did not attempt to set up their own or employ established civil authorities, or even act as an occupation force. The limited numbers of troops available for pacification owing to political and military requirements meant that they did not have the appropriate mindset, training or adequate numbers to pacify the insurgents. The pre-war neglect of counterinsurgency by the German Army made the failure of German anti-partisan strategy inevitable. When dealing with partisans between 1941 and 1944, the commitment to an institutionalised and ideological severity, based on reprisals against the civilian population that targeted 'ideological enemies', such as Jews and communists, meant that the *Wehrmacht* never really sought to win hearts and minds. The 'pacification' operations that killed non-combatants on a large-scale, alleging that they were 'bandits and bandit accomplices', went far beyond the dictates of military necessity.[39]

The employment of brutal methods as a substitute for a properly co-ordinated counterinsurgency strategy by the Germans proved to be the greatest asset to the mainly Soviet partisans, flooding their ranks with recruits as a result of the atrocities committed against the civilian population. Large numbers of executions in reprisal were expected, not only to remove potential threats and send a message to others not to engage in activities, such as attacks or sabotage, but also to deter the guerrillas by making their operations too costly in terms of human lives. In fact, this policy of terror, notably the public hanging of civilians from lamp posts and gibbets, only further alienated the population already humiliated by the German occupation. The execution of civilians exacerbated the antagonism towards the occupation forces, which in turn provoked even greater reprisals from the Germans, creating a vicious circle. Intermittent attempts to treat the civilian population, both the uncommitted and partisan supporters, more humanely might have proved successful, if they had been employed earlier and as part of a strategy that was applied and refined systematically over a number of years.[40] Otto Kumm, one of the most experienced SS officers in counterinsurgency, noted that the destruction and killing not only 'rapidly destroyed our credibility and increased the resistance against us' but was also indicative of a 'narrow-minded approach' that prevented the Germans 'as a collective military body' from learning lessons that 'should have been learned much sooner'.[41]

Although the Germans fought well and more skilfully than their guerrilla adversaries, their counterinsurgency strategy lacked the basic requirements for suppressing a partisan uprising. Most notable among these were unity of command, ample numbers of well-trained and equipped men, a sophisticated hearts and minds strategy to isolate the guerrillas by gaining the support of the local population and, above all, a co-ordinated political as well as military strategy. The German occupation was neither co-ordinated nor functioned smoothly. There was no coherent command structure, because responsibility for counterinsurgency was divided between the SS, *Wehrmacht* and police and the resulting confusion badly hindered the pacification of the partisans. The Army often clashed over policy and areas of control with the SD (*Sicherheitsdienst*, SS security service police) and *Waffen* SS. The failure to develop a co-ordinated counterinsurgency doctrine created immense problems, forcing individual German commanders to implement their own methods. Even with the development of a rudimentary counterinsurgency doctrine in late 1942, the lack of co-ordination within the high command and the fragmented approach to counterinsurgency warfare endured, handicapping the *Wehrmacht* until its final defeat in 1945. Moreover, with its increasing reliance on the SS to undertake anti-partisan operations, in part due to its wish to focus on the conventional war, the *Wehrmacht* gave full rein to the terror that alienated the population and prevented an effective occupation. The long-standing traditions that shaped the conduct of anti-guerrilla warfare as a whole during the nineteenth and early twentieth centuries influenced and conditioned German military thinking on counterinsurgency, which was particularly brutalising well before the advent of National Socialism. The harsh and often indiscriminate German countermeasures against partisans and civilians during the Second World War were driven not only by the Army's traditional attitude towards insurgents but also by Nazi ideology. It was a combination of the two, a fusion of powerful forces, which had catastrophic results.[42]

Balkans

Following the occupation of Yugoslavia and Greece by Axis forces in April 1941, Yugoslavia was divided. General Milan Nedić led the regime that was allowed to administer German-occupied Serbia while Ante Pavelić became *poglavnik* (leader) of an 'independent' Croatia, which included Bosnia and part of Dalmatia, where he conducted brutal campaigns against Orthodox Serbs and Jews. Resistance broke out in Serbia, Montenegro, Herzegovina, Bosnia and Croatia in the summer and autumn of 1941 as German troops transferred to the Eastern Front. The relatively weak remaining German and Italian forces held only the major cities and outposts and lacked the numbers and central command to suppress the rebels. In Croatia resistance blazed in the Serb-populated areas in response to the massacres of large numbers of Serbs by the Croatian fascist militia *Ustasha*. The bloody political and ethnic conflicts that beset Yugoslavia during the Second World War meant that winning the hearts and minds of the civilian population was vitally important. The communist partisans, the Serb nationalist *četniks* (Chetniks) and the fascist *Ustasha* were the most important in an array of ethnic or political groups that participated in the civil war that split Yugoslavia. In such a confused situation, the only method for the *Wehrmacht*'s overstretched occupation forces to gain and maintain control was to retain the support of the occupied population.[43]

The implications of this were ignored by the Germans whose main objective was to crush the resistance and restore control over the region. The German High Command viewed both the partisans and *četniks*, despite their ideological antipathies, in the same way, as part of a national uprising by the 'Serb conspiratorial clique'. The Germans simplistically regarded the partisan uprising as an unacceptable challenge against the legally constituted authority, referring

to them as 'bands', 'bandits', 'insurgents' and 'rebels'. They saw the conflict as
one fomented by a small number of criminals, believing that the majority of
the population accepted or at the very least tolerated the occupation. The
Germans, like many occupiers, could not conceive of the disaffection for their
occupation felt by the general population. They believed that the majority of
the population was either neutral or favoured the German-sponsored regimes
of Nedić in Serbia and Pavelić in Croatia.[44]

Serious frictions between the Axis powers also prevented a consolidation of
effort against the guerrillas.[45] Pavelić, commenting on the disharmony between
the Germans and Italians, noted that 'they are like cat and dog'.[46] There was
also friction with the Bulgarians over operations in Greece. Constant rivalry
occurred between the Italians and the Croatian *Ustasha* regime. The traditional
hostility between Serb and Croat made the two pro-fascist regimes in Serbia
and Croatia irreconcilable enemies and created an insoluble barrier to a unified
anti-guerrilla front. Attacks by *četniks* upon the Croatian and Muslim population in the Italian and German occupation zones angered both the Germans
and the Croatians. The ineptitude of the Croatian Army in dealing with the
partisans and *četniks* also caused frustration. These vicious rivalries between
the Axis powers and their allies preoccupied the German High Command
and prevented the implementation of a co-ordinated anti-guerrilla strategy.
In the end the failure to form a cohesive alliance unravelled the pacification
of Yugoslavia.[47]

When it suited their short-term purpose the Germans were willing to
recruit the assistance of indigenous forces, notably the Serbian *četniks* who
opposed Tito's communist partisans in the civil war and who knew the terrain
and their enemy well. The *četniks* were especially effective when scouting for
German units, indeed often serving in their ranks, when their leader General
Mihailovic co-operated with the Germans. Although some local German commanders differentiated between the contending guerrilla organisations, many
doubted the usefulness of using the various paramilitary organisations such as
the Serbian Guard (*Srpska Straža*) and Volunteers (*Dobrovoljci*) against the guerrillas. Believing that neither the paramilitaries nor the *četniks* were trustworthy
and should not be supported or armed to fight the partisans, the requests of
Nedić for heavy weapons was still being turned down as late as 1944. German
policy in occupied Serbia was to keep indigenous forces weak. Their puppet
government was unpopular and unable to raise the 15,000 men necessary to
form the National Guard. A Serbian Volunteer Corps provided only 9,000
men by 1943.[48]

The counterinsurgency effort was further undermined by the low priority
given to the south-east by the German High Command, which was increasingly hard pressed on all fronts and thus reluctant and often unable to provide

front-line troops and equipment. The German forces had to be content with whatever was to hand to overcome the deficiencies and to employ more effective organisation and planning to destroy a dynamic and mobile foe. Troops sent to the Balkans were often over-age and lacked equipment – what they had came from the plundered military stocks of France and Belgium and other conquered countries. Lacking the means to deal with an insurgency, the Germans deployed a strategy of terror to maintain their occupation, employing a series of encircling drives in order to wipe out the partisans and setting up a network of *stützpunkte* (strong points) in an attempt to contain and restrict their movements. This concept was similar to the construction of blockhouses and the employment of 'drives' to trap Boer commandos by the British during the Boer War. Owing to the preoccupation with the Eastern Front, suppression of the insurgency was improvised and, although the partisans suffered huge losses, they were never completely eliminated, repeatedly escaping encirclement. Facing a resourceful, elusive partisan enemy who employed guerrilla tactics in difficult terrain, German units often resorted to extreme violence as the result of a frustrated impotence. This brutality is also explained by the poor quality of the troops, who resorted to harsh measures to compensate for their failings and lack of training and equipment.[49]

As early as April 1941 General Maximilian *Freiherr* von Weichs issued orders to shoot all male civilians in areas of partisan activity, stipulating that 'guilt was to be assumed unless innocence could be proven'.[50] Sent to suppress the Serb national uprising in Serbia and Bosnia between August and December 1941, Lieutenant General Hans Böhme also encouraged his troops to terrorise the occupied population into obedience, thereby avenging Serb 'treachery' during the First World War. The massacre at Kraljevo in Serbia in October 1941 was arguably one of the two most infamous carried out by the *Wehrmacht* in southeast Europe (the other was at Kalavryta in Greece in December 1943; both were perpetrated by the 717th Infantry Division). Such incidents encouraged further outrages. In the six months following the outbreak of the Serbian revolt, 20,000 hostages were shot in reprisal for the deaths of German soldiers. These reprisals included the incident at Kragujevac in October 1941 when 2,300 hostages were massacred. These atrocities increased the resentment of the population, many of whom joined the partisans. These consisted mainly of Orthodox Serbians who were driven *en masse* into the ranks of the guerrillas by *Ustasha* massacres, religious persecution and reprisal executions in Serbia by *četniks* and the Germans. Other notable massacres took place at Foca in August 1942, including the burning of thirty-three villages by *četniks* and the killing of some 400 men and 1,000 women and children; at Sandjak by *četniks* in January and February 1943; and at the village of Kosutica in July 1943, where people of all ages were gunned down after the corpse of a *Waffen* SS soldier was discovered.

Under the communist leadership of Josip Broz, nicknamed Tito (the Hammer), a national war of liberation was unleashed against the German occupiers[51] by partisans who were aware that attacks were likely to unleash brutal countermeasures that would gain them more recruits. A report by General Wisshaupt noted that 'even with the most unrestricted reprisal measures' they did not 'restrain the continual growth of the armed revolt'.[52] The perils and frustrations from the German view of suppressing 'a tenacious and well-armed adversary', who neither gave nor expected any quarter while making full use of the mountainous terrain and dense forests of the region, weather and demography, was highlighted by one German commander in the field. He recorded that the troops 'had to fight a malicious enemy who could not be caught, in a wild mountainous country' and, hiding among the population, 'were everywhere and nowhere'.[53] The many atrocities initially intimidated the population but the brutality eventually alienated them irrevocably.[54]

Ambitious anti-partisan operations (Operations *Weiß I* and *Weiß II*, Operation White), between January and March 1943 against their main strongholds in Bosnia and Herzegovina known to them as the 'Liberated Zone' and to the Axis forces as 'Titoland', demonstrate both German brutality and also their inability to suppress the insurgents. These operations resulted in the deaths of large numbers of civilians, who had either been assisting the partisans, were suspected of doing so, or had merely been living in partisan-controlled regions. Instructions issued by the high command were harsh and ruthless, as was traditional in German counterinsurgency warfare. 100 hostages were executed for every German soldier killed by the partisans. Before Operation *Weiß I*, which employed thirty-seven divisions, totalling 90,000 German and Italian troops, and 12–15,000 *četniks*, commenced, the *Befehlshaber der deutschen Truppen in Kroatien* (Commander-in-Chief of German troops in Croatia), Lieutenant General Rudolf Lüters, issued two directives that make clear the severity planned. The first (7 January 1943) stated that mere suspicion, not proof, was sufficient for civilians 'to be summarily shot or hanged and their homes burned' and that subordinates should be given free rein in implementing this ruthless policy. The second (12 January) demanded that '*every* measure that ensures the security of the troops and appears to serve the purpose of pacification is justifiable ... No one should be held to account for conducting themselves with excessive harshness'. Anyone opposing the occupying forces was to be shot or hanged, and any villages or other places identified as partisan strong points would be destroyed. Similarly, the 369th Infantry Division issued an order (6 January 1943) that urged 'ruthless measures against the partisans and the population who support them'. However, having received advance warning of *Weiß I*, Tito used the ten days prior to the attack to destroy roads and bridges in the region, preventing a major encirclement, and, as a result, despite heavy

Axis pressure large numbers of partisans escaped across the River Neretva into the Durmitor mountains.[55]

In May 1943 a much larger operation (Operation *Schwarz*, Operation Black) used 117,000 Axis soldiers to cut off a large number of Tito's troops and to inflict severe casualties on the entire partisan leadership. The harsh methods employed at divisional level are demonstrated by the 'ruthless harshness' of the 373rd Infantry Division's operations in the Cardaci region in July 1943, in which anyone 'found with a weapon in their hands' was shot and any settlements that aided them were 'razed to the ground'. In total, 5,697 Partisans were killed, 2,537 civilians shot and 50 villages burned during the prolonged series of sweeps through the area. But, once again the Germans failed to wipe out the partisans, who broke out and returned to Bosnia. German counterinsurgency had been very close to eliminating the partisan threat, but Tito was given a vital breathing space by the Italian surrender in September 1943 that left German forces with vast new zones to secure.[56]

As the war progressed, however, the Germans began to employ new tactics and in May 1944, somewhat belatedly, OKW issued a manual, *Warfare against Bands*, which reflected many of these experiences and sought to disseminate the best practice. Learning from earlier failures to prevent break-outs from the 'cauldron' by partisans, units would attack partisan strongholds from all directions instead of attempting a total encirclement. Once split up, the guerrillas could be hunted down in operations known as 'cleaning up the cauldron'. Launched on 23 May 1944 to deliver the decisive blow, Operation *Rösselsprung* (Knight's Move) began with a parachute drop by the 500th SS Parachute Battalion on Drvar, which nearly trapped Tito in his headquarters. Tito escaped by air, flown out by the British. Advancing into 'liberated' territory, German units captured airstrips and supply dumps, disrupted command structures and broke up partisan formations. But a follow-up operation was abandoned at the end of August because of inadequate manpower that was the result of German troops having to be rushed to the collapsing Eastern Front. The German counterinsurgency campaign had ceased to function effectively and, gradually, in the autumn of 1944 and the spring of 1945, they were thrown out of Greece and Yugoslavia.[57]

Eastern Front

From the very beginning of Operation *Barbarossa* (the invasion of the Soviet Union in June 1941), German security measures were harsh. This followed the German tradition of dealing ruthlessly with insurgents, whether fighting against *francs-tireurs* in France in 1870, in Belgium and France from 1914 to

1918, or communists in Germany itself between 1919 and 1923. As a result, once on Soviet soil the German Army employed brutality in the name of security, committing numerous atrocities including village burning, mass executions, torture and indiscriminate shooting of civilians during anti-partisan operations, which guaranteed the hatred of the population. For example, in the Ukraine General von Salmuth (XXX Corps) ordered ten hostages shot for each German or Romanian soldier killed by the partisans. Similarly, General Kock (XLIV Corps), ordered the shooting of hostages in retaliation for the sabotage of telephone or railway lines, and in November 1941 the 454th Security Division reported the shooting in Kiev of 800 inhabitants in reprisal for sabotage. When Lieutenant General Georg Braun (68th Infantry Division) and some of his staff fell to partisans in Kharkov on 14 November 1941, the Germans executed fifty hostages on the spot and another 150 the next day. The German authorities also took a further 1,000 hostages, declaring that they would kill 200 for each new act of sabotage. Hitler's order to kill commissars was accepted by the German high command and applied in the field by most commanders, even those who later opposed him as the tide turned against Germany. General Erich Höpner, (*Panzer* Group 4), who was executed for his participation in the bomb plot of 20 July 1944, reported on 10 July 1941 that his men had 'liquidated' no fewer than 101 commissars. Many generals issued their own orders exhorting the troops to show no mercy to the Slavs. The German Army was complicit in the killings carried out by the SS and police, relying on them for the 'elimination' of partisans in order to husband its overstretched manpower.[58]

These ruthless measures ensured that the Soviet partisans were hit very hard but any successes were short-lived. German doctrine of extreme punishment failed to provide commanders and their troops with either an understanding of the political nature of the insurgencies on the Eastern Front or the well-developed policies necessary to win the hearts and minds of the local population. Moreover, with its emphasis on the superiority of the German 'master race' to the Slav 'sub-humans', Nazi ideology underpinned this brutal repression, preventing any significant evaluation of pacification tactics. The German Army therefore failed to evolve and implement an effective counter-insurgency strategy. Twenty-six police battalions (some 12,000 men in total) provided manpower for garrisoning occupied Russia but mainly waged a genocidal war against Jews and communists behind the front, killing about 5 million Jews and other 'undesirable elements' and rationalising their murderous actions in the guise of anti-partisan operations. The unease of some *Wehrmacht* officers at the indiscriminate killing of Jewish women and children was nullified for many by the desire to secure their communication lines and rear areas. While having few scruples about issuing orders to shoot men and uproot whole populations, the German generals worried that executing women and children might cause disciplinary and morale problems among the soldiers, and preferred to leave such unsavoury tasks for the SS and SD and indigenous police and militia units to carry out.[59]

The German High Command, permeated by Nazi ideology that mirrored its own repressive doctrine, saw no reason to change its methods in fighting an anti-Bolshevik 'crusade'. Ultimately, the collaboration of the German Army with the *Einsatzgruppen* (task forces), the murder squads of the SD and SS, in the ideological liquidation of communists, Jews, Slavs and others, coincided with its own doctrine of savagely suppressing opposition. It was also closely linked to its disinclination to participate in counterinsurgency operations. The emphasis on *rücksichtlos Vorzugehen* (ruthless action) in 'mopping up' rear areas led to the mass killing of prisoners of war, civilians who were suspected of supporting the partisans, and communists and Jews, who were handed over to the *Einsatzgruppen,* resulting in a vicious circle of violence. Having sanctioned the indiscriminate killing of Bolsheviks, civilians, Jews, partisans and prisoners, it proved difficult for the Army to control their brutalised troops or to implement a strategy to win hearts and minds.[60]

During the invasion of the Soviet Union, the German Army gave a very low priority to counterinsurgency, allowing the three Army groups (North, Centre, South) only three security divisions each, while the two *Reichskommissariat* (commissariates), *Ostland* (the Baltics) and Ukraine, which were responsible for administering the occupied areas, received an additional security division each. Thus, with a number of *Landesschützenbataillone* (territorial battalions), five small, poorly equipped third-class divisions (5,000 to 10,000 over-age troops each) assumed the task of securing the long communication lines and vast rear areas on the Eastern Front. German security forces in rear areas were given massive regions, which were much too big to control effectively or even patrol. About 110,000 security personnel controlled an area that, by the end of 1941, covered more than 850,000 square miles. For example, the 707th Infantry Division was responsible for an area twice the size of Belgium with only 5,000 over-age men and reserve officers and, in addition, it lacked transport vehicles, which made a fast deployment to troubled areas impossible. These forces needed the assistance of front line units, primarily *Waffen* SS, draining manpower from the embattled front. The shortage of men prevented the stationing of German troops in villages on a long-term basis. This could have provided an armed presence among the population on a permanent basis and separated the insurgents from their support. Even during the low-level partisan activity of 1941, the security forces were unable to deal with the guerrilla bands that roamed the forests and swamps of central Russia. The partisans simply dispersed and disappeared into areas where the few German troops could not hope to find them.[61]

Moreover, attempts to stamp out partisan resistance in Russia were not well planned or co-ordinated. In such circumstances, the German security effort against the partisans was unlikely to succeed. The Germans invaded the Soviet Union ill-prepared to meet a determined insurgency and lacking the prerequisites to implement the required strategy. They had neither a single anti-partisan organisation to co-ordinate policy or cover the entire occupied territory, nor any clear doctrine for the conduct of anti-partisan warfare. German Army doctrine when combined with Nazi ideology meant that it was impossible to achieve a successful hearts and minds campaign or 'to divide and rule'. Instead, whether Ukrainian, Belorussian or Russians loyal to Moscow, the population was treated uniformly as *Untermensch* (sub-human), subject to exploitation, deprivation, humiliation and death. The policy of terror without regard to operational necessity was the 'Achilles' heel' of the German occupation and counterinsurgency effort, infuriating the local population, who joined the partisan ranks. Most of the ultra-nationalist Ukrainians, many of whom had initially welcomed the Germans as liberators, turned against them *en masse* within a year.[62]

Attempts were made to improve the situation but it was too little, too late. The population had been irrevocably alienated from German rule. The partisans, although still disliked as representatives of the Soviet system, became the lesser of two evils. Moreover, subsequent attempts to win over the population were too little and too late. For example, in July 1943 OKH directed that all partisans should be treated as prisoners and no longer shot after capture, making official orders that had already been issued locally in 1942. However, the effects of this were negated because the German war economy desperately needed labourers, and captured partisans and civilians evacuated from the 'desert zones' instead of being killed were now deported instead as forced labour to the Reich. This economic exploitation and the savagery of security troops were not ameliorated by a campaign to win the hearts and minds of the indigenous population, which proved to be a fatal flaw. In securing the rear areas, the Germans were heavily reliant on the eastern volunteers, allowing the employment of the better trained and equipped German fighting troops on the front line. In summer 1942, the Germans also began to form units of Red Army prisoners, who had been recruited almost from the start of the invasion, under the slogan 'Combat against Bolshevism'. By 1943, it has been estimated that there were between 3,000 and 4,000 indigenous troops in each German division fighting on the Eastern front, amounting to a fifth of overall German strength. Indigenous auxiliary troops or *Hiwi* (an abbreviation of *Hilfswillige*, willing helpers), *Landeseigene Sicherungsverbände* (security units), *Schutzmanschaften* (local guard units) and *Ordnungsdienst* (police) were all deployed for security purposes against Soviet partisans. As many as 250,000 former Soviet citizens may have served in Waffen-SS 'Legions' and other front line combat formations raised by the *Wehrmacht*. In 1944 the renegade Russian General Andrei Vlasov was finally allowed to form the Russian Liberation Army but a viable and fully operational strategy of recruiting and 'turning' insurgents was never inaugurated properly. The treatment of the local Russian people was one of unremitting savageness, which served only to alienate them.[63]

The Germans launched the experimental *Selbstverwaltungsbezirk* (local self-government district) in Lokot, south of Bryansk. General Rudolf Schmidt (2nd *Panzer* Army) allowed the local population under its governor, Bronislav Kaminski, to autonomously run schools, police, the local government and the economy without German interference. Indigenous forces, the *Miliz* (Militia), later the *Volkswehr* (People's Defence), fought the local partisans under German supervision. Army Group Centre and the *Reichskommissat Ostland* also established the *Wehrdörfer* (fortified villages) in selected villages. A *Wehrdörf* was armed to protect itself against the partisan threat and given some autonomy of administration. The first stage was to halt the losing of ground, the second to stabilise the situation, and the third and final phase was to expand into areas controlled previously by the partisans. However, by the time this project was in full swing in spring 1944, it was already too late. Nevertheless, the *Wehrdörfer* project was of great interest to post-war armies, notably the French in Algeria and the Americans in Vietnam, who experimented with similar concepts. The US Army in particular was much influenced when developing its post-1945 doctrine by German responses to guerrilla warfare and its emphasis on conducting counterinsurgency as a mere adjunct to conventional war.[64]

From 1942, the Germans employed two forms of fighting, one passive and the other active, against the partisans. The 'passive' consisted of securing communications in the hinterland and using a system of strong points and patrols to defeat any insurgent attacks. Roads and railways were heavily protected and patrolled. All bridges, trains, marshalling yards and airfields were guarded and villages were under constant surveillance. Blockhouses, pillboxes, wire fencing and mine fields were employed to channel the enemy into pre-arranged 'killing zones' within the defence system. Mobile patrols operated as the eyes and ears of the defence. Forests were cleared to prevent partisans from approaching unobserved. This system of static defence was not particularly successful, because the shortage of manpower ensured that all potential targets could not be fully guarded. In the rear areas of 2nd *Panzer* Army alone, 1,100 partisan attacks occurred over a six-month period from May to October 1942.[65]

The Germans grasped very quickly that only an 'active' defence against the partisans was likely to succeed. Large-scale operations commenced in spring 1942 in the rear areas of Army Group Centre, notably *München* (Munich, March–April), *Bamberg* (March–April), *Hannover* (May–June) and *Vogelsang*

(Birdsong, June–July). The so-called 'partridge drive' of Operation *Hannover* took place against the 18,000 partisans of the Soviet Group *Belov* between Smolensk and the front some 125 miles (200km) to the east. The partisans were encircled, pinned against the River Dneiper, and then gradually rolled up by elements of three corps comprising seven divisions, spearheaded by the armour of the 5th and 19th *Panzer* Divisions. Overall, the operation was a great success with 5,000 alleged partisans killed but with 2,000 escaping southwards to Kirov, having inflicted heavy losses on the Germans (2,000). The following year, 1943, saw a continuation of large-scale operations, mainly in the *Reichskommissariat Ostland*, but also in the operational area. Among them was Operation *Zigeunerbaron* (Gypsy Baron, May–June), in which six divisions cleared the rear areas of 9th Army and 2nd *Panzer* Army for Operation *Zitadelle* (Citadel), the last big German offensive on the Eastern Front at Kursk-Orel. Once again, the Germans did not achieve full success and the bulk of the partisans escaped the 'cauldron', leaving the area resembling a 'desert zone'.[66]

Other 'active' operations employed small and flexible *Jagdkommandos* (hunting detachments), mostly consisting of a reinforced platoon deployed at corps, army rear, division and regimental level to beat the guerrillas at their own game, moving by night and laying ambushes for the partisans. These *Jagdkommandos* were made up of four squads each with an indigenous scout and well-armed with automatic and semi-automatic weapons, hand grenades and explosives. Mobility was paramount, with ponies for the summer and skis and sledges for the winter. They formed the basis of anti-partisan operations from 1941, receiving more professional training from 1942 onwards. Despite pioneering modern military counterinsurgency practices and becoming the model for similar units formed by the French in Algeria, the *Jagdkommandos* were always a local solution, rather than part of a well-developed strategy. They failed to eliminate the partisans and, moreover, owing to manpower shortages, each division had only one *Jagdkommando*.[67]

By 1944 new tactics were being employed reflecting the lessons and experiences since 1941. Immense stress was placed upon reconnaissance, the employment of *Jagdkommandos*, the importance of leadership, the proper delineation of authority and responsibility between *Wehrmacht* and the SS, close co-operation with civil authorities and initiative at tactical level. The manual, *Warfare against Bands*, issued in May 1944, differentiated several methods for encircling and destroying the partisans, which prevented them from escaping the cordon. A more flexible refinement was to form a 'cauldron' in a partisan-held area and to split it into sub-cauldrons to be encircled and destroyed in turn. This was a successful technique, which was adaptable to the terrain. One section proposed more humane treatment for the local population, revising the tactics of the past four years. But such measures came much too late.[68]

Conclusion

Operations elsewhere against the Maquis in France, notably in the Massif Central between February and March 1944 and in the mountain massif of the Vercors, near Grenoble in late July and early August 1944, and the Polish Home Army during the Warsaw Uprising between August and October 1944, using tactics employed in Russia, also resulted in atrocities against the civilian population. The German Army was very efficient at a technical level in mounting counterinsurgency operations, having learned much by the end of the war, but failed to learn, at least until it was too late, how to win over a population. The German officer corps never really progressed beyond the belief that terror alone would achieve pacification and failed to offer an alternative to the population other than a strategy that was based on suppression, even the extermination of parts of the population. The Germans defeated the insurgents on the battlefield but never achieved any lasting success in their anti-guerrilla campaigns as, owing to their failure to understand the political dimensions of counterinsurgency, they operated in a vacuum.[69] Thus, there was no concept of using 'the carrot and stick'. Whereas the use of terror and 'scorched earth' by other armies was ameliorated and the bitter pill coated with some sugar, the Germans made very little effort to win hearts and minds. Unlike other armies, notably the British Army in the Boer War, the German Army did not learn from its mistakes and, as a result, failed to develop a coherent doctrine to defeat insurgencies. There has been a long-term legacy as well. A newly reunited Germany, a key ally within NATO, contested the notion that NATO forces were fighting a long-term counterinsurgency campaign in Afghanistan and refused to commit regular combat troops. Moreover, burdened by the war crimes committed by the *Wehrmacht* during the Second World War, the political and military leadership were resistant to becoming involved in 'small wars'. Their reluctance to endorse counterinsurgency as an operational task meant that the *Bundeswehr*, which was organised, structured and equipped to defeat a Soviet onslaught in Central Europe, lacked a comprehensive doctrine for fighting a counterinsurgency, such as has been developed by the Americans, British and French over decades. The focus remained on conventional warfare that hampered severely its ability to adapt to operational realities on the ground in Afghanistan.[70]

2

THE LAST TO LEAVE

THE PORTUGUESE EXPERIENCE, 1961–74

Introduction: Portugal's Empire pre-1961

As the first European colonial power to arrive and the last to leave, Portugal was involved in Africa almost continuously between 1575 and 1974. Portugal established a presence in Guinea and the Cape Verde Islands, Angola and Mozambique by the end of the fifteenth century but only gradually extended its authority from the coast into the interior in a series of colonial campaigns during the late nineteenth century and early twentieth century, following the Berlin Conference of 1885. Thus, almost every year between 1875 and 1924 witnessed military expeditions in all three of the Portuguese colonies, while the last significant uprising, in Guinea, was suppressed as late as 1936. This revival of interest in Africa by Portugal was in response to the 'new imperialism' and 'the scramble for Africa' by more powerful European states, Britain, France and Germany, which looked enviously at the undeveloped and lightly occupied Portuguese enclaves. Just as military imperialism rather than commercialism drove the French conquest of Algeria, Indochina and Madagascar, the Portuguese conquest of the hinterland was mounted to maintain prestige and keep the military busy. The Portuguese empire was run on a shoestring and more than 90 per cent of the 'Portuguese' armies were African.[1]

The 1960s brought the challenge of a growing nationalism and a demand for decolonisation from Portugal's colonies. The Indian invasion of Damião, Diu and Goa in December 1961, when four centuries of occupation were ended abruptly within thirty-six hours, had a traumatic effect on the Portuguese Army, whose grim catchphrase became 'Remember Goa'. Its officer corps did not forgive the 'insult' of the court martial and cashiering of the commander, GeneralVassalo e Silva, and thirteen other officers of the heavily outnumbered and poorly equipped garrison at Goa.They saw it as an attempt by the dictator, Antonio Salazar, to blame them for failing to defend an indefensible colony against impossible odds.[2] General António de Spínola noted that the armed forces had been accused 'of not fighting heroically, while in reality our defeat was only a matter of days'.[3]

The episode increased the misgivings of many officers about the feasibility and desirability of retaining Portugal's colonies given the pressure from African nationalism for independence.The government was criticised for ignoring the implications for the country and the army of its political hard line in refusing to surrender the colonies in Africa, especially when Belgium, the Netherlands, France and Great Britain, encouraged by the USA, USSR and the UN, had already or were in the process of relinquishing their empires.While other colonial powers gradually withdrew from Africa, like the white settlers in Rhodesia, Portugal struggled to retain hers, although increasingly the officer corps questioned this commitment. In April 1961 the Defence Minister (General Júlio Botelho Moniz), the Army Minister (Colonel Almeide Fernandes) and the Army Under-Secretary (Colonel Francisco da Costa Gomes) opposed plans for defending the colonies instigated by Air Force Secretary General Kaúlza de Arriaga. When General Moniz told Salazar that a sustained campaign against decolonisation by the Army was 'a suicide mission' and conspired to remove Salazar, Moniz and Fernandes were both sacked. Costa Gomes and other dissident officers were transferred into sinecure administrative posts. Salazar then assumed responsibility for colonial defence himself and went on television to announce that the Portuguese would remain in Africa 'to defend Western and Christian civilisation'.[4]

Salazar and his successor, Marcelo Caetano, both shared the same fatal weakness – a continuing devotion to maintaining Portugal's empire in Africa, resisting pressures for independence.They regarded Angola, Guinea and Mozambique not as colonies, but as 'overseas territories', part of Portugal itself. Caetano had declared in the 1930s that without Africa 'we would be a small nation; with it, we are a great country',[5] and again in April 1970 that 'without Africa we would be a small nation; with Africa we are a big power.'[6] Behind this was a right-wing authoritarian ideology that emphasised Portugal's 'civilising mission' in Africa and, as Caetano noted in 1968, 'a paternalistic process of government and administration'.[7] Thus, there was very little impetus for change.

The Army's Response to Subversive Warfare

The watershed year was 1961, when, surprised by a bloody rebellion in Angola, Portugal had to deploy a substantial part of its Army. Prior to the African wars, the Portuguese Army was an ossified institution, which had not seen active service on a large scale since the First World War. By the mid-1960s many officers believed that it needed to modernise rapidly in order to face the impending challenges of counterinsurgency campaigns on three fronts in Africa. As a result of some officers being attached to their forces in Algeria, the Portuguese were heavily influenced by the French doctrine of counter-revolutionary war, adopting the counterinsurgency techniques of *la guerre révolutionnaire*. Numerous Portuguese political and military leaders shared the belief held by many French soldiers at the height of the Cold War that all colonial disturbances were orchestrated from Moscow and Peking. As the 1960s progressed it became clear that the communist world was too deeply divided to provide a unified strategy and that nationalism was a more important factor in colonial revolts. The Salazar government, however, continued to portray Portugal as being in the forefront of a global anti-communist crusade.

Gradually, the officer corps rejected the government's ideological claims of defending the West.[8] Although as late as 1973 the right-wing Kaúlza de Arriaga still believed in 'the Communist Grand Strategy' and 'Communist world expansion' which aimed 'to destroy Western civilisation',[9] Spínola probably spoke for the majority when stating in early 1974 that 'we must begin by divesting ourselves of the notion that we are defending the West and the western way of life'.[10] Instead, Spínola believed that 'we must develop and oppose an effective counter-revolution to face the revolution, thus combating ideas with ideas.'[11] The Portuguese Army could present itself as untainted by the brutal torture of insurgents. This had contributed to the defeat of the French Army in Algeria by alienating public opinion in the colonies, at home and internationally. But, in the Portuguese case, the dirty work of interrogation and torture was undertaken by the PIDE (*Polícia Internacional e de Defesa do Estado*, International and State Defence Police, from 1969 DGS, *Direcção Geral de Segurança*, Directorate-General of Security).[12]

Aware of earlier British and French experiences, the Portuguese Army general staff began to formulate a counterinsurgency doctrine in anticipation of trouble in Portugal's African colonies, which were already restless with the emergence of militant nationalism during the 1950s. In 1960, this was repressed savagely, notably at Mueda in Mozambique and Pidjiguiti in Guinea–Bissau. To learn about subversive warfare in 1958–59, Portuguese Army officers attended the British Army's Intelligence Centre at Maresfield Park Camp and the French Army's *Centre d'Instruction de Pacification et Contre-Guerrilla* at Arzew in Algeria. Following a visit to Algeria in 1959 Colonel Hermes de Araújo Oliveira, Professor of Geography and Military History at the *Academia Militar* (Military Academy), gave five lectures that provide a valuable explanation of subversive war from the Portuguese perspective. They were published as *Guerra Revoluncionária* (Revolutionary War) in Lisbon in 1960 through the patronage of the Ministry of the Army. This interest in counterinsurgency culminated in 1963 in the publication of the doctrinal manual, *O Exército na Guerra Subversiva* (*The Army in Subversive War*), written by Lieutenant Colonel Artur Henrique Nunes da Silva (Operations Branch, Army General Staff). This doctrine was heavily influenced by recent British and French experience, and in particular by the doctrine of *guerre révolutionnaire.*[13]

Within Salazar's rationale of retaining Portugal's colonies as an indissoluble empire, *The Army in Subversive War* provided the tactical doctrine for fighting guerrillas. It emphasised that the key to success was winning the confidence and loyalty of the population. In theory it followed the British principle of minimum force, which contrasted with the French practice of *ratissage* (raking over) that terrorised the population or the American practice of conducting counterinsurgency as an adjunct to conventional war. The Portuguese sought to win militarily but in a low-key, cost-effective and inexpensive manner. The traditional five phases of an insurgency evolved by Mao Tse-tung in China and developed by General Vo Nguyen Giap in Vietnam were simplified by the Portuguese into two broader phases: a single clandestine pre-insurrection phase and an insurrection phase. The Portuguese would employ a classic counterinsurgency strategy that employed resettlement of the population, the use of elite airborne troops to hunt down the rebels and the gradual 'Africanisation' of the colonial forces. At a tactical level, the emphasis was on fighting with professional skill in an innovative and irregular fashion, employing light infantry and small units to defeat the insurgent. Much of the Portuguese success was due to helicopter-borne assaults, although the gradual development of effective anti-aircraft defences by the insurgents countered this threat. The Army also addressed the root causes of the insurgency by employing 'civic action', including the manpower to build schools and wells as well as to teach and supply medical, health and sanitation facilities, to win the population's loyalty and support. The war against communism would be fought on all fronts, political, social and economic as well as military.[14] As Lieutenant Colonel Hermes de Araújo Oliveira noted, mobilisation would be national, relying not just on the armed forces, but on 'a country's every resource'.[15]

Also adopted from the French was the use of psychological operations (PSYOP) as an important tool to motivate the Portuguese soldiers and win over the African population to the Portuguese cause, countering the subversion of the insurgents. *The Army in Subversive War* was both comprehensive and impressive in

providing an appropriate and timely doctrine for the Portuguese Army's counterinsurgency operations in Africa based on the lessons learned by the British
and French in the past decade. Unlike the French and British in the 1950s, the
Portuguese in the 1960s had an established doctrine to guide them. Nevertheless,
during the colonial wars, two conflicting approaches emerged. These became
associated with the contrasting personalities of two of Portugal's most celebrated
generals, namely the more conventional Kaúlza de Arriaga in Mozambique and
the less orthodox but equally ambitious Spínola in Guinea–Bissau.[16]

Angola

The 'national revolution' began on 4 February 1961 when the MPLA (*Movimento
Popular de Libertação de Angola*, Popular Movement for the Liberation of Angola)
launched a series of attacks in Luanda, the Angolan capital. Although a tactical failure, they were a psychological success, drawing international attention
to the situation in Angola and prompting a furious reaction from the police,
army and local militia. This backlash included summary executions, leading
to the deaths of some 3,000 Africans, in direct contrast to the theory of using
'minimum force'. This was followed on 15 March 1961 by a rural uprising in
northern Angola initiated by the UPA (*União dos Povos de Angola*, Union of
the Peoples of Angola) that lasted into May and resulted in the massacre of
whites, *mestiços* (mixed race) and blacks. The Portuguese reoccupied northern
Angola during the dry winter season between May and October. General
Venáncio Deslandes, the new Governor-General, then announced that the
military operations had ended and that the policing phase had begun. The
arrival of some 20,000 reinforcements in July and August intensified operations.
A large number of insurgents were killed, but there were some severe white
settler reprisals and also some indiscriminate bombing by the Air Force, which
damaged Portuguese credibility. Although the Portuguese had regained control,
the region had been turned into a desert, and the revolutionary war had only
just begun. The rebels reverted to a guerrilla insurgency using sanctuaries in
the Congo, which had gained independence in 1960.[17] This begs the question
of how far the doctrine formulated by the Army General Staff had been disseminated since in the colonies the Portuguese appeared to lack any doctrine
other than brutality.

In March 1961 the small colonial army was able to suppress, with great
brutality, the insurgents who came across the Congo border. After the successful operations of the second half of 1961 to re-establish control, the security
forces concentrated on securing the economically important areas of northern
Angola in order to protect the capital, Luanda, and to gain the time to make

socio-economic development possible. As part of this strategy the local population were concentrated in strategic hamlets in order to isolate the guerrillas
from their popular support. The Portuguese strategy from the end of 1961 was
to intercept and annihilate infiltration by the insurgents, while keeping up
internal security to cut the population off from the guerrillas and employing
'psycho-social action' to retain the loyalty of the indigenous population. The
military strength in Angola gradually increased from 40,000 in 1962 to 60,000
in 1967, some fifteen to sixteen times the number of European troops that
had been in the country in early 1961, as Lisbon continued to send massive
reinforcements. These troops were supported by the PIDE, 10,000 uniformed
policemen, a corps of some 8,000 volunteers and a black militia. By 1967 some
75 per cent of Portugal's entire metropolitan army was overseas. Such a commitment was only made possible by financial support and equipment gained
through NATO membership. In 1962 the Portuguese forces, mainly soldiers,
were under the command of the Angolan Military Region, which had its headquarters in Luanda and was divided into four operational zones. These in turn
controlled sectors that usually corresponded with administrative districts. This
territorial organisation was to survive, with minor changes, until 1974. By 1965
the Portuguese seemed to have good reasons for being optimistic, having been
successful in re-establishing the situation in northern Angola. They were helped
by serious rifts between the different insurgent movements, the poor leadership
and training of the guerrillas and a lack of popular support for the insurgency.[18]

The Portuguese were able to continue the war in Angola owing to settlement
by a large white population and the disunity of the three separate regionally
based liberation movements. These were the FNLA (*Frente Nacional de Libertação
de Angola*, National Front for the Liberation of Angola), formerly UPA, in the
north; MPLA around Luanda and other central or coastal urban areas, including
many *mestiços*; and UNITA (*União Nacional para a Independência Total de Angola*,
National Union for the Total Independence of Angola) in the central plateau.
They not only confronted tremendous logistical obstacles but also increasingly
as the prospect of independence loomed put more energy into fighting one
another than ejecting the Portuguese. Indeed, UNITA entered into an alliance
with the Portuguese against the MPLA in the last phase of the war (*Operação
Madeira*). The Army was able to employ deserters from the guerrillas – the
Flechas (Arrows) – to attack rebel sanctuaries in Zambia. The operational strategy
throughout the thirteen-year war in Angola was based on the techniques
employed by the French in Algeria. A defensive network of garrisons held a
defined area while mobile reaction forces were employed in offensive search-
and-destroy operations against the insurgents who entered the zone. The
so-called 'static units' were supposed to foster the support of the local population,
gather intelligence, provide security for the local political-administrative

effort and achieve dominance over their specific sector using mobility and offensive operations. One commander complained in 1963, however, that the only objective of many garrisons, which were widely scattered because of insufficient manpower, was to safeguard their own survival. Other weaknesses, which were identified, included inadequate training, a lack of aggression, poor motivation and the shortage of regular cadres. Some civilians were critical of the amenities and lifestyle enjoyed by officers based in the larger towns in *a guerra d'ar condicionado* (the air-conditioned war) and of the intermittent bush warfare known as *a guerra de ginguba* (the peanut war).[19]

As regards internal security, the Portuguese were also able to confine the areas of guerrilla activity and influence. They regained control over the local population, with resettlement playing a decisive role as part of a policy of *reordenenmento rural* (rural reorganisation) in northern Angola, especially to the north of the Dembos area, to separate the inhabitants from the guerrillas. The widely dispersed populace was concentrated either on coffee farms or in specially built *aldeamentos* (fortified villages or strategic hamlets), and, if deemed reliable, armed for self-protection against the guerrillas. Some 150 *aldeamentos*, housing 2,000 people, were built in northern Angola between 1961 and 1964, offering the population some substantial benefits, notably security and jobs. The *aldeamento* system, employing counterinsurgency lessons learned in Malaya, the Philippines and Vietnam, gave the security forces control over the local population, denying its support to the insurgents, and the ability to undertake unrestrained operations against the guerrillas. It also gave the local administration the opportunity to improve the standard of living of the population, building schools, clinics, chapels and stores in the new villages. However, successful resettlement was dependent on good planning, the availability of finance and the quality of the personnel and the facilities. Without these, the movement from poor but traditional settlements would be resented, causing misery and dissatisfaction, which would provide recruits for the insurgents. The sheer size of the programme, which involved approximately a million people, meant that the prerequisites for success were not always achieved. In the early stages of the campaign resettlement was implemented in a somewhat hasty and haphazard manner while later on it was extended from the northern region to other parts of the country, where circumstances were less conducive. Thus, *aldeamentos* became one of the most controversial aspects of the counterinsurgency campaign, especially as it was resisted by white settlers, who perceived it to be a policy of 'appeasement' and 'softness' towards the rebels. Nevertheless, General Joaquin da Luz Cunha, the last Commander-in-Chief in Angola (1972–74), estimated that by 1972 a mere 5 per cent of the population was under insurgent control, 20 per cent was affected by insurgent subversion and 75 per cent was free of subversion.[20]

The Portuguese also used 'psycho-social action' to maintain the motivation and morale of the security forces. It was also employed to secure the loyalty and co-operation of the indigenous population, countering the propaganda of the insurgents, who promised independence, social advances and a bright future. The Portuguese promised equality, multiracialism, security and an improved standard of living. The Army provided civic action in support of the civil administration to reinforce the psychological action, notably education, food, medical, religious and welfare programmes. The garrison troops maintained contact with the local community, establishing schools and sports facilities, but these activities again were hampered by a shortage of qualified personnel and adequate funding. In contrast to American policy of free access in Vietnam, Portugal imposed strict censorship and control over the media in Angola, allowing only small numbers of friendly journalists to enter the war zone. The security police (PIDE, later DGS) also dealt ruthlessly with any resistance in urban centres, which remained comparatively tranquil and unaffected by terrorism. But these achievements could only buy time for more permanent and far-reaching changes, notably on the political front, which were not forthcoming. The Salazar regime was not prepared either to concede political power or to plan for a smooth devolution. Such concessions, too little and too late, would only be made by Caetano in the early 1970s.[21]

In particular, a new command and control system of councils attended by key civilian and military personnel was set up at provincial, district, municipal and rural ward level in 1968. This ensured effective co-operation between the security forces and civil administration and co-ordination of intelligence, logistics, psychological action and resources. The appointment of General Francisco da Costa Gomes, as Commander-in-Chief in May 1970, altered the circumstances in Angola at a critical time. A soft-spoken and shy officer of moderate views, Costa Gomes was a close associate of Spínola and, although a rather grey personality lacking his charisma, was not only very intelligent but also politically astute. He visited his troops regularly in order to gauge their morale and the progress of the campaign. Two years later, when Costa Gomes departed to become Chief of the General Staff, the crisis had been resolved, allowing some room for optimism.[22]

By the end of 1966 there was stalemate in all three Portuguese counterinsurgency campaigns, not just in Angola but also in Guinea and Mozambique. This forced the deployment of 120,000 men, the largest military force in Portugal's history, which consumed 40 per cent of the national budget. Although Guinea was causing concern, the situation in Angola was viewed with optimism as the guerrillas were mostly confined to the Dembos region. However, from 1967 onwards, the MPLA opened up a new front in eastern Angola, which posed a serious threat to the densely populated central heartland, and in mid-1970 the

Portuguese estimated that insurgents were operating in 40 per cent of Angola, an increase of 35 per cent since January 1966. Over the next three years (1971–74), however, the efficiency of the security forces and divisions within the insurgents' leadership resulted in the tide turning against the rebels. A number of improvements were introduced by the Portuguese, who implemented a more effective doctrine, notably better leadership, enhanced command and control, greater co-operation between civilian and military leaders at all levels, and a comprehensive intelligence gathering system that was key to successful operations. Others included the increased availability of helicopters, 'Africanisation' of the security forces, the resettlement of the population on a large scale in eastern Angola, political liberalisation and on-going socio-economic reforms.[23]

By 1972, the MPLA, in contrast to the PAIGC (*Partido Africano Para a Independênica da Guiné e de Cabo Verde*, African Party for the Independence of Guinea and Cape Verde Islands) in Guinea and FRELIMO (*Frente de Libertação de Moçambique*, Front for the Liberation of Mozambique) in Mozambique, had lost momentum. It was suffering a crisis in leadership and a severe loss in morale. The loss of Soviet support in December 1973 was a further major blow. This led to the reduction of operations to a sporadic level and a fight for survival. Indeed, the transfer of significant forces from Angola to Mozambique, where the security situation was deteriorating rapidly, was seriously contemplated by the Portuguese.[24]

Costa Gomes and General José Manuel de Bethencourt Rodrigues, a former Minister of the Army and one of Portugal's 'most brilliant young tacticians', who had been brought to Angola at Costa Gomes' request, were the architects of the successful counterinsurgency strategy. This strategy was aided by increased force levels, better equipment and more helicopters. Portuguese operations were carried out with greater aggression and panache. They employed small groups of black troops to intercept infiltration by insurgents and ambush concentrations of guerrillas in areas such as the Dembos Forest, which hitherto had been sanctuaries for the rebels. Annual dry season offensives, such as *Operação Siroco* (Operation Sirocco) between July and October 1970, were employed against the MPLA infrastructure and principal infiltration routes.[25]

By early 1974 the war in Angola had swung in Portugal's favour. The threat to Portuguese rule had been contained and the challenge to the security forces largely overcome. Despite administrative and military deficiencies, Portugal's strategy had been successful and the Army's counterinsurgency doctrine, employing lessons from the British and French experience, was broad and effective, although there was a gap between theory and practice. The main flaw was the failure to develop a political strategy to meet the aspirations of Africans for independence. General Spínola was correct in arguing that Portugal's colonial wars could not be won solely by military means and that a political settlement was necessary to end the war successfully. As a result, the military *coup d'état* in Portugal in 1974 led to an eventual withdrawal from Angola.[26] As in Rhodesia and Algeria, military success had proved futile unless it was combined with a successful political strategy.

Mozambique

Revolutionary war did not erupt in Mozambique until September 1964. Operating in small groups and using traditional guerrilla hit-and-run tactics, the insurgents of FRELIMO were initially contained by the Army in the north in Cabo Degado and Niassa away from the main population zones. Operating from Tanzania, FRELIMO made little headway, but after 1968, following the resolution of its internal divisions and a change of leadership, it was able to infiltrate from bases in Zaire via Zambia. The strategic but neglected province of Tete provided a 'back-door' for infiltration of Rhodesia, which surprised the Portuguese. Short of equipment, the Portuguese, commanded by Brigadier General António dos Santos, were on the defensive. They relied on resettlement and military fortifications in the north and only occasionally launched operations against FRELIMO bases and sanctuaries. The appointment of Brigadier General Kaúlza de Arriaga as Commander-in-Chief in March 1970 brought a more dynamic counter-insurgency strategy. He emphasised the offensive, notably search-and-destroy operations in pursuit of the insurgents and destroying their bases. They were supported by psychological and social reforms to win over the local population. This change in strategy reflected the increase in Portuguese troops available, from 25,000 in 1964 to 40,000 by the end of 1965 and to some 60,000 by the beginning of 1970. Kaúlza de Arriaga (nicknamed the 'Pink Panther') was a right-wing supporter of Salazar who had made his name as a commander in Mozambique and was a friend and admirer of General William Westmoreland, the controversial American commander in Vietnam. Kaúlza de Arriaga believed that the war in the forests and bush of Africa could be won through determination and leadership. Some of his tactics, such as the employment of highly mobile troops with air support, copied those of Westmoreland. A flamboyant and ambitious character, who was conservative in both military affairs and politics,[27] he declared with an over-optimism reminiscent of Westmoreland that 'in Moçambique itself, the present war cannot now be lost', victory being 'only a question of time'.[28]

In 1970 Kaúlza de Arriga launched an ambitious dry season offensive, *Operação Nó Górdio* (Operation Gordian Knot) followed by *Operação Fronteira* (Operation Border), in the north to destroy the FRELIMO strongholds and infrastructure along the Mozambique–Tanzania border. He employed some 10,000 troops with air and artillery support. Lasting seven months and using modified hammer-and-anvil tactics the operation failed, at great expense, having neither prevented insurgent infiltration nor destroyed the FRELIMO guerrillas, who escaped or hid among the population. Further operations were required in the north, Operations Garrotte and Apio, during 1971. Large-scale dry season offensives were also mounted between May and September in eastern Mozambique in 1966, 1968 and 1972. The Portuguese asserted that Operation Attila in 1972 removed half of the guerrillas operating in eastern Mozambique. Increasingly, however, there was an emphasis on the employment of continuous small unit operations, which were more successful in disrupting insurgent activities than the large-scale operations. The latter merely disrupted them for a limited period and left the guerrilla infrastructure intact. An equally ambitious project was to build a defensive network of *aldeamentos* along the northern border with Tanzania, Malawi and Zambia, creating a *cordon sanitaire* and employing a scorched earth policy, which was then extended to other areas of Mozambique, notably Tete during 1972. By 1974 almost a million inhabitants had been resettled. But, as in Angola, resettlement was handicapped by poor planning, shortages of time, funds and resources, and disruption to traditional customs. These resulted in poor conditions that were counterproductive, providing recruits for the guerrillas.[29]

Although General Kaúlza de Arriaga, in a speech to his soldiers, emphasised the importance of resettlement for 'convincing the minds and conquering the hearts' of Mozambicans, in reality too often the population was simply resettled by coercion with the barest preparation as part of a knee-jerk response to insurgent infiltration. This deferred economic and social development to be implemented at a later date and left the relocated villagers in discomfort and distress. Resettlement was far more ambitious and on a much greater scale than attempted by the British in either Malaya or Kenya, but Portugal simply lacked the resources to implement it to the required standard. Nevertheless, resettlement enjoyed some success and was more successful than Diem's ill-fated strategic hamlets in South Vietnam.[30]

General Kaúlza de Arriaga was one of Portugal's most distinguished public figures, having served for seven years as Under-Secretary of State and then as Professor of Strategy for Higher Military Studies and as Chairman of the Nuclear Energy Board. However, his tenure as the commander of the Portuguese forces in Mozambique was not a success and he retired in 1973. While he was able to prevent FRELIMO from hindering the construction of the Cahora Bassa dam, he was unable to prevent full-scale penetration by insurgents of Tete province. Moreover, establishing his own 'personality cult', he often spoke on the radio and in the press about 'the hundred days war' until the stage was reached in 1973 when few of his announcements were taken seriously. This affected the army, whose morale slumped and, as with American troops in Vietnam, there were problems with poor discipline, motivation and combat performance. Many of the younger conscript officers (*milicianos*) serving in Mozambique became openly sceptical of the High Command during the final eighteen months of Kaúlza de Arriaga's tenure. Certainly the level of preparedness and initiative was inferior to that in Guinea, where conditions were far more severe than in Mozambique. Kaúlza de Arriaga's generalship did not compare to the brilliant tactical leadership in Guinea of Spínola, who was critical of his inept performance. Consequently there was relief when Kaúlza de Arriaga, whose political ambitions and preoccupation with politics drew criticism from some soldiers, was replaced in August 1973. His successor, General Tomas Basto Machado, a quieter, able and imaginative personality, assumed command during the most trying period of the war when the Army's prestige had been seriously eroded both by guerrilla infiltration and allegations of a massacre in December 1972 of some 400 people at the village of Wiriyamu. The Portuguese, and indeed Kaúlza de Arriaga himself, were accused of conducting a counterinsurgency campaign that employed reprisals and torture. It also used forced labour on military projects in war zones together with a scorched earth policy designed to deny food and shelter to the insurgents. This pattern of behaviour undermined Portuguese attempts to win hearts and minds.[31]

FRELIMO relied on a strategy of attrition, employing infiltration of the countryside, hit-and-run raids and the planting of landmines to wear down the Portuguese forces, which remained undefeated on the battlefield. The Portuguese were unable to impede FRELIMO's growing strength. The erosion of the security situation was such that in 1974 white settlers, critical of their ineffectiveness and inability to protect them against guerrilla attacks, clashed with the Army. By 1974 the level of the insurgency in Mozambique, especially in the Tete region, had grown in intensity to the extent that the insurgents were able to operate in some areas unchallenged by the Portuguese security forces. FRELIMO conducted its own campaign of terror against the 'loyalist' population, undertaking 689 assassinations and 6,500 abductions between 1964 and February 1973. During 1971 fifty-five chiefs in Tete Province alone were murdered. By 1974 the Portuguese had lost the strategic initiative in Mozambique but the balance of power still remained generally in their favour and, despite growing problems, the situation was not serious enough to suggest that military defeat was likely.[32]

Guinea–Bissau

In July 1961, the PAIGC led by Amílcar Cabral launched its first guerrilla attacks against the civilian population in the north-west in villages near the

Senegal border. In March 1962 PAIGC began a full-scale revolutionary war with attacks and ambushes on Portuguese troops. During 1962–64, PAIGC gained control of the southern littoral and moved into the north. Between 1965 and 1968 the insurgency led by its new military arm, FARP (*Forças Armadas Revolucionárias do Povo*, People's Revolutionary Armed Forces), extended from the north-east and south-west to the centre, and then into the eastern Fula heartlands. During 1965–66, FARP concentrated on destroying army garrisons, which blocked PAIGC supply lines, and strategic hamlets, which had been built to isolate the population from PAIGC. Although the Portuguese held the towns and villages, the PAIGC controlled the countryside. After *Operação Tridente* (Operation Trident) in February 1964, the Portuguese relinquished the initiative to the PAIGC by retreating into outposts, akin to American fire-bases in Vietnam. They relied on air power, which inflicted heavy casualties and caused 60,000 refugees to flee to Senegal but did not stop FARP. Trident, a conventional amphibious assault in the south by combined arms, was a Pyrrhic victory, failing either to isolate the insurgents from their support or to establish a long-term Portuguese presence in the Como Island region, adjacent to insurgent sanctuaries in Guinea–Conakry where cadres were trained. It was symbolic of the way in which the insurgency was being opposed by conventional methods that failed to provide long-term results.[33]

Within a few years, the Army had its back to the wall and defeat was beginning to look certain. Large-scale sweeps through the swampy terrain by the Portuguese, of the type employed by Westmoreland in Vietnam, had little effect against the guerrillas, who enjoyed sanctuary in Guinea–Conakry and Senegal. By 1967, the Portuguese had abandoned the rural areas and were concentrated in the towns. In 1968, after a fact-finding tour of Guinea, Brigadier Spínola reported to the Supreme Defence Council that the campaign was being lost. He criticised the inflexible and visionless strategy of the rotund and inept Governor-General Arnaldo Schultz (a lacklustre veteran of Angola, a former interior minister and a Salazar loyalist), who had predicted when he arrived in 1965 that the war in Guinea would be over within six months and neglected winning hearts and minds of the population with civic action. Spínola was also critical of the Army leadership, which had glossed over the magnitude of the reverses that had been sustained. Rather than pursuing an all-out conventional war, Spínola's alternative strategy was to adopt a more political approach to win the support of the Africans. This would gain time to allow a political settlement to be reached with the insurgents, although Spínola did not advocate full independence.[34]

Spínola was sent back to Guinea with *carte blanche* to implement his radical ideas. At a time when Portuguese fortunes were at their lowest ebb, he replaced General Schultz, who left with Portuguese authority diminishing fast. In May

1968 Spínola took command of the Portuguese troops in Guinea, at a time when it was estimated that PAIGC controlled 50 per cent of the territory and contested a further 25 per cent while the Portuguese controlled only 25 per cent. In order to minimise the bureaucratic red tape and inter-departmental rivalry, which hitherto had dogged operations in Guinea, Spínola insisted on having control of both the civilian and military administration as both governor and commander-in-chief, following the example of Templer in Malaya of establishing unity of command.[35]

Spínola, an austere but brilliant and stylish cavalry officer, who had won international prizes for his horsemanship, had panache, a 'considerable reputation for eccentricity' and, rather like 'Monty', a love of publicity, cultivating 'an image' with his trademark monocle, brown gloves and riding whip. Spínola's career was not hindered by his father's influence as a trusted adviser to Salazar and his own association with a paramilitary gendarmerie, the GNR (*Guarda Nacional Republicana*, National Republican Guard), responsible for the internal security of the regime. He had been an 'observer' on the Nationalist side in the Spanish Civil War. On the outbreak of the African Wars, Spínola served with distinction as commander of cavalry in anti-guerrilla operations in Angola for three years (1961–64), establishing his reputation not only as a professional soldier who demanded high standards but also for his toughness, riding on horseback in the jungle, and his insistence on visiting the front line. In Guinea, Spínola, known as '*O Velho*' ('the Old One'), emerged as a charismatic patrician conservative from the old Portuguese social elite. He was dismissive of civilian politicians, intolerant of incompetence, reforming and openly critical of the regime. Aware that the campaign in Guinea was futile, Spínola provided Salazar with a famous metaphor, describing guerrillas as fleas that prevented sleep and, being extremely difficult to locate, eventually frustrated the 'host', who died of exhaustion.[36]

Spínola provided dynamic leadership as commander in Guinea; in four years he reinvigorated the conduct of the campaign and restored the morale of the troops. He developed four main essentials. The development of a co-ordinated intelligence organisation was key to success. An extensive hearts and minds campaign, which employed civic action to provide homes, roads, schools, teachers and medical services and integrated Africans into the government and administration, was established under the slogan *por um Guiné Melhor* (for a Better Guinea), to win the support of the population. He introduced an integrated command system. Crucially, Spínola's strategy was underpinned by the realisation that counterinsurgency tended to be of long duration and that 'quick fixes' would not work. Travelling around the country by helicopter and jeep, Spínola got to know the local conditions and population, and his troops, who were unaccustomed to seeing their generals. He also implemented an elaborate training programme, IAO (*Instrução de Aperfeiçoamento Operacional*, Operational Proficiency Instruction) to acclimatise troops and units to the theatre, borrowing methods used by the British in Malaya.

Spínola heavily mined the borders and deployed troops to seal them off. The number of garrisons was reduced drastically but key outposts were retained and reinforced. They formed vital bases on major roads, which were tarred to improve the army's logistical capabilities. With his ability Spínola inspired great loyalty, particularly among many officers, but he could also provoke resentment in those from less fashionable units who felt excluded from his inner circle. He was prepared to make enemies by weeding out incompetent officers and to remove others from cushy support posts to fight in the bush. He built up a battalion of commandos, which was commanded by two white officers but led by black officers and NCOs. He created an African Forces Militia, which was well paid and trained, to provide grass roots support and local defence for *aldeamentos*, which were also increased in numbers. Spínola also exploited the tactical mobility provided by helicopters in counterinsurgency operations to deploy small, aggressive commando units based on the *bi-grupo* (bi-groups) employed by the PAIGC guerrillas. These *caçadores* (hunters) were airlifted by helicopter to hunt down FARP's units and infrastructure, especially its bases and cadres, winning back the initiative for the Portuguese. From 1968 they were able to recover most of the contested regions and now controlled all the centres of population.[37]

As a result of 'civic action' the Portuguese claimed the construction of 15,000 houses, 164 schools, 163 fire stations, 86 fountains and 40 infirmaries. Ninety-five per cent of doctors and 20 per cent of teachers in Guinea were provided by the Army, which also made improvements in agriculture and transportation, by building airfields and roads. In addition to economic development, Spínola attempted without success to establish a political counterforce, FUL (*Frente Unida da Libertação*, the United Liberation Front), recruiting PAIGC's opponents to undermine it. He also attempted to make the government more representative by employing blacks in the administration and ensuring that they now received justice in the courts. To ensure grievances were heard and dealt with, officers met representatives of tribal and religious groups and a system of ethnically based 'People's Congresses' was set up. According to their constitution they were 'to form strong instruments of psychological mobilisation around government policy' and to provide 'dialogue between the government and the people', which would culminate with a 'General Congress' in Bissau. Spínola also presided over a legislative council of fourteen elected members. Spínola's socio-economic reforms stole much of PAIGC's thunder. The adoption of 'psychological warfare' by the Portuguese to counteract PAIGC's political indoctrination was drawn from the experiences of other armies, notably the French in Indochina and Algeria. By 1968 the employment of psychological action to show the merits of a liberal, Western society in stark contrast to the rigid and authoritarian communists, was a well-established component of counterinsurgency strategy. However, authoritarian Portugal faced real difficulties in convincing even its least critical allies that it could provide a real alternative to the liberation movements, especially given its destruction of villages that resulted in heavy civilian casualties. The incident that highlighted the tensions between the differing military and political requirements that often plague counterinsurgency campaigns was the Portuguese invasion, *Operação Mar Verde* (Operation Green Sea), of Guinea–Conakry in November 1970. Although successful militarily, this operation was unsuccessful politically, failing either to topple the regime supporting the rebels or to destroy the PAIGC sanctuaries. It was a diplomatic disaster for Portugal.[38]

Spínola's conviction was that the war could not be ended by a military solution, but that there had to be a negotiated, political settlement. In 1971, having regained enough initiative on the ground to be able to negotiate from a position of strength, the Portuguese were given the opportunity to reach a political settlement with President Senghor of Senegal acting as mediator in exploratory talks. But although in mid-1972 Spínola was allowed to meet Senghor, like Salazar, Caetano refused to countenance a ceasefire followed by direct meetings between Spínola and Cabral leading to a joint Portuguese-PAIGC government in Guinea. Spínola was 'profoundly shocked' to find that Caetano preferred 'an honourable military defeat' rather than 'negotiate with the terrorists' which, according to the 'domino theory' of Cold War strategy, would inevitably open the door to further Portuguese concessions in the other 'overseas provinces'.[39]

Following the breakdown of talks and the departure in 1973 of Spínola, who left morale at an all-time high, the counterinsurgency campaign was continued under General Bettencourt Rodrigues. However, the situation in Guinea began to deteriorate during 1973, and it was clear that Spínola's inspiring leadership had served only to paper over serious cracks in the Portuguese war effort. After his departure attitudes hardened on both sides, especially following Cabral's assassination by PAIGC dissidents in January 1973. In contrast to the Portuguese forces, which suffered from shortages of equipment, PAIGC began to receive a steady stream of arms following the end of the Vietnam War. These included modern ground-to-air missiles, such as the SAM-7, which now restricted the operations of the Portuguese Air Force. This deprived the Army of its air support and, together with the destruction of two important Portuguese garrisons by PAIGC bombardments, the loss of aircraft caused a corresponding and devastating slump in morale. The introduction of sophisticated weapons, including some sorties by MiG fighters from Guinea–Conakry against Portuguese positions, tipped the balance, both militarily and psychologically, towards the insurgents.[40] As one officer, a *miliciano* (conscript), noted, 'We knew then that we were headed for defeat'.[41] The insurgents continued to rest, train and strike from their bases in the sanctuaries of Guinea–Conakry and Senegal, employing hit-and-run tactics to avoid contact with security forces. Neither side could achieve out-and-out victory but PAIGC once again controlled large areas of Guinea, which was Portugal's Achilles' heel. It revived memories of Goa and provided the principal reason for the Army's decision to overthrow the government in the coup of 25 April 1974.[42]

1974 Coup

By prolonging the war, which was increasingly senseless and slipping away from the Portuguese, Caetano encouraged discontent in the armed forces, which resulted in a coup. The regime's growing inability to meet the demands of the African Wars unified the officer corps against it as the Army began to show signs of cracking under the pressure such as low morale, the fatigue of the small regular cadre, a dependence on black troops, the poor training of conscript soldiers and growing tensions between regular and conscript officers. The situations in Guinea–Bissau and Mozambique, which had earlier been well in hand, owing to the poor training and equipment of the guerrillas, was becoming increasingly desperate.[43]

The Portuguese Army was mainly made up of conscripts led by a small cadre of poorly paid regulars who served two-year tours of duty in Africa, separated by six-month breaks at home. The counterinsurgency campaigns became a huge drain on regular junior and middle-ranking officers, who felt increasingly aggrieved. The average conscript was hardy but often illiterate and very insular, being routinely racist. This attitude contributed to the high incidence of atrocities that were committed. Junior officers were also mostly conscripts as a result of the Military Academy's failure to produce sufficient regular subalterns. The shortage of regular junior leaders and NCOs meant that the burden of training the troops fell upon their conscript counterparts, whose morale was suspect from the mid-1960s. They were insufficiently trained, dedicated or motivated to ensure that their troops were properly trained and led. The situation deteriorated when the length of conscription was raised from two to four years in 1967 in order to provide the manpower required to fight the war. The lack of motivation and aggression among conscript subalterns, an indicator of low morale, was a matter of increasing concern to senior officers.[44]

As the war dragged on and the insurgents improved, so the professional standards of the Portuguese Army and the quality of the conscripts declined. Many potentially good fighting men obtained deferments or evaded service by emigrating. Moreover, a 'cost efficiency' system of training and manpower allocation borrowed from the US Army undermined unit morale as the conscripts were trained in large camps rather than by the men who would lead them in battle and who were knowledgeable about the local conditions. All troops in Africa fought as infantry even though artillery, engineer and cavalry units were not trained for such tasks. Officers at the front found that regimental cohesion was seriously undermined by the continual redeployment of soldiers, NCOs and officers. In marked contrast to the tightly knit regimental structure of the kind employed by the British Army in Malaya, Kenya and elsewhere, Portuguese commanders were forced to patch together units in the battle zone. The result was poor tactical efficiency and low morale. The Portuguese, like the Americans in Vietnam, paid a stiff price for this more 'efficient' method of allocating manpower. Moreover, the general staff ignored suggestions that regular cadres should be permanently based in the colonies, choosing instead a system of short two-year tours. As with a similar American system in Vietnam, this gave officers no personal stake in the colonies or the conduct of the war and opened the way to careerism.[45]

This system was exacerbated by Portuguese techniques, copied from French counterinsurgency methods, in which stationary garrisons commanded by *milicianos* undertook civic action, patrols, and gathered intelligence while elite units of commandos, paratroopers or marines moved around the countryside doing most of the fighting against the insurgents. These elite troops, often black, commanded by regular officers or keen *miliciano* volunteers, produced a sharp contrast to the frequently timid stationary garrisons led by *milicianos* who were fed up with the war and felt that they were being used to do 'the dirty work'. These units often did not fight well, evading battle and staying in their bunkers rather than patrolling. The anti-war atmosphere in Portugal was demonstrated by the desertion rate in which some 110,000 conscripts did not report for military service between 1961 and 1974.[46]

The ill-feeling and rivalry between conscript 'temporary gentlemen' and regular officers caused bitter discontent, which led to the coup of April 1974. The final straw was the publication of the so-called Rebelo decrees by the government in June 1973 in an attempt to remedy the continuing erosion of the quality and quantity of candidates joining the regular officer corps by seducing *milicianos* who excelled in combat into joining the regular Army. Unfortunately, these measures deeply embittered regular officers, especially the captains and majors, who would be overtaken in rank and felt that their elite, professional status was being harmed. They complained that the proposed reduction in officer training at the academy wounded the 'dignity, prestige and professional *brio*' of the officer corps. Discontent and resentment in the Portuguese Army, as in the French Army in Algeria and the US Army in Vietnam, was concentrated among the middle generation of fighting officers in their 30s and 40s. While regarding themselves as superior to the non-regulars, they had suffered the most from the Army's declining prestige, although they had borne the brunt of the fighting and isolation of the African wars. In the absence of enough middle-ranking officers to staff overseas posts, they had become increasingly exhausted mentally and physically after lengthy 'dirty service' in the bush and jungle of Africa. War weary and angered by the Rebelo decrees, they formed the backbone of the *Movimento dos Capitães* (Captains' Movement) from which the MFA (Movimento das Forças Armadas, Armed Forces Movement), which led the coup of April 1974, emerged.[47]

No European imperial power fought so long a colonial struggle as that of Portugal between 1961 and 1974. The length of the war put enormous pressures on the nation and the Army in particular. During the African Wars the enormous expenditure on defence amounted to more than 7 per cent of the GNP and 40 per cent of the annual budget, placing Portugal in the same category as Israel. By 1967, some 75 per cent of the metropolitan Army was serving overseas. By 1970, with 150,000 troops in Africa, Portugal had deployed a proportion of its population that was five times greater than that deployed by the United States in Vietnam in the same year. The Portuguese had called up a greater proportion of those eligible for military service than any other country except Israel and the two Vietnams. For the Army of one of the least rich and developed countries in Western Europe to fight three wars simultaneously for such a prolonged period without suffering military defeat was a substantial accomplishment. However, the strains of this effort resulted in political collapse following the coup of April 1974.[48]

Following decolonisation by Belgium, France and Great Britain and the serious deterioration of the situation in Guinea, there was also a growing feeling among Portuguese officers that the time had come to abandon the colonies, which were not only morally reprehensible but also increasingly difficult to defend. Support on the home front for the seemingly endless African wars had collapsed, leaving the professional officer corps to battle on without much support from the conscript officers and men, who were either indifferent or opposed to the war. Aware of the defeats suffered by the French in Indochina and Algeria and the lack of progress made by the Americans in Vietnam, Portuguese officers questioned the commitment to the African wars, contrasting the military assets of the French and Americans to the limited resources available to Portugal. It was clear that the war could now only be won by a political settlement, a point that was made by Spínola in his book, *Portugal and the Future*, when it was published on 22 February 1974, causing a sensation with the sale of three editions in a fortnight. Spínola and Costa Gomes, now respectively Vice Chief of Staff and the Chief of Staff for all the armed forces

(*Estado Maior das Forças Armadas*) were sacked in March 1974 by Caetano, who was under pressure from the hard-right who remained committed to retaining Portugal's colonies, for refusing to endorse publicly the regime's colonial policy. The hard-line, right-wing General Joaquin da Luz Cunha was recalled as Commander-in-Chief in Angola to replace Costa Gomes. Believing that 'the Honour of the Army' was threatened, the officer corps intervened to remove the government. The coup of April 1974 and the death of General Franco in Spain in 1975 signalled the final fall of pre-war European authoritarianism in its Iberian redoubt and the end of the Portuguese Empire, one of the longest lasting of its kind. It brought the independence of its African colonies, leaving Rhodesia and South Africa vulnerable as the last bastions of 'white supremacy'. The officers of the Armed Forces Movement who organised the coup ended the Pyrrhic war that the dictatorship had refused to abandon.[49]

Conclusion

Relations between the white community in the African colonies and the Portuguese Army were damaged as the war dragged on. The colonists often questioned the bravery and martial spirit of the Army, and resented the increasing reliance of the Army on black troops, who made up more than half of the colonial army. One white Mozambican reported that:

> The Army sold us out. They fought well at first, but then it was rubbish. From about 1970, officers began to say that this was not their country, it was not their war. I would estimate that 75 per cent of the officers felt this way.[50]

Like the French in Algeria, the Portuguese Army discovered that civic action to improve the conditions and win the hearts and minds of the African population often resulted in bitter opposition from the white community, who resisted the traditional status quo being overturned. This hostility was increased by the Army's lack of sympathy with the white community. Officers listened to African grievances against local officials and, following guerrilla attacks, intervened to prevent retaliation by white settlers against the local black community. White officers commanding black troops were unsympathetic to the traditional race relations. A clear indication of how poor relations between the Army and some members of the white community had become was given in January 1974. White settlers, who were angry at the failure of the Army to protect them against the guerrillas, who had recently murdered five Europeans, attacked the officers' mess in Beira and other military installations in the central region of Mozambique.[51]

While the Portuguese Army adjusted well to the doctrinal requirements of counterinsurgency, attempts to win the hearts and minds of the local population were less successful. In particular, policy suffered from a division between the military, who regarded any reform mainly as a means of controlling the population, and the civil administration, which perceived reform as a chance to stimulate economic development. Military necessity often won. For example, in Angola spending on roads was six times that on either education or health, since, as one official noted, 'revolt starts where the road ends'. Nevertheless, there were substantial improvements in providing new houses, schools, hospitals and wells, notably in Guinea.[52] More successful was the policy of 'Africanisation' employing indigenous troops. By late 1973 more than 50 per cent of the security forces were black and as high as 90 per cent in some elite units such as the GE (*Grupos Espeçiais*, Special Groups), the airborne GEP (*Grupos Espeçiais de Paraquedist*, Special Parachute Groups) or the *Commandos Africanos* (African Commandos) formed by Spínola in Guinea. The DGS also recruited the *Flechas*, some of whom were former insurgents. There were also large African militia and self-defence forces, guarding *aldeamentos* and villages respectively.[53]

The biggest problem appears to have been a split between the elite units, such as the paratroopers and commandos, who did most of the fighting, and the conscripts who formed the majority of the Army (50 per cent in Guinea, 70 per cent in Angola) and were employed on garrison duties. The resulting ill-discipline and poor motivation of many conscript soldiers meant that the Army often failed to implement basic counterinsurgency tactics to the required standard. In Guinea, Portuguese soldiers were reported taking siesta in the middle of the day while they were accused by the Rhodesians of operating noisily in Mozambique to ensure little contact with the insurgents. Tactically, the Rhodesians and South Africans thought that the Portuguese in Mozambique were clumsy and inept, employing patrols that were too large, consisting of thirty to forty men rather than four-man 'sticks'. They fought from bases like the Americans in Vietnam or remained in the towns, being reluctant either to patrol the bush for any length of time or control it by night.[54]

This attitude can partly be explained by the fact that, having limited resources, the Portuguese were keen to limit casualties. They lost fewer deaths in a decade from 1961 than the French lost in Algeria in a single year. Unlike the FLN in Algeria, the guerrillas confronting Portugal avoided the towns. But this compelled the Portuguese forces to spend twenty-four months in garrison posts in the bush and jungle where they became increasingly frustrated by their inability to crush a foe that could retreat into sanctuaries in neighbouring countries. The Portuguese were constrained by political considerations from striking against these guerrilla sanctuaries or conducting 'hot pursuit' operations.[55]

By 1974, the insurgents in the field in Mozambique and Angola, had much improved in training, adaptability and initiative, although they were still not able to overcome the Portuguese forces.[56] And PAIGC had become a formidable opponent in Guinea–Bissau. The Portuguese had some success in pressurising Malawi, Zaire and Zambia to close down cross-border sanctuaries but failed to do so in either Tanzania, which was absolutely crucial for FRELIMO, or Guinea–Conakry, equally important for the PAIGC. Generally the guerrillas were able to obtain havens, both external and internal, where they could receive financial support, obtain rest and supplies, and undergo training.[57] Huge numbers of Africans had also been relocated – more than 1 million people (20 per cent of the population) in Angola, 1 million (15 per cent) in Mozambique and 150,000 (30 per cent) in Guinea–Bissau – and 'civic action' had been employed to win the hearts and minds of and control of the population. But this strategy failed because, as with a similar French policy in Algeria, 'civic action' was not linked to the establishment of an indigenous government. The punitive actions of the security forces, resulting in indiscriminate civilian casualties, also alienated the population. Moreover, as General Costa Gomes noted in May 1974, the Army had 'reached the limits of neuro-psychological exhaustion'.[58]

By 1974, the war in the three colonies had reached very different stages: in Guinea–Bissau PAIGC as much through adept diplomacy as force of arms had victory in sight and the Caetano regime was actually talking to them secretly in London on the eve of the coup. In Mozambique FRELIMO had opened up new fronts in the central region but was still a long way from 'victory'. And in Angola the guerrilla challenge was very limited because of inter- and intra-liberation movement conflict. But, as in so many counterinsurgencies, the final outcome had been decided politically. Thirteen years of costly warfare in three of its African possessions stretched Portugal's resources to the utmost, creating a climate of 'psycho-political collapse'. Although undefeated on the battlefield, the cumulative drain of fighting three wars, the increasing financial burden and the growing dissatisfaction domestically meant that the political struggle was lost and Portugal cracked in 1974. The coup of 1974 was followed by the abrupt withdrawal from and the attainment of independence in rapid succession by Guinea–Bissau (10 September 1974), Mozambique (25 June 1975), Cape Verde (5 July 1975), São Tomé and Príncipe (12 July 1975) and Angola (11 November 1975) – a decolonisation of indecorous haste, which was overshadowed on the global news front by Watergate, the collapse of South Vietnam and tensions in the Middle East.[59]

3

STATE TERRORISM

THE RUSSIAN EXPERIENCE, 1919–2011

Introduction: Russia's Empire pre-1919

Beginning with the annexation of Georgia, between 1801 and 1864 Russian imperial power was extended southwards into the Caucasus, establishing forts to defend its territory. Russia also deployed a 'scorched earth' policy, including brutal, punitive raids against hostile Muslim mountain tribes. Following repeated failure to crush the resistance in the mountains of Dagestan, Chechnya and Avaria, which was led from 1832 by the charismatic Shamil (1797–1871), the Avar political and religious leader of the Muslim tribes in the Northern Caucasus, from 1846 Russia employed a more patient, methodical approach for limited, achievable objectives that, together with a commitment of larger forces, systematically wore down the mountain tribes with relentless campaigning and denied Shamil any victories. The Russians also employed a cut-and-burn policy to clear areas as bases for future operations, while large numbers of the population were forcibly resettled. Systematic deforestation and the destruction of crops and villages continued in a series of winter campaigns, which denied the guerrillas either food or any respite. Weakened by Russia's scorched earth policy and a massive campaign of forced resettlement, which permanently transformed the central Caucasus, although they only partially affected Dagestan and Chechnya, Shamil was finally forced to surrender in 1859. Nevertheless, localised uprisings continued during the period of 1860–77. Russia at this time was preoccupied with the modernisation of her Army to face the conventional threat from Germany and Austria–Hungary. There was therefore no systematic effort to preserve and disseminate the lessons of the Caucasus, which had little relevance to European warfare.[1]

Nevertheless, an informal, unwritten strategy had gradually been developed by Tsarist Russia to defeat insurgents. This combined a number of techniques. A 'surge' of large numbers of troops was made into the 'infected' region. The area was isolated from outside influences. The towns were controlled and that control extended outwards from them. Lines of forts were built to restrict enemy movement and the size of the territory in which the guerrillas could operate. The enemy infrastructure was disrupted by the destruction of crops, livestock, orchards and settlements. The conquest of the Caucasus set the stage for further Russian expansion, stimulated by rivalry with Britain, into Central Asia between 1864 and 1881. The Russian Army would learn a new style of warfare to cope with the harsh climate and vast deserts of Central Asia. On the Kazakh steppe, Russian commanders, like their American contemporaries fighting the Indians on the plains, during the 1840s learned to raid Kazakh villages and encampments in order to drive off cattle, destroy property and demoralise the population. The defeat of the Turkomans by General Mikhail Dmitrievich Skobelev in Transcaspia (Turkmenistan) in 1880–81 extinguished the last effective resistance to imperial rule in Central Asia. The nomadic Kazakhs and the Turkomans were bold fighters but lacked cohesion and were driven largely by a desire to preserve their traditional way of life. As in the Caucasus, the legacy for the Russian Army was modest and short-lived. Similarly to the Americans in the Philippines and the British in India and Africa, Russian officers saw Central Asia, like the Caucasus, as a useful training ground but of little use for the more serious work of preparing to fight a conventional war in Europe. As an institution, the Russian Army never codified the lessons culled from decades of campaigning in these regions. During the twentieth century the Soviets were to rediscover that irregular enemies could drain the resolution and resources of a powerful state, and that unconventional resistance by a highly motivated enemy is very hard to overcome by conventional methods. The 'sword and samovar' strategy, which combined military force with diplomacy, was also used to divide and rule by setting chieftains and tribes against each other.[2]

The Red Army was forced to wage a number of counterinsurgency campaigns during the establishment of Bolshevik rule in Russia after the First World War. Between April 1919 and the autumn of 1921 these were revolts by the Tyumen in Western Siberia, Tatar and Kirghiz tribesmen in Samara, Finnish guerrillas in Karelia and the mountain tribes of Dagestan in the Caucasus and also insurrections in the Tambov province of the Volga region and the Ukraine. Brutalised by war, the Soviets unleashed a Red terror, including the destruction of property, deportation, rapes, the killing of prisoners and religious leaders, and the taking of hostages. In the Ukraine in April 1919 there was an astonishing variety of insurgents with ninety-three guerrilla groups operating against the Bolsheviks. Although they were led by Simon Petlura, the former chairman of the All Ukrainian Military Committee of the Central Rada (or Council), by far the most significant guerrilla leader was Nestor Makhno, an anarchist. In order to continue the fight against the predominant threat posed by the White forces of Generals Denikin and Wrangel, the Bolsheviks were forced to make temporary truces in 1919 and 1920 with Makhno, who at the peak of his powers had some 25,000 followers. However, once the White armies had been defeated, between November 1920 and August 1921 the Bolsheviks turned on Makhno. In Georgia, Islamic guerrillas from the mountain tribes of Dagestan under the leadership of Muslim clerics, notably Najmudin Gotsinski and Shekh Uzun Haji, were able to put as many as 10,000 insurgents, mainly Avars but also Chechens, into the field. The Chechens would revolt against the Soviets again in 1928, 1936 and 1942.[3]

Tambov, 1920–21

The uprising in the Tambov province of the Volga region, which broke out in August 1920, by some 21,000 peasants or guerrillas (the 'Greens' to distinguish them from the Bolshevik 'Reds' and anti-Bolshevik 'Whites') led by Aleksandr Antonov was typical of a number of rebellions. It developed from small-scale peasant revolts against the requisitioning of food into a full-scale peasant war. The initial Soviet reaction was the implementation of martial law, the burning of villages and the taking of hostages. The failure of such methods to curb the insurgency resulted in a mixture of the carrot and stick. These included temporary political concessions designed to separate the guerrillas from the population and the ruthless suppression of the insurgents themselves by military action. Thus, a politico-military strategy was developed to secure popular support, notably in urban centres, by adapting the Tsarist policy of divide and rule.[4]

Vladimir Antonov-Ovseenko was sent with plenipotentiary powers providing unified command to co-ordinate both the party apparatus and the military strategy in the Tambov. Local units and militia were raised and reinforced with some 47,000 troops. This allowed a systematic domination of insurgent areas with the establishment of 'zones of occupation' in March 1921, although a failure to prevent insurgents slipping back into 'cleared' areas remained a problem. With the military presence well-established, lists with some 10,000 names of Antonov's sympathisers and supporters in the villages were compiled by the Cheka, the feared secret police, who terrorised the population under Orders No 130 and 171 in order to isolate the guerrillas from their supporters. For offences such as concealing weapons and harbouring insurgents the death penalty was mandatory. Hostages were often taken and executed. Entire villages were sometimes burned down. In a crude form of resettlement, the families of known guerrillas were detained in concentration camps and their property seized prior to deportation from the province, leaving them to survive without assistance. Whole village populations were often interned and later shot or exiled to the Arctic Circle. It is estimated that 15,000 people were shot and another 100,000 deported or imprisoned during the repression of the Tambov rebellion. Isolated and deprived of support from the population, the guerrillas were broken up into smaller groups by search-and-destroy operations supported by artillery, armoured cars and aircraft. Poison gas was employed against rebels hiding in the forests.[5]

Mikhail Nikolayevich (later Marshal) Tukhachevsky, an aristocrat who was commissioned in 1912 and purged by Stalin in 1937, arrived in May 1921 to command in the Tambov. He was appointed by Lenin and Trotsky after his ruthless suppression of the Kronstadt uprising. As most of the Soviet forces were employed in static garrison duty, to harry the insurgents he immediately set up a mobile strike force of three cavalry brigades (3,000 men), which included the young officer Georgi K. Zhukov, and special motorised units. Tukhachevsky also used both the carrot and stick to entice Antonov's supporters to desert him, offering an amnesty and protection to peasants who handed in their weapons and turned in the insurgent ringleaders. But they were warned that they would suffer the full consequences if they persisted in sheltering the rebels. Other incentives, such as handing over property confiscated from the guerrillas and their sympathisers, were used to drive a wedge between the peasants and the 'bandits'. Antonov's forces were destroyed by mid-July 1921 with only a few diehards remaining, including Antonov himself, who was finally tracked down and killed in June 1922.[6]

Tukhachevsky can be said along with his friend Mikhail Vasil'evich Frunze to be not only one of the founders of the Red Army but also of modern Russian counterinsurgency theory. He wrote three articles on counterinsurgency in *Voina I Revolyutsya* (*War and Revolution*), which recorded his repression of the Tambov Uprising between May and July 1921: '*Iskorenie banditizma*' (The Eradication of Banditry) in 1922 and '*Voina Klopov*' (War of the Bedbugs) and '*Bor'ba s Kontrevolyutsionnymi Vosstaniyami: Iskoreneniye Tipichnogo Banditizma*' (The Struggle against Counter-revolutionary Uprisings: Rooting Out a Typical Case of Banditism) in 1926. Like Frunze, Tukhachevsky demonstrated a clear understanding of the political dimension of counterinsurgency when opposed by guerrillas who enjoyed widespread popular support. In his articles, Tukhachevsky set out an excellent strategy for suppressing an insurgency. He emphasised the importance of adapting to the local conditions, customs and religion in order to win over both the population and the insurgents, and of raising local forces and using former guerrillas to hunt down the 'banditry'. Army garrisons would secure territory against insurgent attacks while mobile forces pursued and destroyed guerrilla bands. He also advocated a unified command, with the appointment of one individual to co-ordinate not only the military, the police, the intelligence services but also the economic, political and social strategies. The 'carrot' of temporary political concessions would be balanced by the 'stick' of harsh military repression against the insurgents and their supporters. 'Bandit families' would be evicted and detained in concentration camps and their property confiscated on the principle of collective guilt. The Cheka purged the insurgent supporters hiding among the population while the military eliminated the guerrillas in the field.[7] Tukhachevsky concluded that 'the struggle must be essentially conducted not against the [guerrilla] bands, but against the entire civil population'. He also noted that 'the morale of the bandits can be broken only if they know suppression will be conducted consistently and with cruel persistency'.[8] However, Soviet military thinkers were preoccupied with conventional warfare in preparation for the forthcoming Great Patriotic War (Second World War). Tukhachevsky's ideas, like those of Frunze, provided the template for a successful counterinsurgency doctrine but remained undeveloped.

The Basmachi, Central Asia, 1918–33

The most serious counterinsurgency conducted by the Bolsheviks was the complex military, social and political struggle to overcome the *Basmachi* (bandits), who resisted the Soviets in Central Asia for more than ten years. This campaign provided the Red Army with considerable experience of counterinsurgency and foreshadowed later conflicts in the developing world during the late twentieth century. The roots of the insurgency were the disruption of traditional ways of life and bitter resentment of Russian domination that caused social tensions and gave impetus to Pan-Islamic and Pan-Turkic movements. These clashed with the Bolsheviks, who were ideologically hostile to religion and traditional societies. The durability of the resistance was especially remarkable since the diverse and autonomous groups, which operated under the umbrella of the *Basmachestvo* (the *Basmachi* movement), lacked either a cohesive political agenda or a coherent military organisation. It was not a modern nationalist movement with a political programme that commanded mass support. Instead, the insurgents were motivated by the negative demand for expulsion of the infidel and a fanatical determination not to submit. As defenders of traditional society, religious identity and self-determinism, the *Basmachi* provided a dangerous, if fragmented, resistance. They forced the Bolsheviks to play down or modify much of their political programme. Struggling to adapt to the challenges of irregular warfare, which required tactical and logistical changes, the Red Army was unable to control much of rural Central Asia between 1918 and 1933.[9]

During the first half of the civil war, the Bolsheviks fought for survival on many fronts. They had little control over the former imperial borderlands of Central Asia, which were cut off from Moscow by White armies operating in the southern steppe and Siberia. Short of manpower, the Bolsheviks established a tenuous control over the cities and towns in Turkestan. But their lack of discipline, notably the looting, rape and slaughter of 14,000 people following their capture of Kokand, enraged the population. Guerrilla warfare soon spread from the Fergana valley to engulf the countryside in what was essentially a number of unconnected tribal revolts fed by ethnic and religious tensions. Lacking modern weapons and tactical co-ordination, the *Basmachi* bands relied on hit-and-run tactics. They mounted ambushes and small raids against Red Army garrisons, and generally withdrew when faced by superior numbers. Operating in small groups the *Basmachi* were elusive and exploited their superior knowledge of the terrain and mobility. They also had support from the local population, who provided intelligence.[10]

In November 1919, having broken the White forces in Western Siberia, Moscow sent a six-person Turkestan Commission to take over in Tashkent and to re-establish its rule in Central Asia. In February 1920 Frunze (the future 'Soviet Clausewitz' who commanded the Turkestan Front), himself the son of settlers in Central Asia, arrived. The reorganisation of troops into cavalry units capable of intercepting and harassing the *Basmachi* bands was followed by a series of conciliatory policy changes to placate local grievances. These included the reopening of bazaars, improved food distribution and the recruitment of indigenous party members and troops. In May 1920 this good work was undone by the implementation of conscription, which gave the *Basmachi* renewed political support and many new recruits. Although the Soviets believed that the *Basmachi* received weapons from external supporters (especially the British) via Afghanistan, there is little evidence that this aid was significant. A reorganisation by the Red Army, whose attention was no longer focused elsewhere, turned the tide in favour of the Soviets. They gradually wore down the resolve of the resistance over the next decade.[11]

During his brief stay in Central Asia between February and September 1920 Frunze, who believed in total war, outlined the tactics and strategy that he and his successors would pursue against the *Basmachi*. Demonstrating an unusual comprehension of counterinsurgency operations, Frunze set out to annihilate the *Basmachi* bands and to isolate them from the population. By creating a network of forts to occupy the ground and employing flying columns to trap the elusive 'bandits', Frunze hoped to destroy the *Basmachi* in the field. The forts were complemented by the creation of home guard units in the small *kishlaks* (the winter quarters of nomadic tribes). This was done in order to separate the guerrillas from the population and cut them off from their supplies and supporters. This suggests that Frunze had some knowledge of the techniques previously employed by the French and the British in their colonial wars. This was some ten years before Mao Tse-tung delineated his 'fish-in-water' theory and showed an awareness of the dependence of the modern guerrilla on the support of the people. Frunze's tactics were again employed in the Soviet campaigns of 1923 and 1926 against the *Basmachi*, underlining his claim to be the 'Father of Modern Soviet Counterinsurgency'. It is worthy of note that an article in *Kommunist Vooruzhennykh Sil* (Communist of the Armed Forces) written during the Afghan War in the 1980s examined the pacification techniques employed by Frunze in Turkestan.[12]

In 1923 in a secret document, entitled 'System for the Struggle with the Basmachis', Sergei Kamenov, commanding the Soviet armed forces, drew up the most coherent and comprehensive plan for the defeat of the insurgents, which advocated the occupation of important population centres, the defence of the key communications and attacks on *Basmachi* bases. The Soviets sought to isolate and destroy the hostile guerrilla bands through unremitting pressure and to curtail their flight to remote internal sanctuaries or in Afghanistan. In order to provide the flexibility and mobility to operate against the elusive *Basmachi*, the Red Army formed *letuchie otriady* (flying detachments) of light cavalry to patrol between the garrisons protecting the lines of communications, as well as attacking *Basmachi* sanctuaries. The flying detachments were supported by *istrebitel'nye otriady* (raiding detachments), local units that searched for and harassed the insurgents. The Red Army conducted search-and-destroy operations to destroy *Basmachi* bands. Only unremitting pressure could ensure the submission of the guerrillas. Success in the field was merely preparation for the main problem of controlling and convincing the local inhabitants to support the Bolsheviks. The Soviets invested considerable resources in political indoctrination. They encouraged education and used propaganda, notably training indigenous cadres, and staged large-scale political rallies. They also toned down attacks on local customs and Islamic institutions and disbanded units that had committed crimes against the indigenous population.[13]

In 1920, with the arrival of reinforcements and the end of the isolation of Turkestan with the collapse of the Whites, the Soviets were able to overcome the main centre of resistance in the Fergana Valley. Committed to a protracted struggle, the Soviets employed draconian measures. Frunze announced that anyone joining the *Basmachi* would be summarily shot, and the 2nd Turkestan Division was given full powers in the Fergana region (modern Tajikistan) to put down the rebels. But efforts were also made to address the ethnic, political and social origins of the conflict and win hearts and minds. Rebel leaders were offered amnesties and attempts were made to demonstrate a tolerance of Islam, to restore the cotton economy, thus making it more difficult for *Basmachi* bands to obtain food. Although inevitably prevailing in any conventional encounters, the Red Army still faced widespread opposition. Relentlessly hunting down the elusive insurgent groups across great distances and in remote mountainous regions, the Soviets gradually wore down the opposition through unceasing pressure until in 1923 only sporadic resistance remained in the more remote areas of the Fergana Valley.[14]

The Red Army was able to reassert control of Bukhara in August and September 1920, but was unable to transform tactical success into strategic victory. The escape of the Emir of Bukhara, who eventually took refuge in Afghanistan, provided a figurehead for future opposition while, in defiance of defeat on the battlefield, *Basmachi* uprisings erupted in response to the Red Army's advance and alleged atrocities against the population. The *Basmachi* showed an ability to avoid and infiltrate past Soviet columns conducting large-scale sweeps, so although large numbers of insurgents were captured the main leaders and their hard core supporters usually escaped. The appearance in the spring of 1922 of Enver Pasha (a former Minister of War of the Ottoman Empire), who joined the resistance as commander of the armies of Turkestan, Bukhara and Khiva, brought tactical and organisational knowledge. He also brought a sense of purpose and some military cohesion and provided the zenith of success for the *Basmachi* movement. But, despite broad popular support, disunity, poor organisation and shortages of weapons left the *Basmachestvo* incapable of facing the Red Army in battle. Enver's campaign, a premature move to semi-regular warfare, was disastrous. It allowed the Soviets the opportunity to bring the *Basmachi* to battle, in the type of set piece conventional warfare that guerrillas should avoid, and then to track down the survivors relentlessly. With the reoccupation by the Red Army and the death of Enver in August 1922, the *Basmachi* faded away and were dispersed by the inexorable pursuit of the Soviets, and in the spring of 1923 the Lokai Valley, one of the last *Basmachi* bastions, fell.[15]

Although *Basmachi* bands continued the fight in Bukhara into 1924, the cumulative effect of military successes, superior organisation and equipment, a war-weary population and skilful propaganda backed by pragmatic social policies enabled the Soviets to prevail. From 1923 onwards, a key aspect in all Soviet operations was the increasing political disaffection of the local population from the *Basmachestvo* owing to the Soviet political policy, which complemented the military campaign of relentless large-scale sweeps and clearing operations by turning the population against the insurgents. Red Army garrisons secured individual villages while other troops hunted down the guerrillas. Militia and volunteers were recruited to provide invaluable local knowledge of the terrain and the rebels. Harsh measures were employed by the Soviets, including in 1929 the deportation from Turkestan by the Cheka of more than 270,000 people accused of aiding the *Basmachi* and a scorched earth campaign in which four cities and 1,200 villages were destroyed. Nevertheless, this did not mean the end of the conflict in remote rural areas, and the final pacification of Khiva had to wait for another decade. Based in sanctuaries in Afghanistan, rebels launched a revolt in the Lokai Valley in 1926. Stalin's collectivisation of agriculture once again stirred up peasant unrest in Tajikistan in 1929–33 with insurgent leaders taking refuge in Turkmenistan and Iran.[16]

The Soviets understood that the military and political aspects of the war were thoroughly interwoven and so they combined 'soft' and 'hard' strategies, employing the carrot-and-stick approach. While relentlessly engaging and killing as many insurgents as possible, they also attempted to persuade pivotal sections of the population to support their cause. The Soviets pragmatically made economic concessions and moderated their policies on religion and conscription, reversing policies or unpopular decisions that were ill-adapted to local conditions or offended indigenous traditions. While building up their own mass support, they also exploited the disunity of their opponents, splintering the opposition, and dividing and ruling between ethnic and religious factions. For example, the Soviets exploited the urban-rural divide in the Ferghana Valley. They gained the support of the inhabitants of cities, mainly Uzbeks and Tadzhiks, against the peasants and nomads of the region, mostly Uzbeks and Kirghiz, who formed the backbone of the *Basmachis*. Soviet wooing of important Muslim leaders ensured that the *Basmachi* movement never acquired the character of a *jihad* (religious war). The Red Army learned from its own experiences and from studying previous imperial campaigns in order to adapt the best methods of operating against an unconventional foe in the deserts and mountains of Central Asia. The Soviets also succeeded in closing *Basmachi* sanctuaries in Iran and Afghanistan, employing both diplomacy and a policy of 'hot pursuit' across the frontiers.[17]

As a result of Tsarist experiences and their own confrontations with various guerrilla movements during the 1920s and 1930s, the Soviets had developed a formula for dealing successfully with insurgent uprisings, based on five central policies:

fragment the opposition using existing ethnic, religious and social divisions;
win over key indigenous elites, especially the tribal leaders, the traditional religious leadership, and the modern intellectual class;
build up a strong indigenous communist leadership and cadres;
employ indigenous troops;
adapt communist ideology to local customs and traditions making it more acceptable and appropriate to the local population.[18]

From the 1930s, the Soviet Army rarely fought guerrillas and instead concentrated on preparing for conventional warfare against Nazi Germany. Henceforth, the Soviets would be far more interested in developing their conventional capabilities than in codifying their counterinsurgency doctrine. This was done through industrialisation and the operational art. Ironically this was led by thinkers such as Frunze and Tukhachevsky. This left counterinsurgency in the hands of security agencies such as the GUBB (*Glavnoe upravlenie po bor'be s banditizmom*, Head Directorate for Struggle against Banditry) and the Security Troops Directorate of the NKVD (*Narodnyi komissariat vnutrennikh del*, People's Commissariat of Internal Affairs) during and immediately after the Second World War. Later the KGB and Border Guards assumed this role, which was later transferred to the Interior Ministry troops after the collapse of the Soviet Union.[19]

Western Borderlands, 1944–59

The insurgencies in the 1920s, and in particular the long-running struggle with the *Basmachi*, provided the Soviets with considerable experience of counterinsurgency. The Red Army was to apply these methods in the territories gained during 1939–40 and in the guerrilla war, which formed a major obstacle to the sovietisation of the region until the early 1950s. This anti-communist resistance took place in western Ukraine, western Belorussia, Lithuania, Latvia and Estonia and in particular the backbone of the resistance was provided by the Ukrainian People's Army (UPA). The UPA enjoyed popular support and maintained a sophisticated civilian infrastructure in the Ukraine during the 1940s and 1950s and so too did the Lithuania Liberation Army (LLA) between 1945 and 1952.[20]

As the Red Army reconquered German-occupied territory in 1944 and 1945, it faced resistance from nationalist forces. The UPA openly resisted the Red Army and Soviet political administration, ambushing and killing General N.F. Vatutin (1st Ukrainian Front), one of the best Soviet front commanders, in February 1944 and the Polish General Karol Swierczewski in March 1946. Initially, the rebels attempted to deploy large units that suffered heavy casualties but then resorted to smaller units that employed guerrilla tactics. They terrorised and killed Soviet functionaries and their collaborators and controlled rural areas. Moscow was forced to bring in Communist Party administrators from other regions, but they lacked knowledge of either the Ukrainian language or the local area. The resistance to the Soviets continued in the Ukraine and parts of Belorussia until the 1950s, when the Soviet security forces finally won. The Red Army was ill-suited for counterinsurgency and prohibited from fighting insurgents and the counterinsurgency campaign, which was conducted by the police under GUBB. The police maintained law and order and collected intelligence, while NKVD security troops, who were organised into NKVD divisions, participated in larger operations. After the reoccupation of the western borderlands, the security forces were badly overstretched across large areas. The authorities had not only underestimated the insurgent strength but also overestimated the impact of reforms in placating opposition.[21]

It took roughly a decade from the mid-1940s to the mid-1950s before the insurgencies in the Western Ukraine and the Baltic states could be crushed through the use of unstinting violence. Although there was no manual setting out a counterinsurgency doctrine, the Soviets employed the political, military and security measures that had been developed during the civil war and redeployed against the *Basmachi*, employing a carrot-and-stick strategy based on a combination of populist reforms and amnesties to win hearts and minds of the majority and harsh, punitive measures to eliminate all opposition. The Ukrainian leader, Stepan Bandera, was assassinated in Munich in October 1959. Agrarian reform and collectivisation were employed as a pacification tool to win over the poor, peasant majority and utilise class tensions. But the ideological and doctrinaire rush to establish collectivisation during 1944–46 before control over the countryside had been gained proved disastrously counterproductive in the Baltic states and western Ukraine, showing an insensitivity to local needs and exacerbating popular grievances. In Lithuania, collectivisation helped to deny food supplies to the LLA. Another weapon was mass deportation, which had also been developed in the 1920s and 1930s to remove not only active opponents but also potential opposition. It was also designed to intimidate the population from supporting the rebels. Families were punished for the activities of those who served with the insurgents. From 1944, deportations of insurgents' families were used in the western borderlands to break up the rebel infrastructure. But at first they aggravated and fuelled the resistance and therefore prolonged the insurgencies. Between 1945 and 1952 108,362 Lithuanians were deported. The Soviets also put the clergy of various denominations under duress to act as police agents, secure the compliance of the population and undermine nationalist resistance by making appeals for an end to the insurgency. Many priests were caught between state and the insurgents, neither of which would allow them to remain neutral.[22]

The Soviets also attempted to exploit divisions and to persuade rebels to defect by promising a full amnesty. Priests were also pressed to persuade guerrillas to surrender while insurgents who had surrendered also made appeals to their comrades to do the same. Others earned their pardons by enlisting in militias, recruited locally to defend villages. They used their local knowledge to locate the rebels and served in commando units that targeted the guerrilla bands and the rebel leadership in particular. Although often poorly armed, disciplined and trained and susceptible to infiltration by the insurgents, such units alleviated the shortage of manpower. But they committed more atrocities than regular forces, being influenced by local animosities. The security forces lacked linguists and had a bad reputation for ill-discipline, robbing, murdering and raping suspects. They perpetrated atrocities during counterinsurgency operations, including summary executions of suspected guerrillas and their families. The police placed garrisons in rebel villages, which were supported by the militia, to hunt down the guerrillas with aggressive patrolling and the setting up of ambushes. To achieve this, they changed tactics, employing small units to hunt down the elusive and dangerous guerrillas and their hideouts with patrols, ambushes and covert operations that were directed by intelligence against specific targets instead of undirected, larger-scale cordon-and-search operations that left the insurgent infrastructure largely unharmed.[23]

During the period 1947–49 the Soviets also identified the lack of intelligence as a glaring weakness and switched the emphasis in their strategy from military operations against the banditry to action against the insurgent infrastructure. They began to establish networks of informers and agents to carry out surveillance and infiltrate the insurgent infrastructure, and to identify their leaders and key supporters. Many volunteered but others were paid or coerced. In Lithuania, for example, the Soviets had succeeded in penetrating the rebel movement, including the highest echelons, by the end of 1949. But as late as 1951 they continued to complain in western Ukraine about the low quality and poor quality of agents within the nationalist movement and attributed the failure to end the insurgency to this. Covert operations were also undertaken by special forces and rebels who had been 'turned' or 'converted' to sow suspicion among and infiltrate guerrilla units, and especially to eliminate their leaders. In a campaign of particular brutality, in which the NKVD deployed 60,000 men, the Lithuanian leadership was targeted in the same way that the leadership of earlier insurgencies had been singled out for destruction. Such tactics created divisions and tensions within the resistance movements, which were exacerbated by some of the methods used by both Lithuanian and Ukrainian insurgents to root out and punish informers and their families. Torture was the main interrogation method employed by the police. Show trials, public executions and the desecration or public display of the corpses of guerrillas killed in combat were used to intimidate the civilian supporters of the insurgency and 'persuade' them not to help the insurgents.[24]

Although the insurgencies continued for several years, keeping the Soviet administration in the countryside at bay into the mid-1950s, the rebels were eventually suppressed by a carrot-and-stick strategy. The 'stick' employed coercive techniques such as the ruthless but effective operations of the security forces, which shifted from large-scale to small unit operations, and the deportation of the hostile elements of the population. The 'carrot' used agrarian reform – the cornerstone of Soviet counterinsurgency – the church and other measures, such as amnesties, to win the hearts and minds of the peasants. However, some policies, such as the introduction of collectivisation, and a culture of indiscriminate violence among the security forces, which did not grasp the importance of the concept of 'minimum force', prolonged the resistance in the short term. The rural population was also increasingly alienated by the shift from selective terrorism and atrocities by the insurgents, notably the UPA, which became increasingly indiscriminate in targeting 'collaborators' and supporters of collectivisation, and over-reliant on coercion in its recruitment. Both were counterproductive, especially when, increasingly, supporters began to believe that further resistance was foolish owing to the lack of external sanctuaries or support from the Western Allies and opted for accommodation with the Soviet state. Soviet state violence, mass arrests and terror, imprisonment, deportation, torture and rape was all part of a deliberate campaign to disorganise the insurgent infrastructure. It drove a wedge between the guerrillas and their civilian supporters, and cowed the local population into submission by depriving them of any hope of effective resistance to Soviet occupation.[25] However, the Soviets promptly forgot the lessons of these campaigns and lacked a counterinsurgency strategy by the end of the twentieth century. They fought guerrillas in Afghanistan and Chechnya with conventional techniques, relying on air power, armour and artillery, which were more appropriate to the Great Patriotic War. More importantly, the political and military policies in Afghanistan and Chechnya were completely unco-ordinated.

Afghanistan

The Soviet invasion of Afghanistan in December 1979 in response to revolt against the Soviet client state and civil war was a brilliant tactical operation, employing airborne, mechanised infantry, *Spetsnaz* (short for *spetsialnoe naznacheniye*, special purpose forces) and tank units to seize key Afghan installations. The 40th Army's intervention was modelled on similar actions in East Berlin in 1953, Hungary in 1956 and Czechoslovakia in 1968. However, they had not required the Soviets to deploy the full range of techniques developed between the 1920s and the 1950s and rather than stabilising the situation, the Soviets triggered an uprising, which became one of the most successful modern insurgencies. The Russians never committed the numbers of troops required to control the countryside of Afghanistan, which reverted to the rebels whenever Soviet units withdrew. Moreover, large numbers were tied down protecting bases and lines of communications. The intervention in Afghanistan marked the first time in several decades that Soviet forces had fought an unconventional war against a determined resistance movement. They were trained for large-scale, rapid tempo operations on the European battlefield and were unprepared for the intensity of opposition in Afghanistan. Lacking a well-defined counterinsurgency doctrine, wedded to conventional warfare and dependent on large armoured formations with air and artillery support, the Soviets were inflexible. They were reluctant to operate at night, relying on well-planned major cordon-and-search operations in daylight. Classic large-scale offensives employing large, combined arms formations against the *Mujahideen* (Soldiers of God) and massive artillery firepower and air power in order to save Soviet lives and compensate for shortages of infantry, showed little concern for civilian casualties. They were indiscriminate, not very effective and consequently did not win hearts and minds.[26]

For both ideological and military reasons, the Soviet Army, like the American and Indian armies and in contrast to the British, French and Portuguese armies, largely ignored the problem of counterinsurgency and concentrated in its doctrinal manuals on conventional and nuclear warfare. Neither the various editions (1966, 1984 and 1987) of the main Soviet tactical manual, Reznichenko's *Taktika* (Tactics), nor Savkin's *Basic Principles of Operational Art and Tactics* (1973) made any reference to counterinsurgency. Although the exploits of Soviet partisans during the Civil War and the Second World War were widely analysed, the 1920s counterinsurgency campaigns against the *Basmachi* in Central Asia and against partisans in the Baltic and the Ukraine after the Second World War received little consideration. Furthermore, there was little attempt to investigate or develop a counterinsurgency strategy. What interest had been shown was mainly focused on mountain warfare and in particular on large-scale operations by the Red Army in the Caucasus, Carpathians and Manchuria during the Second World War. Once it became clear that a quick, decisive victory was not possible, the Soviets had to develop an ad hoc counterinsurgency strategy that combined not only military but also economic, ideological, social and political elements.[27] As the general staff later noted, Soviet forces entered Afghanistan with 'no practical skills in the conduct of counter guerrilla warfare' and without 'a single well-developed manual, regulation, or tactical guideline for fighting such a war.[28]

One of the weaknesses of Soviet counterinsurgency was the centralised command structure, which was suited to the large European battles envisioned by Soviet commanders but proved to be cumbersome and too rigid to manage the campaign in Afghanistan effectively. The military leaders there had limited initiative and independence, having to obtain the approval of the Soviet leaders Leonid Brezhnev, Yuri Andropov and Konstantin Chernenko, who were too old to provide the necessary dynamic leadership. Operational planning and support elements were grouped at divisional and even army level rather than at battalion or regimental level. This institutional culture discouraged junior officers and NCOs from displaying initiative, which was a prerequisite for success in Afghanistan. There was an over-reliance on motorised rifle and tank units being used in 'hammer and anvil' operations and on set piece tactics, using standard drills that were mechanically applied regardless of the tactical situation. The rugged terrain demanded timely and independent decisions by junior leaders at company and platoon level. The Soviets did make attempts to improve their command structure by decentralising and dividing Afghanistan into seven military districts (each had its own brigade or divisional head-quarters), giving increased responsibility and autonomy at the regimental and battalion level, and providing enhanced training of junior officers and NCOs. But the weakness remained. Such a transition and change of culture and ethos took time to implement.[29]

In their first major offensive of the war in the Kunar Valley during February and March 1980, employing some 5,000 troops supported by armour and air power, the Soviets achieved little of lasting impact. They learned that, although they could go wherever they wished, securing strategic ridges and devastating insurgent groups and villages, the *Mujahideen* simply melted away into the mountains making it impossible to hunt down and eliminate the resistance. The *Mujahideen* learned that large groups of guerrillas presented much too easy targets for a powerful conventional army and instead operated in smaller groups. The Soviets also learned that employing large formations, being slow and ponderous and lacking the element of surprise, and the techniques of con-ventional warfare were ineffective when employed in large operations such as Operation *Magistral* (Highway) and the nine major operations in the Panjshir Valley. The more villages destroyed and civilians that were killed in clumsy conventional operations, the more that the 'godless' Soviets were hated.[30]

Successive but periodic large-scale conventional offensives in the Panjshir Valley reflected the determination of the Soviets to batter the resistance into submission. But they also merely highlighted the fact that Soviet command of an area lasted only as long as its forces remained on the ground. As soon as they left, control reverted to the insurgents, who controlled the countryside by night. Eighty per cent of the countryside remained outside the control of the government. The Soviets never had enough troops to hold the territory captured in their operations. Although hampered by shortages of infantry, the Soviets developed more effective tactics and training as the war progressed. They began to employ night attacks, air assaults and heliborne assaults by para-troopers, *Spetsnaz* and other elite mountain units, developing small unit tactics of more physically hardened and trained formations, while the conscripts, who lacked training and experience as a result of short tours of duty, did little fight-ing. The airborne and air assault forces enjoyed special privileges and uniforms, and a pick of the conscripts. They were also awarded many of the decorations and accolades. These elite units were deployed to attack enemy strongholds and cut off insurgents but there were never enough troops or helicopters to be truly effective. Soviet air power was the crucial element in shaping these new tactics and the crucial dynamic on the Afghan battlefield. In particular, the rapid increase in the numbers of Soviet helicopters during the first half of the war was an indication of their important role in supporting operations on the ground. The introduction of American Stinger and British Blowpipe surface-to-air missiles had immediate and serious consequences by forcing a reduction in Soviet air support for Soviet ground forces and their Afghan allies.[31]

Repeated Soviet operations into insurgent-controlled areas, which were usually supported by massive air and artillery support, resulted in 'migratory genocide' – a population exodus on a massive scale. By 1988, after nine years of war, possibly as many as 1.3 million Afghans had been killed and a further 5.5 million refugees (a third of the pre-war population) had fled to Pakistan or Iran. Unable to destroy the elusive insurgents, the Soviets resorted to a 'scorched earth' strategy, which had been so effective in the Caucasus and Central Asia during the late nineteenth and early twentieth centuries. They targeted other supplies that would sustain the resistance and systematically destroyed crops, granaries, irrigation systems, livestock, orchards, wells and villages in order to deny rural support to the insurgents and destroy the infrastructure and cover for ambushes. These Soviet tactics, aimed at terrorising and intimidating the population, negated the medical treatment and pro-regime propaganda that attempted to win hearts and minds. It turned the rural population into supporters of the insurgents, providing them with intelligence, setting up ambushes and hiding them and their weapons. In this instance, the carrot-and-stick policy backfired badly. The Soviets also attempted to close the borders with Iran and Pakistan in order to deny the guerrillas access to sanctuaries, fighters, supplies, training and weapons but with little success. In just the four years between 1983 and 1987 the assistance given by Pakistan included the training of some 80,000 *Mujahideen* (by mid-1984 more than 1,000 a month), the distribution of hundreds of thousands of tons of arms and ammunition, and several billion dollars of logistical support.[32]

Another weakness, which bedevilled the 40th Army throughout the campaign, was the lack of good intelligence and of qualified personnel, GRU, KGB or MVD, who knew the local customs and languages and were capable of rectifying this problem. Good knowledge of tribal and clan relationships and politics in the theatre of operations was necessary to gather intelligence about guerrilla bases, supply routes and arms dumps and build up networks of agents among the *Mujahideen*. Such information allowed them to hunt down guerrillas and assassinate their leaders, and sow dissension between the various insurgent bands. Lacking any pre-war, colonial infrastructure, the Soviets were largely reliant on their Afghan allies, notably KHAD (Afghan Secret Police). In such a situation security was also a problem with all Soviet troop movements being reported by insurgent agents who had penetrated both the Afghan Army and police widely. The Afghan Army also displayed serious weaknesses, being poorly trained, frequently averse to fighting and well below strength in both equipment and men. The soldiers often deserted and their loyalty was suspect.[33]

In an effort to monitor and control the population all males over the age of 13 were required to register with the Soviet authorities and to carry personal identification documentation at all times. Curfews were also maintained and the movement of visitors regulated. Pro-government defence committees worked at city, village, street and neighbourhood level to monitor the movement and mood of the population. The numbers of police and other security agencies were built up to act in support of the government. Agents infiltrated the guerrilla groups and provided intelligence for search-and-destroy operations. But the Soviets were unable to project their influence into enough of the countryside owing to the shortage of troops, poor infrastructure and the rural society within Afghanistan. This made it very hard to control and isolate the population from the insurgents. The relatively low numbers meant that the troops were too few and spread too thinly over a vast country. This precluded effective control of the countryside and prevented the Soviets from implementing a sound strategy effectively or broadly enough to ensure success. The Soviet forces struggled to secure their base areas and the cities, and to maintain their lines of communications back to the USSR, but failed to pacify the rural areas. Large numbers (50 per cent) of Soviet troops were tied down guarding airfields, installations and lines of communications and securing cities and garrisons. Soviet troops also helped to garrison the provincial security posts protecting roads and securing key areas. This produced a 'bunker mentality' and an over-reliance on artillery and air support rather than manoeuvre. This further degraded the ability to defeat the insurgents.[34]

Above all, substantial progress was hampered by the failure of the Soviet political strategy, which was vital to a successful counterinsurgency campaign. The Soviets recognised the urgent necessity of reconstructing the infrastructure of the Afghan government and Army. Thus, much toil was put into educating and training new party cadres and attempting to create a competent officer corps for the Afghan Army. But the latter was riddled with *Mujahideen* sympathisers who provided detailed intelligence about forthcoming operations. Soviet advisers served with the Afghan Army down to battalion level, but such postings were poorly received because of the bad conditions and the language and cultural problems, and not regarded as stepping stones to promotion. The Afghan government and its Soviet supporters failed to stamp out corruption, ethnic tensions, factionalism within the ruling party and astounding incompetence. The Afghan government was split by the violent and incessant rivalry between the *Khalqi* (Masses) and *Parcham* (Banner) factions, which constantly and murderously conspired against each other, and prevented the creation of a strong party apparatus to run the administration of the country. It also generated disarray within the Army, which, shrinking in size and often defecting to the rebels, proved to be an unreliable burden rather than an asset to the Soviets.[35]

By the late 1980s, Soviet commanders were increasingly aware of the need to win hearts and minds, minimise the destruction that alienated the population and carry out reconstruction. The Soviets were also aware of the isolation and unpopularity of the Afghan government, which had alienated the traditional religious and tribal leadership as well as the intelligentsia. They therefore attempted to bolster their client regime with civic, economic and political initiatives to win over the various elites and galvanise the population into supporting the regime. As well as spending an estimated $78 billion between 1982 and 1986 on aid to Afghanistan, the Soviets undertook various projects such as irrigation, state farms, hydroelectric and pumping stations. They also sent large numbers of Russian advisers to serve in the ministries and factories and with the Afghan Army. In return, the guerrilla movement specifically targeted government supporters, officials and projects in their attacks. For example, it was asserted by the Afghan government that in 1983 the *dushmany* (outlaws) destroyed 1,812 schools and killed 152 teachers across Afghanistan. Prominent and successful government leaders such as Lieutenant Colonel Faiz Mohammed and Muslim clerics who co-operated with the 'godless communists' were unceremoniously liquidated as an object lesson to waverers. 'Collaborators' and 'spies', even whole families, were killed ruthlessly. Although the Soviets were able to secure Kabul and the roads to other cities, they could not guarantee the personal security of the Afghan government's officials against assault by the insurgents, who frequently carried out shootings, bombings and assassinations within the capital itself. Ironically, the Soviet intervention to bolster the pro-Soviet regime in Afghanistan exacerbated the conflict by supplying a focus for a hitherto fragmented and rudimentary resistance movement and rekindling the historic opposition to outside interference and any intrusion into local and religious matters.[36]

Unprepared for guerrilla warfare in the mountains of Afghanistan, the Soviet troops had neither the operational or tactical skills nor the mass support needed to overcome an insurgency. Slow to adapt to the terrain and the hit-and-run attacks by the guerrillas, tactical commanders and troops lacked the initiative, discipline and field craft required in fighting small, indigenous forces that refused to stand and fight and often slipped through Soviet attempts to encircle them. The Soviet Army eventually acquired the necessary skills for combat in

Afghanistan, but the rebels adapted to these changes. During the initial inter-vention, Soviet forces included a large number of troops from Central Asia who were selected because of their closeness to Afghanistan and close ethnic ties with its population. But these units did not perform well and fraternisation with the Afghans led to lapses in discipline and serious clashes between Russian and non-Russian, particularly Uzbek, troops. Soviet Muslim troops became demoralised rapidly when they found themselves fighting their kinsmen. Morale was low and discipline and combat effectiveness declined as soldiers deserted and bartered their equipment and weapons for food, alcohol and drugs. Caught up in a vicious civil war, Soviet conscripts in general suffered from poor discipline and low morale. They suffered too from poor conditions, the bullying of new recruits and alcohol and drugs abuse. They also lacked motivation and train-ing, participating in indiscriminate looting and atrocities. They displayed utter contempt for their Afghan allies and showed a reluctance to leave the security of their bases or to dismount from their armoured vehicles in the field.[37]

It took four to five years for Soviet soldiers to adapt from mechanised war-fare, learning new skills and tactics, notably the essential principles of mountain warfare that had been learned previously in the mountains of the Caucasus, Central Asia and Carpathia prior to and during the Great Patriotic War. At first, reluctant to leave the roads and handicapped by their heavy equipment, Soviet troops lacked the mobility and tactical skills to avoid ambushes or pin down an elusive and skilful enemy. The Soviets enjoyed the advantages of air mobility, superior firepower and modern communications. But these were counterbal-anced by inadequately trained troops and unreliable Afghan allies. With their counterinsurgency strategy fatally undermined, the Soviets withdrew from Afghanistan from 1988, deserting their client regime that, like South Vietnam in 1975, eventually collapsed and fell to the *Mujahideen* in 1992. Ironically it actually outlasted the Soviet Union by four months.[38]

The Soviet counterinsurgency campaign in Afghanistan during the 1980s can be perceived as a key transitional stage between the 'traditional' insurgencies of the twentieth century and the more modern insurgencies of the late twentieth century and early twenty-first century. It included elements of both the older model and also of the new emerging pattern. Like many traditional insurgen-cies, the *Mujahideen* enjoyed the support of the local population, and were also supported by outside funding, notably from the Americans, safe havens in Pakistan, a flood of Muslim fighters from around the world and a global Islamic ideology. Although lacking the Internet to broadcast their message to the world instantaneously, the insurgents conducted a propaganda campaign around the world to attract support, notably funds and volunteers. In this way they also influenced Russian policy-makers and weakened support on the Soviet home front, where public opinion turned against the war, precipitating the eventual dissolution of the Soviet Union. Like the US Army after Vietnam, the Russians preferred to avoid involvement in counterinsurgencies as the main lesson of Afghanistan and to concentrate on conventional warfare. This was a factor in the initial failure of the Russian Army in Chechnya in the mid-1990s.[39]

Chechnya

The Chechens have been waging *Gazavat* (holy war) and guerrilla wars intermittently against the Russians since the 1780s. Following their reoccupation by the Soviets in 1920, Dagestan and the Chechen-Ingush Republic remained the most insecure areas of the USSR. Revolts broke out in Chechnya sporadically during the 1920s, 1930s and 1940s. The region was a bastion of continuing resistance to Russian dominance as a result of the incompatibility of local traditions with Soviet modernisation, notably collectivisation and conscription. This resulted in an intense historical hatred of the Russians, which began with the 'scorched earth' policy employed by General Alexei Yermelov in 1816 and was exacerbated by many further Russian and Soviet depredations. After the Bolshevik seizure of power in 1917, the Chechens had declared their independence, waging guerrilla war against great odds into the 1930s. This culminated with Stalin's deportation of the entire Chechen population to Central Asia and Siberia in 1943–44, resulting in the deaths of a quarter of those deported. The Chechens were eventually allowed to return home in the mid-1950s to high unemployment and high levels of Russian immigration. After the Afghan War, with the collapse of the Soviet Union, this separatism resurfaced and the Soviets and then the Russian Federation were entangled in a succession of brutish, small wars.[40] The most damaging were the First Chechen War (December 1994–August 1996) and the Second Chechen War (August 1999–April 2009).

Like the Soviet forces that invaded Afghanistan, the Russian forces that entered Chechnya and besieged the capital, Grozny, in December 1994 were organised and trained for large-scale conventional warfare. They relied on large numbers of tank and mechanised formations heavily supported by artillery. With indifference towards civilian casualties, the Russians relied on massive force, employing heavy aerial and artillery bombardments of Grozny and other cities. This only produced great carnage and destruction and helped the Chechens to recruit more soldiers and win the support of the population. In response, like the *Mujahideen*, the Chechens avoided open battle, relying on classic guerrilla tactics, ambushes, surprise attacks and attrition to wear down a superior enemy and force the Yeltsin government to abandon the conflict. The Russians badly underestimated the ability of the Dudayev regime to resist even though the conflict came only six years after the Soviet withdrawal from Afghanistan. The Russians lost the propaganda war by default, failing to retain the support of public opinion at home. The Minister of Defence, Pavel Sergeievich Grachev, boasted that only a single parachute regiment would be needed to topple the Dudayev government in a matter of hours. The invading Russians were poorly paid, ill equipped and ill informed about the objectives of the operation, and their excessive and indiscriminate use of force caused heavy civilian casualties and widespread destruction. The Russian mechanised forces were unable to destroy the elusive guerrillas and were not trained to fight in small units in either urban or mountainous terrain.[41]

The Chechens' resistance and their determination to continue the struggle were consolidated. No effort was made by the Russians to win hearts and minds. Instead, they treated the population, including many ethnic Russians, indiscriminately as enemies. Indeed, the troops committed many atrocities, notably arson, beatings, looting, murder and rape. The Russians have also been accused of employing systematic torture, extrajudicial executions and causing the disappearances of male detainees held in so-called 'filtration camps'. The Chechens very effectively used the media to publicise their cause, project themselves as victims of Russian brutality and exploit the weak popular support for the war. Terrorist attacks, such as the dramatic raids into Russian territory, notably against Budennovsk (June 1995) and Kizlyar and Pervomaiskoe (January 1996), provoked indiscriminate responses from the Russians and kept the Chechen cause in the limelight. The Chechen offensive against Grozny, Argun and Gudermes in August 1996 was a huge defeat for the Russians. Like the Viet Cong's Tet Offensive in South Vietnam in 1968, it was a catalyst for change. The Russians were forced to negotiate and to end the first campaign against the Chechens, who gained internal autonomy within but not full independence from Russia. And the question of the republic's legal status was to be deferred for five years. The failure to subdue a former colony demonstrated that the Soviet superpower had been replaced by a weak Russia and paved the way for a new kind of Russian nationalism under Vladimir Putin, the new Russian Prime Minister.[42]

Following the rise of the international Islamist radicals and radical nationalism and the spread of crime and terrorism in Chechnya between 1996 and 1999, and the invasion of Dagestan by Chechnya in August 1999, Russia was faced with a chaotic and lawless state, which was a haven for violence and Islamic terrorists. Thus, fighting was renewed in October 1999 when Russian troops moved into northern Chechnya. Putin ordered the Army to reassert control over Chechnya using 'all available means'. By February 2000 the Russian Army had control of Grozny and by mid-2000 a firm presence in most of Chechnya and all the major towns as a result of large-scale operations that inflicted large casualties and enormous damage. But they did not, however, control the rural areas and the impenetrable mountain strongholds that sheltered the insurgents, who inflicted severe casualties in ambushes, assassinations and terrorist attacks. The police were accused of collaborating with and supplying the insurgents with weapons and even taking part in terrorist attacks. Putin promised to 'wipe out the terrorists and bandits', but the Russians struggled to restore 'normal' life despite making every effort to control the media. The creation of the Russian Information Centre ensured that news was censored, filtered, suppressed and in some cases falsified before being disseminated to the mass media, especially the foreign press. Once again Russian troops had engaged in systematic human rights abuses. Winning hearts and minds continued to have a low priority in the Russian campaign, which in a heavily 'kinetic' approach focused on eliminating the insurgents, notably their leadership both in Russia and in other countries. The Russians employed a carrot-and-stick strategy in which there were few 'carrots' and lots of 'stick'. They also suffered from a complex command structure that lacked the necessary unity of command. An even bigger obstacle was the low morale of Russian conscript troops who suffered from poor training, equipment, pay and conditions, which led to ill-discipline, including widespread alcohol and drug abuse. Russian units in Chechnya were riddled with corruption, notably drug trafficking, prostitution, arms dealing – including with the rebels – and kidnappings for ransom. The Russians suffered some of the world's worst terrorist attacks between 2002 and 2004 in retaliation.[43]

After evacuating Grozny in February 2000 the Chechens relied on hit-and-run tactics, notably the use of ambushes, assassinations, bombings, sniper attacks, suicide bomb attacks and bombings, and IEDs, which became increasingly sophisticated. Terrorist attacks included bombings in Moscow and other Russian cities and the seizure of Middle School No. 1 in Beslan, North Ossetia, on 1 September 2004, which had the hallmark of links with al-Qaeda. This turned Russian opinion against the Chechens and merely made it easier for Putin to depict the war in Chechnya as part of the global war against international terrorism. It precluded any attempt at a political settlement, and thus ensured that the war dragged on for a decade. It led to an increasing radicalisation of the Chechen guerrilla movement, who proclaimed allegiance to Wahhabism, and an influx of foreign 'jihadists' into Chechnya. Russia forged a new relationship with Turkey, its traditional enemy, who supported the Chechens, in order to cut off external support to the insurgents. The pro-Russian government in Chechnya installed by Russian troops in June 2000, like the client government in Afghanistan in the 1980s, lacked the required political leadership, although it offered a means to divide and rule the Chechens. It also provided support from indigenous security forces. Resentment of the government and its brutal and corrupt methods provided recruits for the rebels and ensured that the insurgency continued to smoulder. A new constitution in March 2003 gave greater autonomy but fighting continued as Chechnya

remained part of Russia, which announced in April 2009 that military operations against the insurgents were over. However, sporadic terrorist attacks by
the separatists persisted and levels of violence escalated again in 2011, notably
with suicide bombings in Grozny in August. The American counterinsurgency
campaigns in Afghanistan and Iraq diverted the attention and glare of the media
away from the Russian counterinsurgency against the Chechens and their
heavy casualties during 2008–09.[44]

Conclusion

Forced to wage a number of counterinsurgency campaigns to maintain
Bolshevik rule in the aftermath of the First World War, the Soviets had gained
considerable experience of counterinsurgency. These foreshadowed in many
important ways the many sided character of modern conflicts in the developing
world during the twentieth century. In particular, during their campaigns in
the 1920s they demonstrated a clear understanding of the political dimension
of counterinsurgency when opposed by guerrillas who enjoyed widespread
popular support. The Soviets understood and employed the strategy of carrot
and stick in order to win over and secure the population while ruthlessly hunting down the 'banditry'. Temporary political concessions would be tempered
by harsh military repression against the insurgents and their supporters. They
showed an awareness of the dependence of the modern guerrilla on the support
of the people and their strategy was much more centralised than other states
such as the Nazis in Germany or the regimes of Diem and Thieu in South
Vietnam, where different security and intelligence agencies were rivals working
in a chaotic and unstructured campaign rather than working together in a co-
ordinated way. The totalitarian state, which is under relatively fewer restraints and
constraints than a democratic system, ensured that counterinsurgency operations
were closely monitored and implemented in a coherent way. But the strategy
employed was often too rigid as a result of a stubborn and inflexible adherence to communist ideology, and this limited their options. They overrated the
importance of class theory while underestimating that of nationalism, which
underpinned much of the resistance. Thus, policies such as collectivisation and
the violence employed by the security forces often prolonged insurgencies.
Moreover, the Soviet Army promptly forgot the lessons of these campaigns, and,
from the 1930s, rarely fought guerrillas concentrating on conventional warfare.
As a result, the Soviets lacked a counterinsurgency strategy by the end of the
twentieth century, fighting guerrillas in Afghanistan and Chechnya instead with
conventional techniques, which relied on air power, armour and artillery and
were more appropriate to the conventional warfare of the Second World War.

4

A SLOW, LINGERING DEATH

THE FRENCH COUNTERINSURGENCY EXPERIENCE, 1947–62

Introduction

As the Cold War descended over a Europe recovering from the Second World War, the French colonial empire, *France d'Outre-Mer* (Overseas France), was dismantled in the era of decolonisation. After 1945, France tried to persuade its colonies that being part of 'Greater France' gave them a common purpose, special status and a share of imperial power, which independence and nationalism could not provide. Refusing to relinquish illusions of imperial grandeur and to acknowledge the demands of the colonies for independence, the post-war governments of the French Fourth Republic were particularly stubborn in fighting to retain Indochina and Algeria in long, brutal, costly and ultimately futile wars. In acquiring and defending an empire second in size only to that of Britain, France could look to a long experience of colonial rule, particularly during the second half of the nineteenth century. Although lacking a systematic doctrine, the French Army had developed a well-established counterinsurgency practice. It had tried and tested the strategy and techniques of pacification during a number of insurgencies following the French Revolution[1] in which a peculiarly Gallic approach to counterinsurgency was established and developed.[2]

Historically, the French response to insurgencies was a brutal one. In March 1793 the First Republic introduced the death penalty for all rebels, in support of the Army's operations against the peasant (and royalist) uprising in the Vendée. A 'scorched earth' campaign of suppression or 'ideological genocide' killed some 250,000. Twelve *columnes infernales* (infernal columns) under General Louis-Marie Turreau carried out indiscriminate and summary killings of priests, women and children, deportations, rape, the removal of food and livestock, burnings and confiscations. The removal of 'all inhabitants who had not taken up arms' was also ordered to eradicate the infrastructure supporting the insurgency and to stop loyal Republicans from being intimidated. However, from July 1794 General Louis-Lazare Hoche developed a more sophisticated carrot-and-stick strategy. Hoche attempted to win support by addressing local grievances and religious beliefs in the Vendée and in Brittany, where the *Chouans* (Owls) had also rebelled, offering 'carrots' such as returning confiscated property, promising religious toleration, indulging priests and improving the discipline of his troops. The 'stick' included blocking rebel communications and isolating them from foreign assistance, while implementing a clear-and-hold strategy, in which numerous outposts and mobile columns were used to reduce the guerrillas' room for manoeuvre. The revolts collapsed in 1796 although sporadic resistance continued until 1815.[3]

The pattern of employing repressive measures had nevertheless been set. During the insurrection in Calabria in southern Italy, which having begun in 1806 presaged the guerrilla war in Spain, peasants possessing weaponry were summarily hung. In particular, General Jean-Baptist Verdier became infamous for his savagery. In response, the guerrillas attacked French communications and unleashed a reign of terror against the occupying troops and their supporters. The insurgency was gradually isolated and destroyed by the French, who had abandoned any attempt to win over the population. They burnt villages and massacred their inhabitants in order to control Calabria. Having built new roads and bridges to improve communications, they also employed flying columns from the garrisons in the towns to pursue the guerrillas and recruited locals into the Calabrian Civic Guard and the Calabrian Chasseurs to supplement their manpower. The French also placed strict controls to deny the rebels food and supplies, and the taking of hostages to intimidate the populace. Inheriting a system of terror, which had been implemented during the invasion, the French continued to treat insurgents harshly, routinely executing captured guerrillas. In Spain in March 1810, General Peyrremond employed ferocious reprisals in Andalucía. During October 1810 General Honoré Reille pursued the policy of hanging four guerrillas for every Frenchman killed and substituting civilians if there were insufficient guerrillas in French hands. Reille's terrorisation of the population, which included the burning of villages, and the detention, deportation and summary execution of civilians, cowed Navarre. But the emphasis on terrorising the populace backfired. It failed to destroy the guerrillas, who were able to regroup, recruit and resume operations with a greater strength than before.[4]

One commander, General Louis-Gabriel Suchet, did buck this trend, seeking 'peaceful co-existence' with the population of Aragón. He took a more

conciliatory tone and relied on local officials and raising local forces to supplement his French troops. By keeping his men under control and placating
the population, Suchet was more successful than other French commanders
in obtaining a 'sullen neutrality' but this fleeting success was undermined by
Napoleon's demands for more resources and taxes and by renewed guerrilla
activity in Catalonia and Valencia fuelled by Spanish nationalism. Moreover,
Suchet meted out harsh treatment to captured guerrillas, who were tortured
and then executed. As a result of this experience in the French Revolutionary
and Napoleonic Wars, at least one French soldier, General Christophe Michel
Roguet, showed an appreciation of the importance of subordinating military
action to political action and of developing a strategy to win hearts and minds.
In particular, Roguet suggested employing Suchet's counterinsurgency methods in Algeria, but his proposals were viewed as being overly intellectual and
old habits died hard.[5]

Significantly, Marshal Thomas-Robert Bugeaud, who founded the 'colonial
school' of warfare in France, was a Napoleonic veteran with extensive experience of guerrilla warfare under Suchet in Spain. Bugeaud made his reputation
as Governor-General and Commander-in-Chief in Algeria between 1840
and 1844, fighting the Arabs under Abd el-Kader. Before Bugeaud arrived the
French Army was repeating the mistakes made in Spain. Too many soldiers
were pinned down defending static positions under attack by Arabs, who raided
French supply lines. French reprisal raids by heavy columns, slowed down by
artillery and supply convoys, achieved little except exhaustion. They crawled
across the scorched terrain in search of an elusive enemy who retreated before
them, refusing battle, but swiftly swooped to attack any exposed flanks, isolated
wagon trains or stragglers. Furthermore, as Bugeaud noted, a column that failed
to occupy was 'like the wake of a ship upon the sea'. It allowed the insurgents
to wait until the column passed and then renew their revolt, requiring a second
invasion and permanent occupation.[6]

Yet, in six years, Bugeaud conquered most of Algeria by reinvigorating his
demoralised troops and employing the four principles of mobility, morale, leadership and firepower. In place of manning fixed fortifications, which hitherto
had been the main French technique for controlling the countryside, Bugeaud
favoured mobile columns. These, stripped of heavy, slow-moving artillery and
wagons, lived off the land to improve mobility. They undertook *razzia* (punishment raids), seizing Arab livestock and destroying crops and buildings. Bugeaud
stressed the importance of good intelligence that allowed the mobile columns
to pursue the enemy without respite. Bugeaud was a seminal influence on the
French colonial Army and its subsequent counterinsurgency campaigns. The
Algerian campaign became a nursery for French counterinsurgency experts
and the 'Bugeaud method' was much imitated. It was employed during the

1850s and 1860s in West and Central Africa by Louis Léon César Faidherbe and during France's imperial expansion in the 1880s and 1890s. During the ill-fated expedition to Mexico (1862–66), François Achille Bazaine sent flying columns to attack Republican guerrillas, who opposed the restoration of the monarchy. But a shortage of troops to control a large country, a reliance on conventional tactics and the use of draconian methods meant that the insurgency continued to fester. In 1867 France withdrew its support of Maximilian's regime, which promptly collapsed.[7]

Frustrated by the refusal of the Arabs to stand and fight, Bugeaud had operated against their infrastructure. Hitherto, the French had only sporadically burnt crops and villages, confiscated livestock and cut down fruit trees. Bugeaud's campaign of 1841 set a pattern for the systematic destruction of crops, fruit orchards and villages throughout Algeria. A culture of brutality was established in which massacres of Arab men, women and children often occurred. This ruthless policy had long-term political consequences. It alienated the Arab population, who were driven to participate in a number of uprisings, notably the revolt by Mohammed el-Mokrani in 1871. It also highlighted some important trends in French civil-military relations. Bugeaud's campaigns outraged public opinion in France, as the gulf between French claims of bringing 'civilisation' to Africa and the excesses in Algeria became apparent. Bugeaud considered that the Army was a higher power, not accountable to the politicians, and when Abd-el-Kader received sanctuary in neighbouring Morocco, he did not wait for instructions from the War Ministry but crossed the border to defeat the Sultan of Morocco's Army at Islay in August 1844. This set a trend during the nineteenth century for colonial commanders to ignore the government's wishes. This exacerbated the tensions between the Army and the politicians of the Left, which would plague French defence policy, and underpinned the officer corps' alienation from the rest of the nation. Anti-militarism in France was an important political factor during the Dreyfus Affair during the 1890s, which tore French society apart. The Affair smashed non-partisan support among political parties for the Army, and had a serious effect on its relations with its political masters. Believing that it was fighting for *la gloire* (the glory of France) and that it was being vilified at home, the Army closed ranks against attacks by radical or socialist politicians, including Georges Clemenceau and Jean Jaurès, who denounced 'Algerian generals'. The strategic theories of Bugeaud's successors, Joseph-Simon Galliéni and Louis-Hubert-Gonsalve Lyautey, were formulated in response to the growing anti-militarism in France. Lyautey, in particular, sought to bolster the Army by promoting not only the development of the colonies but also the regeneration of French society and politics, thereby hoping to restore France to a pre-eminent place in the world after the humiliating defeats of 1870–71. This opened the door to the politicisation of the French Army, which had political implications for the *guerre révolutionnaire* doctrine during the 1950s.[8]

During the late nineteenth and early twentieth centuries in Indochina, Madagascar and Morocco, the 'colonial school' of warfare developed by Bugeaud, which employed rapid military thrusts through insurgent territory in Algeria during the 1840s, was refined into a more methodical style of counterinsurgency by Galliéni and Lyautey. They were the two great exponents of a strategy of progressive pacification, known as *tache d'huile* (oil stain), which was used to expand French authority slowly and methodically. Describing the technique in a letter to Galliéni in November 1903, Lyautey believed that subjugation of colonies should be 'not by mighty blows, but as a patch of oil spreads, through a step by step progression, playing alternately on all the local elements, utilising the divisions between tribes and between their chiefs'. The 'Galliéni method' of pacification placed as much emphasis on political and socio-economic measures as on military ones. Attacks by converging mobile columns on rebel strongholds were followed up with the introduction of amenities such as free medical care and subsidised markets, as well as posts designed to protect the population from insurgent attacks.[9]

In every subjugated region, the French built a network of blockhouses commanding the roads, which also served as bases for further operations by light columns. This combination of static territorial control and mobile reserves had a long tradition in the French Army. It had been employed against Royalist uprisings during the Revolution. Any military advance would be followed up by the introduction of French administration; soldier-administrators undertaking a dual role. They acted not only as administrators and police but also in a variety of civil roles. In this way, French control and influence would be established over the countryside while the traditional authority of local political leaders would be upheld. The 'Galliéni method' was codified by his subordinate, Lyautey, in a celebrated article, '*Du Rôle Coloniale de L'armée*' ('On the army's colonial role'), published in the prestigious journal *Revue des Deux Mondes* in January 1900.[10]

The *tache d'huile* doctrine, which emphasised unity of command, the importance of propaganda and of social and economic measures, and a system of close territorial control, provided continuity among French counterinsurgency methods prior to the Second World War and after. It was the strategy applied by the French in combatting the Maoist-style insurgency of the Virtminh in French Indochina between 1946 and 1954. It also formed the backbone of *guerre révolutionnaire* in Algeria during the 1950s. However, there were a number of problems with the doctrine. These resulted in military domination over civilian affairs, although for pacification to progress smoothly close co-operation between civil and military authority was essential. Bugeaud had understood the necessity of establishing a unified military and political response to insurgency. He employed officers of the *Bureaux Arabes* (Arab Bureaux) who knew Arabic to extend French control over the countryside and liaise with the local population. They gathered intelligence, administered justice and services to the population, and employed political warfare (divide and rule) to disrupt the insurgents' cohesion.[11]

The imperial expansion of the nineteenth century meant that France did not have the appropriate instruments in place for the centralised administration of its colonies. Soldiers usually became the first administrators in conquered territories and, given the limited control by politicians, continued to dominate. Going further than Bugeaud, both Galliéni and Lyautey, believed that a division of powers was a weakness. They wished to ensure unity of command and advocated uniting both administrative and military authority in military hands at all levels of the hierarchy. Emphasising that French administration should go hand in hand with a military presence, the 'Galliéni method' placed unified territorial command in the hands of soldier administrators. They retained indigenous administrative systems and relied on 'collaboration' by the population to succeed. In French Algeria, the Army continued to dominate the rural administration, whereas under the British system administrators remained civilians, thus ensuring that there was no real integration of the civil and military administrations. Thus, in large tracts of France's second colonial empire, the *officier-administrateur* replaced civilian control. The French colonial corps lacked a large pool of talented personnel, who could provide local knowledge. Its officials were often mediocre owing to a rapid turnover. Administratators served in a colony for only two years, allowing little time to 'connect' with the indigenous population and to learn their languages and customs. Unity of command was created at the expense of close co-operation with civilian administrators, which was in marked contrast to British counterinsurgency practice.[12]

This reinforced the Army's isolation from *La Nation* (the Nation), which was an important factor during the post-1945 colonial wars in Indochina and Algeria when some officers were increasingly in dangerous conflict with their political masters. During this era of colonial disintegration, successive French governments were held hostage by colonial interests that resisted any political concessions to indigenous groups. This meant that imperial policy was driven by the periphery. During the unstable Fourth Republic with no fewer than seven different political parties operating under a system of proportional representation, the imperially minded parties, notably the Catholic right of centre *Mouvement Republicain Populaire* (Popular Republican Movement, MRP), whose adherents invariably secured important ministerial posts in successive coalition governments, were able to block any changes to the status quo. Moreover, support by the French Communist Party for Ho Chi Minh, the

Vietnamese communist revolutionary leader, made it difficult for other parties to negotiate with the Virtminh. The socialists were driven into coalition with the colonial lobby.[13]

Lyautey had argued that the French Army liberated indigenous populations from despotism, anarchy and poverty, and that only 'minimum force' was employed to attain victory. This was a public relations exercise rather than a workable strategy to win over the population. It was designed to appeal to Frenchmen who were at best ambivalent about the colonies that their soldiers were conquering. Imperial expansion was achieved not only by economic penetration and winning hearts and minds but also by French bayonets and the *razzia*, which forced the local population to succumb from utter exhaustion. The French Army continued to stress the primacy of military action and to employ conventional tactics, such as the heavy column, into the twentieth century. But it remained a very blunt instrument, which, as Lyautey noted, resembled employing 'a hammer to crush a fly'. The French relied on their technological superiority, notably when Pétain crushed the rebellion by Abd el-Krim in the Rif Mountains of Spanish Morocco following its spread to French territory (1924–26). Pétain's excessive methods were denounced by a talented junior officer, Captain Jean-Marie-Gabriel de Lattre de Tassigny. Similarly, General Maurice Gamelin shattered the Druze rebellion in Syria (1926–27), bombarding insurgent positions in Damascus and employing tanks, artillery and aircraft while General Antoine Huré completed the conquest of Morocco (1934) with a force of 40,000, which included tanks, lorries and aircraft. When celebrations to mark the end of the war in Europe in May 1945 erupted into violence between Algerians and French settlers at Sétif, the restoration of order was swift and ruthless with summary executions, air strikes and even naval bombardment of Muslim centres.[14]

The tone for post-war French counterinsurgency was set by a two-year military campaign to suppress an uprising in Madagascar during 1947–49, which attracted very little attention outside France. During the two years of the fighting some 350 soldiers and 200 civilians, mostly French but including other Europeans and Asians, were killed by the insurgents. They had gained control of the south-east quarter of the island and established a provisional government. To restore French authority against the rebels, who were severely handicapped by an acute shortage of firearms and cut off from any international support, France sent in around 18,000 reinforcements, who arrived between April and July 1947. The French forces under the command of General Pellet and then General Garbay mounted an extremely repressive counterinsurgency campaign, which featured excessive retaliation by both the *colons* (French settlers) and the *Tirailleurs Sénégalais*. Mutilation of the living and the dead was carried out by both sides. The suppression campaign included a large number of arrests, the torture of prisoners, summary executions, the punitive burning and destruction of villages, and the rape of women and girls. This led to thousands fleeing into the forests where many died of starvation or disease. Somewhat belatedly, General Garbay exercised stricter control but some 60,000 people were killed as the French Army met force with force.[15]

Indochina

However, as the French were repressing the Madagascan insurgents, another brutal war, against the Virtminh in Indochina, which had broken out in December 1946, was progressing less well. As the Dutch also found out at the same time in modern day Indonesia, the French discovered that the old colonial methods were no longer applicable. The techniques of *tache d'huile* were no match for the Virtminh, who had an ideology and a disciplined political organisation that mobilised Vietnamese nationalism effectively. It became increasingly apparent between 1946 and 1954 that the tactics of pacification, employing the twin elements of static garrisons in fortified posts to intimidate the local population and mobile 'flying columns' to crush any revolt, were no longer sufficient to combat communist revolutionary warfare. This bore little resemblance to previous colonial revolts. Although *tache d'huile* had been applied successfully and extensively in Africa, Madagascar and Indochina, its application prior to 1945 had been in circumstances highly favourable to the French. It was employed against rebels who were divided by religious or tribal allegiance and who were inadequately armed, equipped and organised.[16]

In the changed circumstances of the aftermath of the Second World War, traditional French area-by-area methods of pacification no longer succeeded when faced by continuous political subversion. This ensured escalating popular support, intelligence, recruits, sanctuaries and supplies for the Virtminh. The traditional French reliance on technology and firepower, although occasionally an asset, too often worked against them, as they lacked the mobility to engage the elusive guerrillas. Pacification relied on tactical mobility, but the French lacked the aircraft and helicopters for the transportation of troops or supply by air on the scale that would be later employed by American troops in Vietnam. Helicopters were still being developed and no more than ten were available at any time during the war. Instead, the French in Indochina had to depend upon mobile columns of half-tracks, tanks and trucks moving on roads, which were vulnerable to ambush. As the first European power to confront a communist revolution organised along Maoist lines, the French were outmanoeuvred and outfought. They employed inappropriate and outdated tactics that displayed crucial, notably political, weaknesses in suppressing the new types of insurgency.

The French counterinsurgency campaign in Indochina was later summarised by journalist, Lucien Bodard, who reflected that 'the French Army, like a Louis Quinze armchair, was the masterpiece of an extinct civilization'.[17]

But this was not immediately apparent following the triumphal entry of General Philippe Leclerc de Hauteclocque into Saigon in October 1945. Outwardly, Saigon was again the 'Paris of the Far East', providing a base for Leclerc to re-establish French control and to destroy the Virtminh. However, Leclerc, a devout Catholic aristocrat with a brilliant wartime career, soon realised that the French lacked the men and equipment to achieve a decisive vistory. Leclerc set about the reassertion of French control in the south, despatching columns out into Cochin–China (one column leader was Colonel, later General Jacques Massu), which were told to use minimum force and to avoid brutality or looting. Questioning whether a full military reconquest of Indochina was feasible before he left in July 1946, Leclerc warned that France was heading for a guerrilla war that she could neither win nor afford. But he was ignored. Although the French had superior military organisation and modern equipment, notably tanks and aircraft, they lacked either popular support or local knowledge of the terrain. They had badly underestimated the Virtminh led by Ho Chi Minh and Vo Nguyen Giap, who followed a political–military strategy pioneered by Mao Tse-tung in China. Exploiting the vacuum of power, which followed the defeat of Japan in 1945, the Virtminh had already established their own infrastructure and 'safe bases' in the rural areas of Vietnam, notably in North Vietnam (the 'Viet Bac') and the Red River Delta. These sanctuaries allowed them to mobilise the support of the population and to wear down the French Army with guerrilla operations prior to launching a conventional war.[18]

In 1947 the French celebrated a decisive victory, having removed the Virtminh from the Hanoi area and reoccupied most of Laos. But their euphoria proved to be misplaced. In reality, they lacked the manpower either to prevent a Virtminh withdrawal to their sanctuaries in the north or to pacify Cochin–China, where the situation remained precarious. The uprising in Madagascar delayed the arrival of some badly needed reinforcements; when they did arrive pacification of Cochin–China was seen as their first priority. Thus, although suffering a setback, by withdrawing into the 'safe bases', the Virtminh were able to rebuild their forces and strengthen their hold on the population in Tonkin and Cochin–China. They concentrated on political education, recruitment and training, while maintaining a guerrilla campaign against the French occupation, particularly in Cochin–China. For nearly three years, convinced that the rebels had been defeated, the French did nothing to contest the Virtminh's hold over the population. They ignored the implications of the successes of Mao's identical strategy in China. French operations during 1948 and early 1949 continued to be restricted by limited manpower and indecision in Paris. General Roger-Charles-André-Henri Blaizot (Commander-in-Chief, 1948–49) wanted to lead a powerful force into the Tonkin mountains, the Virtminh stronghold, but the French government refused to denude the economically important Cochin–China of troops. One commander, Raoul Salan, was replaced for being too outspoken in demanding reinforcements.[19]

Some progress was made by the French in the Red River Delta and Cochin–China, where General Boyer de Latour du Moulin enjoyed success against the Virtminh stronghold in the Plain of Reeds. But overall an opportunity was lost in 1948. Inadequate numbers of troops meant that the French were always faced by an ongoing dilemma of where to employ their meagre forces as concentration in one area always risked defeat in others. The French were generally able to maintain control of the cities and roads, although ambushes became an increasing problem. The French occupied the rice fields and jungle by day, but the insurgents ruled them by night. Indeed, when Mao emerged the victor in China in October 1949, the French position in Indochina was compromised. The war was not proceeding well, requiring the French to remedy the situation in some way or face a slow but steady decline. Once supplies began to flow into the Viet Bac from China, the strategic balance tilted decisively against the French. China provided international recognition in January 1950 and an external safe haven to train and rearm troops, military advisors to plan operations and conduct training. Substantial Chinese economic and military aid was provided, allowing the Virtminh to make the transition from a guerrilla to a conventional force, forming regular units that received Chinese equipment and training. By 1954, the Virtminh had about 125,000 regulars and 350,000 militia. Armed and trained by the Chinese, the Virtminh went on to the offensive, deploying some thirty regular units. These were employed at first against smaller frontier posts and then against larger ones, notably on the *Route Coloniale 4* that linked Lang Son and Cao Bang.[20]

When in late 1949 the Virtminh went on the offensive, the French were unprepared, their troops scattered in static defences, which ranged from big complexes of bunkers and trench systems to little forts that housed a squad, a platoon or a company. These had previously successfully controlled and intimidated the local population but now merely tied down large numbers of troops, leaving the countryside in the hands of the enemy. The outposts, as is well described in Graham Greene's novel, *The Quiet American*, were extremely vulnerable to attack and deathtraps for their garrisons, being often isolated and undermanned. Even the more substantial garrisons ('hedgehogs' or well-armed outposts that were considered militarily self-sufficient and impregnable), notably along the Lang Son–Cao Bang Ridge in north-eastern Vietnam, could not be held by the French. For logistical support the French garrisons were reliant upon mechanised transport, either travelling through difficult terrain or along roads that were very exposed to ambushes. As a result, the casualties were high and French morale declined. When the Virtminh switched to direct attacks on the ridge garrisons, the outpost of Dong Khe was captured in September 1950, forcing the French to evacuate Cao Bang while under constant attack. By the end of October 1950 the French had suffered more than 6,000 casualties and had been driven out of the north-east, which now became the strongest of communist bases. By 1950 the French had lost any realistic opportunity for victory.[21]

The French should have reappraised their strategy following this disaster but they refused to admit defeat. Believing that the French were on the run, Giap launched an all-out assault on the Red River Delta, hoping to administer the *coup de grâce* and take Hanoi. This was premature, as the French government responded decisively, for once, to the *débâcle* of the so-called Battle of the Frontiers in 1950–51. They despatched substantial reinforcements to Vietnam and appointed the exceptional General Jean-Marie-Gabriel de Lattre de Tassigny as both commander-in-chief and governor-general. Arriving in December 1950, de Lattre at once withdrew the remaining garrisons into the delta region and prepared to defend the approaches to Hanoi. He reorganised his elite units such as the paras, marines and Foreign Legion, into mobile groups that, with artillery, armour and air support, formed a general reserve. Giap's large-scale conventional attacks between January and June 1951 were badly defeated by superior French firepower. Giap lost approximately 12,000 casualties in these attacks. This victory appeared to exonerate the principles of pacification and set a pattern for future operations, which were designed to exploit French technological superiority, notably the firepower of their artillery and air support. The de Lattre Line – a complicated defensive system of mutually supporting concrete blockhouses – was built around the delta to defend the French base and its population. In accordance with Mao's strategy of revolutionary warfare, Giap reverted to guerrilla warfare, defeating French attempts in late 1951 and early 1952 to break out of the Delta defences with their mobile reserves. By early 1953 Giap had recovered the military initiative and the French were running out of options. A Virtminh invasion of northern Laos in April 1953 forced the French to commit a substantial part of the Army to 'hedgehog' positions on the Plain of Jars, further limiting the reserves available for an offensive. With hindsight, it can be said that pacification had failed.[22]

This, however, was still not apparent to the French. General Henri-Eugène Navarre, a cavalry officer with no previous experience of field command, became commander-in-chief in May 1953. He was instructed to adopt an overall defensive strategy in the north for a year, conducting only minor tactical offensive operations to mop up the delta while preparing for a major campaign to bring the Virtminh to the negotiating table. In November 1953 Navarre dropped paratroopers into the valley of Dien Bien Phu in Tonkin (Operation Castor) to establish a 'hedgehog' across Virtminh communications with northern Laos. French intelligence, both tactical and strategic, was good, notably in providing a good picture of the build-up of the regular formations of the Virtminh, their location and intentions. But this was often undermined by the complacency and overconfidence of the high command. It was reluctant to accept the unpleasant reality that the war might be unwinnable with the limited resources available. At Dien Bien Phu the results of this attitude were disastrous. On 7 May 1954, isolated and dependent upon air supply, Dien Bien Phu surrendered and, although the French regrouped, the war was in effect over. Militarily bankrupt and with *la sale guerre* progressively more unpopular, the government of Pierre Mendès-France ended the French presence in Indochina. The last troopship sailed from Saigon in November 1954, leaving Laos and Cambodia as independent states and Vietnam divided arbitrarily along the 17th Parallel.[23]

Some basic counterinsurgency lessons did emerge from the Indochina War. It was recognised, for example, that the guerrillas gained considerable strength from their support among the local population. Valuable experiments in the resettlement of the indigenous population took place within defended zones in the border areas of Cambodia, notably in Svay Rieng in 1946 and in Kompong Chau in 1951. This deprived the Virtminh of support and left these previously infiltrated areas under French control. The aim was to deny food, shelter and recruits to the Virtminh, and some success was enjoyed. Unfortunately, resettlement was never adopted as a general policy by the French in Vietnam. This was due to the difficulty and expense of implementation in a country whose densely populated arable land sharply limited the possibilities for creating new farming communities. Some local tribesmen – principally from the T'ai mountains in the north – were organised, under French officers and NCOs, into *Groupements de Commandos Mixtes Aéroportés* (GCMA), renamed *Groupements Mixes d'Intervention* (GMI) in December 1953. They were used for anti-guerrilla operations in remote areas. These units, however, were generally distrusted and rarely operated with regular units. But, although given little support or publicity, they demonstrated that local anti-guerrilla forces could beat the insurgent on his own ground and with his own tactics. These new methods contributed to the evolution of the new counterinsurgency doctrine of *guerre révolutionnaire*, which the French were subsequently to apply in Algeria.[24]

Failing to recognise that the insurgent threat was primarily a political one the French made little effort to find a political solution or to counter Virtminh mobilisation of popular support. Information about the organisation and strength of guerrilla forces and, more importantly, the political infrastructure of the Communist Party, was lacking, and nothing was done to rectify this. There were, for example, no policies designed to tempt deserters from the Virtminh or to recruit captured guerrillas into 'pseudo-gangs' for deployment against their erstwhile colleagues. The war in Indochina was essentially a colonial campaign to reassert French rule and France's position as a great, global power. The French solution for Indochina was to build a non-communist alternative to the Virtminh, devolving limited internal self-government, while employing military force to defeat the nationalists. Although the former Emperor Bao Dai was belatedly recognised as head of an 'independent' and unified Vietnam in March 1949, something denied to Ho Chi Minh in 1946, he was discredited in the eyes of his people. The devolution of power to him was restricted by membership of the French Union, which allowed him little freedom to develop either his own policies or international standing. Furthermore, although possessing ability, Bao Dai was a playboy, known as the 'emperor of the night clubs', and his successive governments lacked genuine mass support and administrative capabilities. At the political level, the attractions of continued French domination were undermined by the Virtminh's exploitation of nationalist sentiment, which resented French control over the civilian administration. Seeking support from Bao Dai and minority religious sects such as the Cao Dai and Hoa Hao gave the French little credibility and allowed the Virtminh to establish a monopoly in terms of patriotism and nationalism. To this the French had no ideological reply. French propaganda lacked relevance to the cause of Vietnamese nationalism, promising almost exclusively a continued French dominance.[25]

The French also showed a marked reluctance to raise a truly effective national Vietnamese Army. Although de Lattre eventually coaxed Bao Dai into the creation of such an organisation, the initial target of an establishment of 115,000 men was not achieved until 1953. De Lattre himself supervised the formation of a cadre school and ordered each French unit in Vietnam to raise a second battalion from the local population. But this initial expansion lost momentum under his successor, Salan, who distrusted the Vietnamese Army, which the Virtminh were constantly infiltrating. Instead, French units were encouraged to recruit more Vietnamese into regular French units. For example, when the 2nd Battalion, 1st Parachute Light Infantry landed at Dien Bien Phu on 20 November 1953, nearly half their number were Vietnamese. The French were also reluctant to arm local militias, which might have freed the regulars from defending the population. As a result, a large percentage of French troops remained tied down as garrisons manning static defences. These proved vulnerable to defeat, often being overrun by human wave assaults, and they also reduced the manpower available for mobile operations. The lack of men knowledgeable in the local languages and customs and of good tactical intelligence meant that the troops were largely blind and often brutalised the population in an attempt to extract information.[26]

The French failed to co-ordinate the civil and military efforts. Political instability in France and constant changes of government prevented the implementation of consistent policies. Coming so soon after the defeat of 1940, the war in Indochina was increasingly unpopular in France. Support in polls fell from 37 per cent in July 1947 to 19 per cent by July 1949. This meant that politicians who sought domestic support did little to provide adequate resources. For example, the Budget Law passed in 1950 restricted the deployment of conscripts to the 'homeland' territory of France, Algeria and the occupation zone of Germany. Conscripts could not be sent to Indochina unless they volunteered and very few did. This fact ensured that there was little real commitment to the war by ordinary French men and women. Many socialists and the Radical Party leader, Pierre Mendès-France, opposed the war. The substantial Communist Party was openly hostile, calling strikes in arsenals and ports and even sabotaging equipment and stores waiting for transport to Indochina. Demonstrations and hostility meant that gendarmes were deployed to safeguard the embarkation of troops overseas, the wounded could not return to Paris or receive blood from the national blood donor organisation, and citations for bravery were not published. This caused a backlash from the Right, who contended that if Indochina was lost the rest of the French Empire would soon follow.[27]

French colonial forces, regular soldiers from *Troupes Coloniales* (colonial forces, or *La Coloniale* (The Colonial) for short), the *Légion Etrangère* (Foreign Legion), and North African *Tirailleurs* (Sharpshooters), suffered heavy casualties (3,500, of whom 352 were officers, killed since September 1945) and a drop in morale. Repulsed by the fractious nature of French politics and isolated by the lack of domestic support for the war, the Army believed increasingly that the withdrawal from Indochina had been caused by the political instability of the Fourth Republic. This lack of political direction was reflected in the failure to establish a unified command structure in Indochina similar to that established by the British with the appointments of Templer in Malaya and Harding in Cypus. De Lattre held the posts of both civil and military authority in late 1950, but in fact was not allowed to make either decisions or display initiative by his political masters at home. This dual appointment was not repeated and de Lattre's successors, Generals Salan and Navarre, faced interference not only from Paris but also a civilian governor-general in Indochina. As *la sale guerre* (the dirty war) dragged on, a widely held perception among the officers corps was that poor and irresolute leadership by the politicians was the reason for defeat in Indochina. It gave substance to a 'stab in the back' myth now current within the Army.[28]

Guerre Révolutionnaire

The defeat in Indochina was a seminal trauma, which had a profound effect upon the French Army. It acted as a catalyst for the evolution of the counter-revolutionary theory known as *guerre révolutionnaire* that sought to adapt *tache d'huile* with new ways of countering anti-colonial insurrection. This doctrine had its origins in the writings of an influential group of thinkers within the officer corps, which became obsessed with employing the lessons of the Indochina war to win future revolutionary wars. These were already imminent in other parts of the French Empire. From 1954 French officers studied Mao's principles of revolutionary warfare, the Virtminh's campaigns and evaluated the political-strategic implications of guerrilla warfare. Many of the younger officers who evolved this new counterinsurgency doctrine had fought in Indochina and in some cases had been captured by the Virtminh, where they learned at first-hand about communist insurgency and Maoist revolutionary methods.[29]

As a result of these studies, the French analysed and defined the separate 'phases' of a communist campaign of subversion. These were the infiltration of the population by political cadres, the creation of a guerrilla infrastructure and an alternative government, and finally an all-out offensive to seize political power. One analyst who promoted the new doctrine, Commandant Jacques Hogard, further divided the revolutionary war into five stages:

Propagandists and agitators worked to secure a foothold among the population.
Agitators built up a subversive infrastructure.
Insurgents employed ambushes, agitation, sabotage and terror.
The rebels set up 'liberated zones' or 'no-go-areas' from which the government would be excluded.
Finally, an offensive was launched to defeat the government and seize power.

Within these phases, the importance of popular support, international backing and the steady demoralisation of government forces was emphasised. This was a development of *tache d'huile*, rather than any radical departure. It recognised that the highly integrated and structured Virtminh had employed a new combination of political, military and psychological techniques to win over the population and subvert the French colonial authorities.[30]

Having recognised and dissected the nature of the threat, the theorists of *guerre révolutionnaire* promulgated a doctrine of counter-revolution against what they pinpointed as the inbuilt Achilles' heel of the Maoist model. The insurgency was most vulnerable during its preliminary phases, when popular support had yet to be established and its dependence on a logistic base, often in an adjacent country, remained strong, and it remained weak militarily.

Countermeasures were designed to contain communist subversion. They included the promotion of education and reform and closely monitored the activities of the population. Far-reaching population control, including massive resettlement and the construction of complex barriers along international borders, would be employed to separate the insurgents from both internal and external support. Above all, the French theorists of *guerre révolutionnaire* emphasised that, in contrast to Indochina, the Army should have the complete support of the government, despite the unpopularity or repressive nature of its methods. In short, to defeat the communists' revolutionary strategy, France had to implement its politico-military counter-doctrine with an equal ideological determination, fighting fire with fire, as the only remedy to the revolutionary warfare waged by Mao and Ho Chi Minh.[31]

These ideas were disseminated in books and articles in military journals such as *Revue Militaire d'Information* and *Revue de Défense Nationale* during the mid-1950s. General Lionel-Max Chassin's book, *La conquête de la Chine par Mao Tsé-Toung (1945–1949)*, appeared as early as 1952. Probably the most widely read exposition of *guerre révolutionnaire* was an anonymous pamphlet, *Contre-révolution, stratégie et tactique* (Counter-revolution, strategy and tactics), published in 1957 and distributed widely to officers. In the 1950s and 1960s a number of key texts appeared, including articles by Commander Hogard (*Guerre révolutionnaire et pacification*, 1957) and Colonel Charles Lacheroy (*La guerre révolutionnaire*, 1958) and the book by Colonel Roger Trinquier (*La guerre moderne*, 1961) which was later translated as *Modern Warfare: A French View of Counterinsurgency* (1964). *La guerre révolutionnaire* (1965), a small book by Claude Delmas, a prominent civilian contributor to the *Revue de Défense Nationale*, introduced the new ideas to a wider public audience.[32]

The French Army entered the Algerian War in 1954 with the theory of *guerre révolutionnaire* still evolving. Algeria became a testing ground in which techniques were implemented before they had been fully analysed or gained wide support. Indeed, David Galula, who served in Algeria and later had a big influence on the US Army, notably the counterinsurgency manual prepared by General David Petraeus and his associates in 2005–06, believed that the French Army lacked a single, official doctrine for counterinsurgency. This meant that the troops in Algeria operated in an intellectual void until an official doctrine appeared in December 1959 entitled *Instruction pour la pacification d'Algérie* (Instruction for the Pacification of Algeria). It was better known as the Challe Instruction, named after General Maurice Challe, commander-in-chief in Algeria from 1958. Endorsement of *guerre révolutionnaire* by the Army was slow. This was due to the scepticism of many officers, who considered the enthusiasts to be extremely arrogant. They accepted only the narrower, tactical theories of the doctrine, and ignored or rejected its wider, political implications. But, by

1956–57, an influential part of the officer corps had succumbed to relentless propaganda and the unceasing demands of the Algerian campaign, which was increasingly being waged according to the theories of *guerre révolutionnaire*. For example, the *Centre d'Instruction de Pacification et de Contre-Guerrilla* (Centre for the Teaching of Pacification and Counter-Guerrilla) was set up at Arzew in Algeria in March 1956 to inculcate French officers in the new doctrine and the communist threat, devoting little time to winning the hearts and minds of the Algerians.[33] Galula remained unconvinced that 'psychological action' and indoctrination could control the population. He maintained that a 'very articulate' minority had 'managed to take hold of the professional French Army magazines', giving the impression that their theories were official doctrine. It had become 'the subject of the standard jokes in the Army in Algeria'.[34]

The principles of *guerre révolutionnaire* provided French soldiers with a helpful operational tool for studying revolutionary warfare, which would make them not only professional but also well-versed in Maoist strategy. But these ideas did not probe the more fundamental motivation for subversion. In studying enemy doctrine, particularly the writings of Mao Tse-tung and Giap, French officers were more concerned with obtaining quick results to urgent operational problems rather than objective historical analysis. As a result, their studies tended to be rather superficial. They stressed that Mao's writings provided a general theory of modern war when in fact this had not been his intention. The operational lessons from the Indochina War combined with a deep dissatisfaction within the Army concerning the social and political realities of contemporary France to provide a new doctrine with far-reaching impetus and influence but one that was highly political and rigid in its ideas, eventually leading to a dangerous politicisation of the Army.[35]

At the core of *guerre révolutionnaire* was the belief that the Army should act in the role of the defender of the West and its values against a worldwide conspiracy by dedicated communist revolutionaries to subvert and overthrow existing political structures. Following traditions dating from the 1930s and the wartime Vichy regime, the enemy was now seen as being Soviet communism. Many officers were convinced that France should be mobilising all resources against this threat, believing that emerging Arab nationalism in North Africa was inspired by Soviet Communism. They also believed that it posed a direct strategic threat to France herself even though communism had minimal influence over the nationalists. In Algeria, this belief in a communist conspiracy blinded many French officers to the real causes of the conflict, which was chiefly Muslim discontent with political and economic inferiority.[36]

Just as *tache d'huile* as defined by Lyautey had repercussions, so the new doctrine of *guerre révolutionnaire* also had serious political ramifications, in that it advocated that the French government and people should support the Army with the required ideological vigour and resolve unconditionally in order to resist any future insurgency. A counterinsurgency campaign had to be fought without restraint, as a total war. It should employ all available military and psychological means, including semi-legalised brutality, including the deportation or internment of communities, and the detention and torture of individuals. Proponents of the doctrine concluded that the resources of the entire nation, and not simply cadres of the regular Army, must be committed to the struggle against a revolutionary enemy. Another lesson drawn was that the French Army should not be allowed to suffer another ignoble defeat because of a lack of political and national will. Indeed, many officers, who extolled the tenets of *guerre révolutionnaire*, also revealed a deep suspicion of politicians. This mistrust had been exacerbated by the 1956 Suez Crisis and ongoing squabbles with Paris over resources. These officers came increasingly to believe that the doctrine could not be realised satisfactorily in the multi-party, liberal France of the Fourth and Fifth Republics. They argued that, if the politicians failed to achieve the required ideological cohesion and unity of purpose, then the Army had the right to impose it.[37] As an article in *Verbe* argued, 'the Army's concept of the common good was superior to that of the state'.[38]

This ideological demand for a centralised command and advocacy of ruthlessness when combined with a tendency to see communism behind every insurgency that excluded any flexibility when dealing with either politicians or the enemy meant that, although the methods of *guerre révolutionnaire* had some successes during the Algerian War, it also led to deep divisions within the Army and France itself. That many officers, including General Maurice Challe and Colonel Charles Lacheroy, who had developed the doctrine, later rebelled against the state, notably during the 'Week of the Barricades' in January 1960 and the 'Generals' *Putsch*' of April 1961 in Algeria, is a serious indictment. For these adherents the war in Algeria was a crusade for France itself. Not only was Algeria to be rescued from communism and Pan-Arabism, but France was to be revived from its inefficiency and corruption and restored as a disciplined, progressive world power. Such theories led to disaster, plunging the Army into the treacherous waters of open involvement in French domestic politics and by doing so temporarily ruined its reputation and effectiveness.[39]

Nevertheless, French methods and the doctrine of *guerre révolutionnaire* had a big influence on the counterinsurgency campaigns of other countries. Portuguese Army officers attended the French Army's *Centre d'Instruction de Pacification et Contre-Guerrilla* at Arzew. Following a visit to Algeria in 1959 Colonel Hermes de Araújo Oliveira, Professor of Geography and Military History at the *Academia Militar* (Military Academy), gave five lectures that were published as *Guerra Revoluncionária* (Revolutionary War) in 1960. Portuguese doctrine was thus heavily influenced by the French doctrine of *guerre révolution-naire*. General Paul Aussaresses lectured on revolutionary war and the use of torture to 'break' insurgent cells while an instructor and adviser at the School of Special Warfare at Fort Bragg in the United States and at the military college in Buenos Aires in 1962–63. French thinkers have continued to influence the writers of American counterinsurgency doctrine, as was demonstrated in Vietnam and Iraq. The officer corps in Argentina later went on to fight a 'dirty war' against left-wing terrorist groups. Other leading French ideologues, notably General André Beaufre and Colonel Roger Trinquier, veterans of Indochina and Algeria, lectured at Argentina's Superior War School and their books and articles were translated into Spanish during the 1970s. Argentinian officers studied counter-revolutionary war at the French Superior War School. The South African Defence Forces were also heavily influenced by French experience in Indochina and Algeria, notably the counter-revolutionary warfare theories of Beaufre and his concept of 'total strategy', which put great emphasis on an all-encompassing mobilisation of civilian and military resources by the nation. These ideas formed the basis of lectures at the Joint Defence College in 1968.[40]

Algeria

The Algerian War began on 1 November 1954 with a series of unco-ordinated attacks against forty-five French targets. These included the hijacking of buses, bombing department stores and killing Europeans in Bône and Algiers, as well as in smaller towns. This war of terror was designed by the FLN (*Front de Libération Nationale*, National Liberation Front) to force France to grant independence. Owing to inadequate intelligence and poor co-ordination of the available information, the French were unprepared. However, the French regarded Algeria as not just a colony but as an integral part of metropolitan France with the election of representatives to central government. It had enjoyed the same status as the Dordogne or Provence since December 1848, eighteen years after the French arrived in North Africa. Thus, while the government of Pierre Mendès-France contemplated independence for Indochina, Tunisia and Morocco, it was difficult to fulfil the FLN's demands. They would change the structure of France itself and also trigger violence from the notoriously anti-Muslim *colons* or *pieds noirs* (French white settlers). When the revolt began in 1954, Mendès-France stated, '*Ici, c'est la France!*', while his socialist Minister of the Interior, François Mitterand, declared that 'the only possible declaration is war ... for Algeria is France'. As a result, the Army was increasingly caught between the conflicting demands of the French government, the *colons* and the FLN.[41]

With no political solution on the horizon, the military situation steadily deteriorated during the summer of 1955. From the beginning, recognising that it was too weak to secure a firm base area inside Algeria, the FLN sought to create a climate of fear internally. It used hit-and-run tactics and avoided major conventional clashes with the security forces. The targets selected were large *colon* estates, French military personnel and security posts, and Muslim loyalists. Violence and sabotage proliferated across North-East Algeria. In response, Army commanders increasingly ignored *répression limitée* instructions, turning to bombs, rockets, and the destruction of Muslim settlements, with occasional reprisal executions of their inhabitants. The cycle of rebellion and repression gathered momentum. The FLN established a six-person cell structure in settlements, villages, and small towns and some of the cities, termed the OPA (*Organisation Politico-Administrative*, Political-Administrative Organisation) by the French. It administered justice, stockpiled funds, supplies and intelligence and disseminated propaganda, forming a local, shadow government in opposition to French colonial rule. At the same time, the FLN leadership appealed externally to world opinion for support.[42]

The French did enjoy some initial success by deploying a substantial number of troops in the mountainous Berber areas of the Aurès and Kabylia, where they dispersed the FLN guerrillas. Furthermore, the new Governor-General, Jaques Soustelle, who arrived in Algiers in February 1955, promised reforms, notably a policy of 'integration' in which Algerian interests would be respected and improved. The continuing guerrilla campaign against military strongpoints and Muslim 'collaborators' had been highlighted in August 1955 when French settlers, including women and children, were brutally murdered at el-Halia near Phillippeville. This was a turning point, as hitherto the FLN had been perceived by the French as being more of a nuisance than a serious threat. But in an era of Muslim nationalism throughout the Middle East and North Africa the French were unable to totally eliminate the insurgents. Moreover, the Algerian rebels received external support and set up 'sanctuaries' when Guy Mollet's socialist government granted independence to the neighbouring protectorates of Morocco and Tunisia in 1956.[43]

Soustelle abandoned his political reforms following the massacre near Phillippeville, and Mollet's government similarly dropped planned reforms in February 1956 following settler demonstrations in Algiers. The liberalisation under Soustelle, which had engendered such hostility from the *pieds noirs*, who bitterly opposed limited Muslim representation in the Assembly, was now halted in a major change of policy. The French felt obliged to respond with a military commitment that alienated the population and generated further support for the FLN. The Mollet government now endorsed a 'military solution' by passing a Special Powers Act in March 1956 that surrendered civilian authority to the generals. The government also took the drastic and radical step of providing 500,000 conscripts as reinforcements for Algeria by lengthening military service to twenty-seven months and recalling a whole class of reservists. This escalation of the conflict led to an increased consciousness of the war in metropolitan France, much as the escalation of the American commitment to Vietnam was to do in the United States later. Meanwhile, the FLN was skilfully canvassing support in the Third World, the United States and at the United Nations, turning what France maintained was an internal quarrel into an international affair.[44]

As the campaign progressed to overcome the insurgent threat, the French Army developed a number of techniques, based upon the 'lessons' of Indochina as analysed through the prism of *guerre révolutionnaire*. The first – cutting the guerrillas off from their external supporters – was achieved by the construction of *barrages* (barriers) along the frontiers of both Morocco and Tunisia. In September 1957 the French finished construction of the first complex barrier, the Morice Line. This ran south from the coast to the Sahara some 200 miles (360km) along the full length of the frontier with Tunisia, which had provided bases for the FLN's 'army-in-exile' and logistical support for the guerrillas inside Algeria. It consisted of two rows of electrified fencing and barbed wire, surrounded by minefields and supported by floodlights, radar and blockhouses, and was patrolled constantly. The line withstood many attempts by the FLN to breach it. Any guerrillas who did penetrate the line were usually eliminated by patrols, supported by mobile reserves of tanks, artillery, aircraft and helicopter-borne quick reaction units. A similar, though not so extensive, system, the Pedron line, stretching some 90 miles (145km) inland from the coast, guarded the Moroccan frontier. To increase the effectiveness of the barriers, large numbers of the population were evacuated from some areas, including 300,000 inhabitants near the Tunisian border in the Constantine area, who were transported to 250 inland settlements. The vacated areas were designated forbidden zones where civilians could be shot on sight. This allowed the French Army to conduct operations unconstrained by political and social considerations. These fortified lines successfully closed the Algerian borders against FLN incursions from both Morocco and Tunisia, although logistical support from their external sanctuaries was never completely cut off. But they cost a great deal in money and manpower and could make no direct contribution to securing the population. The Morice Line alone required 40,000 men to garrison it. The French were unable to destroy either the sanctuaries on the Algerian border that ensured that the rebellion would survive despite heavy losses, or the FLN's regular forces. These acted as an 'army in being' inside Tunisia and provided Algerian nationalism with international recognition.[45]

Secondly, the 'surge' in manpower initiated by Mollet, a steady flow of American helicopters to move troops swiftly, new leaders with Indochina experience such as General Beaufre and innovative tactics improved the Army's ability to counter the insurgents. In addition, The Centre for Training and Preparation in Counter-Guerrilla Warfare (CIPCG) was established at Arzew in June 1956 to 'acclimatise' the conscripts arriving from France as the troops in Algeria swelled from 50,000 at the end of 1954 to more than 400,000 by mid-1956. The soldiers were taught counterinsurgency tactics although, following the instructions of Salan, the Commander-in-Chief, the main emphasis was on training in psychological warfare. This was seen as being the best weapon to defeat the revolutionary war in Algeria, which it was alleged was being 'directed systematically from Moscow'.[46]

Success in the field could not compensate for the fatal weakness of the French counterinsurgency campaign. This was the political price in the bitterness aroused among the local populace, which lost homes and crops and was forced to resettle in inadequate camps. At first, Algerian communities were resettled as a result of local decisions but by 1957 large-scale *regroupement* (resettlement) had become general policy with some 3 million people being moved into 1,840 *auto-défense* (self-defence) villages by 1960. This created an internal refugee problem and acute housing shortages, notably in Algiers, Oran and Constantine. Resettlement secured control over the population and, in theory, provided social and educational programmes. But, in reality, it demanded an investment of personnel, finances, and staff work that the Army could not sustain. Too many resettlements were poorly prepared, resulting in cold, overcrowded and insanitary camps, which caused malnutrition. In one model resettlement, some 300,000 people in the Blida region were moved from their villages and farms into compounds surrounded by barbed wire, watchtowers and guard posts, thus securing them from insurgent political activity and terrorism. Conditions improved after criticism in the press but ensured that it was unlikely that the Army would win the hearts and minds of the Muslim population. This was especially true as the French consistently underestimated the nationalist appeal of the FLN. Moreover, it was difficult to prevent insurgent infiltration into the resettlement centres, which proved fertile ground for subversion as a result of the hatred and frustration engendered by the poor conditions. By 1959, these shortcomings were apparent and the emphasis was changed to developing existing settlements rather than creating new camps. The Army, however, continued to resettle people in large numbers until the truce of June 1961, believing that separating the population from the guerrillas was more important than placating an alienated population. In theory, repression was inflicted only on the FLN rebels but in practice the security forces employed indiscriminate brutality against the whole population. This drove many to join the insurgents. Another problem was the inability of the Army to protect those still loyal to France. This was highlighted in June 1960 by the kidnapping and killing of twelve pro-French Muslims near Oran.[47]

In order to apply the 'carrot' rather than the 'stick', in September 1955 Soustelle set up the SAS (*Sections Administratives Spécialisées*, Special Administrative Sections) and the SAU (*Sections Administratives Urbaines*, Urban Administrative Sections). They formed part of the *Section d'Action Psychologique* (Psychological Action and Information Service). This was an attempt 'to re-establish contact with the population' in the areas where the FLN had destroyed the existing French administration. Modelled on the *Bureaux Arabes* formed in 1844 by Marshal Bugeaud, they acted as a link between the Army and the local people. By the end of 1959, there were 1,287 officers serving in some 660 sections, assisted by 661 NCOs and 2,921 civilian specialists in public health, agriculture and education. These teams attempted to win the hearts and minds of the population and counter the FLN's infrastructure by providing 'civil action'. The *képis bleus* (blue caps) undoubtedly gave substantial aid to the local Muslim population, handling an estimated 940,000 medical visits every month. But there were fundamental weaknesses. The project was under-financed and spread thinly with seventy-three SAS teams in the Kabylia, which covered some 4,000 sq km and had 900,000 inhabitants. Success was too dependent on the qualities of individuals and there were always insufficient officers and civilian specialists, whose recruitment was limited by low pay, the high qualifications required and the dangerous conditions of service. They remained representatives of an imposed alien culture and were targets for the FLN, suffering the heaviest casualties of any category of administrator. Moreover, they faced opposition from the *colons* and elements within the Army to helping Muslims. In any case, brutality or forced resettlements could ruin months or years of painstaking nurturing of the population.[48]

At first, the French repeated the early errors of the Indochina war, employing tanks and heavy equipment in 1945-style conventional operations, which were inappropriate against guerrillas. The Commander-in-Chief, General Paul 'Babar' Cherrière, who had no experience of guerrilla warfare, believed that he was dealing with only a small-scale tribal revolt, sending his available units into Kabylia and Aurès during 1954–55. A formal State of Emergency for the Aurès and Kabylia was declared in April 1955 and was extended to all Algeria in 1956. Until 1959, lacking good intelligence, the French Army relied on *bouclage* (large-scale manoeuvres of encirclement) and on great sweeps or drives with extended fronts – *ratissage* (raking over) – at divisional or corps level. These ponderous and large-scale operations proved futile. They failed to eliminate the elusive insurgents, who were always one step ahead of the cumbersome French columns. The guerrillas lost heavily in these conventional operations but were flexible enough in tactics and organisation to avoid total destruction. Furthermore, these operations were linked to the concept of collective punishments in retaliation for insurgent activity in which villages were burnt down and their communities summarily executed, and subject to the indiscriminate bombardment of villages. The Army's brutal 'pacifying' operations, notably by Senegalese troops, who were known for their ferociousness, alienated the population. There was no clear distinction whether the French strategy of 'pacification' was to win the hearts and minds of the inhabitants or to repress the revolt brutally. Mitterand and Soustelle had attempted to restrain the Army's use of firepower and draconian methods, but local commanders often continued to use these punitive methods.[49]

In 1956 General Henri Lorillot, who had replaced Cherrière, introduced a refined version of the traditional system of *quadrillage* to contain the insurgents. This had been revived by Beaufre in 1955, drawing on his experience of Morocco during the 1920s. It used a chequerboard of French outposts backed by mobile reserves that differed little from the previous tactics of pacification, deploying the same blend of static garrisons and nomadic pursuit groups. Large numbers were involved: some 300,000 men were employed in the outposts, while 60,000 *harkis* (French-led Muslim troops) and a further 30,000 elite French troops drawn from the *Légion Etrangère* or *La Coloniale* were used in the *Reserve Générale* (General Reserve). The greater part of the country north of the Sahara was covered with a network of fortified posts, manned by detachments of sixty to 100 men, usually metropolitan conscripts or North or Black African units. These secured the countryside and applied strict controls on the rural population. They controlled movement, and kept lists of inhabitants, issuing an identity card to all inhabitants. Gradually, *quadrillage* became very effective. It allowed garrisons to establish personal contact with the local population and this provided a constant flow of intelligence that restricted the activities of the guerrillas. In contrast to Indochina, greater emphasis was put on intelligence and winning hearts and minds on the assumption that conscripts, as 'civilians in uniform', would be able to bond better with the local population. Placing small garrisons across the countryside gave the French a permanency of control that could not be achieved by remaining based in the cities and strategic centres, or by launching search-and-destroy operations that gave only temporary control of the rural areas.[50]

The pursuit groups, the so-called *commandos de chasse*, whose name derived from *Jagdkommandos*, German anti-guerrilla formations that fought on the Eastern Front and in the Balkans during the Second World War, were formed of groups of sixty to eighty men. As in Indochina, they were mostly drawn from the elite *La Coloniale* and *Légion Etrangère* and carried out 'hot pursuit' of the guerrillas that could last for days, sometimes weeks. These mobile operations often relied on *harkis* to locate the insurgents. While much publicity went to the paratroops, they comprised no more than 5 per cent of the Army in Algeria. They only became totally effective in the difficult terrain with the appearance in 1958 of big troop-carrying helicopters. These pursuit groups pulled no punches. They burnt down villages or settlements, sometimes with their inhabitants still inside, and killed whole communities in reprisal for soldiers killed.[51]

The division of duties within the system of *quadrillage* created problems of morale and *esprit de corps*. Local garrisons, executing dangerous but dull and unglamorous responsibilities, resented the publicity and decorations showered on the elite units that fought major battles and then returned to their bases. Many conscript troops displayed neither enthusiasm nor talent for unconventional warfare, because they had received only a rudimentary training in counterinsurgency techniques before arriving in Algeria. These conscripts made up 90 per cent of the troops undertaking static guard duties under the *quadrillage system*. By contrast, the paratroopers were differentiated by *esprit para* (para spirit) and by their distinctive uniform. In claiming that they were winning the war, the elite force commanders confused the ability to achieve local military successes with the actual suppression of the rebellion.[52]

The French used indigenous troops, who made a considerable contribution, to supplement their forces. By 1956 at least a quarter of the 250,000 'French' forces were North Africans and the numbers who volunteered or were drafted for service in regular units increased from 20,000 at the beginning of 1957 to 61,500 in June 1960. North Africans fighting in auxiliary units were even more numerous. Besides the village self-defence forces there were three main types of auxiliaries. Organised in squads with their own corporals but commanded by French senior NCOs and officers, the *harkis* served as combat troops in both the *quadrillage* and the mobile reserves. The *makhzan* were recruited as guards, orderlies, and messengers by the Army's civil affairs teams. Thousands were also hired by the civil administration as auxiliary policemen, the *Groupes Mobiles de Sécurité*, among whom were the *Para Bleus*, recruited from former FLN guerrillas. By 1957–58 60,000 *harkis* and 35,000 other levies had been enlisted. However, the use of either independent North African units in action alongside French troops or the arming of Muslim villagers for self-protection was limited. *Harki* leaders complained that the French refused to provide automatic weapons.[53]

Nevertheless, the *quadrillage* system, when supported by large-scale, well-co-ordinated sweeps of selected areas, proved effective in destroying insurgent gangs. The most impressive, the Challe offensive, was carried out to a set pattern against the FLN mountain strongholds during 1959. A key facet of Challe's plan was the reduction of static garrisons in Algeria's cities, towns, and villages, and an increase in mobile operations to seek out the insurgents. Living off the land, *Harkis* would operate in an area of known FLN activity, fighting the guerrillas at their own game and pinpointing their locations. Regiments from the general reserve would then be deployed to destroy or split up large FLN groups. The mobile *Commandos de Chasse*, operating in small highly mobile units and trained to live like the insurgents, learning their habits and hideouts, would then pursue the small parties of surviving insurgents relentlessly by day and night. They were guided by *harkis* with local knowledge. The *Commandos de Chasse* were exceedingly effective but the high command remained reluctant to create counterinsurgent units that would stay in the field for operations over long periods. Instead, elite units flitted in and out by helicopter rather than remaining to hold and control territory. By doing so they left the local population at the mercy of the guerrillas. Despite their numerical superiority, the French lacked the manpower to implement fully both the *quadrillage* system and the large-scale sweeps, owing to the large number of men needed to man the static garrisons.[54]

The manpower employed by Challe, in one of the classic counterinsurgency campaigns, was colossal. The new tactics, which reinvented the *tache d'huile* technique, were applied from February 1959 in a series of major operations that squeezed the guerrillas against the Morice Line. There were also numerous smaller local sweeps to break up not only the insurgent units but also their infrastructure (OPA cells). For example, during Operation *Jumelles* in July some 25,000 troops were deployed in addition to the troops already operating in the area. Only two or three major British operations in Malaya employed 12,000 troops, while in Kenya the largest sweeps deployed only 5,000 troops. For even the smaller sweeps ten or more battalions were employed, often for two weeks at a time supported by helicopters that provided the most innovative technological development of the Algerian War. By the end of 1959 the French had arguably won the war, inflicting heavy blows at a low cost in casualties to themselves and defeating the FLN in the field. The fighting seemed virtually over, although the insurgents were still able to mount occasional attacks on individual *colons* or Muslim loyalists and engage in sporadic acts of sabotage.[55]

The strengths and weaknesses of French counterinsurgency were highlighted in the 'Battle of Algiers'. In January 1957, following the failure of the gendarmerie to master the urban terror campaign of the FLN in Algiers, Resident-Minister (Governor-General) Robert Lacoste and the new Commander-in-Chief, Salan, sent in the 10th Colonial Parachute Division (General Jacques Massu). Massu's division had just participated in the Anglo-French Suez Expedition. The Casbah was cordoned off and searched carefully in a *nettoyage* (cleansing) operation. Rapid surprise raids and arrests were carried out by the paras. An eight-day general strike called by the FLN was broken within two days. The serious battle then began as the paras moved in to attack the FLN's infrastructure in a series of ruthless and systematic operations. Total military control was imposed, with constant patrols, house-to-house searches, and a network of street checkpoints. Using the Special Powers Law of March 1956 to revoke legal restrictions on detention, Massu seized the police files and instituted large-scale arrests, after having already constructed a detailed intelligence picture of the FLN's order of battle. Informers known as *bleus* (blues) were also employed, while the *Dispositif de Protection Urbaine* (Urban Security Service) of Colonel Roger Trinquier divided the city into sectors, monitoring population movement and behaviour. By March 1957 the FLN infrastructure had been broken up although the hard core rebel leadership remained at large until early October. Similar tactics and techniques were also carried out in Constantine.[56]

Following its defeat, the FLN found it increasingly difficult to supply its forces inside Algeria. In purely military terms, Massu had carried out his orders efficiently, freeing Algiers from terrorist activity and destroying the FLN urban network. But politically the campaign was a disaster both in France and internationally. As the full scale of the paras' brutality and ruthlessness seeped out, France was condemned for adopting policies that were little different from those of the Nazis. The intelligence battle was won primarily by the systematic torture, euphemistically referred to as 'special measures', of some 24,000 men, women and children. At least 3,000 Muslims arrested during the 'Battle of Algiers', including the FLN leader Ben M'hidi, 'disappeared' while in detention. Conditions in ten internment camps, with some 8,000 detainees, added to international criticism. This allowed the FLN to gain recognition from the Non-aligned Movement in 1955 and from the UN in 1960. In the United States, the Eisenhower administration wanted the conflict resolved before it weakened NATO. During the 1960 presidential election Senator John F. Kennedy, sympathetic to the FLN, demanded a French withdrawal. The media played a pivotal role in eroding both domestic and international support for the war. It ensured that from now on events in hitherto 'isolated local wars' had repercussions around the world. In France reports of torture resulted in considerable protest. Public opinion was swung around in favour of a political settlement or eventual Algerian independence. Foreshadowing the US experience in Vietnam, the Army in Algeria was subjected to an intense barrage of anti-war propaganda from France. By 1959 relations between the military and politicians, who had begun a fundamental reappraisal of French attachment to empire, had sustained irreparable damage as a result of the Battle of Algiers. Few politicians were prepared to back the Army and a continuance of French rule. Algeria was seen now as a liability rather than an asset.[57]

The war had been lost on the diplomatic and political fronts. In political terms the FLN continued to be unbeaten because, as in Indochina earlier, the

French had failed to address the nationalism of the population by offering an alternative political solution. This undermined any reforms that had been implemented. Some 350,000 Algerian migrant workers in metropolitan France, at that time the largest 'colonial' community living in Europe, were also fully engaged in the independence struggle. They raised large sums to finance the FLN cause but also waged an urban terrorist campaign in Paris itself. The FLN saw more clearly than the French that the war had never been primarily about killing people but about gaining public support, particularly in the *métropole*. Its political offensive at home and internationally had turned the tables. This essential truth was ignored by the counter-revolutionary warfare school of *guerre révolutionnaire*. The Army's spectacular operations during 1959 solved none of the political problems and by July 1962 Algeria was independent, the FLN in control and the French hurriedly withdrawing.[58]

The frustrations of the Army boiled over during the events of 1958 and 1959, bringing an end to the Fourth Republic. The war in Algeria, which lasted nearly eight years, toppled six Prime Ministers, five governments and the Fourth Republic itself in April 1958. It also came close to toppling General Charles de Gaulle, who formed a government in June 1958, bringing the newly formed Fifth Republic to the brink of civil war. The Army and *colons* expected de Gaulle to retain Algeria but he did not have the same attachment to the empire as some soldiers, having served only one tour in Syria in the 1920s, and offered the Algerians 'self-determination' in September 1959. In November 1960, de Gaulle announced a national referendum in both France and Algeria on self-determination. In January 1961, while recording an overwhelming no vote among *colons*, it delivered a massive vote for self-determination by French voters. The failed 'Generals' *Putsch*' against de Gaulle in Algiers in April 1961 exposed the Army's politicisation and internal divisions. It also forced de Gaulle to negotiate with the FLN. The conscripts and the majority of the Army's officers, who were not prepared to contemplate civil war, supported their president. Nevertheless, some 14,000 officers and men had been implicated in the revolt and its effects were to last long beyond the Algerian War. The *Organisation Armée Secrète* (OAS, Secret Army Organisation), made of *colons*, ironically modelled partly on the FLN, launched a terror campaign against Muslims in Algeria. It hoped to provoke a violent reaction that would force the Army to reassert French control. The OAS also tried to assassinate de Gaulle and blow up the Eiffel Tower. The deaths of some conscripts and civilians in OAS bomb attacks turned French opinion against the dissidents.[59]

Conclusion

By the time the Algerian War ended in July 1962, most of French Africa was independent. The painful disengagement from Algeria was in fact the last major act of French decolonisation, leaving France only a few very minor possessions, islands or small enclaves. But the metropolitan Army continued to maintain permanent garrisons totalling close to 8,000 troops in Francophone Africa, supplemented from 1983 by an intervention force, FAR (*Force d'Action Rapide*, Rapid Deployment Force) of 47,000 men in a corps of five airborne divisions. The French often propped up Francophile leaders such as François Tombalbaye (Chad), Ahmqadou Ahidjo (Cameroon) and Sédar Senghor (Senegal) and, between 1963 and 1988, launched twenty military interventions into troubled African states. They included Gabon (1962 and 1964), Niger (1963 and 1973), the Central African Republic (1967 and 1979), Zaire (1978), Togo (1986), Rwanda (from 1990) and Mali (from 2013). But they showed a marked (and understandable) reluctance to become too deeply involved and just provided aid, advisers, air support, garrisons and training instead.[60]

The French have not enjoyed a great deal of success in counterinsurgency since 1945. In fighting insurgency they evolved and adopted a rigid 'theory' of *guerre révolutionnaire*, which was essentially reactive, trying to 'catch up' with events. Ultimately it proved not only unsuccessful in practice but also led to deep rifts in French society and the Army itself. Since 1962 little has been heard of *guerre révolutionnaire*. The lesson is clear, namely that the adoption of rigid, structured approaches to the complex subtleties of modern insurgency is fraught with dangers. The subsequent rejection by the French Army of all aspects of a counterinsurgency doctrine was unfortunate, as *guerre révolutionnaire* undoubtedly contained many useful techniques, which have been applied elsewhere.[61]

Once Algeria had become independent, de Gaulle set about the total restructuring of the French Army. It was revealing that France's first atomic bomb was detonated at Reggane in the Algerian Sahara just two weeks after the 'barricades week', while a second bomb exploded as the 1961 *Putsch* collapsed. One of de Gaulle's principal motives in wanting to disentangle the Army from Algeria was to modernise and transform it into an updated version of the *armée de metier* (professional army), which he had advocated during the 1930s. The atomic *Force de Frappe* (strike force) was an essential component, as after decolonisation nuclear rather than imperial power was now the basis for France's place in the world. Although there was some bitter infighting concerning which of the three services was to be its ultimate beneficiary, the formation of France's new atomic-age Army undoubtedly was a vital sop for the loss of Algeria. Thus, the Army accepted the role of the *Force de Frappe* as primarily one of political prestige, rather than as a purely military accessory, at the same time as, on the whole, it supported de Gaulle's withdrawal from NATO for the military independence and extra authority granted to it.[62]

The Algerian War 'created a deep wound in French society and a deeper one within the Army', which remained bitterly divided.[63] The officer corps was extensively reshaped from 1962. At least one-sixth of the officer corps resigned, including the majority of those officers who believed that Algeria should remain as a part of France. Internal divisions and recrimination within the officer corps were, however, to endure for most of the 1960s. The elite para and Foreign Legion regiments, which had been deeply implicated in the *Putsch* of 1961, were broken up and their officers purged. *La Coloniale* returned to its pre-1900 title, *troupes de marine*, and was given the lead role in the new intervention forces. In Indochina and Algeria there had already been an increased democratisation through promotions from the ranks and, although remaining conservative, the officer corps of the Army has gradually become less aristocratic and more bourgeois in its composition. By the 1970s a new, streamlined and highly professional Army, with better pay and conditions, and based entirely on home soil with a simplified new mission of defending France, had emerged. By the beginning of the twenty-first century French Marines had amassed more than forty years of experience of stability and support operations (SASO) in Africa. They are now taught about the diverse African cultures, traditions and fighting methods in order to train national forces and win the hearts and minds of local villagers in rebel areas.[64]

5

BANANA WARS AND OTHER SMALL WARS

THE AMERICAN COUNTERINSURGENCY EXPERIENCE, 1898–1975

Introduction

Like the British Army, the US Army lacked an extensive official, written counterinsurgency doctrine before the Second World War. It relied instead on an informal, unwritten tradition and accumulated experience for the conduct of 'small wars'. This had evolved pragmatically and was passed from one generation of soldiers to the next haphazardly. Ironically, during its War of Independence America produced guerrillas leaders such as Francis Marion (the Swamp Fox), Andrew Pickens (the Wizard Owl) and Thomas Sumter (the Gamecock). Apart from short conventional wars against the British and Mexicans, the US Army spent the majority of the eighteenth and nineteenth centuries policing the western frontier against Native Americans, accumulating a wealth of experience. Nevertheless, it remained an orthodox force attempting to control an unconventional enemy by conventional methods. It learned to employ columns to pin down an elusive foe and bring about a decisive engagement by attacking their villages and crops and forcing them either to abandon property and prestige or to stand and fight against superior discipline, organisation and firepower. The column was employed by General John Sullivan against the Six Nations in 1779, against the Shawnee in 1786 and 1790, by Major General 'Mad Anthony' Wayne in the Northwest Territory during 1794, and against the Creeks in 1813–14. Expeditions over difficult and unfamiliar terrain presented problems, and commanders abandoned equipment to increase mobility. They also developed specialised light infantry and riflemen, and employed irregulars such as frontiersmen and Indians as guides and auxiliaries. During the Second

Seminole War between 1835 and 1842 the tactics of waging war on the whole enemy population were combined with a cordon of small posts that provided patrols and security for white settlements in northern Florida. The Army also used 'collective punishments' and from 1838 recruited other tribes to kill the Seminole population. The lessons of 'Indian warfare' were codified in lectures given by Dennis Hart Mahan at West Point between 1836 and 1840.[1]

The US Army gained further experience of pacification warfare during the Mexican War of 1846 to 1848. During the war Major Generals Winfield Scott and Zachery Taylor were supplied over roads swarming with Mexican guerrillas, who had been enraged by American atrocities, mostly committed by troops from volunteer units. This was despite Taylor's warning in December 1846 to respect the local population. Taylor employed sweep operations and raids on guerrilla bases and punished his soldiers for atrocities committed against the Mexicans. He also held local communities responsible for guerrilla attacks and threatened to confiscate their property. He also convened military tribunals to try Mexicans, whose support for the guerrillas now began to waver. Scott also developed a carrot-and-stick strategy to win hearts and minds. The economic and social life of the country was to be maintained and he pledged to protect Mexican lives and property. Demanding impeccable conduct from his soldiers, Scott paid for supplies rather than requisitioning them. Free rations and food were distributed and schools, hospitals and other institutions maintained in order to promote good relations with the population. He also placated the powerful Catholic Church by attending Mass himself and ordering his soldiers to salute priests. But at the same time Scott waged a war of extermination against the Mexican guerrillas who were attacking his supply lines. He ordered that captured guerrillas be shot summarily and he treated civilian supporters of the insurgency harshly. The property of local officials who were held personally responsible for insurgent attacks in their districts was confiscated and villages believed to be sheltering guerrillas were fined or burnt down. Scott accomplished his task but the American advance between Vera Cruz and Mexico City was marked by devastation and the insurgents were never completely suppressed. The campaign provided the blueprint for officers during the American Civil War, which would in turn influence the Army's conduct of counterinsurgency operations for the next 100 years.[2]

During the American Civil War Confederate insurgents, often supported by the local population, pinned down large numbers of the US Army. They avoided set piece battles and employed the classic guerrilla tactics of ambushes, raids, surprise attacks on foraging expeditions and supply lines. They also terrorised citizens loyal to the Union and disrupted the federal government in the southern and border states. At first, President Abraham Lincoln and his Army commanders sought to win over opposition through moderate and benevolent policies. They believed, wrongly as it turned out, that most Southerners could be reconciled quickly. But victory remained elusive and, as the US Army continued to be harassed by 'bushwhackers' and government officials and supporters intimidated and assassinated, more severe methods were introduced. Scott's methods in Mexico were resurrected. Commanders, including Major General George B. McClellan in western Virginia in June 1861 and Major General John C. Fremont in Missouri in 1861–62, began to deal with the guerrillas and their civilian supporters 'according to the severest rules of military law'. Fremont announced that he would execute anyone found guilty of bearing arms. He also formed a counter-guerrilla company to locate and fight the guerrillas. Major General Henry W. Halleck advocated a 'no quarter' policy and in December 1861 issued a general order decreeing the death penalty for 'insurgent rebels' caught in the act of sabotage. He made local towns and counties pay for the repair of damage. A no quarter policy was widely employed, and many units, notably the 36th Ohio Volunteer Infantry under Colonel George Crook in West Virginia in the winter of 1861–62, did not take any prisoners. In East Tennessee the summary execution of 'bushwhackers' was common practice as neither side gave any quarter in their attempts to gain control over the region. Charging the 'Yankees' with 'atrocious and systematic violations of the laws of civilised warfare' the Confederates formed partisan rangers in April 1862 to wage a war of 'retaliation' led by leaders such as John Hunt Morgan, Nathan Bedford Forrest and John Singleton Mosby.[3]

Lacking manpower, the Union Army struggled with the problem of locating and eliminating the insurgents, who were mobile, elusive and supported by an intransigent and hostile population. In response, commanders such as Ulysses Grant, Philip Sheridan and William Sherman applied increasingly harsh, heavy handed and widespread measures. These included mass arrests, banishment, demands for 'oaths of loyalty', the implementation of pass systems and the confiscation or destruction of property. Sherman, the main advocate of severity against civilian supporters of the insurgents, believed that the population should be made to 'fear us and dread the passage of troops through their country'.[4] This hard attitude was echoed at a lower level by Colonel J.H. Shankling in Missouri, who believed that guerrillas 'are not only enemies of our country, but of Christianity and civilization, and even of our race, and the only remedy for the disease is to kill them'.[5] The Army levied fines on Southern sympathisers who harboured rebels, knowing that they relied on a largely friendly population for shelter, supplies and intelligence. Those who refused to pay had their property confiscated. Frequently, this policy was regarded as a sanction for wholesale plundering and as such provided recruits for the bushwhackers, increasing rather than decreasing their activity.[6]

Civilians were increasingly held responsible not only for the damage done by the guerrillas but also for failing to notify the Union Army about their presence. Brigadier General John Pope, a veteran of the Mexican War, fined Missouri towns within a 5-mile radius of any guerrilla attack. In Virginia he executed bushwhackers, destroyed their property and forced local inhabitants to repair railways and telegraphs. In Eastern Tennessee, Confederate families were accosted constantly by Federal troops, their homes raided and searched, their possessions stolen or destroyed, and they were often summarily arrested and deported for disloyalty. The Union Army forcibly resettled recalcitrant populations. As early as the autumn of 1862, General Ulysses S. Grant, in the first major resettlement, removed all active pro-Confederate civilians living in western Tennessee and northern Mississippi. A year later Brigadier General Thomas Ewing, brother-in-law of Sherman, deported almost the entire population of western Missouri.[7]

Out of desperation, the Union Army resorted to the destruction of private property in retaliation for insurgent activities. This led to the wholesale burning of towns and farms in order to deny supplies to the guerrillas and also to intimidate their supporters. The most notorious examples of this 'scorched earth' policy were the laying waste of the strategically important Shenandoah Valley by Sheridan and the swathe of devastation cut by Sherman on his march to Atlanta and to the sea in 1864. But similar methods were also employed in Arkansas, Missouri, Virginia and West Virginia.[8] This policy was enshrined in General Orders 100, 'Instructions for the Government of Armies of the United States in the Field', published on 24 April 1863. This was a landmark document as it provided, for the first time, an official code of conduct, which regulated the behaviour of the Army towards enemy forces and civilian population. Influenced by the views of Francis Lieber, a noted legal scholar, it advocated moderation in the treatment of unarmed civilians but recommended severe treatment of irregulars and disloyal civilians. It influenced European and British military thinking and became one of the pillars of the Hague Conventions of 1899 and 1907.[9]

The Union Army established small posts to protect the local population. These also served as bases for patrols that searched the countryside for insurgents. Blockhouses were built to defend bridges and railways. In addition to these passive measures, patrols, raids and sweeps were conducted to locate and destroy the guerrillas themselves. But too often the commanders displayed little initiative and their men stayed on the main roads and in towns rather than venture into the countryside. Nevertheless, some officers, such as Colonels George Crook and Henry M. Lazelle, learned through trial and error how to defeat the guerrilla bands. Adopting guerrilla tactics, the Federal forces used small, flying columns to hunt down the insurgents relentlessly on their own turf.

They burnt disaffected areas, took no prisoners and executed suspected rebel supporters. The provost marshal system was used to restore law and order and gather intelligence. It also administered loyalty oaths, issued passes, seized property, and arrested, tried and imprisoned or executed the disloyal. It recruited troops with local knowledge of the terrain and populace to serve as guides, spies and in the 'Home Guard'. Finally, it resettled the population, setting up self-contained communities of fifty families defended by a fort or blockhouse to protect supporters of the Union and separate the population from the guerrillas, foreshadowing the 'Strategic Hamlets' of Vietnam a century later. By these methods, although never destroying the insurgents completely, the Army contained them sufficiently to allow a conventional victory to be won. Little effort was made post-war to codify the lessons. The Army, which embraced the Prussian military system as its model for professionalism, adopted an 'uncompromising focus' on conventional warfare that became 'deeply ingrained' in its culture for the next century. Apart from the Marines, the American military trained to fight traditional wars against major powers, viewing small wars and insurgencies as distractions. Consequently, whenever conducting counterinsurgency campaigns, it lacked either a doctrine or an institutional memory of how to fight them. Although fighting on the frontier against the Indians, the Army dismissed such operations as police work, preferring to concentrate on training and planning for conventional warfare.[10]

During the Indian campaigns, the Army relied on a ruthless offensive strategy of relentless pursuit to eventually wear down and destroy the Plains tribes, who were acknowledged as superb warriors and masters of guerrilla warfare and 'the best light cavalry in the world'. Supported by civilian irregulars and Indian auxiliaries, columns of troops would advance, ideally in winter, into hostile territory and search for Native American bands or encampments, seeking not only to kill the enemy but also to destroy his supplies and shelter. The mobility of the Indians usually allowed them to evade these thrusts because co-ordination of converging columns was difficult and posed many logistical difficulties. But, such attacks, notably on Black Kettle's Cheyenne village on the Washita River by Lieutenant Colonel George Custer's 7th Cavalry in November 1868, could be very effective, especially when followed by a relentless pursuit that hounded the exhausted and demoralised hostiles ruthlessly until they were worn out and surrendered. A network of forts kept the Indians away from white settlements and communications and provided bases for patrols, which chased the elusive Indians. There was no formal doctrine for this warfare but the Civil War, notably the devastation of the Shenandoah Valley by Sheridan, served as a model for a strategy of striking at the enemy population in order to destroy its ability to carry on their resistance. 'Reservations' were also employed to resettle the Indians into areas where they could be intimidated, monitored and controlled. This was done mainly by supervision of the food supply. Total war against the entire enemy population allowed soldiers to employ the harshest measures since 'savages', like guerrillas, did not obey the rules of civilised warfare. It was regrettable that women and children died but such brutal methods were justified because they ended the war more quickly and decisively and were therefore more humane in the long run. The Army merely refined the methods of the Civil War and continued to wage a conventional war but with a few unconventional techniques in order to devise solutions for the problems posed by the Indian wars.[11]

The main difficulty was to locate the hostile Indians in the vast Western landscape. General Crook was one of the foremost advocates of using Indian scouts to track down hostile bands but he won few converts among his fellow senior Army officers. Crook also made drastic cuts in the equipment carried by his troops and he developed the use of pack mules to gain greater mobility. He also adopted a divide and rule strategy by promoting the disintegration of tribal loyalties and employing Indian collaborators to infiltrate, divide and break up hostile bands. Some units, using just mounted infantry, also cut down on all unnecessary equipment, abandoning their artillery and wagons. Thus, the Army achieved some mobility that, if still inferior to that of the Indians, allowed the soldiers to wear down their foe. In West Texas the strategy of Brigadier General E.O.C. Ord and his subordinate, Colonel Benjamin Grierson (10th Cavalry), was to garrison the handful of waterholes while sending out patrols with Indian scouts from forts and sub-posts to pursue the elusive foe relentlessly. Controlling access to the waterholes and cutting off supplies from the reservations, Grierson outmanoeuvred and outmarched Victorio's Apaches in 1880. General Nelson A. Miles created special elite units of handpicked and specially armed troops, which, when employed with Indian scouts, could provide small, mobile strike units to take on the Indians on their home ground and deprive them of any respite. Similar units would be later formed in the Philippines. In his campaign against Geronimo in 1886 Miles wore down the enemy with relentless pursuit that was combined with resettlement of the Chiricahua population to Florida, thus denying them logistical support. Unlike the British in India and the French in Africa or the Constabulary forces later raised by the Army in the Philippines and the Marines in the Caribbean, no attempt was made to raise large-scale Indian units. Such units might have fought the Indian wars in a less conventional fashion but the Army mostly relied on conventional methods and attrition, and this would remain the norm in future campaigns.[12]

Philippines, 1899–1902

On 30 June 1898, towards the end of the Spanish–American War, American forces landed in the Philippines, a Spanish colony, and after a brief campaign, aided by local Filipino forces whose goal was independence, captured Manila on 13 August. The purchase in December 1898 of the Philippines from Spain by the United States resulted in hostilities with the Filipino revolutionaries who, following defeat in a brief war between February and November 1899 that showed that they were unable to win the war conventionally, resorted to guerrilla war. Supported by a clandestine infrastructure and shadow government in the countryside and towns, the insurgents led by Emilio Aguinaldo sought to wear down the Americans through hit-and-run tactics. They hoped that the anti-imperialist movement in the United States led by the Democrat William Jennings Bryan would win the forthcoming presidential elections in November 1900. American commander Major General Elwell S. Otis and his commanders focused on the pacification of what they saw as routine lawlessness, not realising until early 1900 that they were fighting an insurgency. Indeed, until mid-1900, Otis refused to acknowledge that the insurgency existed.[13]

The US Army's experience during reconstruction in the South following the Civil War and the Indian Wars provided an administrative structure that Otis adopted on 20 March 1900. The Philippines was divided into departments, divisions, districts and sub-districts when the Army's tactical organisation was replaced by the military division of the country into four departments of Northern Luzon, Southern Luzon, Visayas and Mindanao-Jolo. The department and district commanders held dual responsibilities, undertaking not only their military duties but also the civilian administration. They ran the local government, attempting to display the benefits of American rule by building schools and roads. In doing so they had considerable freedom of action. Although the occupation of the entire archipelago tied down thousands of troops, it was the key to controlling the population and in securing intelligence about the enemy's infrastructure. But the US Army was ill-prepared for counterinsurgency operations and spent the next three years mired in a demoralising war.[14]

In the Ilocano Province (northern Luzon), for example, after mid-January 1900 Brigadier General Manuel Tinio's *insurrectos* fought on as guerrillas, ambushing American units and occasionally attacking towns. If attacked the poorly armed insurgents would escape into the *barrios* (urban areas), hide their weapons and blend in with the local population, from whom they received food, intelligence, recruits, supplies and taxes. At first the Americans were optimistic that their programme of re-establishing civil government, building schools and improving public health would win hearts and minds. Amnesties were also offered by Major General Arthur MacArthur (father of General Douglas MacArthur), commander of the Philippines from May 1900, to any *insurrectos* who surrendered. But cultural and linguistic barriers meant that local customs and needs were ignored. The Army, lacking good intelligence, unlike the guerrillas who had excellent information about American movements, was a 'blind giant', unable to use its powerful forces against an invisible enemy. Only gradually did the Americans build up a picture of the guerrilla bands and their infrastructure. This enabled them to understand the links between guerrillas and the towns. They used indigenous officials to identify insurgents and, by placing restrictions on the supply and movement of arms, food, and shelter, cut off their support from the towns.[15]

Local commanders established networks of spies and informants to gather intelligence and identify the enemy infrastructure. This allowed suspected supporters of the insurgents to be rounded up and interrogated. Building on Spanish methods, identity cards, card indexes, fingerprinting, photographic identification, travel permits, census records and the colonial police were all employed to monitor the local populace and create a modern surveillance state. As in the Mexican and Civil Wars, provost courts played an integral part in breaking up the insurgent infrastructure. Local provost officers developed reputations as experts in intelligence operations and were soon moved around the Philippines to disseminate their methods and break up recalcitrant areas. Early successes depended on the initiative of individual officers at a local level but gradually intelligence activities were centralised. Under MacArthur the Division of Military Information (DMI) was formed in December 1900 to undertake the key task, hitherto neglected, of collecting, analysing and disseminating information. In February 1901 DMI issued a comprehensive list of some 560 known rebels operating in south-west Luzon and the following month began keeping records not only of insurgents but also important community leaders, such as priests and municipal officials. In September 1901, Commander-in-Chief Major General Adna R. Chaffee ordered all post and field commanders to appoint an intelligence officer to collect information on his local area for DMI, which could now assume a crucial role, that of a central repository of intelligence. A strategy of divide and rule was also employed. During 1900 the Americans exploited tensions in La Union province between the *insurrectos* and a local Filipino sect, the *Guardia de Honor. Guardia*, previously persecuted by the rebels, was employed to identify and break up the entire clandestine guerrilla infrastructure in the towns and villages. Similarly, in western Mindanao, Catholic rebels were suppressed by local Muslim chiefs.[16]

Based on the experiences of the Civil and the Indian Wars, the Army operated aggressively, employing small columns of 50–100 men on search-and-destroy operations to locate and annihilate the insurgents. It sought to improve mobility by abandoning heavy equipment, by employing mule pack trains and indigenous bearers, and increasing the number of cavalry. Elite detachments of mounted infantry and scouts were created. They bore the brunt of the counter-insurgency campaign. The Army also slowly recognised the importance of using indigenous forces who were familiar with the terrain, local population and languages and were able to provide the means of implementing a divide and conquer strategy. It exploited the internal divisions, geographical and cultural, of the Philippines, which had proved successful against the Native American Indians. In particular, the rivalry between the Macabebes and the pro-guerrilla Tagalog, provided opportunities to undermine Filipino unity by exploiting existing fissures in their society.[17]

By the end of 1899 more than half the US Army in the Philippines was composed of volunteers who had joined up following the outbreak of the Spanish–American War. The Americans augmented their forces from September 1899 by recruiting indigenous troops to work with them as guides and scouts and guard lines of communications. At first, there was strong resistance from conservative officers, who feared atrocities by local auxiliaries or collusion with the guerrillas, which limited Filipino recruitment. Some units, the Macabebe Scouts in particular, were indeed notorious for committing outrages against the population of Batangas. But the return of volunteer regiments to the USA obliged the transfer of the pacification process to indigenous auxiliaries. They proved effective in policing the jungle and destroying armed bands. MacArthur increased the recruitment of auxiliaries, especially police and scout units, in the first half of 1901. That August Governor William H. Taft created the Philippine Constabulary, and the following month Chaffee formed the Philippine Scouts.[18]

As in the Civil War, the main lesson was that the policy of winning hearts and minds would be inadequate without an accompanying strategy of coercion breaking the link between the guerrilla's support in the towns and the insurgents in the field. The initial response by the US Army was to focus on combating the main security problem – the armed guerrillas. But gradually it had identified the importance of the link between the guerrilla bands and their infrastructure. By the end of 1900 American commanders were employing the brutal and oppressive measures used previously in the Mexican and Civil Wars to suppress the insurrection. The re-election of President William McKinley on 14 November removed the foundations of Aguinaldo's Fabian strategy. It also allowed MacArthur, who hitherto had avoided jeopardising this electoral success, to replace the previously benevolent pacification campaign with a much more ruthless one against the insurgent infrastructure. On 20 December 1900 MacArthur invoked martial law throughout the Philippines under the provisions of General Orders No. 100 (the Lieber Code) originally issued in April 1863. This provided the tools, such as provost courts, the arrest and imprisonment of civilian supporters, indigenous troops and intelligence, to separate the guerrillas from their infrastructure. Non-uniformed combatants were treated as 'highway robbers or pirates' and they, along with any civilians who helped them, were subject to the death penalty. MacArthur confiscated the property of some rebel leaders and exiled thirty-eight of them to Guam. Press censorship was also increased. The policy of benevolent pacification, or winning hearts and minds, was now counterbalanced by coercion. This formed part of a carrot-and-stick strategy to combat intimidation, kidnappings and assassinations by the guerrillas. Relentless pressure would also be kept up on the guerrillas by far larger and more effective sweeps of the countryside, when crops and property were systematically destroyed to deny the insurgents food and also to punish their supporters. Pro-American rallies were also held as support for the insurrection began to weaken. The constant combined pressure on the inhabitants of the towns and on the insurgents in the field proved decisive. It culminated in the destruction of both the insurgents and their infrastructure and led to the eventual surrender of many of the major *insurrectos* leaders.[19]

With the capture of Aguinaldo in mid-March and the surrender of increasing numbers of guerrillas, MacArthur believed that 'the armed insurrection is almost entirely suppressed'. Despite resistance from Army officers who disliked a return to civilian government before the insurgency had been eliminated, Taft became Governor on 4 July 1901. Many soldiers wished to retain unity of command and took exception at civilian intrusion, especially when military control had to be reinstated in Batangas province (Luzon) and on the islands of Cebu and Bohol. By September 1901 only three areas were still actively hostile – Luzon, Samar and Cebu. In south-western Luzon popular support for the insurgency meant that resistance in the Tagalog provinces of Batangas, Lagunas and Tayabas lasted longer than elsewhere. The rebel infrastructure on Luzon was the best led, organised and supported. Coercing the local population into acquiescence with fines, arson and assassination, it frustrated American attempts to suppress resistance. Unable to control either the hostile population or the countryside, the Americans compounded the problem by stressing the importance of 'killing armed *insurrectos*' instead of dismantling the insurgent infrastructure in the towns. The Americans also failed either to recruit indigenous auxiliaries or to recognise the key importance of intelligence. Sweeps to destroy the guerrillas produced few breakthroughs and the frustrations of pursuing a phantom enemy took their toll. Many, including Major General Loyd Wheaton, believed that, 'you can't put down a rebellion by throwing confetti and sprinkling perfumery'. Others, such as Colonel George S. Anderson, advocated 'a thorough destruction of all stores that may serve as subsistence to the Insurgent Army'. By June 1900 the Army had burnt 10–15 per cent of the houses in Bataan Province. Although counterproductive, such actions were commonplace as the Americans increasingly viewed the *insurrectos* as 'rank barbarians, not much above our better class of Indians'. They employed the destruction of property, indiscriminate arrests and unofficial interrogation techniques that included tortures such as the water and rope 'cures'. Colonel Benjamin F. Cheatham encouraged his troops 'to burn freely and kill every man who runs'. Reports began to reach the United States that beatings, torture, mock executions and denial of food and sleep were widespread, all integral parts of a highly controversial counterinsurgency campaign.[20]

The attack on the American garrison at Balangiga on Samar on 28 September 1901 proved to be the last straw. It resulted in the decision by Taft and Chaffee to end the insurgencies on Samar and Luzon. Chaffee appointed two of his most capable and experienced commanders, Brigadier Generals Jacob H. Smith and James Franklin Bell, who were successful in suppressing the insurgency but caused great destruction of property, crops and animals. In contrast to Bell's model pacification in the Tagalog region of Luzon, Smith's operations on Samar were highly controversial. They led to a Congressional inquiry to investigate allegations of atrocities. Smith's campaign was one of 'fire and sword', which established 'colonies' on the coast while sending columns to devastate the interior by destroying large quantities of food, livestock and property. Smith ordered Major (later Major General) Littleton W. T. Waller, to 'kill and burn' and take no prisoners (males over the age of ten were to be regarded as combatants) and turn the interior of Samar into a 'howling wilderness'. Smith was court-martialled and retired from the Army while five other officers were also tried for 'war crimes'. Their actions tainted the reputation of the Army as a whole.[21]

An experienced and capable officer with a reputation for being aggressive and ruthless, Bell changed the pattern of previous operations and addressed the particular circumstances of the region. He employed drastic measures in a carrot-and-stick campaign (or 'attraction' and 'chastisement' in the terminology of the time) which, while attempting to treat the population well, forced the guerrillas and their supporters to sue for peace. The soldiers were to be 'considerate and courteous in manner, but relentless in action'. Bell employed provost courts to gather intelligence and separate the population from the insurgents, to investigate the insurgent infrastructure in the towns and arrest supporters of the insurgency, who were then often tortured and beaten. Local leaders were coerced into supporting the Americans. But it was the inept strategy of the insurgents that alienated and lost the population's support. Copying the methods of Generals Valeriano Weyler in Cuba and Herbert Kitchener in South Africa and the 'reservations' during the 'Indian wars', the technique of 'concentration' was employed. The population was resettled into zones of protection where it could be intimidated and monitored, its food supplies controlled and movement restricted. Anyone outside these zones was considered to be hostile. With the towns under his control, Bell hunted down the insurgents. He swept areas of known guerrilla activity with columns in a well planned and executed campaign which destroyed the link between the guerrillas in the field and their urban infrastructure. But many of the 'concentrated' population died of malnutrition and sickness and large areas of Batangas were turned into a wilderness like other insurgency supporting parts of the Philippines. The US Army had won the Philippine War by re-employing the methods of total war from the Civil War and the Indian Wars. On 4 July 1902 President Theodore Roosevelt declared the insurrection had ended, but periodic uprisings, notably by the Moros, continued into the 1930s.[22]

Once again, the US Army made little effort to learn from the Philippines counterinsurgency campaign. A five-volume official history was never published for political reasons. The war had been unpopular, both at home and within the military. The Army was reluctant to discuss the seamier side of counterinsurgency in public at a time when allegations of abuse and torture had besmirched its reputation. Unlike the British, who included a section on 'Savage Warfare' in their 1902 regulations after the Boer War, the US Army ignored irregular warfare in its manuals and field service regulations. The study of counterinsurgency was relegated to the fringes, as reforms were instituted during the late nineteenth and early twentieth centuries. The system of military education introduced by General Sherman and the formation of the general staff by Elihu Root made the Army a more effective and professional institution. But it also reinforced the inclination to focus rigidly on conventional warfare as the Army's true role and to ignore the unconventional. After the Philippines the Army was increasingly reluctant to participate in any other counterinsurgency campaigns, which were seen as unpopular and institutionally unrewarding. Senior officers such as Generals Leonard Wood and John J. Pershing saw small wars in Cuba and the Philippines as preventing the Army from preparing for modern warfare and conventional operations. The Army's traditional inclination to relegate irregular war to a distinctly secondary status was reinforced by its experiences in Mexico and Russia before and after the First World War. Concentrating on conventional warfare, the Army's study of counterinsurgency virtually ceased in 1916 and during the inter-war years it was happy to leave intervention in Santo Domingo (1916–24), Haiti (1915–34) and Nicaragua (1926–33) to the Marine Corps. When the Army and Navy defined their roles in 1927, the responsibility of undertaking expeditions overseas was formally given to the Marines who gained the nickname of 'State Department troops'.[23]

Banana Wars, 1915–40

In the century prior to 1909 the US Marine Corps took part in some fifty minor operations in Central America and the Caribbean. The Marines also participated in the controversial pacification of Samar in the Philippines in 1901, for which Major Littleton Waller had been court-martialled and acquitted, and a more peaceful pacification of Cuba between 1906 and 1909, which was again led by Waller. From 1909 to 1912 the pattern of these interventions changed when the Marines carried out a series of landings on the Mosquito Coast of Nicaragua that presaged the 'Banana Wars'. These were to involve large expeditions and led to occupations that dragged on for years. Between 1915 and 1932 the Marines fought three small wars in Haiti, the Dominican Republic and Nicaragua, the experiences of which were to form the basis of a counterinsurgency doctrine published in 1934 and again in 1940 as the famous 'Small Wars Manual'. This advocated breaking up guerrilla bands and then pursuing them with small units, winning hearts and minds and recruiting a local constabulary. Commanded by Colonel Waller once again, the Marines landed in Haiti in July 1915 following the murder of the Haitian president. They became embroiled in a small war against the Cacos and undertook two campaigns in 1915–16 and 1918–20 to suppress the rebels, not departing until 1934. The Cacos reverted to classic hit-and-run guerrilla tactics from bases in the mountainous interior. They employed good intelligence and used knowledge of the terrain to retain the initiative, as well as terrorising the urban population of Port-au-Prince. Alongside military operations the Marines undertook 'civic action', constructing roads with forced labour and providing medical services. There were never enough Marines to occupy all the country so the Haitian Gendarmerie was created in 1915. It was commanded by Captain Smedley Butler with 120 Marine officers and it helped to garrison countryside villages, pacify the rural population and gather intelligence to locate the Cacos. Internal passports were also issued to monitor and control population movement. At night large mobile columns were deployed in search-and-destroy operations to attack Cacos camps and forts. Small Marine patrols, which used captured Cacos as guides, then systematically pursued and harried the surviving insurgents, who were also strafed and bombed by aircraft. The Marines destroyed houses and burnt down villages to deny the rebels shelter. Allegations of the 'indiscriminate killing of natives' became an issue during the American presidential election of 1920.[24]

Between 1916 and 1921, the Marines also occupied the Dominican Republic, where a growing insurgency in the countryside opposed the president. They divided the country into two military districts and disbanded the Dominican Army. As in Haiti, the Marines employed a combination of civil and military measures to pacify the population and defeat the guerrillas. These included reorganisation of the government, the provision of medical services, educational and agricultural reform and road building. Garrisoning the countryside to provide security for the population against the rebels, they used search-and-destroy operations followed up by small patrols to locate and destroy the elusive insurgents. The Marines protected the rural population in the Dominican villages by recruiting and training indigenous 'Home Guard' units. These mixed units, composed of ten to fifteen Dominicans and two or three Marines led by a Marine officer, were effective in bringing security to rural communities and anticipated the Combined Action technique employed in Vietnam half a century later. The *Guardia Naçional Dominicana* (Dominican National Guard), later renamed the *Policia Naçional Dominicana* (Dominican National Police), was created in 1917 and proved very effective in patrolling and fighting the rebels. Operations were undertaken at night and during bad weather in order to gain the element of surprise. Property and supplies that would aid the enemy were routinely destroyed. Cordon-and-search operations were introduced in order to round up adult males and identify insurgents. National identity cards were issued to the whole population and central intelligence offices set up in each district. By 1922 only a few small bandit groups remained in the field. The lessons learnt in Dominica and Haiti, notably the key importance of an indigenous constabulary in the successful campaigns, were disseminated in the *Marine Corps Gazette*.[25]

In the summer of 1927 the Marines faced an insurrection in Nicaragua led by Augusto Sandino, an excellent guerrilla leader operating from his base in the northern jungles of the province of Nueva Segovia. But by that time, American military intervention in the Caribbean and Central America was facing growing criticism, both domestically and also internationally, and President Calvin Coolidge wanted a quick victory with the minimum cost and controversy. Brigadier General Logan Feland, the Marine Commander in Nicaragua, therefore emphasised that winning over the local population was essential, especially given Sandino's popularity and the unpopularity of America's Nicaraguan allies, principally mine owners and businessmen. Even before the insurgency had begun the Marines formed a constabulary called the *Guardia Naçional de Nicaragua* (National Guard of Nicaragua), which garrisoned the major cities and towns but failed to secure the countryside. It had a growing reputation for corruption and ineffectiveness. In January 1928 the Marines located and stormed Sandino's mountain fortress, El Chipote, but he escaped to continue his guerrilla war against the Marines from deep within the jungles of Nueva Segovia. Sandino's insurgents were well trained and armed and also had an excellent local intelligence network to warn of Marine patrols. The Marines, who found the local populace hostile, used the well-tried combination of garrisons, indigenous constabulary forces, small-scale patrolling, large-scale search-and-destroy and night operations, intelligence networks, together with the innovative employment of aviation to locate and defeat Sandino. The *Guardia* released the Marines from garrison duty and freed them to hunt down the insurgents. Long roving patrols, notably along the Coco River with the aid of the local Miskito tribe, pursued the rebels. Property was again destroyed and the emphasis on 'civil action' was minimal. In May 1928 the Marines offered an amnesty to insurgents who surrendered and then launched an offensive, breaking up the guerrilla bands and forcing Sandino into exile in Mexico. The *Guardia* hunted down the remaining rebel bands and, by the autumn of 1932, the insurgency had collapsed.[26]

Greece, 1946–49

At the end of the Second World War, the US Army had a considerable body of experience to fall back on when undertaking small wars, although relatively little of this doctrine was either recent or included in official manuals. During 1939–45, the US Army mainly undertook a conventional war and did not participate in any significant counterinsurgency campaigns. As a result of the post-war emphasis on conventional warfare there was little incentive to develop pre-war small wars doctrine, which was largely neglected by the service schools and by the writers of Army doctrine. Together with the death or retirement of the Army's pre-war veterans who had experience of counterinsurgency, this ensured that the service lacked the doctrine or organisational expertise for counterinsurgency operations. Much of the doctrine that emanated post-war concentrated on lessons gleaned from studies during the 1950s of German responses to fighting against partisans who had supported the Red Army, which was seen as the main threat during the Cold War era. Guerrillas were therefore regarded as an adjunct to conventional warfare rather than requiring a separate concept and strategy. The emphasis remained on small unit tactics, mobility, surprise and good intelligence. The population would be won over through the good conduct of the troops, the introduction of government reforms to redress grievances, and the provision of modern economic, social, and political institutions as part of a carrot-and-stick strategy. This blended aggressive operations, punitive measures and enlightened administration to separate the insurgents from their supporters.[27]

Lieutenant General Sir Ronald Scobie's British troops liberated Greece during October 1944 and, remaining until 1946, were quickly engulfed in the savage Greek Civil War (1944–49) between the communists and their rivals. Following the election of a right-wing government in 1946, the communist ELAS (*Ethnikòs Laikòn Apelevtherotikòs Stratòs*, National Popular Liberation

Army) began classic guerrilla operations, employing ambushes, raids, sabotage and terrorism. ELAS was supported internally by a formidable clandestine organisation (the *yiafka*), which had developed during the Second World War. It was aided also by Greece's communist neighbours, Albania, Bulgaria and especially Yugoslavia, which provided training, equipment and sanctuaries. The guerrillas established bases along the mountainous northern border, where they formed 'liberated zones'. The Greek population was caught between the cruel repression of the government, whose police had a reputation for brutality, and the insurgents, who terrorised the rural areas, press-ganged recruits and seized supplies by force. Both the police and the Greek Army, which did not have an efficient staff and was handicapped by a factional officer corps, lacked the training and equipment to deal with the guerrillas. The Army had dispersed its troops across the country to provide security against communist raids and terrorism, leaving no central reserve to conduct offensive operations or regain the initiative. By the end of 1947, more than half the Greek Army was pinned down in static garrisons, while the guerrilla infrastructure was largely untouched despite a series of sweeps and encircling operations. Planning and execution had been poor, allowing the insurgents, warned by their excellent intelligence network, to escape.[28]

In December 1947, the Joint United States Military Advisory and Planning Group, Greece (JUSMAPG) was set up first under Major General William G. Livesay and then Lieutenant General James A. Van Fleet to provide operational guidance for the Greek Army. In practice Van Fleet commanded the Greek Army. With JUSMAPG planning, he implemented operations, personally inspected and assessed the strengths and weaknesses of the Greek Army, and insisted that American recommendations were adopted. The Americans also provided advisers to train and administer the Greek Army, introducing their own techniques and intense battle-focused training and doctrine, which emphasised aggressive small unit tactics to defeat the less well equipped but elusive insurgents. Special platoons were raised to demonstrate these techniques to troops in the field but it was recognised that it would be a mistake to turn the Greek Army into a clone of the US Army. Thus, JUSMAPG formed up a division for the Greek Army in 1949 that was especially designed to meet conditions in Greece. Since tanks and heavy artillery were of limited usefulness in mountainous terrain, the Greek Army was given more suitable weapons such as machine guns, mortars and pack artillery. JUSMAPG also reduced the motorisation of infantry battalions so they would not become road-bound. Proposals to equip the Greek Army with helicopters were rejected and instead horses and mules were provided as a means of improving mobility. Nevertheless, the Greek forces retained ingrained habits, difficult to change, notably a lack of aggression and an over-reliance on artillery and air support.[29]

Improvements in the Army freed the police to focus on the key tasks of maintaining law and order, providing security for the rural population, and breaking up the communist infrastructure. Economic and social aid was provided in areas cleared of guerrillas by the Greek Army. It shifted aid from cities to the countryside where the battle to win hearts and minds was critical to success. An armed militia – the TEA (*Tagmata Ethnofylackha Amynhis*, National Defence Corps) – was created in the autumn of 1947 to replace existing ad hoc organisations that supported rural pacification. Evolved from the wartime right-wing paramilitaries, it implemented the 'White Terror' of 1946–47 in local areas. It controlled the population, freed the Greek Army from providing static garrisons to carry out offensive operations, supported its cordon-and-search operations and helped to hunt down insurgents. In 1948 a Home Guard was also formed to provide security for villages and outposts. Wishing to control the local populace and cut off the insurgents from their supporting infrastructure, the Greek government developed a harsh but effective strategy of mass arrests, rounding up tens of thousands of suspected supporters of the insurgency. By November 1947 310,000 had been forcibly evacuated, mainly from the north where the guerrillas were strongest. At the start of Operation Pigeon in the Peloponnese during December 1948, 4,297 suspects were detained, often without trial. Some were executed while many were held in remote island internment camps. The government also removed entire populations from areas infested by insurgents, draining the 'sea' in which the 'fish' swam and thus eliminating the guerrilla intelligence network. By 1949 there were 700,000 refugees (about 10 per cent of the Greek population), the majority of whom had been relocated by the government. While some American civilian officials were uncomfortable with these methods, believing that they could be counterproductive, the soldiers, notably General Van Fleet, approved of the mass arrests and resettlement and did not enquire too closely into the treatment of internees. Large areas of the Greek countryside were laid waste, food and shelter in the mountain villages destroyed, all in order to force the guerrillas either to retreat into sanctuaries across the border or to fight out in the open.[30]

Although the government offensive of 1948 was no more successful than that of 1947, American aid and advice began to shift the operational initiative towards the Greek Army, helped by other developments in 1948–49 that now tipped the situation in the government's favour. Internal divisions within the Greek communists were exacerbated in early 1949 by the decision to support an independent Macedonia, perceived as treason by most Greeks. This policy had a number of important consequences. It alienated not only internal supporters, allowing the Greek government to rally public opinion against the communists, but also their main external supporter, Tito. He abruptly closed the Yugoslav border in July 1949, which removed the main source of support for the Greek communists, denying them sanctuary during the government's offensive of 1949. The insurgents also attempted prematurely to transform the guerrilla war into a more conventional conflict (Mao's third stage of an insurgency). They established fortified bases along the frontier and attempted to capture towns in an ill-fated move to seize and hold territory. However, this allowed the security forces to deploy their superior firepower, which destroyed the bulk of the rebel forces.[31]

In January 1949 the Greek Army was galvanised by the appointment to the new post of supreme commander of General (later Field Marshal) Alexandros Papagos. He enjoyed a good relationship with Van Fleet and ensured that the offensive of 1949 was much more effective than those of 1947–48. Special commando units, LOK (*Lókhoi Oreinòn Katadroméon*, Companies of Mountain Rangers), trained in anti-insurgent warfare by the British and paid for by the Americans, who overcame their institutional dislike for elite forces, to take on the guerrillas. They pursued and harassed the communists as never before, while at the same time the population was protected and controlled in the towns and villages. Having cleared southern and central Greece, the Army overran the communist bases in the north and soon the war was over. The Army had been helped by the enemy's mistakes but it had also improved sufficiently to achieve victory. This had been largely as a result of employing the 'stick' rather than the 'carrot'. It had combined conventional operations, mass arrests of communist sympathisers, and resettlement and security measures in the countryside to remove the communists' control over the population and deprive them of their intelligence. But, ultimately, the insurgency failed because the majority of Greeks were opposed to it. As the first major confrontation of the Cold War, the Greek Civil War was a testing ground for the counterinsurgency tactics and techniques that would be employed during the next four decades.[32] As with future operations in the Philippines and Korea, those in Greece showed the importance of working with and supporting local regimes and of limiting the numbers of foreign troops committed on the ground.

Philippines, 1945–55

Between 1946 and 1950 the President of the Philippines, Manuel Roxas, and his successor Elipidio Quirino, waged an ill-conceived and ineffectual campaign against a pre-war peasant movement, which had formed an alliance with an urban-based communist party and other groups to oppose the Japanese occupation during the Second World War. The government wavered between coercion and unconsummated promises of reform to remedy peasant grievances, notably poor economic conditions and oppression by the land-holding class. The guerrillas (the *Huks*, an abbreviation for *Hukbalahap*, in turn short for *Hukbo ng Bayan laban sa Hapon*, People's Anti-Japanese Army) fought not only the Japanese and their collaborators but also anti-communist guerrillas led by Americans. The *Huks* were supported by a clandestine infrastructure in the villages, the Barrio United Defence Corps, which provided them with food, intelligence, recruits and shelter among the rural population. The campaign against the 'bandits' was led by the security forces of the Ministry of Interior (the Military Police Command and its successor the Philippine Constabulary), which were supported by undisciplined and poorly trained paramilitary private armies recruited from former Japanese collaborators and pro-American guerrillas by landowners who feared the *Huks*. The police acted ruthlessly against the insurgents and their supporters, modelling their techniques on the counter-insurgency methods employed by the Japanese during their occupation. These included cordon-and-search operations, neighbourhood watch organisations and the seizure of hostages. The police also tortured *Huk* suspects and destroyed and burned villages and crops to deny food to the insurgents and their supporters. Large numbers of refugees were created by these methods. Combined with the government's flagrant rigging of the 1946 and 1948 elections and failure to fulfil pledges to redress peasant grievances, this drove the population into the arms of the *Huks*.[33]

Hiding among the population and expanding outwards from bases in central Luzon ('Huklandia'), the *Huks* at first employed classic guerrilla tactics, launching ambushes and raids against the security forces. But by 1950 they were beginning to capture larger towns in preparation for a transition to conventional warfare – the final phase of a successful Maoist insurgency. As a result, responsibility for the counterinsurgency was transferred to the Army from the ill-disciplined Constabulary, which had alienated the population. When opened in Manila in 1947, the Joint US Military Advisory Group (JUSMAG) provided only logistical assistance and organisational advice as it was forbidden to visit Philippine units and bases. But in 1950 under Major General Leland S. Hobbs it was permitted to be more pro-active. It employed successful techniques from the recent Greek Civil War and provided inspirational leadership to shake up the Philippine security forces from their lethargy. Scattered garrisons were consolidated into larger units to take the offensive. Security forces were streamlined and the police returned to their more traditional duties. The creation in December 1950 of four Military Area Commands (MACs) provided centralised control while Army units were rapidly expanded from two battle-ready infantry battalions in 1950 to twenty-six battalions, restructured to fight in the varied and rugged terrain of the Philippines. Hobbs and his subordinates were resolutely opposed to either giving these battalions an American divisional structure or to the introduction of US troops and advisers. These would bruise

Filipino pride, Americanise the war and deprive the Philippine Army of its initiative and self-sufficiency. They also attempted to reduce the over-reliance on indiscriminate firepower and large-scale search-and-destroy operations. Instead, to regain the initiative, they advocated large-scale operations to break up the larger groups of insurgents, which were then to be followed up by constant, saturation patrolling and relentless pursuit by specialist units. These included newly created elite Scout Rangers, who undertook long-range reconnaissance, intelligence and raiding roles, and were often disguised as *Huks*, and small and mobile 'hunter-killer' units, which employed assassination and counter-terror in a 'no-holds-barred' counter-terrorist approach. Suspects were 'snatched', interrogated and sometimes 'disappeared'.[34]

Overemphasizing the communist influence over the insurgency, the Americans provided aid to stabilise the deteriorating internal position and applied pressure on the Quirino regime to instigate political reforms to undermine support for the *Huks*. But, initially, JUSMAG had only limited success in persuading Quirino to implement these reforms. Then, from late 1950, the tide began to turn as, under pressure from the *Huk* insurgency and American threats to withhold aid, Quirino's regime introduced some significant economic and political reforms, holding free elections, thereby restoring faith in the democratic process. Above all, the appointment of Ramon Magsaysay as Secretary of National Defence in September 1950 provided the charismatic and strong leadership, with the slogan of 'All-Out Friendship or All-Out Force', which was required. Magsaysay understood that a co-ordinated political and military strategy was the only way to defeat the guerrillas. This meant expanding and restructuring the command, intelligence systems and the Army. He revitalised the officer corps with surprise inspections, removing more than 400 officers and sending hundreds of others to train in the United States. US Army manuals and training were adopted as the basis of Filipino doctrine. He also built up a propaganda machine with the help of Lieutenant Colonel Edward G. Lansdale, chief of the intelligence and unconventional warfare section at JUSMAG, with whom he enjoyed a close relationship. Lansdale went on to play a major role in Vietnam.[35]

Magsaysay employed a carrot-and-stick strategy, wielding the 'stick' of pacification and the 'carrot' of 'civic action', a term coined by Lansdale. The treatment of civilians and prisoners and the discipline of the troops were all improved. Magsaysay attached troops to the private security forces to ameliorate their behaviour, although ill-discipline remained a problem. He raised an additional 10,000 civilian commandos who were trained by the Army to secure the local population and separate them from the insurgents. They were also to collect intelligence, identify and arrest *Huk* supporters, and guide and support the Army's operations. These auxiliaries released units of the Philippine Army to undertake offensives and they also consolidated areas liberated by such operations. Simultaneously, the Army attempted to win hearts and minds by carrying out public works. Efforts were also made to counteract the *Huks*' slogan 'land for the landless' by reforming agriculture and rehabilitating captured *Huks* on land cleared from the jungle as part of the Economic Development Corps (EDCOR) project.[36]

By 1952, despite some continuing deficiencies in the counterinsurgency campaign and the somewhat cosmetic aspects of the 'civic action' and socio-economic reforms, the initiative had clearly moved from the *Huks* to the Filipino government, helped by the capture of their top political leadership in October 1950. The key to defeating the guerrillas was to separate them from their support among the population, making the civilian 'sea' no longer hospitable to the guerrilla 'fish' in Mao's analogy. The population was controlled and monitored by the police and civilian paramilitaries, who garrisoned the rural areas, providing the intelligence necessary to dismantle the rebel infrastructure and isolate them from their supporters. Secure from retaliation by the guerrillas, the flow of intelligence about the *Huks* increased as, persuaded by propaganda, 'civic action' and rewards from the government, the people provided information. As in Greece, some harsh policies were employed, notably mass arrests without trial, the destruction of crops and property in *Huk*-held zones, the use of reprisals, the taking of hostages and the resettlement of the civilian population. Luckily, the insurgents had alienated the populace with their own brutal methods, notably the murder of the widow and daughter of the popular President Manuel Quezon in August 1949. Following the election of Magsaysay as president in 1953 and the surrender of the *Huk* leader, Major General Luis M. Taruc, in 1954, the tide turned against the *Huks*. By 1955 fewer than 1,000 remained as fugitives in remote mountain regions. The dynamic leadership of Magsaysay suppressed the *Huks* with a strategy that combined relentless military pressure and well-crafted political reforms that matched local conditions.[37]

Korea, 1948–55 and 1966–69

The regime of Syngman Rhee assumed power in South Korea in August 1948, replacing the US military government. It had a tenuous hold on power owing to political disunity, peasant unrest and opposition from communist insurgents who already controlled large areas of South Korea. The communist SKLP (South Korean Labour Party) opposed Rhee's government. It infiltrated the government apparatus and disseminated propaganda. Controlled by the communist North and supported by an infrastructure among the peasants in the rural countryside, SKLP guerrillas used remote mountain bases to attack the police, raid villages, extend their influence and terrorise any opposition. The insurgents were particularly strong in the Chŏlla and Kyŏngsang provinces and by early 1949 the government had lost control in South Chŏlla. South Korea's counterinsurgency strategy was hampered by the corruption and disunity of the government, which resulted in political and economic instability and an inability to address grievances. Poorly paid and disliked because of their brutality and extortion inherited from the Japanese colonial police in a war of extreme harshness, with scattered small garrisons across the country, the National Police were vulnerable to insurgent assaults and had few reserves to suppress large insurrections. A rivalry with the Korean Constabulary, renamed as the South Korean Army in 1948, prevented unity of command in implementing counterinsurgency strategy. The Army also suffered from mutinies and had to be purged of political undesirables. At the same time, it was being expanded and reorganised into divisions. Having been raised hastily in 1946, both the Army and the police lacked training and battle experience, and they suffered heavy casualties in fighting against the insurgents in 1948–49.[38]

Following the first sustained insurgent uprising on the island of Cheju-do (Jeju) in 1948, which was brutally suppressed, the American Korean Military Assistance Group (KMAG), as in Greece and the Philippines, recommended economic development, better and more responsive government and social change, notably land reform, as the best remedy to the insurrection. Improvements in the security forces, which had regularly employed torture during interrogations, used forced labour, killed suspects without trial, seized food and property and sometimes massacred peasants, were also required. Communists were purged from the Army and the 14th Regiment was disbanded and its colours burned for mutinying at Yosu in October 1948. KMAG emphasised the training of Korean officers as leaders and the removal of incompetent officers. KMAG trained and equipped the South Koreans for internal security duties rather than for conventional operations and exercised much influence in planning and executing the counterinsurgency campaign, which was harsh but increasingly effective. The South Koreans blended American organisation and small unit tactics with techniques learned from the Chinese Nationalists and the Japanese. These included the use of identity cards and the establishment of a civilian defence corps in each village. Between November 1949 and March 1950 this strategy broke the backbone of the guerrilla movement, which now could no longer sustain large-scale operations. By early 1950 relations between the security forces and the rural population had improved to the extent that villagers were providing information about local insurgents. There had been a number of important successes in 1949 when leaders of the insurgents on Cheju-do and of the Yosu Rebellion were killed. Large-scale operations, such as Operation Ratkiller (December 1951–March 1952) and

Banana Wars and Other Small Wars

Operation Mongoose (July–August 1952), broke up the remaining guerrilla groups. These operations often took place during the winter when there was a lull in conventional operations and the guerrillas were most vulnerable. Once these major operations finished smaller units hunted the surviving guerrillas ruthlessly. A 'scorched earth' strategy was used to destroy insurgent supplies and shelter and break the will of their supporters, draining the 'sea' that sustained the 'fish'.[39]

Under American pressure, the government changed tactics, providing some 'carrot' to complement the 'stick'. It opposed the indiscriminate killing of the rural population, offered amnesties to communists and mobilised public opinion with propaganda and by reforms that provided land and jobs. Schools and political re-education were provided in areas dominated by the insurgents. The population and rural areas were controlled and monitored by the police and neighbourhood watches and by the forced resettlement of the rural population into fortified villages, which were garrisoned by self-defence groups and militias, as a means of attacking the communist infrastructure. Granted sweeping powers to arrest and detain suspected communists, the police and paramilitaries ruthlessly destroyed the communist party organisation and disrupted the command and supply of the insurgents. Some 30,000 suspects had been arrested by the end of 1949. That same year some 100,000 civilians had been resettled from Cheju-do, an area controlled by the guerrillas, while prior to the Winter Punitive Operation of 1949–50 another 100,000 were relocated. The increasingly war-weary population was caught between the excesses and terror tactics of the guerrillas and the harsh and ruthless conduct of the government, which was increasingly combined with just enough incentives to win their hearts and minds.[40]

As a result of its commitment to internal security the South Korean Army was unprepared for conventional warfare when Kim Il Sung launched a major conventional invasion of the South in June 1950. The action indicated that insurgency had failed to subvert the South Korean government but it reinforced the belief of many Americans, whose response had been disjointed and mainly traditional, that counterinsurgency diverted resources away from the key priority, conventional warfare. Indeed, the lightly armed South Koreans and their KMAG advisers had borne the brunt of the counterinsurgency campaign while the Americans concentrated on conventional operations. Numerous South Korean communists came out in the open during the North's invasion and temporary occupation of the South and were arrested when the South Korean authorities returned. The communist infrastructure was wrecked and never fully recovered. The arrival of extra American materiel and troops did not change this campaign radically. It remained based around the relocation of the civilian population and the destruction of supplies and property in order to clear insurgent bases. Between October 1950 and February 1951 American troops participated in some counterinsurgency operations, notably the 'Pohang Guerrilla Hunt' (January and February 1951), employing large-scale sweeps and small unit patrols. The Americans used the destroyed food crops and villages to deny them to insurgents. But in March 1951 Lieutenant General Matthew B. Ridgway was forced to ban further wanton destruction of towns and villages. General Van Fleet, who arrived in the spring of 1951, brought valuable coun- terinsurgency experience to Korea. By 1952 the insurgency had been reduced to a low level of activity and three years later had been virtually wiped out. This was achieved mainly through coercion and without making any significant economic, political or social reforms.[41]

A communist insurgency however continued to be supported by North Korea, although it never properly recovered from the defeat of 1948–55. It lacked too the means to eject the staunchly anti-communist regime of President Park Chung Hee. Kim Il Sung escalated infiltration by sending complete units of guerrillas into South Korea, hoping to reinvigorate the underground move- ment and drive out the Americans. By doing so he hoped to achieve the reunification of Korea as a single communist state. Following the ambush of an American patrol in November 1966 the infiltration increased, leading to further clashes during the spring of 1967. General Charles H. 'Tick' Bonesteel III, the intellectual if uncharismatic American commander of the United Nations forces in Korea, was blessed with a number of advantages over his fellow gen- eral William C. Westmoreland, the US commander in Vietnam. Like Malaya, South Korea was a peninsula with a long coastline and a short, well-guarded land border. The unified command system, a legacy of the Korean War, gave Bonesteel the capability to shape and direct the South Korean Army's coun- terinsurgency operations. As a more homogenous nation than South Vietnam, South Korea had a more capable military and a stronger system of government. It had a track record of successful counterinsurgency between 1948 and 1955 and the insurgent movement in South Korea was a much less formidable opponent than the Viet Cong.[42]

Bonesteel employed a two-prong strategy to overcome the insurgency. The first was to close South Korea's borders to infiltration by erecting a new defen- sive barrier behind the Demilitarised Zone (DMZ) with sensors, minefields, observation posts, patrols and quick reaction forces. A system of co-ordinating the coastal defences was also developed. The second was a counterinsurgency campaign in the interior, the burden of which as in the first insurgency of 1948–55 was left to the Koreans to bear together with KMAG advisers and a few special forces training teams. Once again the Korean police were respon- sible for maintaining interior security, especially in the Taebaek and Chiri mountains, where the guerrillas had their bases. Civilian paramilitary forces guarded villages, provided coast watchers, gathered intelligence and mobilised support for the government. KMAG provided advice, equipment and training. 'Civic action' such as the construction of schools and wells was also undertaken. By the end of 1969, the second Korean insurgency was over. It had failed to subvert the government, which retained strong support from the people.[43] Success in the Philippines and Korea proved harder to replicate in Vietnam.

Vietnam, 1965–75

When in January 1961 Russian leader Nikita Khrushchev announced the Soviet Union's support for wars of national liberation, he signalled an apparently deliberate strategy to undermine the West in the Third World. President John F. Kennedy took up the challenge, telling Congress in May 1961 that the crusade to rescue the world from communism was a 'battle for minds and souls'.[44] Kennedy believed that the defeat of the insurgency and subversion of Maoist revolutions required not just military assistance but also economic development and political reform. But future Army Chief of Staff General Earle Wheeler was to contradict him in November 1962, saying that 'the essence of the problem in Vietnam is military'.[45] Kennedy's attempts to restructure the Army, which included the introduction of mobile light infantry, to meet the new 'threat' were opposed by the Army leadership under Army Chief of Staff General George H. Decker. While wishing to defeat guerrillas, the Army was planning and training to win a conventional war in either Europe or Asia, using the unfettered and massed firepower, which had proved so effective during the Second World War. It also concentrated on the conventional threat posed by North Vietnam rather than insurgency in South Vietnam. Senior officers, led by Generals Decker and Lyman L. Lemnitzer (Chairman of the Joint Chiefs of Staff), questioned Kennedy's assumption that soldiers trained for conventional warfare could not defeat insurgents. They also resented what they regarded as political interference in the Army's internal affairs. This led to specialists in guerrilla warfare, including Major Generals William B. Rosson and William P. Yarborough, being sidelined. Yarborough was even sacked as commander of the Army's Special Warfare Center and School because of his closeness to Kennedy. He only received promotion after serving in more conventional posts in Korea and Washington, DC. The conventional warfare 'top brass' were still able in 1962 to get one of their own – Lieutenant General Paul D. Harkins – appointed as the first chief of Military Assistance Command, Vietnam (MACV). But that same year, both Lemnitzer and Decker paid the price for their opposition to the president when their tenures in office were not renewed. The appointment of Harkins, who had no direct experience of counterinsurgency, exemplified the over-optimistic 'can-do' mindset of the US Army.[46] That attitude was summed up by an anonymous senior American officer serving in Vietnam:

> I'll be damned if I permit the United States Army, its institutions, its doctrine, and its traditions, to be destroyed just to win this lousy war.[47]

The implication that specialisation in counterinsurgency did not enhance an officer's career prospects was reinforced by the allocation of unconventional warfare missions to the Central Intelligence Agency (CIA). Following the assassination of President Kennedy in November 1963, traditional, conventional methods prevailed by default. Training remained largely conventional with most of the instruction relating to armour and mechanised units. Little priority was given to counterinsurgency as the Army focused on a possible war with the Soviets in Europe. The result was that the Army entered the Vietnam War in 1965 with little interest in developing a counterinsurgency doctrine or producing a suitable literature in its manuals and professional journals. It was more 'career-enhancing' for officers to serve with conventional forces in Europe and Vietnam rather than with irregular forces in small wars or as an adviser attached to indigenous units.[48]

During the 1960s British officers observed the inflexibility of the American chain of command. Subordinate officers were severely restricted in their freedom of action and this tight command and control adversely affected the US Army's capacity to develop the tactical flexibility central to the ability to fight insurgency. The tendency of American generals to compartmentalise political and military operations as separate entities was a severe handicap to conducting a successful counterinsurgency, in which political rather than military goals should have dominated. The lack of knowledge of Vietnamese culture and language was also a major problem. The adoption of the twelve-month tour lowered the efficiency of tactical units, undermined discipline and handicapped the ability to learn and adapt to conditions in Vietnam. These tours were too short, giving commanders and troops little time to get to know the area of operations and ensuring that there was little coherence in the long-term planning of the campaign. Above all, it encouraged 'careerism' while discouraging any questioning of policy within the officer corps, which supported easy, short-term solutions rather than the development of long-term projects.[49]

The first two Chiefs of the Military Assistance Advisory Group (MAAG), Vietnam, Lieutenant Generals John W. 'Iron Mike' O'Daniel (1954–55) and Samuel T. 'Hanging Sam' Williams (1955–60) pursued a policy similar to that of previous American advisory groups in Greece and Korea during the 1940s and 1950s. The main difficulty was creating a viable army out of the various units inherited from the French and an officer corps that was highly corrupt and politicised. Rather than being a clone of the US Army, the South Vietnamese Army was given a lighter, leaner structure capable of undertaking both conventional and counterinsurgency operations. Nevertheless, little attention was given to counterinsurgency and the American training was mainly conventional in orientation. Indeed, after independence in 1954, O'Daniel wanted to use Vietnamese divisions in conventional operations to clear the Virtminh from Vietnam within two years, re-employing a plan developed by Van Fleet for the Greek Army in 1949. Similarly, Williams opposed the creation by President Ngo Diem in 1960 of rangers as special anti-insurgent units because they diverted resources away from the main priority, conventional forces. Gradually, believing that the main threat was still a conventional invasion and remembering that in South Korea having been trained for internal security the Army had faced a massive cross-border assault, the Vietnamese divisions were 'beefed up' to become a 'mirror image' of the US Army, trained to fight an airmobile and mechanised war against a North Vietnamese offensive. Lacking aggression and reluctant to leave their bases, operate at night, or venture into the 'bush', the South Vietnamese Army remained, like the French had been, road-bound and over-reliant on air and artillery support. They preferred ponderous large-scale operations rather than patrolling the countryside with small units to locate the elusive guerrillas, allowing the Viet Cong to control the countryside and its population.[50]

Meanwhile, the Viet Cong was growing in strength during 1959–65 and North Vietnam was supplying regular troops and modern Soviet and Chinese materiel. By the mid-1960s the South Vietnamese Army was not only out-thought and out-manoeuvred but also increasingly out-fought and out-gunned. It put up a dismal performance in clashes with the Viet Cong, which were harshly criticised by American advisers attached to South Vietnamese formations. By the spring of 1965, the government in Saigon was losing the war with about half of South Vietnam controlled by the Viet Cong and no improvement in sight. With the approval of President Lyndon B. Johnson, Westmoreland committed the first US ground combat units in Vietnam during March 1965 to provide security for American support bases. American forces were then committed to active operations in June 1965 and succeeded in stabilising the situation. Heavy units with limited mobility, trained and equipped for conventional warfare, rather than the special forces that could undertake a counterinsurgency role, were employed.[51]

Key to the suppression of the insurgency was the provision of security for the rural population and mobilising them politically, but the importance of the police and militia in achieving this was largely ignored by the Americans. The internal security forces that had been inherited from the French were woefully inadequate, lacking equipment, experience or training. Furthermore, being absent from the hamlets and villages, hopes that South Vietnam's police would assume the burden of controlling the countryside, as such forces had in Greece and Korea, were frustrated. The police alienated the population with their corruption and inefficiency while failing to destroy the insurgent infrastructure. Diem (and his successors) saw the security forces not only as a means of retaining his grip on power by employing them against his opponents, but also as possible threats to his regime. The principle of divide and rule within the Vietnamese government created chaos, resulting in the formation of three separate national police forces. MAAG had little influence over the training of the police, which became increasingly militarised during the 1960s and generally ignored internal security. Attempts to build up the police's security role were rejected by Lieutenant General Lionel McGarr (MAAG) and Westmoreland (MACV). Paramilitary forces, notably the Self-Defence Corps (later Popular Forces) and the Bao An (Civil Guard, later Regional Forces), which recruited local units to provide rural security at village and province level respectively, were poorly led, trained and equipped and, like the Army and police, alienated the population with their brutality and ill-discipline. The American decision in 1957 to withhold assistance from these paramilitaries was a serious mistake politically, and resulted in the local security forces not receiving crucial assistance during the first critical phase of the insurgency. They were subsequently unable to protect the local population from intimidation by the Viet Cong. In the two years from 1957 more than 400 South Vietnamese local government officials were assassinated, with 110 being killed in the last four months of 1959. The paramilitaries were also later neglected by Westmoreland, who did not realise their importance in providing the intelligence required by the government to locate and destroy the guerrilla infrastructure. There was, however, an improvement between 1968 and 1972 as their competence, numbers, equipment and leadership improved. In a vicious circle of instability, the South Vietnamese Army employed its limited resources to provide internal security, hampering its professional development, yet failing to provide security for the rural population.[52]

The Advisory Group recognised the need for a co-ordinated politico-military strategy and civic action to support reforms and economic development, to improve behaviour by the soldiers whose heavy handed methods had destroyed crops, livestock and villages. General O'Daniel asked for Colonel Edward Lansdale to be sent to Vietnam to co-ordinate civic action and psychological operations (psyops) to counter the Viet Cong infrastructure. Lansdale linked the lessons learned during his service against the *Huks* in the Philippines with local Vietnamese methods, built on the experience of the French and Virtminh. But his unconventional ideas received little official support from the conventional American high command. Lansdale sent civic action teams in with the troops into communist-controlled areas to win hearts and minds through propaganda and small-scale socio-economic projects that provided medical facilities, roads and wells. Although General Williams was unenthusiastic, his successors increasingly supported the employment of South Vietnamese soldiers to provide such 'services', and from 1960 the Army sent its own civic action teams. This forced the Saigon regime to rejuvenate its own projects. From 1963 the Army provided civic action and psyops advisers down to division and province level. It also started MEDCAP (the medical civic action program), in which teams of American and Vietnamese personnel toured the countryside giving free treatment to the population.[53]

Lacking sufficient forces to garrison all villages, the South Vietnamese government employed resettlement, relocating the population to sites where they could be controlled, monitored and defended. This fulfilled the twin aims of increasing security for the local inhabitants while destroying the infrastructure that supported the guerrillas. In 1959, without consulting the Americans, Diem started to employ defended hamlets ('agrovilles'), based on French colonial experiences. Three years later Diem launched the Strategic Hamlets, an over-ambitious scheme inspired not only by British resettlement in Malaya but also by the French in Algeria. In advocating this approach Robert Thompson, head of the British Advisory Mission (BRIAM) in Saigon, emphasised using the police and local security forces to secure the local population rather than mount offensive operations to destroy the guerrilla forces in the field. However, to gain the support of MAAG the British concept, which downgraded the importance of conventional forces and stressed the primacy of the police and paramilitary forces, was replaced by more conventional operations. However, these failed to control the rural population. The attempt to use resettlement on a vastly greater scale than the more efficient British administration had done in Malaya was doomed to failure. The Strategic Hamlets and a succession of similar schemes proved counterproductive. They were poorly planned, badly executed and lacked sufficient resources. The resettlement process was hugely unpopular with the local population, who endured hardships through it, especially as it often meant leaving their sacred ancestral lands. Little was done to prevent infiltration by the Viet Cong, who continued to recruit and expand. Thompson believed that victory could only be achieved through securing the population and denying the enemy any access to the local populace. This was to be done through 'saturation patrolling' of the populated areas and attacking Viet Cong bases in inaccessible regions by long-range patrols, forcing them out into the open where they could be pursued relentlessly. The advice was controversial because a British officer was entering into the decision-making loop, and there were also doubts about the validity of the Malayan experience. In any case, the Americans were largely uninterested in learning from experts in counterinsurgency.[54]

Instead, the US Army roamed the countryside in search of an elusive enemy, attempting to clear areas of guerrillas and mostly thinking in terms of 'body counts'. As in the Philippines and elsewhere the Army advocated an aggressive counterinsurgency strategy in which, having been freed from garrisoning duties by police and militia, the Army could now drive the insurgents away from the population, thus allowing the gradual expansion of government control across the country. Faced by an insurgency in its third phase, that of conventional warfare, the Americans now made the very same mistake for which their predecessors had criticised the Greeks and South Koreans. They persisted in mounting large unit operations, deploying massive firepower, in the hope of catching and destroying the insurgents. But in most cases the elusive Viet Cong demonstrated an uncanny capacity for escaping the net. Although some supplies would be discovered and a few guerrillas killed, these large-scale search-and-destroy operations rarely resulted in decisive victories. Furthermore, the conventional tactics employed were self-defeating. They often resulted in heavy civilian casualties that further alienated the population and gained the Viet Cong many recruits. Pressing social issues were not addressed and little support was given to the locally raised paramilitary forces, which might have made a real difference.[55]

Eventually, the prevailing belief that firepower provided by aircraft, helicopters and artillery 'in larger quantities than ever before' in support of the infantry 'outweighs maneuver as the decisive element of combat power' began to be questioned.[56] Brigadier General Ellis W. Williamson of the 173rd Airborne Brigade thought that, although 'a lot of effort, a lot of physical energy' was being expended, 'a major portion of our effort evaporates into the air.'[57]

The Army was over-reliant on air and artillery support. Apart from the vast expenditure of ammunition, such methods sapped the initiative of the soldiers, creating a 'fire-base psychosis'. It also had serious ramifications politically, when the population was forcibly removed from communist-dominated areas in order to create 'free-fire zones' where civilians were treated as enemy combatants, deny the enemy supplies, and break the spirit of supporters of the guerrillas just as Sherman had done in the Civil War. Major General Frederick C. Weyand (25th Division) admitted that his conventional operations in Hau Nghia Province had no effect on the local population because the hamlets were controlled by the local Viet Cong, who continued to collect taxes, impress labour and assassinate government 'loyalists'. Captured Viet Cong documents suggest that they were happy to engage American forces in remote areas to draw them away from pacification of the population centres on the Coastal Plain. Frustrated officers such as Brigadier General Willard Pearson (1st Brigade, 101st Airborne Division) proposed more subtle methods. These included the extensive use of patrols, ambushes and guerrilla tactics. Such small unit tactics became increasingly prevalent but the conventional ethos and poor training of the American troops hindered progress. The Army created specialised long-range reconnaissance patrol (LRRP) and ranger units and recruited Viet Cong defectors into units such as the Kit Carson Scouts and Tiger Scouts to try to locate the elusive insurgents. But specialised units remained under-utilised by the Americans, who ignored their potential use in hunting the guerrillas. For example, LRRPs were mainly employed for surveillance and intelligence gathering, although they were sometimes used for 'direct-action combat', especially small unit ambush, 'hunter-killer' or 'snatch' missions.[58]

Consequently, many villagers were reluctant to help Americans troops, who only stayed for short periods, for fear of communist retaliation. This ignored the basic requirement of counterinsurgency, which was to provide greater security for the population in rural areas and separate them from the insurgents. The US Marine Corps were advocates of this approach, influenced in part by a long history of fighting numerous small wars and in part by British experiences in Malaya. In contrast to Westmoreland's search-and-destroy operations, Marine Corps Major General Lew Walt implemented a clear-and-hold strategy from mid-1965. Importantly, he 'sold' this idea to his Vietnamese counterparts, who supported its initial implementation and subsequent expansion. Walt employed the Combined Action Platoon (CAP) consisting of a Marine rifle squad linked with a platoon of either Regional Forces or Popular Forces to live, work and fight in a village until the Vietnamese were able to defend it themselves. The Marines then relocated to another village to repeat the process, which gradually extended government control like an oil stain. In tandem with this garrisoning of the villages, search-and-destroy operations at battalion level in the surrounding area destroyed large insurgent units and were then followed up by small unit patrols. Thus, the Marine strategy worked from the 'bottom up' at village level, whereas Westmoreland's worked from the 'top down' by destroying the enemy's large units first. Westmoreland refused to allow Army units to participate in a similar scheme, objecting to dispersing his soldiers in penny packets across South Vietnam. He believed that he lacked the manpower to adopt this strategy and feared that it would also result in higher casualties. With hindsight, these problems now appear not insurmountable. Indeed, the casualty levels were lower than those of units undertaking search-and-destroy operations. Although only a small percentage of the Marine forces were used in CAPs and much depended on the quality of the troops employed, the results were impressive. They gave security, denied supplies to the VC and harried them with constant patrols and in ambushes. The CAPs were successful in wresting control of the village of My Thuy Phuong from the Viet Cong. But when troops from the 101st Airborne Division replaced the Marines just before the 1968 Tet Offensive, the villagers complained of their disruptive and destructive behaviour, especially when they started forcibly relocating families.[59]

Clearly, a pacification strategy did not fit in with Westmoreland's misguided large unit search-and-destroy strategy nor with the Army's impatience to achieve quick results through conventional means. The Army's quick fix approach to counterinsurgency was illustrated by the fact that the four-man MATTs (Mobile Advisory Training Teams), the Army version of CAPs, worked with village paramilitary forces for only about a month before moving on. Westmoreland had enjoyed a successful career but had no experience of counterinsurgency. He also displayed little grasp of political nuances. He tended to promote fellow airborne officers to key posts and ignored the advice of experienced officers such as William Yarborough.[60] The overemphasis on conventional operations diverted attention away from the pacification process, which was left either unprotected or under-resourced. Robert Thompson condemned the Americans for ignoring the first principle of counterinsurgency and in failing to realise that 'the insurgent political subversive organization should be the primary target'.[61] From 1965 onwards North Vietnam received massive aid from Russia and China. Some 170,000 Chinese soldiers, many in anti-aircraft units, were in the country by 1967, allowing the Hanoi regime to release large numbers of regular troops to fight the Americans and their Vietnamese allies. This meant that Westmoreland was unlikely to win a conventional war of attrition.[62]

Population security was improved by the introduction of a Revolutionary Development policy by Saigon in 1965 and the establishment of CORDS (Civil Operations and Revolutionary Development Support) to manage MACV's pacification but was handicapped by a lack of funding and an over-reliance on statistics in the evaluation of progress. Although ultimately a failure, the massive communist offensive that began on 30 January 1968 was a turning point in the war. But, just as in Korea eighteen years before, the communists in coming out of hiding suffered enormous losses that seriously weakened their infrastructure. Beginning in 1968 the *Phuong Hoang* (Phoenix) programme established committees at national, regional and provincial level throughout South Vietnam to co-ordinate attacks on the Viet Cong leadership and infrastructure. But the programme proved to be a controversial one, both domestically and internationally, as it allowed the use of assassinations to eliminate the Viet Cong. Aided by a more compliant South Vietnamese leadership after the Tet offensive, the Americans got some significant measures, including major land reform and improvements in the South Vietnamese Army, implemented. But, although government control over the countryside increased after 1968 while support for the communists declined as more insurgents were eliminated between 1969 and 1971, control over the rural population remained highly contested. In July 1968, General Creighton Abrams replaced Westmoreland as American commander in Vietnam. Abrams had a more sophisticated understanding of the political threat posed by the Viet Cong than his predecessor. He changed the emphasis from search-and-destroy to clear-and-hold operations. He also switched from the 'war of the big battalions' to small unit operations made up of patrols, ambushes and night actions to harass the enemy. There was a prohibition on 'kinetic' weapons, such as heavy artillery and aircraft, and a new emphasis on a greater contribution by the Vietnamese, notably the Regional Forces and Popular Forces, that had been 'cold shouldered' by Westmoreland. But these changes were only partially successful, not only because of the Army's rigid culture but because they came too late. Even in 1969–70, Lieutenant General Julian J. Ewell, commander of II Field Force, and Major General Melvin Zais, commander of 101st Airborne Division, were still mounting large-scale operations supported by air strikes and artillery. Although a military defeat, the Tet offensive was a political victory for the North Vietnamese. The war became increasingly unpopular with the war-weary American public, leading eventually to the withdrawal of American troops in 1973 and to the fall of South Vietnam in April 1975. Ironically this was to be a North Vietnamese invasion of the conventional type the Americans had always feared.[63]

Although notable progress had been made in suppressing the insurgency after 1968, South Vietnam still faced many problems. Its leadership remained weak, Diem's fall in 1963 had ended civilian rule and initiated an era of constant coups and instability, and little progress had been made in coming to grips with the many social ills as many of the reforms were purely cosmetic ones. Poor administration remained a problem and this was exacerbated both by inflation and the huge numbers of refugees that the war had dislocated. 'Vietnamisation' prior to the withdrawal of US troops highlighted the fact that without American support South Vietnam was extremely precarious. Any successes were achieved by the Americans rather than by the government's security forces. The ultimate test – creating a stable regime – had been failed. Despite progress, the population did not support the Saigon government.[64]

Conclusion

After Vietnam, American involvement in small wars was constrained by public opinion and Congress, which showed little interest in 'imperial policing' and supported minimum involvement on the Latin American model of the 1960s. Defeat in Vietnam had reinforced the US Army's unwillingness to either engage with or develop a comprehensive doctrine for counterinsurgency. It continued to have a low priority as the Army preferred to fight conventional conflicts and exhibited a marked aversion towards involvement in any further counterinsurgency campaigns. The service once again left counterinsurgency with its disagreeable connotations of a 'dirty war' to the CIA. Army and Marines participation in 1980s low-scale counterinsurgency interventions, including El Salvador and Guatemala, was strictly limited. They were undertaken under the euphemisms of 'low-intensity conflict' and 'operations other than war' as 'counterinsurgency' was too closely associated with failure in Vietnam. Training of indigenous security forces was left to military advisers and special forces, at the School of the Americas at Fort Benning, Georgia, where instruction in the use of torture was controversial. Reluctant after Vietnam to be drawn into fighting revolutionary warfare in the Third World by the Reagan Administration, which doubled the size of special forces, the Army developed the Weinberger–Powell Doctrine. This marginalised and perpetuated its dislike of counterinsurgency as an irrelevant aberration to the main priority of fighting a conventional war on the Central Front in Europe. During the 1980s the Army concentrated instead on the creation of the AirLand Battle doctrine to win a mechanised war against the Soviets. Analysis of the Arab–Israeli war of 1973, rather than Vietnam, underpinned the Army's subsequent restructure. Success in Desert Storm during the 1991 Gulf War was seen as a vindication of the Army's conviction that future wars would be conventional. Contradictory experience in Somalia, Bosnia and Kosovo had had little impact on the mainstream Army. Iraq and Afghanistan would shatter these illusions, highlighting the failure to prepare for counterinsurgency campaigns prior to 11 September 2001.[65]

6

THE EMPIRE STRIKES BACK

THE BRITISH COUNTERINSURGENCY EXPERIENCE, 1900–94

Introduction

From the eighteenth century, the British Army's main role was to defend the Empire unless called upon to maintain the balance of power in Europe. Conventional wars in Europe were less frequent and the Army was very active in and proficient at colonial warfare. Nevertheless, until late in the twentieth century, it had an aversion to having a well–defined, written counterinsurgency doctrine. One was at last published in 1969 but then remained unrevised until 1997. The Army preferred a more pragmatic and flexible evolutionary process. It relied on the skill and experience of its leaders, supported by a close relationship with policy–makers, rather than specific guidance to provide operational effectiveness. It is arguable that this pragmatic ad hoc approach was a strength allowing flexibility and preventing doctrine from becoming too artificial or theoretical. Moreover, there was plenty of unofficial material for officers to consult, notably W.C.G. Heneker's *Bush Warfare*, T. Miller Maguire's *Strategy and Tactics in Mountain Ranges* and *Guerilla or Partisan Warfare*, and Francis Younghusband's *Indian Frontier Warfare*.[1]

The main strength of this ad hoc practice was that the British were not tied to a specific approach. The weaknesses were that traditional methods were often forgotten or ignored and had to be relearned. There was also no proper dissemination of successful techniques from previous insurgencies. The 'colonial' techniques that succeeded in Malaya and Kenya proved less effective against urban terrorism in Ireland, Palestine, Cyprus and Aden. Moreover, the Empire was run on a shoestring. This meant that effective police forces and intelligence services were rarely well established at the beginning of each insurgency. This resulted in an absence of organisation and good information for the security forces, which were nearly always initially caught unawares. This is a recurring theme during British counterinsurgency campaigns. In 1956 the Colonial Office implemented standardised colonial intelligence techniques but continued to resist a more centralised imperial intelligence bureaucracy.[2]

In 1896, *Small Wars: Their Principles and Practise* was published by Major (later Major General Sir) Charles E. Callwell, the 'Clausewitz of colonial warfare'.[3] *Small Wars*, by which the 1906 edition is better known, became the standard manual in English. A comprehensive synthesis of the lessons of past experience in colonial warfare, Callwell provided the basis not only for British doctrine but also for future discussion and development. This was not just for the British but for many other nations, too, since he assimilated the experiences of other armies and chose his examples from many periods and many armies. In *Small Wars*, Callwell clearly distinguished between colonial, small wars 'against savages and semi-civilised races' and regular campaigns between organised, conventional armies. 'Imperial policing' of the colonies provided a rich legacy of experience against irregulars along the frontiers of the Empire during the late nineteenth century. This was separate from the theory and practice of the home armies prior to the Second World War. Callwell recognised the importance of vigorous leadership, good intelligence and public opinion while openly supporting 'scorched earth' tactics. The seizure of livestock and the destruction of crops and villages weakened insurgency, although they might 'shock the humanitarian'. Callwell also explicitly justified reprisals and punitive actions against civilians, including collective punishments. In doing so he failed to fully realise the negative impact and counterproductive implications of these techniques on the population. Reflecting his influence, British counterinsurgency campaigns continued to use these methods both before and after 1945. Although their counterinsurgency methods could be brutal, seeking to defeat the insurgency by destroying its civilian infrastructure, the British generally behaved with greater moderation than some other colonial powers and totalitarian states. Certainly, Bernard Montgomery ('Monty') when serving in Palestine in February 1939 noted that the French in Syria were much more ruthless. The Amritsar Massacre, 1919, and Ireland, 1919–21, reinforced the need for restraint and 'minimum force'.[4] But the punitive habits of the colonial era continued throughout the twentieth century.

South Africa, 1900–02

During the last phase of the South African War, although beaten conventionally, the Boers fought on. Led by a new group of able leaders, notably Louis Botha and Jan Smuts, who refused to submit, the Boers hoped to undermine British support for the war and thus compel the British government to make peace. Undertaking guerrilla warfare in December 1900, they split their forces into small highly mobile and well-armed bands of mounted infantry (commandos). They adopted hit-and-run tactics against 'soft targets' such as isolated garrisons, communications, convoys and railways. Having successfully invaded Orange Free State and Transvaal in 1900, Field Marshal Earl Roberts lacked sufficient troops to simultaneously pursue the Boer commandos, maintain the occupying field army and garrison the conquered country. His troops resorted to the burning of farms and the collection of fines in reprisal for Boer attacks on their communications and the railways. Deploying reinforcements, the British made lavish use of blockhouses (as had been used by the Spanish in Cuba), wire barriers and the removal of the population to pin down the elusive insurgents by restricting their mobility and room to manoeuvre. Simultaneously they employed mobile columns in sweep operations to harry the guerrillas constantly.[5]

Good intelligence was crucial to the success of the mobile columns. The British were handicapped by an over-centralised command, poor scouting and a dearth of good information while the Boers were provided with supplies and information by the local farmers. Colonel David Henderson reorganised and expanded the Field Intelligence Department. Each column had its intelligence officer who organised black scouts, guides and spies to gather information. Colonel Aubrey Woolls-Sampson, an *Uitlander* (a British emigrant to the Transvaal) and the ex-commander of the Imperial Light Horse, was one such intelligence officer. He achieved several coups for Lieutenant Colonel George E. Benson against the Boers, based on local knowledge and good intelligence, notably from Bantu scouts. Captured Boers were 'turned' and persuaded to serve the British as guides or trackers. 'Loyal' Boers served against the commandos on the veldt in the National Scouts in the Transvaal and the Orange River Colony Volunteers in the Orange Free State. Amnesties and rewards weakened the resolve of the guerrillas to fight on and enticed them to surrender, turning the tide against the Boers.[6]

It quickly became a 'dirty war'. The Boers responded ruthlessly towards those who surrendered, court-martialling and sentencing as traitors anyone who collaborated with the British. One emissary to Christian de Wet was flogged and then shot by General C.C. Froneman, while the Secretary of the 'Burgher Peace' Committee was executed. Nevertheless, some 5,000 burghers fought for the British, who failed to fully exploit their usefulness. The South African Constabulary (SAC), including some 400 'Loyal' Boers, was raised by Major General R.S. Baden-Powell in September 1900 to police the Transvaal and Orange Free State. It was used as a military force by Kitchener to sweep the countryside and man blockhouses when it might have been

better employed, as the War Office noted, to occupy specific areas, contribut-
ing to the pacification of the country. Large numbers of natives also played
a central role as non-combatants on both sides, often as forced labour. They
were also deployed as armed troops by the British, defending the railways
and blockhouse lines and providing local knowledge as scouts and guides
for columns. Skilled horsemen and trackers from 'Coloured Corps', like the
Namaqualand Scouts, Bushmanland Borderers and Northern Border Scouts,
helped to defeat the commandos. Whether armed or unarmed, these black
soldiers were often killed by the Boers if captured and the rebellion by 'Kaffir
tribes' was one factor in the Boers eventually seeking peace. The British also
resorted to coercion, ruling the Cape Colony by martial law, which allowed
the imprisonment and execution of rebels. Boers caught in British uniforms
were also shot summarily. British forces, notably the colonial contingents from
Australia, Canada, Cape Colony and Natal, entered a bitter 'spiral of reprisals' with the Boer guerrillas. Consequently, six officers from the Bush Veldt
Carbineers were court-martialled for the murder of Boer prisoners, and two
Australians, Lieutenants Harry 'Breaker' Morant and Peter Handcock, were
executed. Their defence had been that the shooting of prisoners as a reprisal
was accepted practice.[7]

   The British failed to pin down the elusive Boer commandos and stop their
hit-and-run raids. This was despite employing mounted infantry, mounted
columns, chains of blockhouses and attempting to decentralise their command,
supply and transport. When these more conventional methods failed to bring
the stubborn Boers to the negotiating table, Roberts' successor, Lieutenant
General (later Field Marshal Earl) Kitchener, resorted to harsher 'scorched
earth' tactics. These included the destruction of crops, farms and livestock as
a collective punishment to deter support for the guerrillas and destroy the
Boer support base, notably food and supplies, such as fodder for their horses.
It has been estimated that 30,000 farms were burnt and more than 3½ million
sheep slaughtered by peacetime. The British also used the Spanish technique of
'resettlement', known also as 'reconcentration'. This had been employed against
Cuban insurgents by General Valeriano Weyler during the 1890s and by the
Americans, as 'concentration', in the Philippines from 1899. Boer women and
children were moved from their farms on the veldt and placed in 'concentration
camps' to cut them off from the Boer commandos in the field. African families were held in separate 'concentration camps'. This strategy eventually wore
down the Boers but at a heavy cost in civilian lives. The deaths of thousands of
Boer women and children from disease in the concentration camps caused both
domestic and international outrage. Although such methods were often used
against indigenous populations around the Empire, it was felt that they should
not have been employed against those of European descent.[8]

Official estimates of the deaths vary from between 18,000 and 28,000 Boer civilians and between 16,000 and 20,000 black inhabitants. Public opinion, both in Britain and overseas, forced the British government to act. It reined in some of Kitchener's demands that even harsher methods, such as the deportation of women and children, be employed. It also sought to improve the poor conditions and rations in the internment camps in order to reduce the death rate, which promptly fell. Schools were also provided to teach the Boer children. Although successfully ending the war in May 1902, the policy of scorched earth, treating the guerrillas as 'simple bandits', together with the many deaths in the concentration camps proved to be an expensive policy politically. Little effort was made to counterbalance the 'stick' with the 'carrot' by winning the hearts and minds of the population, whose freedom was heavily restricted. This embittered the Boers, prolonging the war, souring future British–Boer relations and having long-term political effects, notably the division between the British and Boer populations in South Africa, that are still felt today.[9]

Ireland, 1919–21

A Norman intervention in Ireland during the twelfth century marked the beginning of more than 700 years of British involvement in Ireland, although the English Crown did not assert full control there until 1541 when Henry VIII assumed the title of King of Ireland. The subsequent arrival of protestant settlers from England and Scotland dislodged the existing Catholic landholders, sowing the seeds for centuries of sectarian and military conflict. The wars of the seventeenth century, notably William III's victory over James II at the Battle of the Boyne (1690), cemented the protestant ascendancy. Following the Irish Rebellion of 1798, which was bloodily suppressed, the Irish Parliament was abolished under the Act of Union of 1801 and Ireland became part of the United Kingdom of Great Britain and Ireland. The nineteenth century saw continued demands by Irish nationalists for Home Rule, which was finally restored in 1914 but soon suspended on the outbreak of the First World War.

The situation in Ireland deteriorated following the Easter Rising of April 1916, which was brutally dealt with by the authorities, allowing a Republican recovery. The British failure to find a political settlement and use of repression allowed the extremists to grow in influence as nationalist support drifted away from the Irish Parliamentary Party. It had neither delivered Home Rule nor prevented conscription. *Sinn Féin* (Ourselves Alone) won most seats in the December 1918 general election in southern Ireland. Its members of Parliament refused to take their seats at Westminster and instead proclaimed an Irish Republic. The arrest of the *Sinn Féin* leadership in May 1918 and the closing down of the *Sinn Féin* organisation in July 1918 had driven the Republicans underground. The British government attributed *Sinn Féin's* success in local and national elections in the south to intimidation. They failed to understand that repression, which did not root out the opposition yet alienated potential supporters, was mobilising the population behind the rebel cause. The rebellion was launched with the murder of two constables at Soloheadbeg on 21 January 1919, but the British were slow to react and did not officially recognise the insurgency until April 1920. Between 1919 and 1921 the Irish Republican Army (IRA), the military wing of *Sinn Féin*, with 3,000 activists, made Ireland ungovernable despite the presence of 80,000 British troops. Unable to defeat the British Army in the field, the IRA resorted to guerrilla tactics of hit-and-run. They developed the 'flying column', a small mobile unit based on the Boer commandos of the South African War, to harass and attack the security forces. While the guerrilla campaign was proceeding, a shadow government built up its own administration and judicial system to win the support of the population while subverting British rule. A complementary campaign of sabotage and terrorism was carried out in the United Kingdom during 1920–21, which was countered by arrests, imprisonment, deportation and internment. British credibility was undermined by atrocities, such as the shooting of 8-year-old Annie O'Neill and the murder of Mayor Thomas MacCurtain of Cork City. These made headlines not just in Ireland and Britain, causing a domestic backlash, but also internationally, notably in the United States. Failing to create the Propaganda Department until late in the campaign (August 1920), the British lacked the means either to counter this adverse publicity or to highlight the atrocities of the IRA's terror campaign.[10]

Unable to suppress the insurgency, the British exacerbated the situation by venting their frustrations. Although by the end of 1920 the Army had restored discipline, the damage had been done. Instead of winning the hearts and minds of the population by offering incentives to support the government, the British resorted to the collective punishment that was often employed in colonial wars. They attempted to make a prolonged insurgency prohibitive for its supporters by using curfews, collective fines and the destruction of homes, crops and livestock. This only further alienated the population by penalising the community in an indiscriminate fashion. As in South Africa, the British policy of burning the houses of suspected rebel supporters was disastrous. It drove the local population into the republican camp and resulted in retaliation in which loyalists' homes were burnt. This formed part of the IRA's campaign of intimidation against 'collaborators' with the British, such as the Royal Irish Constabulary (RIC), magistrates and civilians. By 1926 the protestant population in the Irish Free State had fallen by a third owing to 'ethnic cleansing' during 1920–22. The British failure to provide security for loyalist lives and property was a decisive factor in allowing the IRA to gain control of the rural countryside. Moreover, as it was considered impossible to forcibly remove a 'white' European population, no effort was made to resettle the population, separating it from the rebels.[11]

More important was the failure to create the central command or organisations, such as joint committees at different levels of the hierarchy, to ensure co-ordination of the counterinsurgency strategy and civil-military co-operation. Above all, the British lacked an effective police force working closely with the Army to provide local knowledge about the population and intelligence about the activities and infrastructure of the guerrillas. From 1916 onwards, following the Easter Rising, there were communal pressures on the RIC. Its constables faced intimidation and sectarian violence between the Catholic and protestant communities. The recruitment and influx of English recruits, who lacked training and local expertise, into the RIC and its Auxiliary Division exacerbated the situation. Lacking training and discipline, these Auxiliaries treated the Irish population as the enemy. They committed reprisals and terrorised inhabitants during raids and searches, notably the attack on Croke Park (November 1920) and the burning down of part of Cork (December 1920) in retaliation for the murder of comrades. Unresolved tensions between the Army and the Auxiliaries, whose reprisals made the soldiers reluctant to work with them, aggravated the lack of co-operation between the police and military. This hampered intelligence collection, one of the key ingredients of any successful counterinsurgency campaign. Bureaucratic rivalries between Special Branch and MI5 in Whitehall also hindered the gathering of intelligence during 1919–21.[12]

A paramilitary force that was very different from the English police, the RIC was by 1916–19 in a state of disarray. It lacked cohesion and leadership, and was increasingly isolated from the population. The government had failed to reform the RIC after the Easter Rising of 1916 or to bring in outside experts, such as Charles Tegart, a police officer from India with experience of revolutionary terrorism in Bengal, until too late. Demobilised soldiers, the notorious 'Black and Tans' and Auxiliaries, were recruited by Major General Sir Hugh Tudor, the head of the RIC from May 1920, to solve the shortage of policemen. Former Army officers were also recruited to staff the Special Auxiliary Division of the RIC. There were allegations of assassinations of prominent republicans, summary executions and abuse of suspected rebels during interrogations. The British failure in Ireland was largely self-inflicted, the result of a lack of restraint in a war that was fought in the full glare of world attention.[13]

From July 1919 the well-informed IRA terror campaign attacked police stations and killed informers, policemen, magistrates and officials. In doing so, it broke the key link between the police and the community. This prevented intelligence gathering. The police (including Special Branch) and the civil service had been infiltrated, allowing the IRA to target intelligence personnel and informers. The 'hit squads' of Michael Collins (Director of Intelligence, IRA) then killed virtually the entire undercover branch of the Dublin Metropolitan Police on one 'Bloody Sunday' in November 1920. The intelligence system, which was reasserting itself, collapsed and had to be rebuilt during 1921. To remedy the lack of police intelligence soldiers were deployed in both Cork and Dublin during 1920 to protect police posts. The Army gradually realised, as in Malaya later, that information and local knowledge was available only if the Army and police worked closely together for a lengthy period to win the trust of the population. Slow to appreciate their importance, the Army did not appoint specialist intelligence officers at brigade and divisional level until November 1920 and at battalion level until December 1920. Nevertheless, in mid-1920, a new Central Intelligence Bureau was set up in Dublin and local intelligence centres in each police district collated and sent intelligence there for analysis. A Raids Bureau was also founded in September 1920 to collect and analyse the intelligence received from raids. This built on the Army's first offensive in early 1920, which accumulated much information from captured documents and the interrogation of prisoners and allowed it to gradually compile the IRA's 'order of battle' and 'black lists' of IRA members. Having already lost the hearts and minds of the population, the British never succeeded in infiltrating the IRA, but improved intelligence did reduce the organisation's capacity to operate during 1921. In the ensuing stalemate, the IRA conducted a 'war on informers' that targeted protestants, ex-soldiers and other minorities, while the British mounted a parallel 'dirty war' to terrorise the republican movement and its supporters. Both sides employed political murders and 'death squads' until eventually a truce was signed in July 1921. This was followed by the Anglo-Irish Treaty in December 1921 that resulted in a divided Ireland.[14]

At first, facing a sophisticated political and military insurrection, the British failed to grasp the full challenge presented by the complex situation in Ireland. They made every cardinal error that security forces opposing an insurgency can make. Civil-military co-operation and the collection of intelligence, which was the key to success, were poor. Above all, failing to offer any political option other than Home Rule, they could not provide a 'carrot' with which to win the hearts and minds of Irish moderates, who rejected the IRA's violence. The Army and the police were over-reliant on the 'stick'. The Army employed conventional tactics that were no match for the more mobile rebels. In using motorised transport that was too road-bound and widely dispersing their forces to protect police stations, they provided easy targets for the IRA and surrendered the initiative to it. Large-scale 'drives' to encircle and destroy the insurgents were employed to maintain the 'offensive spirit' but were largely

futile. They usually allowed the rebels to escape. Although never entirely abandoning large-scale operations, as when exploiting good intelligence they forced
the guerrillas to break up into small groups, the British started by 1921 to
employ small 'hunter-killer' units led by junior officers. Their success was reliant on good information. This allowed them not only to retain the element of
surprise but also to target and ambush specific IRA units. They often adopted
a retaliatory policy of 'shoot to kill'. Wearing rubber-soled shoes instead of
hobnailed boots and employing superior discipline, equipment and training,
the British beat the guerrillas at their own game. Augmented by constant
patrols and air support, these tactics regained the initiative. They harassed the
insurgents, lowered their morale and gave them a feeling of insecurity. Many
British officers were convinced that they were close to defeating the insurgency but the discovery of the right tactics came too late, although the losses
sustained by the IRA strengthened Lloyd George's position in the negotiations
that ended the war.[15]

Unprepared to deal with the complex military and political problems posed
by the new style of guerrilla warfare, Britain learned from its mistakes. It had
developed an effective counterinsurgency strategy by the spring of 1921 but
it is doubtful if by this stage victory could have thwarted Irish independence.
The Anglo-Irish War was the first truly modern revolutionary insurgency. The
IRA was able to win a 'people's war' some thirty years before Mao's victory
in China in 1949. It provided the template for modern revolutionary warfare
developed later by Mao, Tito, Ho Chi Minh, Che Guevara and Fidel Castro.
Inspired by the Boers, the Irish themselves would provide a model for the
Jewish leadership in Palestine in the 1940s and George Grivas in Cyprus during
the 1950s. Yitzhak Shamir, a member of the Stern Gang and a future Israeli
Prime Minister, adopted the *nom de guerre* 'Michael' in honour of *Sinn Féin's*
Michael Collins, while the only Jewish member of the Irish Volunteers played
an active part in organising *Irgun Zvei Leumi* on the IRA model in the 1930s.[16]
The IRA proved that:

> Well-organised violence not only can pin down large armies of occupation,
> but can create a world opinion so hostile to the occupier that he is forced to
> listen to rebel demands.[17]

Skills gained during the inter-war period would be invaluable in fighting
insurgents during the post-war period. Revolts in Iraq (1919–22), Ireland
(1919–21), the Malabar Coast of India (1921–23), Lower Burma (1930–32), and
Palestine (1936–39) and continuous operations in the Middle East and along
the North-West Frontier of India provided much experience. As a result, the
British entered the 'counterinsurgency era' with more practical expertise than

any other nation. Britain adopted the use of 'minimum force' as a key principle following the Hunter Commission's report on the Amritsar Massacre in 1919. It accepted that brutality and excessive force created greater problems in the long-term and backfired politically. During the Moplah Revolt in India (1920–21) and the Burma Rebellion (1932–36) the British employed a clear-and-hold strategy, which was crucial for a successful counterinsurgency campaign. In Burma it used 'civic action' in combination with resettlement. Emphasising the co-operation of colonial administrators, police and Army, they became models for future campaigns. This culminated in the elaborate committee system used during the Malayan Emergency and provided the foundations for a coherent post-war counterinsurgency doctrine, which would be applied in Kenya, Borneo and Oman.[18] This nascent doctrine would be tested during the Arab Revolt in Palestine.

Palestine, 1936–39

The Arab Revolt between 1936 and 1939 was a watershed because many of the modern counterinsurgency techniques that were employed after 1945 were already emerging, building on the lessons of imperial policing. Lacking troops, security arrangements and good intelligence, the British were ill-prepared for the Arab Revolt. It was fuelled by Arab nationalism, anti-British unrest and resentment of Jewish immigration and aspirations for an independent state in Palestine. Caught in the middle, between Jewish demands for an increase in immigration and Arab ultimatums for restrictions, the British political position was untenable. This left them unable either to win hearts and minds or to divide and rule. The rebellion, known semi-officially as 'The Arab Troubles', started on 15 April 1936 with Arab attacks on Jewish settlers. It was followed by an Arab general strike and attacks on oil pipelines, railways, roads and telephone communications across the country. It was a major peasant rebellion, which lacked a central command, an articulate cause, or a charismatic leader of Michael Collins' stature. The insurgents relied on small, self-contained gangs, which were poorly equipped but employed traditional guerrilla tactics very effectively in attacks on Jewish settlers, British officials and security forces, and Arab 'collaborators'. The police were unable to cope and Sir Arthur Wauchope, the High Commissioner, requested reinforcements from Egypt that began to arrive in May 1936. By mid-June 1936 two brigades were controlling two regions around Haifa and Jerusalem. But operations were hampered by Wauchope's refusal to implement full martial law or allow the Army to take the offensive, although some units were committed against Arab guerrillas in the Nablus–Jenin–Tulkarm 'triangle' in Samaria.[19]

These restrictions were lifted as the violence increased and special emergency regulations (introduced in June) allowed the Army to conduct searches of villages and reassert control over Jaffa. Air Vice Marshal (later Air Chief Marshal Sir Richard) R.E.C. Peirse, Air Officer Commanding Palestine Transjordan Command, employed 'drives' on an unprecedented scale in a large area of northern Palestine to trap the elusive rebels. However, after the initial success of breaking up the larger insurgent groups, the campaign degenerated into guerrilla warfare. Villages that had been identified as supporting the insurgents were cordoned off and searched. Lacking good intelligence, the Army alienated the population, providing recruits for the insurgents. A request by Peirse to be allowed to bomb villages was refused by the Cabinet. On 2 September, the Cabinet agreed to send an entire division to Palestine and the implementation of martial law in order to reimpose British rule by sheer weight of numbers. Lieutenant General (later Field Marshal) Sir John Dill, General Officer Commanding British Forces in Palestine and Transjordan, preferred to employ small company-sized columns to scour the countryside. Intelligence gathered from local villages that were protected by permanent detachments was used to target and surprise the guerrillas. As a result of these methods the level of violence fell.[20]

This persuaded the Arabs to call off the general strike in October 1936 and accept the offer of a Royal Commission chaired by Lord Peel. This political solution, backed up by preparations to deploy a substantial force and the full power of martial law, had broken the deadlock. However, the Peel Commission's recommendation, published as a White Paper on 4 July 1937, to create two separate states – one for the Jews in the north and one in the south for the Arabs – satisfied no one. Immediately, the level of Arab unrest increased. It included attacks on Jewish settlements and sabotage against the oil pipelines and the railways and culminated in the assassination on 26 September 1937 of Lewis Andrews, the District Commissioner of the Galilee, by Arab gunmen. Again, poor intelligence about Arab aims and intentions meant that the British were caught off-guard. By the end of 1937, with a full-scale insurgency under way, the British had lost control of large areas of Palestine to the Arabs. Emergency regulations, including extraordinary powers of arrest and detention, the death penalty for possession of firearms, the levying of collective fines and house demolitions, were introduced. British troops took reprisals against Arab villages, destroying them as punishment for attacks. Villagers were held in cages in poor conditions, beaten, tortured and killed. Some regiments gained a reputation for being overly 'robust' in their treatment of the Arabs.[21]

Both Ireland and the Arab Revolt demonstrated the importance of winning the support of the local population and obtaining good intelligence from the police, who served among the community. These were key factors in maintaining internal security, since without local knowledge and intelligence about the rebels the security forces were blind and operated at a disadvantage. Thus, the police and other officials became targets for terrorism as the insurgents sought to break down public security, deprive the security forces of their ability to function and demonstrate the government's competence. As throughout the Empire, the backbone of the British administration in Palestine was the district officer. But, as Wauchope complained to London in November 1936, there was a severe shortage of qualified officers. In times of inter-communal tensions, the Arab district officers were afraid of the guerrillas, becoming unreliable and unwilling to operate within the villages. Arab terrorism caused a severe breakdown of law and order as the Palestine Police Force disintegrated. Composed of Arab, British and Jewish policemen who were segregated from each other, the police force was not of a high standard. Both Arab and Jewish policemen were suspect. Forming in December 1935 more than half of the police force, the Arabs faced the same intimidating environment that had paralysed the RIC in Ireland twenty years before. Some, according to the War Office, had betrayed details of operations to the guerrillas. The hostile Arab population and, in particular, the assassination of Arab officers in the Criminal Investigation Department (CID) severely limited the vital collection of intelligence and eventually led to the total collapse of the Arab section. This meant that intelligence, never adequate since the warring intelligence departments seldom functioned as a cohesive unit, completely dried up. The lack of fluent Arabic or Hebrew speakers forced both the British police and Army to rely on information from Jewish organisations, notably the Haganah (*Irgun Haganah Hav'ivrith Be Eretz Israel*, Hebrew Defence Organisation in Israel), to locate Arab insurgents.[22]

Sir Charles Tegart, a distinguished Indian policeman who arrived in December 1937, advocated greater use of the Arab police. During the next three years he built fifty police garrisons, 'Tegart forts', to guard the frontier, roads and railways and the countryside. In Arab villages and Jewish regions they provided security for the police and denied the guerrillas sanctuaries in Syria. Tegart also established 'Arab Investigation Centres' where suspects were subjected to 'waterboarding' and other tortures by the CID. To offset reliance on the Arab police, some 19,000 Jews were recruited, trained and armed to form the *Notrim* (Guards, later the Jewish Settlement Police), a paramilitary Jewish police force, which protected Jewish settlements and provided intelligence. From May 1938 Special Night Squads, consisting of Jewish settlers commanded by British officers and NCOs were led by Captain (later Major General) Orde Wingate, an intelligence officer with strong Zionist sympathies. They patrolled the Haifa oil pipeline, taking the battle to the insurgents on their own ground by ambushing the Arab guerrillas and raiding Arab villages. They provided useful intelligence from Arab informers recruited by Wingate, but also earned a reputation for abuse, torture and execution of Arabs. Wingate was eventually sent home amidst concerns about the impact of his methods on inter-communal relations. Nevertheless, the British would develop this 'counter-gang' strategy later in Malaya, Kenya and Oman.[23]

The tide turned during 1938 as the British changed their strategy from conciliation to coercion, systematically keeping the Arab rebels under surveillance, harassed and pursued. The British relied on a two-pronged approach of targeting the rebel fighters while separating them from their supporters among the rural population. For the first time, there was firm political co-ordination of the campaign. Fortunately for the British the Arab insurrection lacked not only the political and military organisation of the IRA earlier in Ireland and of the later Jewish insurgency, but also any significant external support. Army control of the police, taking over all police districts, allowed unity of command and improvements in intelligence gathering. In continuation of the policy of 'village occupation' started in 1937, twenty key villages in areas of unrest (mainly in Samaria and Galilee) were occupied in May 1938 to isolate the guerrillas from their support. Police posts in other villages prevented the rebels from using them as bases for their raids. This denied them intelligence, food, recruits and shelter, while protecting the local rural population from intimidation. Occupation of the villages not only allowed the government to restore its authority but also to extend its control over the countryside and to win over their inhabitants. 'Punishments' such as heavy collective fines, forced labour, protracted and 'punitive' searches and the destruction of property, houses and whole villages were also employed to cow the population. Brutal cordon-and-search operations often involved unofficial destruction of property, looting, acts of violence and summary executions. Simultaneously, mobile columns and patrols hunted down the elusive guerrillas who had been flushed out of the villages and urban areas relentlessly, denying them security and rest. As a result, terrorism declined as the insurgent bands were forced out into the open and destroyed.[24]

Blockhouses were built to deny the rebels access to the population and also to limit their movement, which was also restricted by the introduction of identity cards, checkpoints and curfews. Attempts, not entirely successful, were also made to close the borders with Transjordan and Syria by erecting blockhouses and Tegart (barbed) wire to prevent supplies reaching the guerrillas and restrict their access to sanctuaries. After October 1938 there were two divisions in Palestine, employing troops from the United Kingdom that had been released by the signing of the Munich Agreement. By November 1938 the insurgency was in decline, with many of the gang leaders killed or captured. With war looming in Europe, the British made their military victory more palatable to the Arabs with political concessions in the White Paper of May 1939, limiting

Jewish immigration and settlement, regulating land sales and promising independence without partition to an Arab Palestine within ten years. Many of the basic techniques of a successful counterinsurgency, which would be developed into a proper doctrine after 1945, had been employed in this campaign.[25]

However, the crucial importance of political factors in overcoming an insurgency continued to be overlooked. Books written during the inter-war period, notably *Imperial Policing* by Major General Sir Charles Gwynn (1934) and *British Rule, and Rebellion* by Colonel H.J. Simson (1937), built on the work of Callwell. They summed up admirably the traditional strengths of 'imperial policing' (the use of 'minimum force', maintaining the rule of law, and civilian control of the military), which formed the core of British counterinsurgency theory from 1945. But both Gwynn and Simson tended to ignore the more brutal realities of actual practice on the ground, emphasising the importance of co-operation between the civil, military and police authorities and good intelligence while overlooking the increasingly political nature and importance of guerrilla warfare. In an article on 'Modern Problems of Guerrilla Warfare' published in 1927, Major B.C. Dening predicted the growing prominence and political character of counterinsurgency and highlighted the need to remedy any grievances underpinning a rebellion. This was the exception and most soldiers, like Gwynn and Simson, ignored the wider political issues of insurgency. The War Office did not issue a pamphlet, *Notes on Imperial Policing*, until 1934. It provided guidance on suppressing rebellion but focused mainly on the military rather than the political dimensions of counterinsurgency. This left the British Army reliant on its 'small wars' traditions and unprepared for the insurgency in Palestine between 1945 and 1948. The Arab Revolt of 1936–39 set the pattern for a number of post-war insurgencies, notably Palestine, Aden and Cyprus, in which the British attempted unsuccessfully to suppress the insurgents by using 'colonial' brutality.[26]

Palestine, 1945–48

In the Zionists, the British were faced by 'white' insurgents who were not easily placated, having the support of the Jewish population, which had been radicalised by the Holocaust and refused to collaborate with the security forces. Moderate Jews had been angered by the detention of illegal Jewish immigrants in 'Nazi-style concentration camps' in Cyprus. The British quickly lost the propaganda war in Palestine, Britain and internationally. Supported by an effective propaganda machine and aided by the revelations of the concentration camps, the Zionists had world opinion on their side. In contrast to the Arab Revolt, British actions were scrutinised closely and other Western nations, particularly the United States, applied pressure to prevent any brutal suppression of the revolt. Lacking a clear political aim, the British could neither offer the carrot nor brandish the stick in suppressing their first real insurgency since the Second World War. Wishing to retain its bases in Palestine and unwilling to offend either the Arabs or the Jews, the British government failed to formulate a clear policy despite requests from General Sir Alan Cunningham, the High Commissioner, for a political solution. Once again confronting the issue of Jewish immigration, the British faced hostility from both Arabs and Jews. Torn between offending the Palestinian Arabs and their supporters, the Arab states on whose oil Britain depended and upsetting the Zionists and their American allies, the British dithered, paralysed by political indecision. Cunningham resisted implementing sterner measures as being counterproductive but without political concessions it proved difficult to divide and rule the moderates and extremists. Cunningham was unable either to win over the Jewish community, or to separate the more moderate Jewish Agency, which had co-operated during the Arab Revolt and the Second World War, from the more hard line and violent splinter groups. Most notable among the latter were *Irgun* (*Irgun Zvai Leumi*, The National Military Organisation), led by Menachim Begin, and *Lehi* (*Lohamei Herut Israel*, Fighters for the Freedom of Israel), led by Abraham Stern and often called the Stern Gang, which in 1944 had assassinated Lord Moyne, the Minister-Resident in the Middle East, in Cairo.[27]

The emergency began on 31 October 1945 with a large campaign of sabotage against the railways and other installations, notably the Haifa Oil Refinery. This was followed up with a general strike and serious rioting, raids on bases and banks to acquire weapons and funds, and the use of bombs and mines to harass the British occupiers. In response the British searched Kibbutzim and villages for weapons and insurgents. Lacking a strategy to deal with the insurgency until the spring of 1946, they lost the initiative to the insurgents by pinning down its troops in static garrisons. Cordon-and-search operations were employed randomly in reaction to sabotage or ambushes on troops, which were hurriedly deployed in Tel Aviv to deal with various disturbances. Most notable was a riot lasting two days during which soldiers opened fire, killing six Jews and injuring another sixty. These 'colonial' tactics were disastrous, increasing American support and sympathy for Zionism. As the situation deteriorated, the government, under pressure from Field Marshal Bernard Montgomery, the CIGS (Chief of the Imperial General Staff), in June 1946 permitted large-scale cordon-and-search operations. These further alienated the population with invasive and abusive searches of houses and screening of civilians that often developed into large-scale confrontations.[28]

Operation Agatha, a huge 'cordon-and-search' operation, was mounted between 29 June and 1 July 1946. Planned in great secrecy, it searched simultaneously the Jewish Agency buildings in Tel Aviv, Jerusalem and Haifa and also twenty-seven settlements. During the extensive searches, valuable documents and an arms cache were secured and 2,718 people detained but three were killed and seventeen wounded. These were quite light casualties considering the violent resistance from some settlement residents. But, as General Sir Evelyn Barker (commander of British forces in Palestine) noted in his operational report, the British lost their few friends within the Jewish community. The Commissioner for the Haifa District reported widespread resentment at the arrest of prominent Jewish settlers and searches of settlements. Moreover, the most dangerous members of *Irgun* and *Lehi* had escaped. After the blowing up by *Irgun* of the King David Hotel on 22 July 1946, the large-scale Operation Shark was launched on 30 July to cordon-and-search the whole of Tel Aviv. It achieved little, partly because inexperienced troops, rather than the police, did the house-to-house searching. Small-scale operations, based on good police information, would have been more effective but the absence of reliable intelligence made them impossible. The number of terrorist incidents declined but any successes gained by Operations Agatha and Shark were removed by the release of Jewish detainees in November 1946 under political pressure. This did not mollify Jewish bitterness and led to a revival in terrorism. In short, the imposition of martial law and a plethora of covert special operations, raids and searches had failed to destroy either the insurgents or their infrastructure.[29]

During the last year of the Mandate the British were increasingly frustrated with the intractable situation in Palestine. They escalated the use of collective and punitive punishments under the Emergency Regulations, which gave the security forces wide powers. Ignoring the political context in Palestine, this further alienated an already sullen population. 'Notes for Officers on Internal Security Duties', a training pamphlet issued by GHQ Middle East Forces, failed to acknowledge the crucial importance of the political dimension. The brutality of the Stern Gang and other incidents, such as the flogging of the Brigade Major of the 2nd Parachute Brigade as a reprisal for harsh treatment of Jews by the 6th Airborne Division and the hanging of two British sergeants by *Irgun* in July 1947, bred a growing anti-Jewish feeling. This brought a backlash from the security forces, whose ill-discipline included the beating up of local people and the deaths of five inhabitants when police armoured cars opened fire on cafes and shops in Tel Aviv. Equally controversial was the employment by Brigadier Bernard Fergusson (Assistant Inspector General of the Palestine Police) of a special unit of 'Q Squads' to hunt down insurgents. This copied the Special Night Squads commanded by Orde Wingate during the Arab Revolt. The scandal of Captain Roy Farran, seconded from the Army to the Palestinian Police, being tried and acquitted for the murder of 17-year-old Alexander Rubowitz, a member of the Stern Gang who disappeared in

Brigadier Mike Calvert, Commandant of the SAS Brigade, at a ceremony in Tarbes in southern France to mark the transfer of 3 and 4 SAS (2 and 3 Regiment de Chasseurs Parachutistes) from the British to the French Army, October 1945. (© IWM, B 15783)

Civilians being interrogated by officers from the Palestine police during a sweep in Tel Aviv (following the bombing of the King David Hotel) for members of Jewish terrorist organisations, 30 July 1946. (© IWM, E 31979)

Following the disturbances in Tel Aviv in November 1945, a Palestinian constable acts as an interpreter for a British Army patrol advising people to leave the area. (© IWM, E 31467)

Members of the Palestine police examine the remains of a truck which was used to carry a bomb into the motor transport yard of the police headquarters in Haifa, January 1947. (© IWM, E 32177)

Damage to the Allenby Bridge on the main route between Palestine and Trans-Jordan following an attack by Jewish saboteurs during the night of 16–17 June 1946. (© IWM, E 31945)

The explosion of a second bomb at the King David Hotel in Jerusalem. The hotel housed the military headquarters for all armed forces in Palestine as well as the offices of the Palestine government. The attack on the hotel was the biggest blow struck against British rule in Palestine. It was carried out by the Jewish *Irgun* organisation. (© IWM, E 31969)

A Bren gunner trains his gun on a car as it is searched in December 1946 during a British road block on the main Jaffa highway outside Jerusalem. In the background, a soldier of the Irish Fusiliers checks suitcases on the top of a bus. (© IWM, E 32160)

Soldiers removing a Bren gun from a hidden cache of weapons found in the Jewish settlement of Doroth near Gaza, September 1946. (© IWM, E 32043)

The United Nations Honour Guard presents arms on the arrival of Field Marshal Sir John Harding at K14 Airfield in Seoul, 1955. (© IWM, MH 33067)

Sikorsky H-5 helicopter of the United States Air Force Rescue Service lands at a forward British ration point to collect supplies needed by units of the Commonwealth Division cut off by flooded roads in Korea, c.1951. (© IWM, MH 31802)

The British High Commissioner in Malaya, General Sir Gerald Templer meeting the chief of the Kadabu tribe, Ratu Lewar, during a visit to the 1st Fiji Infantry Regiment at Bahau in Negri Sembilan, 1951. (Photograph supplied by Department of Information, Federation of Malaya, Ref WH-5506-MAL © IWM, K 13972)

The remains of a Sikorsky S55 Whirlwind HAR.21 helicopter of 848 Naval Air Squadron, which was forced down through engine failure, being recovered from a swamp near Layang, Malaya, in May 1955. The servicemen recovering the aircraft are from the Royal Engineers Gurkha Regiment from Kluang and the Royal Naval Air Station at Sembawang. The helicopter is being wheeled along a bridge that was built over 200 yards of swamp for the salvage operation. (© IWM, A 33203)

British Army soldiers on patrol in Kenya during the Mau Mau uprising, c.1952. (© IWM, MAU 545)

Captured Mau Mau guns, c.1954. These home-made weapons were often more dangerous to the person firing them than to the intended target. (© IWM, MAU 443)

British soldiers on patrol searching for Mau Mau fighters, c.1954. (© IWM, MAU 547)

A Daimler armoured car and crew from the East African Independent Armoured Car Squadron returning from patrol, Kenya, 1953. (© IWM, BF 11141)

A group of Samburu trackers armed with their traditional weapons of spears, bows and arrows, 1954. (© IWM, BF 10954)

Squairs Farm in the Aberdare Ranges, c.1953. The farm was raided by the Mau Mau and subsequently used as the headquarters of 'I' Force. (© IWM, BF 10946)

A British 3-inch mortar team in action during operations against the Mau Mau, c.1953. (© IWM, MAU 221)

Evacuation of injured soldiers on horseback by members of the King's African Rifles, c.1953. (© IWM, MAU 326)

British Army patrol crossing a stream, c.1954. The soldiers carry a variety of weapons, including the 7.62mm X8E1 Self Loading Rifle (first and second soldiers from right); the 9mm Sten Mk 5 (third soldier); and the Lee–Enfield .303 Rifle No. 5 (fourth and fifth soldiers). (© IWM, MAU 587)

Members of a British Army patrol search a captured Mau Mau suspect, c.1954. (© IWM, MAU 552)

Two members of the Kikuyu tribe arrested by security forces, c.1954. The man on the right is a witch doctor suspected of administering Mau Mau oaths to new recruits. (© IWM, BF 10958)

Mau Mau document stamp acquired by Lieutenant M.J. Cornish of the Royal Welch Fusiliers during his service as Intelligence Officer on attachment to the 2/3rd Battalion, The King's African Rifles, in Kenya between 1953 and 1954. (© IWM, EPH 1237)

At the Naivasha Rifle Range men from the 1st Battalion, Royal Irish Fusiliers instructing members of the Rift Valley Home Guard on the use of the Vickers machine gun, c.1954. (© IWM, MAU 867)

Loyal Kikuyu tribesmen, who continue to support British rule, man a Bren gun while on patrol in dense forest, c.1954. (© IWM, MAU 726)

Major General William A. Dimoline, Colonel Commandant of the King's African Rifles, inspecting some of his troops. (© IWM, MAU 253)

Lieutenant General Sir George Erskine, Commander-in-Chief, East Africa Command, greets local tribesmen, 1954. (© IWM, MAU 780)

Lieutenant General Sir George Erskine, Commander-in-Chief, East Africa Command (centre), observing operations against the Mau Mau, c.1954. From May 1953, General Erskine took control of all military units operating in Kenya including auxiliary troops and the police. (© IWM, MAU 821)

Lieutenant General Sir Gerald Lathbury who succeeded Lieutenant General Sir George Erskine as Commander-in-Chief, East Africa Command, in May 1955. Lathbury made greater use of a wider range of tactics to locate and destroy the Mau Mau, notably employing captured insurgents for intelligence gathering and to infiltrate the rebel infrastructure. (© IWM, MAU 261)

The Governor of Kenya, Sir Evelyn Baring, inspects troops of the King's African Rifles during a ceremony to present them with new colours, 1957. Sir Evelyn Baring had declared the State of Emergency in Kenya in October 1952. (© IWM, MAU 240)

1,455 refugees are evacuated from Haiphong, North Vietnam, to Saigon by Royal Navy aircraft carrier HMS *Warrior* during Operation 'Passage to Freedom', on 4 September 1954. (© IWM, A 32999)

The baggage of Vietnamese refugees is searched for firearms and other weapons as they arrive on board HMS *Warrior* at Haiphong. (© IWM, A 33000)

General Sir Harold Alexander, Commander-in-Chief, Middle East, and Major General John Harding survey the battlefront from an open car, September 1942. (© IWM, E 16458)

General Sir Harold Alexander (right), with Lieutenant General Oliver Leese and Lieutenant General John Harding, inspect one of the German Panther tank turrets which formed part of the Gothic Line defences in Italy, September 1944. (© IWM, NA 18349)

Field Marshal Sir John (later Lord) Harding, c.1955. (© Crown copyright: IWM HU 103757)

Portrait of Field Marshal Sir John (later Lord) Harding. (© Crown copyright: IWM MH 20310)

National servicemen of the 1st Battalion, Royal Ulster Rifles prepare to stop a local bus for a search while on road block duty during the EOKA Emergency in Cyprus, c.1958. (© IWM, HU 52030)

Men of the 1st Battalion, the Royal Ulster Rifles, search loads carried on a donkey at a road block during the EOKA Emergency in Cyprus, c.1958. (© IWM, HU 52033)

A patrol from the Queen's Own Highlanders searches the jungle around Seria in Brunei for rebels in hiding and for arms and ammunition, January 1963. Following their landing to free European hostages held by rebels in the police station in Seria, the Highlanders patrolled by boat and on foot throughout the area in an attempt to round up the remaining members of the rebel army. (© IWM, TR 18614)

Supplies being dropped by parachute into a clearing in the Borneo jungle during the Indonesian campaign, 1965. (© Crown copyright: IWM RAF-T 5315)

H Squadron, 5 Royal Tank Regiment carries out routine patrols in Saladin armoured vehicles and Ferret scout cars during operations in Sarawak, North Borneo. In addition, the squadron engaged in 'hearts and minds' exercises to win the support of the local civilian population and gain a source of local intelligence. (© Crown copyright: IWM FEW 66/27/20, FEW 66/27/33, FEW 66/27/35)

A helicopter-borne Royal Marine commando patrol returns to base in the Sarawak jungle with a captured terrorist, brought in for interrogation, 1964. (© IWM, A 34816)

Plastic washbag used to conceal a pistol which had been employed in the murder of two Aden police officers in March 1966, during the disturbances which preceded the independence of South Yemen. The pistol was discovered buried in the washbag by members of the Somerset and Cornwall Light Infantry. (© IWM, (EPH 9986))

Stencil cut from metal sheet with Arabic inscription and word FLOSY in capital letters. FLOSY (the Front for the Liberation of Occupied Southern Yemen) and the NLF (the National Liberation Front for Occupied South Yemen) were bitter rivals in seeking to gain the support of the population for their efforts to eject the British from Aden, 1965–66. (© IWM, FEQ 885)

Men of 'B' Company, the 1st (Norfolk and Suffolk) Battalion, the Royal Anglian Regiment, cordon off the village of Hiswa in order to search for dissidents and illegal weapons during the Radfan campaign. A Royal Navy Westland Wessex helicopter of 848 Naval Air Squadron from the carrier HMS *Albion*, patrols overhead. (© Crown copyright. IWM ADN 65-196-13)

The last British soldier to leave Aden, Lieutenant Colonel 'Dai' Morgan, Commanding Officer of 42 Commando, Royal Marines, arriving on board HMS *Albion* on 29 November 1967. (© Crown copyright: IWM MH 30869)

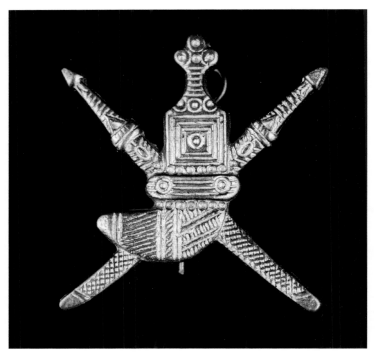

Headdress-badge worn by members of the Dhofar Force, which had been formed in the 1950s to control the Dhofar. Infiltrated by the insurgents, some of its members tried to assassinate the visiting Sultan of Oman in 1966. It became the Dhofar Gendarmerie in 1971 and later the Dhofar Guard. (© IWM, INS 8274)

Trooper J. Scott, serving with 1 Patrol, 3 Troop, 'A' Squadron, 22 SAS Regiment, providing medical treatment to villagers in Falige on the remote Yanqul Plain of Oman, 1970. (© Crown copyright: IWM MH 30626)

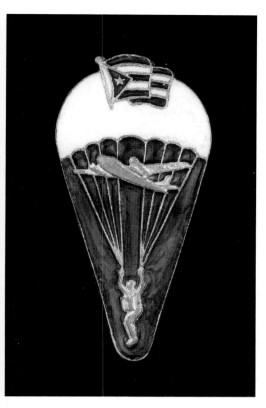

Parachute badge worn by Soviet-funded Cuban advisors who intervened in Angola in 1975 when it gained independence from Portugal. (© IWM, INS 8223)

A Sikorsky CH53 Sea Stallion helicopter used by the German Army, which was photographed during a NATO exercise in Europe, August 1977. (© Crown copyright: IWM CT 737)

A paratrooper serving with 1st Battalion, The Parachute Regiment jumps from an American C130 Hercules on to a drop zone in the Kuwait desert during preparations for operations against Iraq, 2003. (© Crown copyright: IWM OP-TELIC 03-010-20-077)

An RAF Chinook helicopter in Kuwait photographed during preparations for Operation Telic, the invasion of Iraq, 2003. (© Crown copyright: IWM OP-TELIC 03-010-18-286)

Four soldiers of 3 (UK) Division Headquarters and Signal Regiment conduct a foot patrol on a road lined with tanks on the outskirts of Basra during Operation Telic, 2 September 2003. Pictured (left to right): Sergeant Billy Morris, Corporal Kevin McLean, 2nd Lieutenant Ewan Watson and Lance Corporal Tony Gailes. (© Crown copyright: IWM HQMND(SE)-03-053-009)

A Lynx helicopter fires a rocket whilst providing air support to 1st Battalion, Irish Guards during the advance into Basra, 2003. (© Crown copyright: IWM OP-TELIC 03-010-37-013)

Iraqi civilians walk past British soldiers lying prone by the roadside in Basra, 2003. (© Crown copyright: IWM OP-TELIC 03-010-37-075)

At Bridge Four on the outskirts of Basra, a soldier of the 1st Battalion, Irish Guards, looks for possible Iraqi enemy positions as Royal Engineer technicians prepare to cap one of the burning oil wells within the city of Basra, 3 April 2003. (© Crown copyright: IWM OP-TELIC 03-010-34-009)

A British soldier exercising while off duty in Basra, 2003. (© Crown copyright: IWM OP-TELIC 03-010-36-111)

Soldiers of 1st Battalion, The Royal Green Jackets wait for a Royal Navy Sea King helicopter to land in October 2003. The helicopter was to transport the troops to remote areas to set up vehicle checkpoints on roads during a twenty-four hour operation to deter terrorism and prevent the smuggling of weapons and other goods into Basra. (© Crown copyright: IWM HQMND(SE)-03-053-234)

The Tactical Support Unit (TSU) of the Iraqi police, supported by Iraqi, Danish and British military forces, search a house near Basra as part of a counter-insurgency operation during 2006. The TSU searched for arms, ammunition and bomb-making equipment. Thirteen suspected insurgents were detained as a result of this search-and-arrest operation. (© Crown copyright: IWM HQMND(SE)-06-008-072, HQMND(SE)-06-008-097, HQMND(SE)-06-008-127, HQMND(SE)-06-046-040)

Marine K. Rademan of 40 Commando, Royal Marines, is on guard duty at Bagram Airfield, Afghanistan, February 2002. 40 Commando deployed to Bagram in November 2001, ahead of the main force which followed some weeks later. Their initial role was to provide support to Major General John McColl and the Afghan Interim Authority once the latter took office on 22 December 2001. (© Crown copyright: IWM LAND-02-012-0818)

British forces return to base after completing patrol in Lashkar Gah, Helmand Province, Afghanistan, 2006. (© Crown copyright: IW HTF-2006-010-082)

Soldiers of 3rd Battalion, The Parachute Regiment and D Squadron, Household Cavalry, wait to board a Chinook helicopter after completing operations to secure the town of Nowzad, July 2006. British forc had been serving in Nowzad since May, during which time they had been virtual besieged by Taliban insurgents during fier fighting. (© Crown copyright: IWM HTF-2006-043-427)

May 1947, damaged British prestige in Palestine. This demonstrated that 'state counter-terrorism' if discovered and publicised provides excellent propaganda for the insurgents. Similar units were later employed in Malaya, Kenya, Cyprus and Northern Ireland.[30]

Strong differences of opinion about the correct strategy to pursue arose between Montgomery, a key believer in employing repression to crush the rebels, and Cunningham, who was more aware of the political realities in Palestine. The latter recognised the need to appease the Americans and moderate Jews, who could provide the intelligence needed to eradicate the insurgency. Ignoring the implications of an aggressive counterinsurgency, Montgomery and Barker believed that the Army was being forced to fight with its hands tied behind its back. They sought to regain the initiative and to employ more 'robust' methods against the terrorists. Cunningham argued correctly that the only solution was a political one and that sterner measures would only alienate further both local and world opinion unless directed against specific targets by good intelligence. Moreover, the British lacked the security forces to suppress the revolt by sheer weight of numbers. This had been the reason why the imposition of martial law in March 1947 had failed.[31]

The system of committees at the central and district levels, which ensured good civilian–military co-operation, formed the cornerstone of successful British counterinsurgency in the post-war era, notably in Malaya and Kenya. But, as in the Arab Revolt of 1936–39, the local civilian administration, which was based on the district officer and the police commissioner, was missing in the Jewish Revolt. This key source of local knowledge and intelligence dried up if the population became alienated, as it did once again between 1945 and 1948. By the end of 1946 the British had failed miserably to win over the population and the campaign in Palestine was clearly lost. The pre-war doctrine of 'imperial policing' had successfully suppressed insurgencies by clearly identifiable tribal forces in remote areas where there were few political considerations. But it was not so successful against a clandestine and sophisticated Jewish insurgency in the politically sensitive environment of Palestine. The use of 'robust' methods as part of a strategy of coercion, which had crushed the mainly rural Arab Revolt (and worked later in Malaya and Kenya) failed when applied against largely urban terrorism (as it would also later in Cyprus, Aden and Northern Ireland).[32]

The lack of good intelligence, crucial for a successful counterinsurgency, and a failure to establish a central intelligence bureau to co-ordinate the myriad intelligence-gathering agencies were the most significant factors in the failure to defeat the insurgency. The British were unable either to identify the insurgents or to mount successful operations against them. The report of Sir Charles Wickham, formerly Inspector-General of the Royal Ulster Constabulary

(RUC), and County Inspector Moffat (RUC) in December 1946, highlighted police inadequacies. These included the inability to establish friendly relations with the public, who refused to co-operate, and to obtain intelligence. As a result, there was adequate information on the semi-clandestine *Haganah* but little on *Irgun* or the Stern Gang. The Palestine Police lacked a Special Branch, and CID, which was supposed to gather political intelligence, neglected this important aspect, lacking the organisation and manpower. Without the co-operation of the *Yishuv* (Palestinian Jews), the police were short of Jewish officers who, being intimidated by assassinations and infiltrated by the insurgents, could not be trusted. Many sided with the insurgents rather than with the government, acting as spies for the rebels and, when asked to translate documents obtained during Operation Agatha, destroying incriminating evidence. There was also a shortage of the necessary language skills, as only five British officers could speak Hebrew. British intelligence officers were also assassinated by *Irgun* and the Stern Gang. The Inspector-General, Colonel W. Nicol Gray, emphasised military policing rather than intelligence gathering. The force became militarised with the formation in 1943 of a special paramilitary Police Mobile Force, which was closer in behaviour to the Gestapo rather than the British 'Bobby' on the beat. Poorly trained, it was not based in the districts but instead operated from a central location, driving around in armoured cars and treating the local population heavy-handedly as the Black and Tans had in Ireland in 1919–21. Prisoners were tortured with waterboarding and the beating of the soles of the feet.[33]

By August 1947 the security forces had been unable to reverse the rising violence and the loss of internal security, and the British government admitted defeat, announcing that it would surrender the Mandate by July 1948. The British had been levered out of Palestine by a skilful campaign of terror. The main lesson of Ireland, and both the Arab and Jewish Revolts, was that there was no substitute for the support of the population, which provided good intelligence. This was the key to a successful counterinsurgency campaign and without it the best civil-military co-operation and counterinsurgency techniques were unable to defeat the insurgents. Vacillating between concessions and severity the government had failed to win over the population. Henceforth, British counterinsurgencies should be seen in the context of decolonisation and the subsequent retreat from Empire during the era of the Cold War. Britain often failed to detect the rebellions that sprang up in Palestine, Malaya, Kenya, Cyprus and Aden and attempted to hold on to its colonies, often using brutal and repressive methods. It was only in the late 1950s and early 1960s that the pace of decolonisation accelerated. The campaigns in Malaya, Kenya, Cyprus, Borneo and Aden were ultimately fought pragmatically to make sure that, where Britain could not retain sovereignty, post-colonial regimes acceptable to British political and commercial interests were established. The British were successful in this, except in Palestine and South Arabia, although Israel remained pro-Western.[34]

Malaya, 1948–60

Overwhelmingly from the minority Chinese population, which minimised the task faced by the security forces, the communists launched a Maoist insurgency in Malaya in April 1948. They exploited the racial tension and communal violence resulting from economic hardship and alienation of the Chinese peasants from a Malay-dominated society. Surprised by the 'communist insurrection', the British failed during the first two years of the Emergency either to win the hearts and minds of the population or protect them from intimidation. Passing draconian Emergency Regulations, which gave extraordinary powers of arrest and detention without trial and stiff penalties, notably the death sentence for possessing arms, the government employed repression not only against the Chinese communists but also the Malay Left and leading trade unionists. The entire population over 12 years old was registered, fingerprinted, photographed and forced to carry identity cards at all times. This made surveillance and control of the population possible. Unprepared for this crackdown, many communists were detained before they could go underground. But such heavy-handed methods were controversial, drawing much criticism for suspending constitutional rights and intimidating the population. In late 1951 some 6,000 people were being detained without trial amid complaints about the despotism of the 'police state' in Malaya and the loss of civil liberties, notably by the law-abiding Chinese population.[35]

The brutal strategy of coercion allowed the security forces to destroy houses and sometimes whole villages to punish the insurgents' supporters. In October 1948 the inhabitants of Batu Arang were expelled and in November 1948 a large section of Kachu village in Selangor destroyed. In December 1948 the Scots Guards summarily executed twenty-four Chinese peasants in the village of Batang Kali in retaliation for an ambush. Such incidents provided martyrs for insurgent propaganda. Initially, the under-strength security forces were overstretched and dispersed in static defences throughout the country. There was little co-operation between the Army, police and civilian administration. Lacking good intelligence, the police were ill-prepared for anti-guerrilla operations while the Army's organisation, based on the battalion, was too centralised. Viewing their opponents as 'bandits', there was an overemphasis on large-scale sweeps through the jungle rather than on small unit tactics. The British were reluctant to abandon large-scale operations because in the summer of 1948,

Chen Ping, the communist leader, switched prematurely from guerrilla to conventional warfare in accordance with the Maoist model of revolutionary war. This resulted in some significant British successes following the arrival of reinforcements and a temporary lull in terrorist activity after the communist reversion to guerrilla warfare, which led to over-optimism that the insurgency had collapsed. Large-scale sweeps prevented the troops from occupying areas long enough to get to know the local terrain and population, and from gathering intelligence. This made it impossible to identify the enemy 'order of battle'. Appropriate when the insurgents were operating in sizeable formations and the security forces lacked training in small unit tactics, large-scale operations became increasingly redundant with the reversion by the rebels to guerrilla warfare. Gradually, there was a shift to small-scale operations by units located within specific areas. This allowed the British to use good intelligence gathered from the inhabitants to deny food and supplies to the insurgents and to hunt down the guerrillas. Decentralisation of command down to company and platoon level allowed junior officers to employ the appropriate tactics.[36]

The Jungle Warfare School (JWS), officially the Forces Training Centre (FTC), was established at Johore under Lieutenant Colonel Walter Walker by Lieutenant General Sir Neil Ritchie (Commander-in-Chief, Far East Land Forces) in mid-1948. It trained the troops in basic skills, such as field craft, small unit tactics and operating in the jungle, which had been learned in Burma during 1943–45. New tactical innovations employed in Malaya were outlined in a manual, 'The Conduct of Anti-Terrorist Operations in Malaya' (ATOM), written by Walker at the JWS in two weeks. This established for the first time a counterinsurgency doctrine for future British and Commonwealth campaigns. It heavily influenced Australian, South African and Rhodesian manuals of the 1960s and 1970s, and other armies, notably the Portuguese, around the world, providing the groundwork for modern counterinsurgency. For example, two American graduates from JWS were part of the 'Counterguerrilla Warfare Study Group' that set up the 'guerrilla warfare training program' for the 3rd Marine Division in Okinawa in 1961.[37]

The turning point came with the arrival in April 1950 of Lieutenant General Sir Harold Briggs as 'Director of Operations' to improve civilian–military cooperation and the collection and dissemination of intelligence that were key to successful counterinsurgency. The 'Federation Plan for the Elimination of the Communist Organisation and Armed Forces in Malaya' (the 'Briggs Plan') maintained that counterinsurgency requires primarily a political rather than a military solution. Building on years of British experience of imperial policing and irregular warfare, Briggs set up an elaborate committee system at district, state and federation level to provide regular and effective liaison between the civilian and military personnel. This was based on a similar system used in the Burma Revolt between 1930 and 1932 in which Briggs had served. The District War Executive Committee (DWEC) chaired by the district officer, who was the lynchpin of colonial administration, was also attended by the local battalion commander, police superintendent, RAF officer, home guard commander and some key leaders of the local community. It was responsible for daily operations in local areas. DWECs answered to the States War Executive Committees (SWECs), which in turn answered to the Federation Executive Council. Chaired by the High Commissioner, they were responsible for policy and finance. Briggs allocated specific areas of operations to battalions and sub-areas to companies, ensuring that units got to know the terrain and won the confidence of the local population.[38]

Comparing the clearance of 'CTs' (Communist Terrorists) from Malaya to the eradication of malaria, Briggs recognised that the 'breeding grounds' in Malaya were the Chinese subsistence farmers. The 'Briggs Plan' focused on the forced resettlement of more than 400,000 Chinese peasants from 'squatter' villages along the fringes of the jungle into resettlement areas. Employing a carrot-and-stick strategy the British deployed both coercion to cut off the communists from their infrastructure among the rural population and rewards to win hearts and minds. Population control and severe restrictions on movement broke the links between the guerrillas and their cells in the 'squatter' villages, isolating the insurgents from supplies of food and intelligence. The Chinese peasants were also provided with land, running water, schools, medical care and citizenship in the Federation of Malaya. These were things the communists could not provide although, in the short term, the standard of living for many Chinese declined as it took time to install the new amenities and for poor conditions to improve. Rewards were given for information, which led to the death or capture of the insurgents, sowing dissension in insurgent ranks. Pamphlets, newspapers, leaflets, the radio and 'voice aircraft' were employed by the Emergency Information Service (EIS, later the Department of Information) to win the propaganda war. Criticisms of the government and communist propaganda were increasingly counterbalanced by the EIS, which gradually won the 'battle for the mind'. After 1952 the tide turned against the insurgents in the jungle and their morale cracked, leading to increasing numbers of defections that culminated in the mass surrenders of 1958 to take advantage of the *Merdaka* (independence) Amnesty offered by the government.[39]

Ultimately the huge 'surge' of British and Commonwealth troops, supported by a vast expenditure, eliminated the insurgency. Briggs developed the full spectrum of security forces previously deployed by the British on the North-West Frontier of India. These included a special constabulary recruited from Malays to defend the rubber estates and tin mines, a home guard recruited from the Chinese population to secure the resettlement areas, and the Chinese auxiliary police in larger towns, to provide security for the population, gather intelligence and make progress in defeating the insurgency. The *Orang Asli*, jungle aborigines, who served with the Perak Aborigines Area Constabulary (PAAC) and the Police Aboriginal Guards (PAG, a Home Guard unit), and Iban and Dyak trackers from Sarawak were also recruited to gather intelligence and track the insurgents for troops based in the jungle. The reformed Special Air Service (SAS) won over the aborigines by providing medical care and other facilities. They also hunted down the insurgents, driving the communists further into the jungle and disrupting their communications network. The first helicopters made this possible by carrying troops and supplies. Special forces would form an important element in future counterinsurgency campaigns.[40]

As Briggs noted, lack of intelligence was the British 'Achilles' heel'. In particular, the acute shortage of Chinese-speaking officers and Chinese constables remained a serious weakness throughout the Emergency. The intelligence agencies were unprepared for revolutionary guerrilla warfare, notably the Malayan Security Service (MSS), which was disbanded. It was replaced by a newly created Special Branch that handled all tactical intelligence and counter-subversion. Moreover, Colonel W. Nicol Gray (Commissioner of Police) neglected the ordinary police in the local districts, who were intimidated and alienated from the population, placing too much emphasis on police jungle squads and a Special Frontier Force. He repeated his mistake as Inspector-General of the Palestine Police, and created factionalism by bringing 500 constables from Palestine. Following Gray's removal, Colonel Arthur Young (Commissioner of the London Metropolitan Police) was seconded to Malaya in April 1952 to revitalise Special Branch and create a police force that was capable of defeating the insurgents and maintaining order after the British departure. He formed a training academy and recruited Chinese officers, although many Chinese remained reluctant to join the police, who were traditionally composed of Malays and Indians. The intelligence agencies were unprepared for revolutionary guerrilla warfare, notably the Malayan Security Service (MSS), which was disbanded. It was replaced by a newly created Special Branch that handled all tactical intelligence and counter-subversion. In August 1950 Sir William Jenkin was appointed Director of Intelligence (DOI) for the police. However, poor co-ordination with military intelligence remained a major weakness which, with 'turf wars' between competing bureaucracies and clashes of personalities, led to his resignation in October 1951. It was not until General (later Field Marshal) Sir Gerald Templer arrived that Jack Morton (Director of Intelligence) undertook the role of co-ordinating all intelligence agencies, providing a model for Kenya, Cyprus and Borneo. A system of intelligence committees alongside the executive committees and the appointment of Military Intelligence Officers (MIOs), who liaised between Special Branch and the Army, improved the gathering and dissemination of intelligence. During 1950–54, Special Branch was central to the defeat of the insurgents. It provided the intelligence to build up a communist 'order of battle' by 1957. It also recruited hundreds of agents and surrendered enemy personnel (SEP) to infiltrate and destroy the insurgent organisation. Interrogators employed harsh interrogation methods, including torture, financial rewards and blackmail to 'turn' suspects.[41]

However, these reforms required time to become effective and during 1950 and 1951 the situation in Malaya deteriorated. The number of incidents reached a peak at the end of 1951 with the insurgents, who targeted police stations and civilians, committing 100 murders per month. The main problem lay in the command structure, which with its federated system required a strong executive. Briggs (Director of Operations) did not command either the police or the troops and, although he developed the policies for success, lacked the authority to energise the elaborate committee system into implementing them. The turning point was the death of Sir Henry Gurney, ambushed and killed in October 1951, and the appointment of Templer as his successor. By combining the offices of High Commissioner and Director of Operations and control of both civilian and military matters, Templer provided unified command. Adopting the Briggs Plan, which had already begun to turn the tide, Templer travelled widely round the countryside, seeing things for himself, 'electrifying' the security forces and infusing the bureaucracy with a sense of urgency. When Templer left Malaya in 1954 the tide had definitely turned with the number of terrorist incidents falling from the peak of mid-1951 and the insurgency declining in strength, although the Emergency would not end until 1960. Templer gets most of the credit but others, notably Briggs and Tunku Abdul Rahman (Malaysia's first Prime Minister), also contributed to the success of the campaign.[42]

The British achievement in Malaya was not just in mounting a successful counterinsurgency campaign but also in leaving a post-colonial nation-state that was acceptable to the West, gaining independence in August 1957 under Tunku Abdul Rahman. This outmanoeuvred the insurgents politically and avoided the damaged relations that were the legacy of the exit from empire by the Dutch in Indonesia and the French in Indochina and Algeria. Tunku's ruling Alliance Party, having won elections held in July 1955, provided a moderate, constitutional alternative to the communist insurgents. It remained in power long after the British departure, retaining a number of authoritarian powers, such as detention without trial, which had been introduced during the Emergency. This process also included the reform and speedy 'Malayanisation' of the Army and Civil Service prior to independence. By independence all SWECs and DWECs had Malayan chairmen.[43]

The communists' failure to win support from the majority Malay population allowed the British 'to divide and rule', exploiting ethnic divisions. The insurgent leaders played into British hands by failing to adopt a strategy suitable to Malayan conditions by retreating into the jungle, isolating themselves from their support, rather than employing urban terrorism, which proved so effective in Palestine, Cyprus and Aden. They were also isolated geographically by the Royal Navy's blockade of the Malayan peninsula and by the Thai–Malay agreement of 1949, which sealed the border and allowed joint operations against the guerrillas in their sanctuaries, and politically, by the failure either to internationalise their struggle or to mobilise any significant external support, notably from communist China. The rising prices for rubber and tin, a consequence of the Korean War, provided revenue to finance the counterinsurgency and raised the prosperity of the population. Developed from their colonial traditions, British methods were based on a close integration of civil and military authority. They relied on 'minimum force', the supremacy of the police over the Army, good intelligence from Special Branch, the resettlement of civilians, effective local forces, and small and highly skilled units in well-planned, targeted operations instead of the ill-directed use of large operations and heavy firepower. The same techniques would be employed successfully in Kenya but were less successful against a powerful organisation such as EOKA in Cyprus that enjoyed popular support.[44]

Kenya, 1952–60

Beginning in the summer of 1952, the Kenya Emergency was mainly limited to one tribe, the Kikuyu (Gikuyu), which dominated the anti-colonial Mau Mau, and to the Central Province, which included the capital, Nairobi. A nationalist agenda that sought independence and could sustain an effective insurgency was missing. Instead, the main grievance fuelling the insurrection was land shortages. The acquisition of land by European settlers in the White Highlands and government agricultural reforms had affected Kikuyu tenant farmers. It resulted in growing numbers of landless peasants and increasing militancy among the embryonic working class and tribal youths squatting in Nairobi slums. The Mau Mau terror campaign against white farmers and their families, informers, tribal chiefs and other 'loyal' Kikuyu also undermined their credibility as a national-liberation movement. Although the Colonial and War Offices criticised their initial neglect of propaganda, this allowed Nairobi to portray the rebels as 'savages'. The Lari massacre in March 1953 of more than 100 Africans, mostly women and children, by the Mau Mau, was a turning point that the British exploited. Consequently, the security forces fought a particularly 'dirty war',[45] which resulted in Mau Mau veterans bringing a lawsuit against the British government in October 2006 for compensation for abuses committed during the emergency. The government paid an out-of-court settlement.

Following the assassination of Senior Chief Waruhiu wa Kungu (one of the government's main Kikuyu supporters), Sir Evelyn Baring (the Governor) declared a State of Emergency in October 1952. This granted extraordinary and draconian powers, allowing communal punishments and control of population movement. Operation Jock Scott by the police and troops to overawe and crush the rebellion made some 8,000 arrests. It also controversially detained some leading figures, notably Jomo Kenyatta, of the Kenya African Union (KAU), which sought reform through constitutional means. This demonstrated Special Branch's inability to identify that the Mau Mau leadership came from the more radical Kikuyu Central Association (KCA), which had been outlawed in 1940. This swoop surprised the rebels but allowed the extremists to gain influence and recruits by depriving the nationalist movement of its moderate leadership and alienating many Kikuyu.[46]

Following the Malayan model, committees provided the formal organisation to foster civilian–military co-operation and to gather and disseminate intelligence. At the highest level the Internal Security Working Committee formed in 1950 rarely met at first because Sir Philip Mitchell (the Governor) downplayed the danger of a Kikuyu rebellion. This meant that the necessary drive was lacking until the arrival of Sir Evelyn Baring in September 1952. From 1953 the Colony Emergency Committee, chaired by the Governor, initiated policy, while at district, provisional and divisional level emergency committees integrated the Army, police and civilian administration. A 'chief staff adviser to the Governor' (later director of operations) was appointed in January 1953 but no director of intelligence existed until 1955. From June 1953 General Sir George 'Bobbie' Erskine (Commander-in-Chief, East Africa) provided dynamic leadership and knowledge of counterinsurgency, but, unlike Templer in Malaya, did not control the civilian administration. At the beginning of the Emergency, the Kenyan Special Branch had only four officers and few Africans in the ranks and remained extremely under-staffed as late as 1954. The Kenya Intelligence Committee was formed in March 1953 to oversee intelligence and in May 1953 established military intelligence officers at the provincial and district level. They were aided by field intelligence assistants. These were European sergeants recruited from the white-only Kenya Regiment, who provided local expertise and language skills. Each district was divided into areas. From early 1954, each area had a field intelligence officer to recruit informers and agents, gather intelligence on the Mau Mau and its infrastructure, and liaise between civilian, police and military personnel.[47]

Captured guerrillas were employed innovatively to infiltrate Mau Mau units. These 'pseudo-gangs' gathered intelligence and located insurgents who were too elusive for conventional Army units to catch. They also learned about their organisation and arrested their leaders and members. Initially employed in the settled districts, these tactics were given official sanction in June 1954 by General Erskine and their use was expanded by his successor, Lieutenant General Gerald Lathbury, who in May 1955 set up five special forces teams. A training centre, the Special Methods Training Centre, was established and the technique was employed widely during offensive operations to hunt down guerrilla bands in the forests. This use of counter-gangs had considerable impact on later counterinsurgency operations in Malaya, Cyprus and Northern Ireland.[48]

The British attempted to win hearts and minds by protecting the loyal Kikuyu from intimidation. They also introduced development and reconstruction programmes to provide roads, hospitals and housing. Labour and agrarian reform improved job opportunities and pay in urban areas while land use and tenure was expanded in rural areas. But these uses of the 'carrot' were overshadowed by brutal use of the 'stick', notably detention and deportation, to suppress the insurgency. Between 70,000 and 100,000 Kikuyu were forcibly evicted or fled from the Central and Rift Valley Provinces in 1952–53 and 2,609 Mau Mau suspects were tried. Of these, 1,090 were hanged, compared to eight hanged in Palestine, nine in Cyprus and 226 in Malaya. At one stage they were being hanged at the rate of fifty a month after 'drumhead trials'. Detention was employed on a far larger scale in Kenya than in any of the other British post-war counterinsurgencies. By the end of the Emergency some 80–90,000 detainees (including Barack Obama's grandfather) had been held officially, often suffering appalling conditions and torture. By 1954 there were some 78,000 detainees compared to just 1,200 at the height of the Malayan Emergency. This widespread use of detention disrupted the insurgent organisation, notably in Nairobi, and provided much intelligence from the interrogation of suspects and captured documents. But it created huge numbers of suspects and placed a heavy burden on CID and Special Branch, who had to screen them for rehabilitation. It also further alienated detainees, who experienced poor conditions, forced labour, torture, rape and summary execution.[49]

On the Malayan model, the concentration ('villagisation') of the Kikuyu, who lived in scattered dwellings, into 'protected villages' separated the rural population from the insurgents, who were denied their support and recruits. Carried out with greater brutality, problems with the implementation of resettlement, as in Malaya, resulted in deaths from starvation and disease in unsanitary conditions. During 'villagisation' of all Kikuyu areas between June 1954 and October 1955 more than 1 million people were forcibly resettled into 854 new villages where the population was closely policed and often mistreated. Mirroring the successes of Malaya, resettlement played a major part in disrupting the insurgent infrastructure, forcing them on to the defensive and isolating them in the forests. Collective punishment, notably confiscations, fines, evictions, forced labour, movement restrictions and identity checks, controlled the resentful population.[50]

In a poor state and short of manpower at the beginning of the Emergency, by the end of 1954 the police had been reinvigorated by an expansion of a restructured Special Branch, the number of police stations and auxiliaries, such as the Kikuyu Home Guard and the tribal police. The police gained a reputation for brutality and Special Branch, nicknamed 'Kenya's SS', was notorious for torturing suspects, notably at the Mau Mau Investigation Centre at Embakasi, near Nairobi. Unsurprisingly, few Kikuyu served with the police. The auxiliary forces included white settlers, policemen hired in the United Kingdom and a home guard created from 'loyal' Kikuyu. As in Ireland and Palestine, they provided valuable supplementary manpower and local knowledge of terrain and languages. Mobilised as the Kenya Police Reserve (KPR) and the Kenya Regiment, they served as trackers, guides and interpreters for Army units. But, like the *colons* in Indochina and Algeria, the white settlers from the highlands of Kenya created problems, opposing reform and behaving like vigilantes. They were difficult to control and, ignoring the concept of 'minimum force', employed brutality against the indigenous population. Beatings, torture, mutilation and summary executions of suspects were widespread. The KPR was labelled as a group of 'undisciplined sadists' by one newspaper. The recruitment of local levies, which was successful in Malaya, made the technique of divide and rule possible. The Kikuyu Home Guard was mainly employed to defend the population within the Kikuyu tribal reserve, allowing the Army and police to undertake more aggressive operations, but was also employed to hunt down Mau Mau bands. Lacking adequate training and good leadership, the Home Guard carried out local vendettas and massacres against suspected rebel supporters. It used summary executions and torture to extract confessions in the detention camps. The gruesome nature of a small number of Mau Mau atrocities, notably the murder of thirty-two Europeans, including a 6-year-old boy and his pregnant mother, unleashed a backlash from the racist settlers. They called regularly for extermination of the Kikuyu. There was a greater toleration of vigilantism, producing greater abuses of power than other post-war insurgencies and perpetuating cycles of violence. British and King's African Rifles units also committed atrocities. Attempts by Arthur Young (Commissioner of Police), who had served Templer in Malaya, to stop this ill-discipline were obstructed by the 'hawks' who saw 'counter-terror' as an essential weapon. In the end, Young and Donald MacPherson (Head of CID) resigned.[51]

Poorly armed with *pangas* (machetes) and crude, home-made firearms, Mau Mau gangs were much less sophisticated than the communists in Malaya or

Vietnam. They lacked any revolutionary or guerrilla warfare training and made no attempt to sabotage vulnerable installations. The Army operated in 'special areas' within the reserves where the movement of the local population was circumscribed and established the forests of the Aberdares and Mount Kenya as 'prohibited areas' where only the security forces were allowed and where they could shoot Africans on sight. During 1953 the Army's troops were widely dispersed, helping the police and protecting European farms and loyal Kikuyu. This left few troops for the offensive against the Mau Mau bases in the forests. Nevertheless some large-scale sweeps were conducted in the Fort Hall and Nyeri districts between June 1953 and March 1954, breaking up large gangs. As the police was rebuilt and paramilitary units developed into effective forces, troops were freed to take the offensive. Loss of morale amongst the Mau Mau was indicated by more than 800 surrenders by January 1955 in response to the 'Green Branch' amnesty, which was one of successive incentives to surrender offered to the forest fighters by Erskine.[52]

In April 1954, having limited the spread of Mau Mau, Erskine conducted a massive urban cordon-and-search operation (Anvil) to 'clean up' Nairobi, the centre of Mau Mau activity. The operation screened the entire African population, detaining some 19,000 people, and deporting around 30,000 more back to the Kikuyu Reserves. This was the turning point in the campaign, severely disrupting the infrastructure of the Mau Mau, which never recovered. A 50-mile fortified ditch 18ft wide and 10ft deep was also dug by African labour to separate the Reserves from the Mau Mau in the forests. Cutting off supplies to the Mau Mau forest fighters and sealing their fate, Anvil provided a model for future urban operations in Cyprus, Aden and Northern Ireland. Its success was followed up by large-scale operations against the guerrillas in the Aberdare Forest (Hammer I) and Mount Kenya (Hammer II) in January 1955 and First Flute, which during February and March 1955 broke up the larger gangs. These drove many Mau Mau into the settled areas, where they were destroyed. Relentless pressure in the forests, when combined with food denial, persuaded many gang members to surrender, providing intelligence for further successes. By late 1955 the emphasis had changed from large-scale to small-scale operations by units deployed in specific areas to hunt down the remaining guerrillas. Districts were divided into four company areas, which corresponded to the police divisions, allowing troops to get to know the local terrain and inhabitants. As in Malaya decentralisation of command provided the tactical flexibility to meet local requirements. By January 1956, lacking a coherent strategy, external support and the undivided support of the Kikuyu, the Mau Mau were defeated.[53]

The military campaign had been mirrored by a political one. In 1954 the first African minister was appointed, followed in 1957 by the first African general election. In 1960 delegates went to London for a conference to discuss African rule. In 1963 Kenya was granted full independence. Britain preserved its 'vital interests' with Kenya remaining in the Commonwealth and firmly in the Western camp during the Cold War under the leadership of Kenyatta. He co-operated with British intelligence agencies and employed the security apparatus left by the British to keep his political rivals under surveillance. Two manuals, *The Conduct of Anti-Terrorist Operations in Malaya* (Singapore, 1952) and *A Handbook of Anti-Mau Mau Operations* (Nairobi, 1954), outlined the lessons of the Malayan and Kenyan Emergencies. Nevertheless, the British continued to conduct each insurgency in an ad hoc way, reinventing the wheel for each new insurrection. There was little study of counterinsurgency at either Sandhurst or the Staff College. Dissemination of techniques and lessons was hampered by the British Army's decentralised, regimental organisation and dislike of formal doctrine. The Colonial Office failed to create and disseminate a doctrine. This hindered the learning process, which was conducted by individuals rather than through official manuals or pamphlets.[54] In Cyprus, these weaknesses would be more fully exposed.

Cyprus, 1955–59

The campaign started on 1 April 1955 when bombs exploded in government offices, military facilities and police stations across the island. The strategic objective of the insurrection was to force the British to allow *enosis* (union with Greece). Both Archbishop Makarios, leader of the Greek Cypriot community, and Colonel George Grivas, leader of EOKA (*Ethniki Organosis Kypriou Agoniston*, the National Organisation of Cypriot Fighters) realised that 'total defeat' of the security forces was impossible. Instead, subjecting the British to relentless harassment, EOKA waged a terrorist campaign in tandem with external diplomatic and political activity, ensuring that occupation carried a price while keeping *enosis* on the international agenda. Underestimating the internal and external support for *enosis*, the British government had rejected the overtures of the Greek government to engage in bipartite negotiations over the future of Cyprus. This not only alienated the Greek Cypriot population but guaranteed that the insurgency enjoyed the active support of neighbouring Greece, one of Britain's NATO allies. Greece took the issue of Cypriot self-determination to the United Nations General Assembly five times between 1954 and 1958.[55]

Britain's insistence that colonial rule would continue reflected her wider strategic interests. Whereas Britain was able to strengthen its counterinsurgency campaign in Malaya and Kenya by promising independence, it was much less flexible in Palestine, Cyprus and Aden, which were considered essential to its Imperial defence strategy. Cyprus provided the General Headquarters (GHQ) for Middle Eastern Command and a base in the Eastern Mediterranean and the Middle East for its heavy bombers (and their atomic weapons) and Anglo-American Signals Intelligence. The reluctance to leave was reinforced by demands from Conservative MPs, who were dismayed by previous retreats from Palestine and Egypt, to make a stand for the Empire in Cyprus.[56] The announcement in July 1954 by Henry Hopkinson, Minister of State for the Colonies, that 'there can be no question of any change of sovereignty in Cyprus',[57] led to riots throughout the island. The limited political compromises that Britain offered did not satisfy the Greek Cypriot community, which was overwhelmingly in favour of *enosis*. No moderate leaders emerged with whom Britain could deal, on the model of Borneo, Malaya or Kenya. The majority of the population remained sullen and sympathetic to EOKA, rejecting government enticements and defying the security forces' harsh countermeasures. As a result, the government could not protect the minority who supported the security forces. Greek Cypriot support for the insurgents was mobilised through a combination of political appeals and intimidation. Indeed, the Church's political and financial support was a significant factor in EOKA's survival.[58]

Influenced by the Jewish revolt in Palestine during 1945–48, EOKA employed a campaign of terrorism in heavily congested urban areas rather than a full-blown, rural insurgency on the Maoist model. During 1955–59 the security forces fought very few 'engagements' with EOKA. Until late 1955 the British were on the defensive with the troops dispersed in static defences in an attempt to provide security, protect property and support the increasingly demoralised police force. The escape of sixteen EOKA terrorists from Kyrenia Castle and the burning of the British Institute in Nicosia in September 1955 awakened the British government to the seriousness of the situation. Field Marshal Sir John Harding, the retired CIGS, was appointed Governor of Cyprus, arriving on the island on 3 October, closely followed by reinforcements. The insurgents were supported by the National Liberation Front of Cyprus (EMAK), which organised a network of sympathisers who provided intelligence, supplies, recruits and safe houses, and conducted the propaganda offensive.[59]

EOKA's campaign consisted of sabotage, riots, street clashes between demonstrators (often schoolchildren) and security forces, bomb and grenade attacks against bars and the homes of Army personnel and senior officials, and hit-and-run attacks on installations such as the Cyprus Broadcasting Station's radio transmitter. It also included passive resistance and economic boycott by the populace, propaganda advocating self-determination and anti-colonialism, and attacks on the government apparatus. These included the assassination of Greek Cypriots in the police force, communists, 'collaborators' within both the Greek and Turkish communities, and senior British officials, including the Governor. EOKA killed more Greek Cypriots than Britons. The police were so intimidated that they remained in their fortified stations. The cell structure of EOKA and the use of couriers made infiltration by Special Branch very difficult, while EOKA's own highly effective intelligence apparatus allowed it to penetrate every level of the government, including the police force, the civil service, and various British military facilities. The British failed to eliminate EOKA agents among the Greek Cypriot personnel of the police force. They warned of security force operations, targeted colleagues for assassination, betrayed informers, provided safe houses and, on occasion, even carried out attacks themselves. Insurgents also acquired information from various church and nationalist organisations. As a result, the insurgents were often forewarned of security force intentions, notably large-scale cordon-and-search operations that, lacking any intelligence to direct them, failed to achieve tangible results.[60]

By contrast, the security forces lacked the intelligence to accomplish significant breakthroughs until the administrative, police and intelligence-gathering apparatus in Cyprus was rebuilt. This took most of 1956. As Governor, Harding was also Director of Operations, co-ordinating the military, the police and civilian administrations. Prior to the reorganisation of the Cyprus command structure in 1955, there had been little co-ordination of police and military intelligence staffs. As existing intelligence collection and distribution was inadequate, this made the operations of the security forces either ineffective or counterproductive. The new post of Director of Intelligence was created in May 1955 to co-ordinate all intelligence services, improving the availability of intelligence considerably. These changes were made shortly *before* a State of Emergency was declared on 26 November 1955, suggesting that important lessons had been learned from the experiences of Malaya and Kenya. However, the lack of local expertise, notably of the Greek language, remained a serious weakness, ensuring that the British were unable to exploit fully the new intelligence-gathering system. Throughout the Emergency the British lacked an adequate number of trained interrogators who spoke Greek. As late as June 1956, there were only five Greek Cypriot-speaking interrogators on the island. In April 1955, Special Branch had only twenty-two gazetted officers. Men brought in from other locations to provide 'professionalism' lacked the local expertise to exploit the available intelligence, which was thus never sufficient to defeat EOKA completely or eradicate terrorism totally in the towns. Thus, although the integration of civil, military and police intelligence made it possible to disrupt guerrilla operations severely, an excellent intelligence organisation could neither compensate for the local population's lack of co-operation nor win hearts and minds. The counterinsurgency campaign in Cyprus was only partially successful because, although EOKA had suffered serious setbacks in the mountains, operations continued in the towns where there was no parallel decline in violence.[61]

To regain the initiative Harding launched a series of cordon-and-search operations of varying size during 1956 to clear EOKA 'mountain gangs', notably the large-scale operations Pepperpot (May 1956) and Lucky Alphonse (June 1956). Although Grivas managed to escape, these operations were successful, eliminating the guerrilla groups in the Troodos mountains and capturing thousands of documents. This was a real intelligence breakthrough, which, for the first time, gave the British an insight into EOKA's organisation. Operation Sparrowhawk (October 1956) cleared the Kyrenia mountains. The tide was turning, forcing Grivas to concentrate EOKA's efforts in the towns. From December 1956 the security forces achieved further successes against EOKA. The British had taken more than eighteen months to recover momentum. Fighting against elusive guerrillas the security forces learned to replace cumbersome large-scale cordon-and-search operations with more flexible and subtle small-scale operations. Some success was achieved with six-man teams, acting on good intelligence, to fight the insurgents on their own ground. The troops were allotted areas, employing 'soft-shoe' patrols by night and day with blackened faces, removing their boots and wearing gym shoes. In close co-operation with the police, stop-and-search operations on vehicles and snap road-blocks caught terrorists and their weapons. In the 'snatch' a small party would raid specific houses to seize terrorists from their beds. Special, so-called 'Q Squads', heavily armed undercover squads based on the 'pseudo-gangs' in Kenya, were recruited from captured terrorists who had been 'turned' by the security forces, and also pro-British Turkish Cypriots, criminals recruited from the Greek Cypriot community in Britain and British personnel. Q Patrols located and infiltrated EOKA gangs, using local knowledge of the terrain and EOKA. Grivas recalled that the system of rural patrols and 'counter-ambushes' meant that the guerrillas were always in fear of the 'lurking cat'. The effectiveness of these new tactics eventually encouraged Makarios, EOKA's political leader, to compromise with the British.[62]

When 'carrots' such as the £38 million development plan for the island announced by Harding in November 1955, failed to produce results, the government resorted to the 'stick'. It attempted to convince the population of the folly and the cost of supporting the insurgents. A State of Emergency on 26 November 1955 greatly increased the security forces' Emergency Powers but little consideration was given to the negative aspects of coercion despite limitations displayed in Ireland and Palestine. They further alienated an already sullen population. Lengthy curfews and large fines were imposed on villages as punishment for either hiding guerrillas or failing to co-operate with the security forces. Rarely resulting in arrests or the discovery of arms caches, large-scale cordon-and-search operations invariably upset inhabitants, who were roughed up, detained for screening or had their houses searched. Executions of Cypriot terrorists, even though small in number, provided martyrs for the cause of EOKA. Widely criticised as being counterproductive, collective punishments, such as the destruction or requisition of property, were often arbitrary. Far from undermining support for EOKA, they gave ammunition to the Greek propaganda machine and made Greek Cypriots more hostile to British rule. The British had lost the propaganda war. Belated attempts to improve public relations and win over a hostile population floundered on the sophisticated and well-organised campaign by Radio Athens. The use of torture, including beatings, the removal of fingernails and waterboarding, by the security forces was common knowledge among journalists, who jokingly called the interrogators 'HMTs' ('Her Majesty's Torturers'). This resulted in frequent allegations of torture and brutality. By late 1956, the Greek Cypriot population was largely behind EOKA. This allowed the insurgency to continue despite suffering 'hard blows' between December 1956 and March 1957. Increasingly frustrated by their inability to catch the elusive guerrillas, the security forces became more brutal towards the Greek Cypriot population. The most notable incident was in Famagusta on 3 October 1958. An EOKA gunman shot two wives of British non-commissioned officers (NCOs) who were shopping. Within two hours troops had detained 1,000 Greek Cypriot men for interrogation. Of these, 256 required medical treatment, sixteen were seriously injured and two were killed. Major General Darling (Director of Operations) admitted in 1958 that his tactics were based on the realisation that EOKA now had mass support and that to suppress it would require the kind of force used by the Russians in Hungary.[63]

Although unable to defeat EOKA, the British did succeed in preventing the achievement of its main aim, *enosis*. In the aftermath of the Suez debacle, the retention of Cyprus as a 'fortress colony' was no longer regarded as a necessity, and the British agreed to end the conflict. They relinquished sovereignty but retained bases on the island. Urban terrorism and propaganda added an entirely new dimension to insurgency. Although long involved in suppressing colonial rebellion, the Army had little experience in countering insurgents whose organisation was clandestine and whose tactics were political in intent and criminal (rather than military) in method. The implications and lessons of British involvement (and failures) in Ireland and Palestine had been overshadowed by later successes but the techniques employed successfully in Malaya and Kenya were less successful against urban insurgencies that retained popular support. Campaigns in isolated locations, such as Borneo and Oman, where the guerrilla enemy could be identified and unrestricted use of force was possible, became the exception after 1945. More usually, as in Cyprus, it was very difficult to distinguish the opponent, who operated clandestinely, from the population as a whole, especially when the political context and media access made unrestrained use of force unacceptable and inappropriate. By the end of the 1950s the British had more than a decade of experience of counter-insurgency and, despite some failures, notably in Palestine, had an impressive record of success. This was based on the realisation that counterinsurgency was predominantly a political problem. The defeat of an insurgency depended upon developing close civilian–military co-operation and winning the hearts and minds of the population. This made intelligence gathering, the key to a successful counterinsurgency campaign, possible. Campaigns in Borneo and Aden in the 1960s would confirm these strengths and weaknesses.[64]

Borneo, 1963–66

The opposition of President Sukarno of Indonesia to the inclusion of the British protectorate of North Borneo on the island of Borneo into the new Federation of Malaysia led to 'Confrontation' with Indonesia. Unable to defeat the British militarily, the Indonesians employed subversion, diplomatic pressure and military incursions from the jungles of Kalimantan, the Indonesian section of Borneo. The goal was to annex Brunei, Sabah and Sarawak and break up Malaysia. A revolt in Brunei in December 1962 was crushed within forty-eight hours by the British, although it took six months to mop up the rebels. Confrontation began in earnest in April 1963 when Indonesian raiders attacked the police post at Tebedu in Sarawak. The insurgency was led by the Clandestine Communist Organisation (CCO) supported by SUPP (Sarawak United People's Party), which recruited from the disgruntled Chinese community. It opposed the formation of the Malaysian Federation, fearing anti-Chinese measures ('desinicisation') as a result of ethnic tensions with Dayaks and Malays. The guerrillas of TNKU (*Tentara Nasional Kalimantan Utara*, National Army of North Borneo) formed with Indonesian fighters the core of PGRS (*Pasokan Gerilja Rakjat Sarawak*, Sarawak People's Guerrilla Movement), which infiltrated from Kalimantan. They received training and arms from the Indonesian Army. From February 1964, the Indonesian Army committed regular troops to support large-scale raids by the guerrillas. They set up guerrilla bases to intimidate and subvert the local population, notably the Iban, who straddled the border. From January 1965, in an expansion of the conflict, Indonesian regular units were committed to incursions, whose frequency increased. This resulted in large-scale attacks on Commonwealth positions and landings on the Malaysian peninsula, which failed to gain any support from the local population.[65]

The British were unable to close the 1,000 mile frontier between British North Borneo and Indonesian Borneo (Kalimantan), which was largely unmapped jungle, mountains and swamp with few roads and no railways. This provided a ready-made haven for the insurgents, who mounted cross-border raids. Major General (later General Sir) Walter Walker was appointed Commander British Forces, Borneo on 19 December 1962 and Director of Borneo Operations soon thereafter. An expert in jungle warfare who had served during the Malayan Emergency and outlined its lessons in a ground-breaking manual (ATOM, 1952), Walker waged the campaign in Borneo on the principles that had been so successful in Malaya. It once again demonstrated the vital importance of unified command and co-ordination of the civilian–military effort. The command structure above Walker was cumbersome, being directed by the Malaysian National Defence Council in Kuala Lumpur, which was linked to London through the British high commissioner. Nevertheless, it worked because of the good relations between Walker, his superior, Admiral Sir Varyl Begg (commander-in-chief, Far East,) and Sir Claude Fenner (inspector-general of Police). In North Borneo, both Walker and Major General Sir George Lea, his successor, were in supreme command of all three services, thus maintaining unity of command. Walker was a member of the State Emergency Executive Committees formed in Sarawak and Sabah and of the Sultan of Brunei's Advisory Council. The theatre of operations (17th Gurkha Division) was sub-divided into four brigade areas. Unfortunately, the police and Special Branch had its own separate organisation responsible to their inspector-general in Kuala Lumpur. But this violation of unified command did not severely hinder operations as the internal threat from the CCO was far less than the external threat posed by the Indonesians.[66]

Security against intimidation had to be provided for the local population, which was divided into two different communities. The Chinese lived in the towns of Sarawak and on the rubber plantations of Sabah while the indigenous peoples straddled the frontier with Indonesia. The military sought to defeat the external threat by providing security for the population and winning their hearts and minds. The police and Special Branch, which had been strengthened by Walker, dealt with the internal threat posed by the CCO infrastructure, detaining Sarawak Chinese suspected of communist subversion. Medical treatment and the improvement of hygiene and sanitation was provided for the local population. The heavy-handed treatment by the Indonesians gave the local population an added inducement to support the security forces. Auxiliary indigenous troops, such as the Border Scouts, composed largely of Ibans, provided local expertise and information. As in Malaya, loudhailers in long boats, voice flights over guerrilla bases and leaflets were employed to encourage the insurgents to surrender. Some resettlement took place. Entire villages were relocated by helicopter away from the border areas where they were vulnerable to raids.[67]

A network of forts garrisoned either by platoons or companies was set up to provide bases for patrols to dominate the jungle. Units stayed in particular sectors for a lengthy duration so they would understand both the terrain and the local people well. Constant small patrols operated from company bases, which were usually positioned in or near *kampongs* (villages) to prevent intimidation and extend the government's influence. They dominated the local district, observing enemy activity, setting up ambushes and cutting off infiltration routes. The defences of the bases were built up gradually. Most were provided with a small airstrip and helicopter pad, allowing troops to operate in the jungle. The British emphasised small-scale operations in which most of the fighting was at platoon and section level. Such operations required excellent junior leadership, battle drill, training and physical fitness of the troops. As in Malaya and Kenya command was decentralised to the junior leadership. They participated in the constant and active patrolling of the jungle, which maintained the initiative and hunted down the elusive insurgents. The jungles of Burma and Malaya had provided the British Army, notably the Gurkhas, with plenty of experience of jungle warfare. Learning from past experiences the British avoided indiscriminate use of superior firepower. They employed limited air strikes and artillery support. This ensured the safety of civilian life and property. Walker insisted that the tactics should be to 'clear, hold and dominate' rather than to 'search and destroy' as employed by the Americans in Vietnam.[68]

By the end of the Confrontation indigenous, Indian and British units provided a defensive shield very similar to the one that had protected the North-West Frontier of India, dealing with a range of conflicts that stretched from banditry to a full-scale war. Having been revived (though playing a relatively small part) during the Malayan Emergency, the SAS arrived in North Borneo in January 1963, three months prior to the Tebedu raid in April 1963, which signalled the commencement of Confrontation. Initially envisaged as a reserve to act as a 'fire brigade' to deal with incursions, the SAS instead were employed as a 'border watch', acting as Walker's 'eyes and ears'. They patrolled the key areas permanently, mapping the frontier, living among the indigenous population and winning hearts and minds with medical care. They worked with the security forces in the villages along the border, notably the Gurkha 'border watch patrols' and the Border Scouts, to provide the intelligence that was vital in order for the platoons and companies operating from the permanent bases to defeat the Indonesian raids. The successful campaign in Borneo was made possible by the employment of helicopters, which played a central role for the first time, although they had been deployed in the later phases of the Malayan Emergency and in Cyprus. Helicopters provided security forces with the speed, mobility and flexibility with which to cut off and ambush Indonesian incursions. This reassured the local population that the security forces could provide a quick and effective response.[69]

Needing to seize the initiative and end Confrontation, the British instituted Operation Claret. This allowed clandestine incursions by the SAS and Gurkhas up to a depth of 10,000 yards across the border into Indonesian Kalimantan, which took the battle to the Indonesians from August 1964. These platoon, company or battalion level operations, which ambushed and harassed the Indonesians, were linked closely to negotiations between Kuala Lumpur and Jakarta. They were very carefully planned and implemented in order not only to ensure success and drive the Indonesians away from the border, but also to maintain the secrecy of the operations. This made it difficult for the Indonesians to carry out offensive actions across the frontier and imposed unacceptable losses on them, while allowing them to negotiate politically without 'losing face'. They required a high level of junior leadership and small unit tactics by the security forces. They also needed good intelligence, including signals intelligence (SIGINT), and air support, notably the use of helicopters. The remoteness of Borneo and its undeveloped state meant that the Confrontation was not 'a television war' and that the security forces were able to keep strict control over the media and maintain security. Unlike Palestine, Cyprus, Aden and Northern Ireland, the campaign was not the centre of media attention. This spared the British unwelcome coverage of the more draconian counterinsurgency techniques, notably detention without trial. A well-run government and police force supported by a mainly loyal population guaranteed internal security. This allowed the security forces to concentrate on defeating the rebel raiders. Such features are rarely fully replicated in modern campaigns. From 1964, Britain and Malaysia, aided by America, also covertly supported rebel movements in Indonesia itself, notably the Celebes, with limited training and weapons. They sought to undermine the Indonesian campaign in Borneo by overthrowing Sukarno.[70]

Meanwhile, the British and Malaysian governments won the campaign to win hearts and minds on the international front, courting the support of the United Nations, United States and South-East Asian countries who had a stake in the conflict. This meant that the inclusion of Sabah and Sarawak in the Federation of Malaysia received international recognition. Britain also avoided the accusation of colonialism. Crucially, in contrast to Palestine, the British were able to obtain American support, portraying their campaign, like that in Malaya, as an anti-communist crusade to keep Malaysia and the Straits of Malacca in pro-Western hands. This, together with the low casualties and cost, ensured that the campaign had bipartisan support in Britain, in contrast to the acerbic division over South Arabia. The Labour victory in the 1964 election brought no radical change in policy. Malaysia was able to portray Indonesia as the aggressor and pro-communist. This, aided by Sukarno's withdrawal from the UN in January 1965 and alignment with communist China, persuaded the United States to withdraw financial aid. This encouraged the anti-communist Indonesian Army under General Suharto, which colluded covertly with the Malaysians, to overthrow Sukarno, slaughtering some 500,000 communists, and begin the negotiations that ended Confrontation in August 1966.[71]

South Arabia, 1962–67

The Federation of South Arabia faced an insurgency by guerrillas, led by the National Liberation Front for Occupied South Yemen (NLF) formed in June 1963. NLF enjoyed support internally from the population and externally from the republican government of Yemen, which overthrew the royalist regime in September 1962. Using bases in the Yemen the rebels employed a combination of classic guerrilla tactics in the protectorates and urban terrorism in Aden to defeat the British. As in Borneo, it was impossible to close the border with Yemen and prevent dissident tribesmen being supplied with arms and advisers. Granted independence in February 1959 the Federation was perceived to be a puppet state of the British, who retained considerable powers. These included the right to issue compulsory 'guidance' to the Federal authorities and sovereignty over the airfield and the military base in Aden. Having replaced Cyprus as GHQ, Middle East Land Forces, Aden was of great strategic importance to Britain, providing with Singapore (GHQ, Far East Land Forces) its global presence. Aden was also the site of a large refinery. The 'Aden Group' of hawkish Conservative ministers, direct descendants of the 'Suez Group', supported the defence of British interests in South Arabia just as they had previously pushed for the retention of Cyprus. This prevented negotiations until the position had become untenable, resulting in an undignified withdrawal. Whereas the successful war in Borneo received little attention, a very difficult political situation and hostile media coverage made the contemporaneous counterinsurgency campaign in South Arabia a very different proposition. Egypt and the Soviet Union provided support for the insurgents in a clash between Arab nationalism (and anti-colonialism) and British imperialism.[72]

The lack of unity of command was a major problem that plagued the campaign. There were too many civilian and military organisations involved, notably the Colonial Office, the Foreign Office, the Aden State, the Federal Government, Aden Command, Middle East Land Forces and Middle East Command. Rivalry between the Federal National Guard (FNG), controlled by the local emirs of the seventeen states that made up the Federation, and the Federal Regular

Army (FRA), formed from the old protectorate levies raised by the British and also known as the South Arabian Army (SAA), made co-ordination of operations problematic. Another major problem was the reliability of the SAA. It proved to be both professionally inept and untrustworthy owing to the infiltration of its ranks, who mutinied and deserted to the NLF on the eve of the British withdrawal. This was in marked contrast to Malaya and Kenya where the employment of local troops had been a key component of the successful counterinsurgency campaign. Furthermore, although Brigadier James Lunt, as commander of the FRA, was answerable to the Arab Federal Minister of Defence, he was also subordinate to the Commander, Middle East Land Forces in Aden as a seconded British officer. The instructions and intentions of these two superiors were often contradictory. No satisfactory solution was found and unity of command and civil–military co-operation, which was crucial to success in Malaya, Kenya and Borneo, was never achieved. Indeed, it was not until early 1965 that a committee system was set up to provide an effective civil–military structure, bring together the key military, police and intelligence personnel at the different levels of the hierarchy, and oversee internal security. As in Palestine earlier and Northern Ireland later a director of operations, who could provide unity of command and the 'drive' to overcome bureaucratic inertia, was not appointed because of the political sensibilities of the Federal and Aden authorities. As in Borneo, Kenya and Malaya, Aden was divided into battalion areas but there were never enough troops on the ground to secure the large and lawless protectorates and the British resorted to the use of air power as a cheap but indiscriminate alternative that provided the NLF with recruits.[73]

After the Federation invoked its defence treaty with Britain, between January and October 1964 'Radforce' moved into the Radfan, the difficult mountainous region near Aden, to suppress the Quteibi, the main tribe in the area. The tactics were basically those of the colonial period since, although helicopters enhanced the mobility of the security forces, which also enjoyed air support, there were never enough in theatre. 'Proscribed areas' were created and the local population were prevented from returning to their land by patrols and air attacks until the recalcitrant tribesmen came to terms with the government. Collective punishments on the colonial model were also employed. The Army resorted to the destruction of property, crops and livestock to punish and terrorise the guerrillas and their supporters into submission. Civilian casualties mounted in spite of attempts to evacuate women and children and the provision of medical and food supplies. Reliant on firepower and air power, Radforce was not mobile enough to catch the elusive insurgents, who used guerrilla tactics in difficult terrain and crossed the porous border from safe havens in the Yemen. Instead of suppressing a tribal rebellion the British merely added fuel to a revolutionary war fought by well-armed insurgents.[74]

Although militarily successful, politically the campaign in the Radfan was a disaster. It resulted in intense media coverage. In contrast to Oman later, the media were poorly managed and there was also little effort to exploit the 'godless' Marxism of the insurgents. At this stage soldiers were still naive about the increasing power of the media and changed attitudes towards Empire. Although the casualties were small, Britain's international standing was damaged by the image of a colonial power and its federal puppets suppressing a nationalist rebellion and bombing villages. Britain was condemned in the United Nations and exhorted to cease its repression and bombing of villages. The British lost the propaganda war with Radio Cairo (later 'Voice of the Arabs'). The situation was made worse by the failure of the British to conduct a development programme to win over the population. The lack of response to the suggestion by Brigadier Lunt that some wells be dug on the Radfan to achieve the support of the local tribal population was indicative of the neglect of development.[75]

The Radfan operations also stirred up trouble in Aden, where ethnic divisions and the lack of political rights provided widespread support for the NLF. They switched their attention in 1964 from the Radfan to Aden itself, confronting the British with their first urban insurgency since Cyprus. The insurgents employed hit-and-run tactics, including assassinations, grenade attacks, snipers and sabotage. As in Nicosia in Cyprus, they struck and then disappeared into the crowded streets, bazaars, cafes and mosques. On the defensive, the security forces could not protect either the pro-government civilian population from intimidation or the many vulnerable military targets. Handicapped by the absence of a unified command and policies to win the loyalty of the population, they suffered additionally from the scrutiny of the media. This produced not only adverse publicity but also made it difficult for the government to deport or intern migrant Yemenis who were radicalised and recruited by the insurgents. Instead of applying the pseudo-gang technique, 'turning' captured guerrillas against their erstwhile colleagues, the SAS employed counter-terrorism or 'hunter-killer' tactics. Dressed as Arabs they snatched suspects to interrogate or kill, but this did not stem the tide. The British also provided covert support for Royalists and tribal dissidents who opposed the Egyptian presence in Yemen. Corrupt and implicated in arms smuggling, the police, like the SAA, were badly infiltrated by both the NLF and its rival, the Front for the Liberation of Occupied Southern Yemen (FLOSY), and could not be trusted. The armed police mutinied in June 1967.[76]

In the final analysis, as in Ireland, Palestine and Cyprus, the British lost the population's support and thus the ability to gather the intelligence on the organisation and intentions of the insurgents required to beat the insurgency. Lacking Arab speakers and a clear organisational structure (there were more than ten separate intelligence services in South Arabia) and bedevilled by inter-agency rivalry (a Director of Intelligence was not appointed until July 1966, which was far too late), the British exacerbated the problem. The cell structure adopted by the terrorists made it very difficult to infiltrate their groups and to obtain intelligence about their activities, although 'Q Squads' as in Palestine and Cyprus 'turned' captured rebels. Moreover, between 1964 and 1966 the NLF deliberately eliminated the local, Arab-speaking section of the Special Branch through a series of assassinations in which sixteen were killed. As in Cyprus, even when improved the intelligence organisation could not function without the co-operation of the local population, which was either apathetic or hostile towards the Federation. Without either an effective intelligence-gathering capability or the co-operation from the local population, operational intelligence 'dried up'. This left the British unable to exploit divisions between NLF and FLOSY and the various tribes of the Yemen and without an effective counterinsurgency. The security forces were forced to rely on routine patrols and searches, which yielded little information, and the internment and interrogation of suspects without trial. Racist abuse and brutality by the troops alienated hearts and minds, as did brutality and torture, including electric shocks and sleep deprivation, at the Aden Interrogation Centre (AIC) and the detention centres in and around Aden, pioneering the interrogation techniques that became controversial later in Northern Ireland. Adverse publicity culminated in a demand by the UN General Assembly in December 1966 for the release of political detainees.[77]

Aden became 'the nadir of British COIN'.[78] The Labour Party, which was in opposition, echoed such criticisms at home, ensuring that the campaign in South Arabia lacked bipartisan support. The writing was on the wall once Labour, which had never been happy with the creation of the Federation and the South Arabia campaign, assumed power in October 1964. This meant that the security forces lacked the time and support to develop a campaign to defeat the guerrillas. It also proved difficult to win hearts and minds when Britain's commitment to South Arabia was doubtful. Following a Defence Review, the Labour government announced in February 1966 that Britain would neither retain the Aden bases beyond independence, which would be achieved by 1968, nor renew the defence treaty with the Federation. In June 1967, British troops were withdrawn from the protectorates, which were overrun and replaced by a Marxist regime. This lead in turn to a withdrawal from Aden in October 1967. This was the most humiliating defeat since the withdrawal from Palestine almost twenty years before. The British had failed to secure a friendly government in power. This allowed Soviet and Chinese fleets to use Aden as a naval base in the Gulf and provide support for the insurgency against the Sultan of Oman in neighbouring Dhofar. As in Palestine and Cyprus, the counterinsurgency methods developed in Malaya, Kenya and Borneo were of limited

value in a less favourable environment. Experiences during the 1970s and 1980s
would reinforce this lesson.[79]

Oman and Dhofar, 1967–75

The war in the Dhofar, a province of Oman, was waged by the Sultan's Armed
Forces (SAF) supported by British officers and the SAS, who had participated in
the Jebel Akhdar campaign of 1958–59, operating as the British Army Training
Team against communist guerrillas based in South Yemen. The new regime in
South Yemen, following the successful campaign against the British in South
Arabia, provided Marxist ideology and Chinese and Soviet arms to the rebels.
The Jebalis, who were ethnically different from the populations of Muscat and
Oman, had revolted against the reactionary Sultan of Oman, who opposed
modernisation. The insurgency of the People's Front for the Liberation of the
Occupied Arabian Gulf (PFLOAG) made rapid progress and, by 1970, the SAF
was losing the war. Controlling only the coastal towns of Dhofar it made only
brief forays into the interior, which was dominated by the Jebel plateau, to
enforce collective punishments and burn villages.[80]

On 23 July 1970 the political situation was dramatically changed by the
coup of the Sultan's son, Qaboos bin Said, which was probably instigated by
the British. With Sultan Qaboos, who had trained at Sandhurst, providing the
necessary political leadership, the SAF, led by and advised by the British, now
undertook a classic counterinsurgency campaign based on the lessons of the
Malayan Emergency. The emphasis was on winning the hearts and minds of
the dissatisfied population on the Jebel. The tide was gradually turned. The
new era of mass communications meant that the propaganda war was waged
on an international front and not just domestically. Oman gained recognition
by joining the United Nations and the newly formed Arab League. This meant
that it was difficult for the rebels to label the new Sultan as a British puppet.
Instead, the conflict was portrayed as a war against communism, gaining the
support of both Jordan and Iran who sent forces to support Oman. Although
the SAS is often credited with the successful campaign in Oman, large numbers of Omanis, Baluchis, Jordanians and Iranians did most of the fighting.
'Omanisation' within the SAF was also introduced.[81]

Forming the Dhofar Brigade Headquarters at RAF Salalah and a joint
civilian–military committee, the Dhofar Development Committee, chaired
by the Wali of Dhofar, the British and Omanis created an infrastructure to
co-ordinate the winning of hearts and minds, intelligence collection and
operations into one coherent campaign. The undeveloped nature of Dhofar
precluded the elaborate committee system that had been employed in Malaya.

In Dhofar the British lacked district officers, who had been central to the counterinsurgency campaign in Kenya and Malaya. Many of the functions carried out by the District Executive Committees in Malaya were undertaken in Dhofar by The Civil Action Teams (CATs) in a greatly simplified system. The decisions were taken by a small group of people, notably by the Sultan and the commanders of the SAF and the Dhofar Brigade, rather than by ponderous committees. This less formal structure was made possible because Dhofar was much smaller and less developed than Malaya.[82]

The Sultan launched a holy war against what he condemned as godless communists, who had by 1970 created widespread disaffection by their brutal methods in imposing their Marxist ideology and suppressing both Islam and local tribal customs. A split between the communists and Islamic traditionalists led to infighting and mass defections by insurgents to the security forces. A propaganda war employing leaflets and Radio Oman turned the ideology of the insurgents against them. Amnesties and cash incentives also induced many disgruntled guerrillas to surrender. The greatest effort was concentrated on winning hearts and minds. Conditions for local inhabitants were improved using the slogan 'Securing Dhofar for civil development'. CATs were established to develop shops, schools, medical clinics and mosques, and dig wells once the engineers had provided roads. The creation of the *Firquat Salahadin* (Home Guard), formed out of loyal Jebalis and 'Surrendered Enemy Personnel' (SEPs) and supported by SAS 'advisers', provided security and a permanent government presence. In contrast to the indifference in the Radfan, the Army's Chief Engineer, Richard Clutterbuck, a counterinsurgency expert, dug wells enthusiastically in the Dhofar, which became central to 'civic action'. SEPs, the *Firqat* and CATs won the support of the local population, gathering valuable intelligence on *adoo* (enemy) objectives, positions and tactics. The 'carrot' was backed up by the 'stick'. This included an economic blockade, artillery and air strikes against the civilian population in areas controlled by the PFLOAG, and the widespread destruction of crops, food stores, wells and villages that sustained the revolt.[83]

Engineers constructed a series of fortified lines (the Leopard, Hornbeam, Hammer and Damavand Lines) between 1971 and 1974 along the South Yemen border. Similar to the blockhouses employed during the Boer War in South Africa and the defences built by the French along the frontier during the war in Algeria, they cut off the movement of supplies and insurgents. These lines systematically secured the Jebel for development and forced the insurgents into the open. They helped to regain the initiative but did, however, tie down a large portion of the limited manpower. Between December 1974 and March 1975 the rebels were destroyed as a military force, making the task easier by abandoning guerrilla tactics and opposing the offensives by Omani forces and their allies conventionally. The Sultan of Oman's Air Force (SOAF) and SAF forces attacked the large concentrations of guerrillas with their superior firepower in a series of clearing operations, inexorably eliminating any resistance. Patrolling their tribal areas from CAT bases, the *Firqat* dealt with any small parties of *adoo*, hunting down the rebels. With supplies from South Yemen cut off it was possible to clear the insurgents gradually from the Jebel and extend the development to the liberated areas. The establishment of permanent bases on the Jebel was essential to break the rebel hold on the population. Raids across the border into South Yemen by Mahra tribesmen, who had been financed and supplied with arms by MI6 and trained by the SAS, similar to 'Claret' operations in Borneo, harassed the insurgents in their havens and disrupted their logistics. As in Borneo and the Radfan, helicopters were very useful for moving supplies and troops.[84]

Like Malaya and Borneo the Dhofar was a 'model' campaign. It supplanted the Malayan Emergency as a case study taught at the Junior Command and Staff Course. Outlining the main principles of a successful counterinsurgency, notably the employment of clear-and-hold operations and the combination of selective force with winning hearts and minds, it served as a pattern for future campaigns. These insurgencies took place in a confined and well-defined area and were overwhelmingly rural, allowing the security forces to deploy their firepower in the jungle and Jebel respectively. It also isolated the guerrillas and kept the war away from the scrutiny and glare of the media, in marked contrast to the acrimonious debate in Palestine and South Arabia. Moreover, the war was fought on a very small scale and in very favourable circumstances, employing overwhelmingly superior force and resources, which were paid for by the Sultanate's oil revenues, against the rebels. Ultimately, success was the result of the ability to develop an intelligence machinery. This allowed the round-up of insurgent supporters and provided timely operational and tactical intelligence, mainly from informers ('Freds') and SEPs alienated by the harsh methods of PFLOAG.[85]

Northern Ireland, 1969–97

The war in Northern Ireland was more complex because the province was part of the United Kingdom. The unrest began as a protest against institutionalised discrimination against Catholics, who were treated as second class citizens by a protestant state, but developed into a Catholic working-class uprising. When the demonstrations turned violent during rioting in the summer of 1969 the Stormont government asked on 14 August for troops to support the police. Between 1969 and 1972 the British Army was sucked into a contest, for which it was unprepared both in terms of riot control and understanding of the conflict. As in Palestine, the British government faced the dilemma of choosing between two hostile groups. It wished to placate the Catholic minority, which had been politicised by the events of 1969, with reforms, but also to keep rule from Stormont and Unionist support, fearing 'the protestant backlash'. The Army's traditional methods of dealing with rebellions and terrorism in the colonies proved counterproductive, aiding the emergence of paramilitaries. Welcomed at first by both Catholics and protestants, during 1970 the troops alienated both communities with heavy-handed methods, persuading many Catholics to join to the Provisional Irish Republican Army (PIRA). Lacking numbers and good intelligence, the Army implemented inappropriate lessons from Palestine, Cyprus and Aden. These included use of curfews, internment, brutality and torture during interrogation, house and vehicle searches, raids, shootings and large-scale cordon-and-search operations. They further alienated a sullen Catholic population, which established 'no-go areas' in Belfast and Londonderry.[86]

This alienation was confirmed by subsequent events. The most notable of these was the introduction on 9 August 1971 of internment, which included the use of 'robust' interrogation techniques on suspected PIRA members and supporters, and 'Bloody Sunday' on 30 January 1972 when the 1st Battalion, Parachute Regiment shot dead thirteen Catholic demonstrators during a civil rights march in Londonderry. These were disastrous. They further disaffected the Catholic minority and led to widespread international condemnation of Britain. The introduction of direct rule from London on 24 March 1972 further fuelled the insurgency. The deployment of the SAS in a 'hunter-killer' role similar to that tried unsuccessfully in Aden led to further public relations disasters when three of the ten people killed during the first two years of a regular SAS presence in the province had no connection with a terrorist organisation. Allegations of a 'shoot-to-kill' policy by the security forces, linked to incidents such as the killing of insurgents by the SAS at Loughgall in May 1987 and Gibraltar in March 1988, were controversial. They supplied fodder for PIRA's propaganda machine on both the domestic and international fronts, and helped with recruiting. Thus, during the 1970s and 1980s, successful infiltration of PIRA by informers was counterbalanced by accusations of a covert 'dirty war' against its supporters in collusion with loyalist 'death squads', which fanned the flames of the insurgency.[87]

The counterinsurgency techniques, which had been so successful in colonies such as Malaya and Kenya, were disastrous in Northern Ireland. They allowed PIRA to build up support within the Catholic community. Normal civil government during the 1970s was ineffective, failing to win the hearts and minds of the Catholic minority and losing the competition in government that occurs between the insurgents and counterinsurgents as they battle for the support of the population, which is the essence of counterinsurgency. With the aim of forcing the British out of Northern Ireland, PIRA emulated the tactics used in Palestine, Aden and Cyprus by groups such as *Irgun* and EOKA. A containment strategy to reduce the conflict with the PIRA to an 'acceptable level' was gradually developed during the late 1970s, which eventually allowed the British to achieve a settlement. In July 1972 Operation Motorman employed 28,000 troops and 5,300 Ulster Defence Regiment (UDR) in the biggest British operation since the Suez Crisis of 1956 and the largest deployment of infantry since the Second World War. Guided by improved intelligence gathering, this provided a significant success, clearing the 'no-go areas' in Belfast and Londonderry in a display of overwhelming but reasonable force. The result was a loss of military momentum for the IRA, breaking up its organisation in Belfast and Londonderry. This severely undermined the PIRA's operational capacities and led to a lowering of fatalities, which never again reached the levels of 1972. The IRA was forced to adopt a 'long war' strategy in the mid-1970s, resulting in a stalemate that lasted two decades.[88]

Co-operation between the Army and the police began to improve, leading to the introduction of sounder counterinsurgency tactics. Identical boundaries for the operational areas of the RUC and the Army were established, enhancing liaison at all levels with Special Branch and intelligence gathering. Networks of agents and informers and covert surveillance of PIRA activity run by specialist units such as 14 Intelligence Company and E4A provided the main successes against the paramilitaries and severely hampered their activities and their ability to function. Nevertheless, the security forces faced a long battle with PIRA for intelligence. The cell structure of PIRA made it difficult to infiltrate or gather information about their activities, particularly in South Armagh. It proved impossible to seal the ill-defined, porous frontier with the Irish Republic, which was leaking like a sieve, even with the co-operation of the Irish Army and police force. In 1978 two Tasking and Co-ordination Groups (TCGs) were introduced to integrate intelligence gathering and dissemination by the CID, Special Branch and Army. 'Ulsterisation', the greater involvement of local, indigenous forces, was hampered by the overwhelming protestant composition of the Royal Ulster Constabulary (RUC) and the Ulster Defence Regiment (UDR). This reinforced their image as defenders of the Protestant ascendancy. There were allegations of collusion with Loyalist murder gangs that killed Catholic civilians, notably the solicitor Pat Finucane.[89]

A series of security committees attended by military, RUC and Special Branch officers co-ordinated operations in Northern Ireland on the Malayan model at provincial, regional, divisional and sub-divisional level, gathering and disseminating intelligence. Organisational difficulties, notably the failure to make Special Branch, RUC and military boundaries conform before 1990, and the reluctance of the RUC and Special Branch to share intelligence with soldiers, who served only six-month tours, made the system less than satisfactory on the ground. As in South Arabia, political considerations, such as the desire to preserve an air of normality and deny the insurgents a propaganda coup, prevented the appointment of a director of operations to co-ordinate the activities of the security forces. Two training teams prepared and trained units for service in an urban environment prior to departure to Northern Ireland. On arrival in Northern Ireland, the troops were also trained in riot control, patrolling, cordon-and-search operations and street fighting.[90]

Supported by Irish–Americans and Libya, PIRA was never defeated outright. There was no military solution and after the initial disasters of the early 1970s the security forces' strategy was to contain PIRA until a political solution could be found. This involved close control and surveillance of the population by patrols and the construction of bases, communication and watch towers, including an extensive network of bases along the 300-mile border with the Irish Republic, and infiltration of PIRA. The security forces operated within self-imposed restraints having learned that policies such as internment, 'shoot-to-kill' and cross-border 'hot' pursuit, were counterproductive politically. They switched from overt, reactive operations to more covert, preventative operations. 'Success' in Northern Ireland was reliant on keeping violence at an acceptable level while the leadership of PIRA and the Unionists were persuaded to accept peace negotiations and power-sharing as the solution to the military stalemate. Simultaneously, money was pumped into economic development of Northern Ireland in order to alleviate poverty as one cause of Republican dissatisfaction and undermine PIRA and *Sinn Féin*. Although undefeated militarily, PIRA was constrained by the security forces' counter-terror operations and forced to abandon violence in order to achieve its political goals. It was increasingly difficult for PIRA to operate in Northern Ireland and its ability to strike at prestigious targets, such as members of the security forces, declined. Its operations were increasingly thwarted. The PIRA ceasefire between August 1994 and February 1996 paved the way for the Good Friday Agreement of April 1998 and the establishment of the devolved executive in November 1999, which allowed *Sinn Féin* to participate in Northern Ireland's government. Both the republicans and the British had been forced to compromise on their objectives in order to restore peace.[91]

Conclusion

As a result of its colonial history the British Army was ideally suited to waging counterinsurgency campaigns. For most Western armies large-scale

conventional warfare was the main *raison d'être* and counterinsurgency was
perceived as an unwelcome distraction from the real war. By contrast, for most of
its history, the British Army saw its main role as that of an imperial police force
protecting the Empire. The Army's highly decentralised command structure
and tactical flexibility made it very effective in developing counterinsurgency
techniques to fight guerrillas. From past experience, the British developed a
military and colonial infrastructure ideally suited to defeating insurgency, but
they relied on informal methods to preserve and disseminate these techniques
and principles. This risked the loss of expertise and experience if complacency
set in or the connection with the past was broken. As a result, many valuable
lessons had continually to be relearned while bad habits including the often-
disastrous misuse of ill-disciplined auxiliaries such as the 'Black and Tans', could
linger with surprising tenacity.[92]

Until the 1960s British counterinsurgency techniques were passed down
informally as traditional methods that had not been formalised into a written
doctrine. These were based on three broad principles of the use of 'minimum
force' (being highly selective in the use of force), close co-operation between
civil and military authorities, and tactical flexibility (notably decentralised, small
unit tactics, which delegated authority to relatively junior officers). Another
important characteristic of British counterinsurgency throughout the twentieth century was the deployment of units to specific areas for an extended
duration so that they could get to know the terrain and the local population.
During the Malayan Emergency this informal doctrine was codified and fully
developed for the first time, guiding British and Commonwealth methods
for many years. Thus, although some expertise and standard procedures were
imported from one counterinsurgency to the next, commanders on the ground
in Malaya, Kenya, Cyprus and Aden tended to develop their own strategies
and tactics without reference to previous campaigns, being influenced by
'local conditions' rather than a formulaic doctrinal response. Books such as
Defeating Communist Insurgency: The Lessons of Malaya and Vietnam, published by
Sir Robert Thompson in 1966 and *Counterinsurgency Campaigning* published by
Colonel Julian Paget in 1967 tended to rationalise and draw out general principles from the era of the British retreat from Empire. These had been codified
between 1949 and 1963 in various Army manuals, notably *Imperial Policing and
Duties in Aid of the Civil Power* (1949), *Keeping the Peace: Operations in support of
the Civil Power* (1957 and 1963), *Anti-terrorist Operations in Malaya* (1952, 1954,
1958) and *Handbook of Mau Mau Operations* (*c.*1954). They were written just
as this era was concluding, giving the impression that the British Army had
developed a doctrine for defeating insurgencies based on universally applicable techniques. The dissemination of ideas and lessons was often hindered by
the British Army's decentralised organisation and dislike of a formal doctrine.

Similarly, the Colonial Office failed to create and disseminate its own doctrine. Nevertheless, this meant that the British Army was flexible in its response in a way that many other armies, often focused on conventional warfare, were not. Thus, it was able to implement a successful counterinsurgency doctrine in Malaya, Kenya, Borneo and Oman where the circumstances fitted in with lessons from its previous colonial experience. When a political solution proved difficult to find, as in Palestine, Cyprus, Aden and Northern Ireland, the British were less successful.[93]

During the Cold War the British Army's main objective was to face a Soviet invasion of northern Germany. In the 1960s and 1970s British defence policy was focused upon the British Army of the Rhine (BAOR) and defence of NATO's Central Front, where the best equipment and the greatest concentration of British troops were deployed and trained almost exclusively for BAOR operations. 'Small wars' were now a distraction as the Army concentrated on its continental commitment, fighting a very specific, major war in northern Germany. The concept of manoeuvre warfare was formulated during the 1980s to conduct a corps level conventional air–land battle against the Soviets on the central plains of Europe. The first Army-wide operational doctrine since the Field Service Regulations of 1909 was published in 1989 just as the Cold War was ending. The concept of manoeuvre became institutionalised during the 1990s, remaining the central tenet for the Army ever since. Such methods were transferable to the Falklands in 1982 and to the Gulf in 1991 but required the troops, staff and command to make significant adjustments to doctrine and tactics to adapt from large to small-scale warfare when engaged in 'small wars' in Northern Ireland, Afghanistan and Iraq. The Cold War dominated planning and during the 1980s and 1990s the British Army focused on manoeuvre warfare while counterinsurgency became a fringe activity. Moreover, the British Army had little experience of fighting insurgencies as a result of the withdrawal from Empire, which had largely been completed by the end of the 1960s. In practice, it had lost much of its expertise in the practical application of key counterinsurgency principles, as events in Afghanistan and Iraq would confirm.[94]

7

DIRTY WARS
AFGHANISTAN AND IRAQ, 2003–12

Introduction

In both Iraq and Afghanistan there was no government in place that could ensure success by providing either a strong administration or a political settlement with the insurgents. Seen as occupying forces and handicapped by sanctuaries the insurgents enjoyed in Iran and Pakistan respectively, the American and British armies were also unable to compensate for the lack of a central bureaucracy. The British had been able to take this for granted in previous counterinsurgency campaigns from Malaya to Northern Ireland. This framework of intelligence-gathering, doctrine and bureaucratic structure was missing in both Basra and Helmand. In Malaya, Kenya or Northern Ireland the British constituted the government and, as a result, the Army had no need to undertake state-building. Moreover, in Iraq there were no serving officers who had either practical experience of serving in counterinsurgency campaigns such as Malaya, although many had first-hand experience of Northern Ireland during the 1980s and 1990s, or knowledge of the counterinsurgency theory required to implement a war-winning counterinsurgency strategy.[1]

The essential element in successful British counterinsurgency campaigns after 1945 had been civil-military co-operation, although its importance was rarely formally acknowledged. By the 1990s the colonial officials and district officers who had been the crucial factor in that success were missing, as were the structure and local expertise that they had provided. Instead of being the colonial masters with credibility and natural authority, the Coalition forces were intervening in countries where they were regarded as occupiers lacking either legitimacy or local knowledge. As a result, young officers in Iraq and Afghanistan would find themselves having to perform various administrative, economic and political duties, which had been previously performed by colonial and district officers. The British counterinsurgency manuals that appeared in 1977, 1995 and 2001 drew heavily on Malaya and other decolonisation experiences. The doctrine in them failed to recognise that these earlier successes had been achieved when operating in a colonial or a post-colonial environment that no longer existed. Such successful methods would not necessarily be applicable when transferred to a campaign where these conditions had ceased to exist. The civil administration that had formed a key factor in earlier successes had not been replaced effectively. The main principles of counter-insurgency had been set out but there was very little on recent developments, especially the challenges presented by new information technology and Islamic fundamentalism. Counterinsurgency had been neglected and not taught at the Staff College and, as a result, hard-fought lessons had been forgotten and much of the former British expertise lost.[2]

Historically, 'bribes' or so-called 'subsidies' had been employed widely by political officers on the North-West Frontier of India to make sure that tribes such as the Afridis complied with British interests and wishes. As the experts on the spot, the political officers knew the men with political power who were worth influencing. Such expertise and knowledge was largely missing within the security forces deployed in Afghanistan or Iraq. In Iraq and Afghanistan the cultures are very different from the West, so the issue of 'cultural knowledge' has become prominent. The necessity of understanding the population of a non-Western country that is supporting insurgents has once again become clear. Traditionally 'cultural knowledge' of the enemy was an asset the British enjoyed as colonial masters in colonies such as Malaya and Kenya and were able to draw upon administrators and soldiers who knew the local languages and customs. But the British Army in Afghanistan and Iraq no longer enjoyed access to such expertise. Shortages of Arab speakers and of good interpreters, who were under constant threat – more than sixty Iraqis employed by the British were killed – meant that attempts to identify the insurgents by searching houses and questioning suspects were likely to be counterproductive. These alienated the local population, especially as some units such as the Parachute Regiment had a poor reputation for winning hearts and minds. Recognising this problem the Americans set up 'Human Terrain System' teams to provide each brigade commander with the capability to react to these cultural factors.[3]

The classic counterinsurgency tactic is to separate the people from the guerrillas. The unpleasant reality that is rarely addressed is the problem of achieving this when the insurgent is able, as in Afghanistan, to control the population. Even in Northern Ireland, following the early disasters of the campaign, notably

'Bloody Sunday' and internment without trial, the British Army developed in familiar surroundings over the subsequent thirty years a very competent organisation to collect and analyse intelligence. This system was based on the support and local expertise of a very capable police force that evolved over a lengthy period and employed surveillance, central databases and urban warfare techniques. This operating method was not transferable to other theatres such as Iraq and Afghanistan, where the local security forces lacked sophisticated expertise and intelligence gathering was poor. The successful counterinsurgency campaigns such as Malaya and Northern Ireland, which were reliant on good intelligence, had taken a long time to set up.[4]

The training and weaponry for a conventional war against the Soviet Army on the North German plain were not appropriate in complex 'small wars' in Iraq and Afghanistan. Like the British Army, the US Army in leadership, organisation and doctrine was prepared for a high-intensity conventional war. Most of the American high command spent the first decade of the twenty-first century adapting to low-intensity non-conventional war against insurgents, having devoted their careers preparing for a mechanised war against the Soviets. As a result, at the turn of the century the US Army lacked an effective counterinsurgency strategy and arrived in theatre with little knowledge of fighting insurgents. Battalion officers in the field read classic counterinsurgency texts bought via the Internet as a substitute for an official doctrine. They tended to rely on conventional 'kinetic' methods, which were inflexible and took no account of local culture and sensibilities, thus perpetuating the insurgencies. In both Afghanistan and Iraq these failures were exacerbated by the shortages of personnel who understood local languages and culture, which hampered the successful collection and analysis of intelligence. There were also delays in developing indigenous security forces, whose loyalty and morale remained questionable. The Americans became entrapped in a coercive approach to counterinsurgency that exacerbated the cultural chasm and generated a vicious cycle of violence. By 2009–11, however, most of the generals had gained much experience from service in Iraq and Afghanistan and were implementing very different strategies.[5]

Iraq, 2003–09

On the capture of Baghdad in 1917, General Sir Stanley Maude proclaimed that the British came 'to liberate Iraqis from the evil rule of the Turks', but the collapse of the Ottoman Empire in 1918 had resulted in British occupation of Mesopotamia and their ruthless defeat of the 1919–23 uprising. This potent imperial legacy bred scepticism about American and British promises in 2003 that they had not come as invaders but as liberators. American support for Kurdish and Shia leaders fuelled Sunni suspicions that they were seeking the destruction of Iraq. Furthermore, when the British 7th Armoured Brigade first entered Basra in April 2003 it lacked the numbers to stop the looting by Iraqis of houses, shops, hospitals and schools as well as government buildings. The British occupation struggled to re-establish law and order and basic services. The forces had to rely on ad hoc planning while lacking adequate contacts to gauge local views and the possible consequences of their decisions. The collapse of the Ba'ath local administration and the dismissal of 16,000 police and many civil servants left a vacuum the British were unable to fill. Another problem was the detention in poor, over-crowded conditions of large numbers of Iraqi soldiers, militia members and looters. Their treatment, using techniques such as hooding, noise and sleep deprivation, stress positions and verbal abuse, was highly controversial not only in Iraq and Britain but also internationally.[6]

In contrast to the highly successful Allied planning for the occupation of Europe and of Germany (Operation Eclipse) in 1944–45, very little planning or preparation had been made for the occupation of Iraq (Phase IV) following the American and British 'intervention'. One of twenty-seven nationalities that formed the Multi-National Force – Iraq (MNF-I), Britain took control of Multi-National Division (South East) – MND(SE), which consisted of the four southern provinces of Iraq – Basra, Dhi Qar, Maysan and Muthanna – some 60,000 square miles or about a quarter of Iraq with a 600-mile border with Iran, Kuwait and Saudi Arabia. Badly overstretched, the Army lacked the resources, notably funds for reconstruction and manpower, to either control the deteriorating situation in Basra or fulfil promises of providing a better life for the Iraqi population. This undermined its credibility. Over the last half of 2003 the British forces dwindled in strength from 43,000 to 8,000 men as both the British government and the Army wished to vacate Iraq as soon as possible. Expecting to undertake peacekeeping duties rather than fight an insurgency, troops were spread thinly over the four provinces, occupying five major bases, a logistics base and the Airport in Basra. The brigade deployed in Basra was too small for its mission and lacked the combat power to take on the local militias.[7]

As the situation deteriorated in Basra, the British simply lacked the troops to launch clear-and-hold operations to drive the militia from the city, as a result of political constraints implemented by an unsympathetic government. It failed to provide the support and resources necessary to implement an effective counterinsurgency strategy and protect the population from intimidation. The civilian and military leadership in Whitehall failed to understand the political dynamics on the ground in Iraq, while insisting on running operations from London. Owing to the unpopularity of the war, the leadership were obsessed by short-term goals and maintaining the political and military relationship with the Americans. They were reluctant to commit the necessary resources and thus consequently damaged Anglo-American relations. By refusing to admit that there was an insurgency, to be realistic about what could be achieved, and to adopt the correct counterinsurgency strategy, they doomed their troops to fail.[8]

All army mechanised and armoured brigades were locked into a fixed cycle of six-monthly deployments to Iraq that, as part of Operation Telic, provided a new divisional headquarters to act as Headquarters MND(SE) and a new manoeuvre brigade. Each new deployment brought a new interpretation of the threat, the military problem faced and the response required. This resulted in much debate about the correct strategy and tactics, especially as the initial perception was that the campaign was one of reconstruction rather than counterinsurgency. These new units were also often ad hoc formations, in stark contrast to the Americans, who employed only fully formed and trained corps and divisional headquarters and moved to twelve-month and later eighteen-month tours. British arrogance about American operations in Iraq, while somewhat smugly undertaking 'peacekeeping' patrols in Basra in soft hats and unarmoured vehicles as they had done in Bosnia and Kosovo, created tensions with their American allies. However, while the Americans seemed perplexed by the tribal structure in Iraq, the British employed it to their advantage, having learned from their colonial experience the importance of adapting to indigenous cultures. The American troops complained of the lack of local expertise and understanding of local culture, which hampered their ability to operate.[9]

American leaders in Washington were as unprepared as their British counterparts in Whitehall. The brilliant conventional invasion of Iraq brought down the Ba'ath regime of Saddam Hussein, allowing a triumphant President George W. Bush to announce on 9 April 2003 that the mission had been accomplished. However, the occupation forces were never large enough either to secure the country or to contain the Sunni insurgency in central Iraq, which began unexpectedly in May 2003. The Sunnis were disgruntled by the invasion and the dismantling of the Ba'athist regime, which threatened their dominance of Iraq. They were further alienated by the subsequent occupation and by the abuse of suspected insurgents (notably at the Abu Ghraib Detention Facility), the clumsy and brutal American search-and-destroy sweeps, and American-led 'kinetic' offensives against rebel centres, notably Fallujah in April and November 2004. Growing sectarian violence brought Iraq to the brink of civil war by the end of 2006, derailing American plans for a quick disengagement. Law and order had collapsed and there was widespread destruction, looting, sectarian violence and terrorist attacks on the occupying forces. The failure to deal with the lawlessness and to restore essential services convinced many Iraqis that the occupiers could not or would not protect them. Waging the type of urban terror campaign seen previously in Cyprus, Ireland and Palestine, the insurgency was supported internally by the Sunni population and externally by some foreign fighters and, over a porous frontier, by Iran and Syria.[10]

By the autumn of 2003 the situation in southern Iraq had deteriorated badly. Without effective intelligence the British provoked local communities and failed to identify the real threat, the radical and well-armed Shi'a militias. These stepped into the power vacuum, eliminating the Ba'athist supporters of Saddam Hussein, journalists, policemen and politicians and supporting endemic corruption and crime. Crucially the British did not provide security for the Iraqi population of Basra, notably the secular middle-classes, against the religious fanatics of the Shi'a death squads. Often serving with the police, they undertook wholesale ethnic cleansing of Sunnis and Christians. During 2003–04 the British also failed to develop either reconstruction and development or a strategy to win the hearts and minds of the local population. They were unable to provide basics such as education, electricity, healthcare and water. This was epitomised by the failure to immunise the cattle of the Marsh Arabs against tuberculosis (TB) until the second half of 2008 although this group was crucial to sealing the border with Iran. Good relations would have increased significantly the likelihood of success for the British counterinsurgency campaign. Having sustained heavy casualties from direct confrontation with British troops, the insurgents resorted during 2004 to classic guerrilla hit-and-run tactics. After April 2004 the Shi'a militias in Basra and Al Amara, mortared bases, probed positions, restricted movement with improvised explosive devices (IEDs) and inflicted casualties. The British had lost control of Basra by the end of 2006. Claiming 'success' in southern Iraq the British transferred responsibility for maintaining security to the newly established Iraqi police and Army, and withdrew, leaving the city to the tender mercies of the militias. Unbeaten in battle but badly served by their own commanders and political leaders, the troops were subjected to a humiliating retreat, evacuating Basra Palace in September 2007 in a manner reminiscent of the retreat from Aden in November 1967.[11]

The original intentions of employing the Iraqi Army to help to rebuild Iraq had been overturned with disastrous consequences. The decision by the American-run administration in Iraq (the Coalition Provisional Authority, CPA, nicknamed 'Can't Provide Anything') to cleanse Iraq of all links to Saddam's regime was a disaster, alienating half the population. They purged the government of middle- and high-ranking members of the Ba'ath party, which had already been outlawed. This ignored the fact that most mid-level government officials and bureaucrats were compelled by Saddam to join. Moreover, the removal of the four highest Ba'ath cadres removed the most able of the government's bureaucrats, causing the collapse of the hitherto efficient local governance and complicating reconstruction. CPA also disbanded the Iraqi Army (CPA Directive Number 1 of May 2003), which was the country's second largest employer. The disbandment meant that many ex-soldiers were unemployed and without any incentive to support the new regime. The British acquiescence with CPA Directive Number 1 set back the reconstruction in Basra by months and meant they lost the support of influential Iraqis.[12]

The police in Iraq under Saddam Hussein had been notorious for brutality and corruption. American and British attempts to create a new police force showed the divisions between their prospective approaches to such matters. Whereas the Americans attempted to increase the numbers of police very quickly, sacrificing quality for quantity, the British tried to introduce the British model of community policing, building up links with the local population. But it was difficult to change the police culture quickly. Lacking indigenous support, the British were faced with endemic corruption and militia infiltration of the police, who were the source for much of the violence and linked to many of the sectarian deaths that occurred in Basra. Few Western reporters dared to enter Basra, afraid of the risks involved following the death of Steven Vincent, a freelance American journalist, who was kidnapped and killed. The British were also suffering casualties from bombs while on patrol in the city. However, they lacked the manpower, equipment, local knowledge of the political dynamic and intelligence network in Basra to gather the information required to locate and deal with sectarian 'death squads' who were killing Basra's Sunnis, a small community that fled the Shia terror.[13] The release in February 2006 by the *News of the World* of a video of British soldiers beating Iraqi rioters in Amarah, a Marsh Arab town, in 2004 enflamed the situation, resulting in protests in the streets and mortar and rocket attacks on British bases. The growing lawlessness engendered a response in which, as in Palestine, Cyprus, Aden and Northern Ireland before, prisoners and detainees were widely mistreated and tortured during interrogation. The most notorious example was the torture and death of Baha Musa while detained by the Queen's Lancashire Regiment when 'the gloves came off' following the death of an officer.[14]

The twin problems of brutality, notably the torture and murder of prisoners, by the Iraqi police and the intimidation and infiltration by the militias of the police were ignored as the British prepared to transfer to Afghanistan. Paramilitary death squads that supported sectarian violence speeded the descent into civil war between Sunni and Shia. A small force of British troops supported a small Iraqi border force, which attempted to intercept the money and weapons flowing into Iraq from Iran and Syria, giving the insurgents the ability to train, equip and reinforce their forces. Eventually the British formed Military Transition Teams (MiTTs) to train and mentor the Iraqi Army in 2005, as the Iraqi security forces were reformed and built up to replace the British troops while the pressure grew for a quick withdrawal from Iraq in order to redeploy to Afghanistan. The importance of this had been underestimated previously and the British had not followed the American example of employing embedded advisors, ignoring lessons from their past experience of training indigenous forces.[15]

As in Northern Ireland and Bosnia, the British brigades and battalions in Iraq and later in Afghanistan served six-month tours of duty. These were too short, giving commanders and troops little time to get to know the area of operations, and meant in the long run that the campaign lacked coherence. In under three years no fewer than eight generals served in command of British troops in southern Iraq. In Northern Ireland the duration of tours could be up to two years but some were only six months, while in Iraq the Americans insisted on deployments of one year. Two-year tours for each brigade, with its components rotating in a staggered fashion after a six-month period, would have brought an improvement but the British Army was too overstretched to be able to implement it. These short tours resulted in careerism as senior commanders, seeking to make their mark, wrote optimistic summaries at the end of their short six-month tour, which claimed that the situation had improved, rather than adopting a more realistic, long-term view.[16] The lessons from similar short tours of duty employed by the Americans in Vietnam had not been learned.

During 2005–06 there were not enough British troops on the ground to be effective as insurgent control of Basra grew. With a large 'tail' of logistic and administrative staff in their bases, the British lacked the troops on the ground to secure and hold any area of Basra for more than a few hours. Operation Motorman in Northern Ireland in July 1972 to remove the Provisional IRA's no-go areas in Belfast and Londonderry had required some 21,000 troops. The manpower required to employ the same secure-and-hold strategy in Basra or other areas of southern Iraq was simply not available. British manpower resources were stretched over two theatres – Iraq and Afghanistan, which were 1,000 miles apart. Once the British forces had restored order and completed quick impact projects in one area of Basra they moved on to another location. There were not enough reliable, local security forces to maintain any momentum. Under-resourced, lacking good intelligence and without support from the Iraqis, British operations merely stirred up the local militias while lacking the means to suppress them properly. The insurgents continued to attack British bases in the city with mortars and rockets. Basra had become so dangerous that any British civilians leaving the bases to pursue the reconstruction effort were liable to be ambushed, kidnapped or killed. The British PRT (Provisional Reconstruction Team) was evacuated to Kuwait. Any Iraqi civilians working for the British were also in grave danger from retribution by the insurgents.[17]

A small number of civilian administrators worked alongside the military as *de facto* governors of each province in southern Iraq. The civil servants required for such posts, blending an expertise in the languages and political dynamics of local tribes with the toughness to work in isolated and hazardous places, had not been produced by the British since the last days of the Empire. There was little co-ordination between the civilian and military staff on the ground with the result that priority tasks for reconstruction, such as the provision of water and electricity, were never carried out. Whereas British generals complained of the lack of civilian support, some American generals attempted to provide the systems themselves. General Peter Chiarelli (1st Cavalry Division in Baghdad, 2006–07) sent officers to Austin, Texas, to discover how the services of a town are provided and emphasised the importance of re-establishing essential services for the local population.[18]

When invading Iraq in 2003, the British Army was employing equipment and an organisation that had been created to defeat the Warsaw Pact in high-intensity conventional warfare of short duration on the plains of Northern Europe. In lengthy low-intensity operations against insurgents some rather different equipment and organisation were required. Until mid-2006, owing to the lack of resources provided by the government, the Army lacked the correct equipment and strategy to maintain the counterinsurgency campaign. This financial stringency condemned the troops in Iraq to impotence since, given the choice between winning the war in Iraq in the short-term and maintaining the Army's long-term capability for conventional war, the high command preferred to preserve the latter.[19]

Forced on to the defensive by IEDs, the Army retreated from contact with the local community, becoming a distant and isolated occupation force, besieged in their bases. Routine movement between bases was mainly by helicopter and any rare movement on the roads was in heavily armoured columns. Soldiers no longer patrolled on foot in berets and were more intent on defending themselves against constant attack than in protecting the local population from sectarian violence. The announcement of forthcoming British troop withdrawals in December 2005 encouraged the extremists of the militias in their violence and sounded the death knell for the moderates. The British troops were forced to rely on sallies by 'quick reaction forces' and the occasional large-scale raid, which lacked the intelligence to counter the 'hit-and-run' tactics of the insurgents, who remained well-hidden within the civilian population.[20]

Although some British commanders viewed American doctrine with a degree of smugness, the deteriorating situation in Basra and the escalating commitment to Helmand posed disquieting questions about the validity of British doctrine and whether it too should be re-evaluated. This doctrine had been updated by Brigadier Gavin Bulloch in 2000, drawing heavily on the Malayan Emergency to articulate a few clear principles. These included a clear political aim, an information campaign to win over the population and gather intelligence, and the primacy of the police over the military. But it provided little of substance on more modern challenges, such as handling of the media in a new era of information technology and the twenty-four hour cycle of news or dealing with insurgencies led by Islamic fundamentalists. British doctrine was out of date, being more suited to fighting in relatively uninhabited jungles during a withdrawal from Empire than to the kinetic, urban battles being fought in Iraq. Critics of British doctrine felt that the emphasis in the teaching at Sandhurst and the Staff College was on the former successful campaigns in Kenya, Malaya and Northern Ireland. The case studies that were employed stressed the winning of hearts and minds and omitted the more brutal aspects, such as the forced relocation of villages to internment camps and summary executions. This meant that the British lagged behind intellectually in addressing current counterinsurgency thinking. British pride in the strong legacy of counterinsurgency experience tended to prevent either internal reflection and self-criticism or recognition that adaptability to circumstances was required. The Army was gripped by an 'insular, conformist culture'.[21] Defenders of British strategy maintain that historical examples of British failures, as well as successes, in addition to regular reference to the brutal side of counterinsurgency as practised by British, French and American forces were also provided.

The importance of consulting local inhabitants as part of the planning and preparation for operations became apparent. One of the criticisms of British campaigns in Iraq and Afghanistan is that recognition of the importance of understanding the local, indigenous point of view was frequently ignored. 'Influence' tended to be a one-way street rather than a two-way exchange of information. Similarly, career-conscious officers were discouraged from questioning policy, keeping any doubts to themselves and becoming obsequious in their behaviour. Criticism was often stifled by the fear that speaking out would attract disapproval and be 'letting the side down' by spreading defeatism and undermining morale. One officer, Captain Leo Docherty, who spoke Arabic and Pashtu, was dismissed for making comparatively mild criticisms about his service in Afghanistan during 2006.[22] The British Army 'was inhibited by a resistance to external criticism, a sense of anti-intellectualism, conservatism in its approach to change'. It had failed to learn and adapt and needed to re-establish its position 'as a highly professional army'.[23]

Prior to 2003 the British Army portrayed itself as being especially successful in fighting small wars. It emphasised that this success was based on the ideas and lessons learned from an extensive colonial experience, the 'classic period' of counterinsurgency during Britain's withdrawal from empire in the 1940s and 1950s. This formed the basis of Britain's military reputation for having a distinctive approach to counterinsurgency operations. The campaigns in Iraq and Afghanistan were therefore initially perceived as being a chance for the British to demonstrate their skills. There was a tendency for all political and military problems to be viewed in terms of the experience gained in Malaya and Northern Ireland and to neglect the lessons of Iraq. One veteran British aid official, Sir Suma Chakrabati, believed that the constant emphasis on past experience, notably on Malaya, was irrelevant. Visitors to military headquarters in Iraq and Afghanistan would be given PowerPoint presentations by British Army officers seeking to relate the techniques employed in places such as Malaya, Kenya and Northern Ireland to their current operations. Indeed, such was the tendency for 'complacent' and even 'smug' senior British officers to lecture their American counterparts on the relevant lessons of Malaya and Northern Ireland that they were eventually forbidden even to mention either of these insurgencies in American headquarters. Lectured condescendingly on the pre-eminence of the British 'small wars' tradition, by 2007 America's senior commanders in Iraq were increasingly frustrated with their 'useless' British counterparts, believing that their supposed expertise was increasingly irrelevant. In the early years of the insurgency the British had felt rather superior to the inexperienced Americans, but fell silent once the 'surge' implemented by General David Petraeus radically transformed the situation.[24]

In mid-2004 the British made the decision to deploy from southern Iraq, resulting in the major transfer of troops to Afghanistan in 2006, where it was assumed operations were largely completed to southern Afghanistan, reinvigorating the campaign that had stalled during the invasion of Iraq. They viewed Afghanistan as an opportunity to restore their lost reputation as well as ending an unpopular commitment of troops before the British became involved in a protracted counterinsurgency campaign in Iraq. In February 2007, the Prime Minister, Tony Blair, announced a withdrawal of 3,000 British troops from a total of more than 7,000 in Iraq for redeployment to Afghanistan. The replacement of Tony Blair with Gordon Brown as Prime Minister escalated the British withdrawal. At the same time as Basra was handed over to the Iraqis in September 2007, the Americans began the 'surge', deploying an extra five brigades in Operation *Fardh al-Qanoon* (Establish the Law), to restore control in central and northern Iraq. This coincided with a movement known as *Sahweh* (the 'Awakening') in which local Sunnis turned on al-Qaeda, upset by their methods, notably the kidnap of local women and suicide bombings. The Americans encouraged the Sunnis to provide security for the local population, gaining time for an effective government to emerge. It was left to the Iraqis and Americans in Operation Charge of the Knights (*Saulat al-Fursan*) to clear Basra in March and April 2008 and to restore normality in an initiative taken by the Iraqi Prime Minister, Nouri al-Maliki. Handing over Basra airport to the Americans on 31 March 2009, the British left for Afghanistan hoping to retrieve a much diminished reputation in the eyes of their American and Iraqi allies following their perceived defeat in Basra. In a complete role reversal, it was now the Americans who were the experts in counterinsurgency and the British the students, although ironically in the 1980s the Americans, obsessed with 'manoeuvre warfare', believed that the British imperial tradition crippled its ability to master modern warfare.[25]

Confronted with heavy casualties inflicted by the Sunni insurgency in Iraq and by the summer of 2006 sectarian violence that resulted in heavy civilian deaths and was escalating into civil war, the Americans began to seek new solutions and methods. Until 2007 the Americans had relied on a highly 'kinetic' strategy and became trapped in a vicious cycle of violence. Seeing all Iraqis as a threat, they remained isolated on bases located away from the population and thus unable to identify potential collaborators and protect them from the insurgents. The British thought that their American allies were overly belligerent and reliant on technology while ignoring the sensitivities of Iraqi culture.[26] One US Marine officer agreed that the US military tended to 'reward obedience over creativity' and to 'look at the world through a Western cultural lens' that blinded it to 'the complexities of tribal hierarchies and loyalties'.[27] These faults were summed up by Brigadier Nigel Aylwin-Foster in late 2005 in a controversial article that stated American soldiers were 'incapable of effective counterinsurgency' and 'at times their cultural insensitivity, almost certainly inadvertent, arguably amounted to institutional racism'. He also argued that the American Army was over-centralised, conformist and inflexible.[28] Aylwin-Foster's article was written to provoke debate and changes in American strategy at the request of General Petraeus, who left Iraq in September 2005 to take up command at the Combined Arms Center, which co-ordinated the US Army's training and doctrine.[29]

Although British claims of excellence in counterinsurgency based on their superior experience undoubtedly annoyed some Americans, others saw the article, as Petraeus hoped, as a challenge. It dared them to learn from not only the British and French but also from their own mistakes in Vietnam and now Iraq. Petraeus undertook jointly with the Marine Corps, represented by Lieutenant General James N. Mattis, to produce a new counterinsurgency manual that would identify and codify best practice. This was co-authored by Dr Conrad Crane, Director of the US Army Military History Institute, and Lieutenant Colonel John Nagl, the influential author of the book, *Learning to Eat Soup with a Knife*, comparing British success in Malaya with the failures of the Americans in Vietnam. In February 2006 Petraeus gathered counterinsurgency experts for a workshop, which was attended by experts such as Nigel Aylwin-Foster, Montgomery McFate, David Kilcullen and Kalev Sepp. This manual, which reflected tried-and-tested British principles and the theories of Galula, was an entirely new departure for the Americans. It changed mindsets just as much as the direction of the Iraq campaign. It laid out a doctrine of protecting the population rather than targeting the insurgents, being heavily influenced by Galula. The mantra 'clear, hold, build' was the basis for this new doctrine, whose first draft was published in March 2006 and the final version, Field Manual 3-24, *Counterinsurgency*, in December 2006.[30]

By the autumn of 2006 defeat in Iraq was staring the Americans in the face. General George Carey, the American commander, had been unable to prevent the sectarian killings in Baghdad. Casey established a counterinsurgency warfare centre in Iraq modelled on the jungle warfare school set up by the British in Malaya in 1949. However, it failed to change American tactics, which continued to target insurgents rather than protecting the local community. They employed indiscriminate cordon-and-sweep operations that detained thousands of Iraqis in Abu Ghraib prison and swamped the interrogation system. American troops returned to their huge bases after each operation. This lost not only momentum but also the support of the local population, who were left unprotected against insurgents who *did* remain within the community. The American occupation was breaking all the basic tenets of a successful counterinsurgency. President George W. Bush was faced with two options. He could either withdraw from Iraq, as suggested by a non-partisan Iraq Study Group in a report to Congress in late 2006, or gamble on making a startling shift in strategy as indicated by the counterinsurgency doctrine newly issued by General Petraeus. Petraeus advocated moving the American troops out of their bases into the local community and increasing the number of American forces to deal with the insurgents.[31]

Some senior officers had adopted successful counterinsurgency tactics prior to the 'surge' in 2007, realising that success in establishing a permanent presence and starting reconstruction relied on winning over the local population and working in close co-operation with the Iraqi security forces. In Baghdad during 2004 Major General Peter Chiarelli (1st Cavalry Division) had trained and employed Iraqi security forces that were then integrated into all operations. The combined operations of Major General John Batiste (1st Infantry Division) with the Iraqis in Samarra during 2004 also served as a model for the rest of Iraq. In northern Iraq, Petraeus (101st Airborne Division) had worked with the local civilian population in Mosul during 2004–05. Brigadier General Herbert R. McMaster in Tal Afar during 2005–06 and Colonel Sean MacFarland in Ramadi during 2006–07 also demonstrated the value of clear-and-hold operations. In early 2004, critical of Army tactics, notably the employment of artillery and air strikes, the Marines eschewed the use of heavy weapons. Under Major General Mattis (1st Marine Division), a respected and perceptive intellectual, in May 2004 the Marines revived the Combined Action Platoon (CAP) that had been employed in Vietnam and lived among the local population. They also provided training for the newly raised Iraqi Civil Defense Corps, later the Iraqi National Guard. However, implemented on only a small scale and without the formal and standardised structure of the CAP platoons, this revival lacked the number of troops, language skills, trained Iraqi security forces to co-operate with and a lengthy occupation to be successful. Many CAP Marines were withdrawn to fight in the Second Battle of Fallujah in November 2004 and the scheme was abandoned. The special forces were also critical of the Army approach, emphasising that conventional troops should keep a lower profile, abandon their large, fortified bases to live among the population and make a greater effort to train the Iraqi security forces. As regards training, one problem was that, unlike the Marines, the Army did not give high priority to advisory duties or allocate 'high-flying' officers to advisory teams.[32]

During 2006 and 2007, the American military were quicker than the British to conclude that a change of direction was required. The 'surge' was designed by Petraeus to buy time for a political solution to be found, but he was given a very limited amount of troops and time to achieve results, given the lengthy duration of counterinsurgency campaigns and large numbers of troops usually required to achieve success. Somewhat fortuitously the American 'surge' coincided with a backlash by the Sunni tribes against the brutal Islamic fundamentalism of al-Qaeda and its foreign fighters and the resurgent Shias. Imitating Colonel McMaster in Tal Afar and Colonel MacFarland in Ramadi, who had recruited local tribes to fight the insurgents, the Americans were very successful during the 'Awakening' in Western Iraq in bribing tribal leaders to fight al-Qaeda. Less easy was winning over the leaders of the Sunni insurgency in the urban sprawl of Baghdad, where tribal bonds were weaker and it was harder to identify the insurgents. It was also difficult for General Graeme Lamb, Petraeus' British Deputy Commander, to persuade senior American commanders and the Iraqi government of Nouri al-Maliki to negotiate with insurgents, but this proved crucial during the 'surge'.[33]

Appointed as commander of the Multi-National Force to implement a new strategy from February 2007 with five brigades of additional American troops, Petraeus planned the 'surge' to protect the population from the insurgents and retrieve Iraq from the brink of civil war. The bulk of the Americans were transferred from their big bases into smaller outposts within the local communities. Responsibility was delegated from corps, division and brigade to battalions and to company officers and NCOs who lived in outposts (joint security stations) with indigenous security forces, maintaining daily contact with the local population. The key principle of this decentralisation, based on the ideas of David Galula (the French counterinsurgency expert), was to secure the support of the populace. Operating in small groups and located permanently in particular areas in partnership with Iraqi Army and police units, which understood the local terrain and communities, the Americans could interact with these communities. This allowed them to stabilise the situation and acquire accurate intelligence about the insurgents. American patrols collected more than a million fingerprints and retinal scans from Iraqis in order to provide a census of the local population and begin to identify insurgents. This enabled special forces teams to launch raids to capture or kill key insurgent leaders. Simultaneously, a series of conventional operations cleared insurgents from their strongholds, notably in Baghdad. Petraeus and his staff visited units, 'coaching' them as they adapted to their new deployment, environment and tactics. By July 2007 American losses had begun to decrease. In September 2007 Petraeus reported to Congress that sufficient progress had been made to allow the reduction of troops and a return to the levels available prior to the 'surge'. This coincided with the contrasting final and humiliating withdrawal of British troops from Basra, who never possessed either the political support or the necessary equipment and manpower resources to make an equivalent 'surge' feasible. Unlike Bush, neither Blair nor Brown had any appetite for making the investment in manpower and resources required to restore the deteriorating situation. Nevertheless, although the 'surge' brought Iraq back from the abyss of total collapse, it was merely the beginning of a long journey to achieving stability, especially as the invasion of Iraq had left a highly toxic legacy. A major criticism of the 'surge' was that it was not followed up with any political reforms to win hearts and minds but had merely established the sectarian Shiite government of Nouri al-Maliki in Baghdad while the civil war in the countryside continued.[34] The subsequent collapse of the Iraqi government following the withdrawal of US troops that began in December 2007 and ended in December 2011 have confirmed this analysis.

Afghanistan, 2000–12

In response to the attacks on the World Trade Center on 11 September 2001, the American government launched Operation Enduring Freedom, which brought down the Taliban regime in Afghanistan and its al-Qaeda allies during October and November and established Hamid Karzai's government in December 2001. Al-Qaeda's infrastructure had, however, not been destroyed and many Taliban and al-Qaeda had escaped across the porous border into neighbouring Pakistan to regroup. Little was done to establish Karzai's control as the invasion of Iraq meant that Afghanistan was starved of attention and resources, both civilian and military. A low-level insurgency developed during 2002–04 at a time when there was no Afghan Army or National Police. The Americans were slow to introduce a counterinsurgency strategy, concentrating on pursuing the remnants of al-Qaeda and Taliban rather than on nation building and developing Afghan security forces. Initially, there was no Afghan participation in the counterinsurgency strategy, which was being implemented by America and her NATO allies with little understanding of Afghan politics or society. In 2003 the Afghan Army launched some operations against the Taliban, which from 2005 began to be more aggressive. The Afghan government's fragile stability began to crumble as the insurgents increasingly dominated the rural areas. The Taliban employed classic guerrilla tactics while gaining outside support from Iran and employing sanctuaries in Pakistan to train recruits. The police were institutionally corrupt and ineffective (even selling arms to the Taliban), being badly trained and poorly equipped. Rotated every three months, they were unable to either establish law and order or protect local communities against the Taliban. The Afghan intelligence service (the National Directorate of Security) also had a bad reputation for the mistreatment of the population, beating and torturing suspects. The Afghan Army was able to co-operate competently in major counterinsurgency offensives in 2005–06 but lacked modern equipment. Owing to the gap between what British forces were resourced to do and what they were being asked to do (committed to a campaign in Iraq while embarking on another in Afghanistan), the forces deployed to Afghanistan lacked the resources to succeed, notably large numbers of troops to undertake the clear-and-hold operations that could reintroduce a government presence and regain control of the countryside. Instead, they employed ineffective and counterproductive search-and-destroy operations, alienating the population with heavy-handed searches of homes and round-ups of suspects.[35]

Helmand, the southern province of Afghanistan, was similar in size and population to Bosnia and Kosovo where the British had served during the 1990s. But whereas they had then been part of 60,000 'peacemakers' in a non-hostile environment in which no soldiers were killed, in 2006 they deployed only a very weak force (16 Air Assault Brigade) of 3,300 in the far more hostile environment of Helmand. In Malaya the British had deployed some 40,000 troops alongside 100,000 mainly reliable police and auxiliaries, whereas in Helmand the maximum total of British troops was 10,000 and the police were largely untrustworthy. Reinforcements arrived only intermittently. In contrast to Bosnia and Kosovo, there were few metalled roads that were easily blocked and long lines of communications. This policy of minimal manpower meant that, as in Iraq, the British lacked the troops and resources to control Helmand, causing friction with the Americans. They also lacked the financial assets to maintain the very well-funded American civil aid programme, which had been established in the province. Owing to a concentration of the majority of funds on strengthening the central government in Kabul, the British failed to provide sufficient reconstruction and development in Helmand and thus to win hearts and minds. The local government was corrupt and interlinked with the local drug barons. In contrast to the pragmatic American response, the

British pursued aggressive anti-drug policies that threatened the main income of many Afghan farmers and prevented strong support being built up among the local population.[36]

To some Pashtuns (the Pathans faced by the British Empire during the nineteenth century) the British presence (Operation Herrick IV) in Helmand, which was the 'Taliban heartland', simply repeated their past invasions of 1839–42, 1878–80 and 1919. The implications of this and Afghan memories of their victory at Maiwand in 1880 meant that, with hindsight, it was a 'strategic mistake' to use British troops among the Pashtun tribes of southern Afghanistan who looked upon the occupation as 'round four' in their long-term conflict with the infidel British. This legacy of the previous British involvement in Afghanistan was compounded by their support for an unpopular and dysfunctional government, whose administration and police had a reputation for corruption and incompetence. Karzai's government was estimated to be one of the most corrupt in the world (Afghanistan was ranked 172 out of 180 in July 2007). It was incapable of addressing the problem of an endemic culture of political corruption that pervaded the Afghan political system. Most of the government were from northern tribes, of which the Pashtuns were suspicious, especially of the Tajiks. Providing an effective, legitimate government presence in some 40,000 villages in Afghanistan was a major problem. Corruption and misrule by local officials and the weakness of central government that lacked legitimacy at village and district level allowed the insurgents, militias and warlords to step into the vacuum. The progress of the Operational Mentor and Liaison Team (OMLT) in improving the Afghan Army was very slow, owing to ill-discipline, poor logistics and inexperienced commanders. The quality of the training provided also needed to be improved. The mentors were sometimes poorly selected and there was no organisation either to train them or to provide a doctrine to guide them.[37]

Following a reconnaissance the SAS had advised that the small British force should not attempt to garrison the northern and southern zones but remain within the populated central area of Helmand to secure the local population's support. However, under pressure from the Afghan government and the need to justify themselves within a climate of inter-service rivalry, which exacerbated their own 'can-do' ethos, the Army undertook the disastrous strategy of creating 'platoon houses' in support of Afghan police and Army units. Having dispersed too widely, their forces were overextended across the province rather than being concentrated in key areas. This stirred up a hornets' nest and provided magnets for attacks by the Taliban, who provided unexpectedly fierce resistance and, seizing the initiative, forced the British to cling on grimly. The British were not only unable to contribute troops to Allied operations, upsetting their American colleagues, but also failed to secure and hold areas and thus retain control of the population. Far from protecting and separating the people from the Taliban these tactics had increased their support for them. Only a massive 'surge' by US Marines in 2010 finally turned the tide. Having joined the war for essentially political reasons, the British had become embroiled in a long counterinsurgency campaign for which they were unprepared. They then ignored the changed reality on the ground and its implications, refusing to rethink their strategy for two years.[38]

Moreover, the British employment of heavy weapons and 'kinetic' firepower, notably artillery and air strikes, to defend their dispersed garrisons created heavy civilian casualties from 'friendly fire' and displaced thousands of people in a way reminiscent of Vietnam. This revealed a total misunderstanding of the concepts of population control and security. The high level of 'collateral damage' among civilians in Helmand increasingly alienated the population, which was driven into the arms of the insurgents as the traditional British ethos of 'minimum force' was eroded. This increase in the level of counterproductive violence was in response to the perception of failure in Basra and the damage done to the British reputation in American eyes and the high levels of violence experienced in Afghanistan. Owing to the vast panoply of destructive weapons available to the troops, the philosophy of restraint that had been a feature of British military culture in successful counterinsurgency campaigns was felt by some critics to have been largely absent from Iraq and Afghanistan,[39] although there was more restraint by British forces than was often portrayed.

As in Iraq, the syndrome of successive brigades operating on short tours on the Vietnam model exacerbated the situation, ensuring that it was very difficult for any continuity of strategy or tactics to develop. Between 2006 and 2009 six British brigades served a six-month tour, producing a culture of short-term decision-making and institutional procrastination as commanders of units attempted to impress during their brief tour by undertaking inappropriate 'kinetic' operations. These high-profile operations imported from Iraq to Afghanistan attempted to fix and destroy the Taliban, looking for a high 'body count' even though this operational rationale should have been discredited by French and American experiences in Vietnam. While neither holding ground nor building up levels of security for the local population, such operations provided good material for the media war. Any opposition was seen as 'defeatism' in the 'culture of spin' that dominated Whitehall during the New Labour era. The British had little control outside their bases as these operations to 'clear' areas, which the troops called 'mowing the lawn' or 'mowing the grass', allowed the insurgents who had scattered into the hills to return to control and terrorise the population. The shortage of troops to hold ground meant that the policy persisted, resulting in a continued over-reliance on firepower and further civilian casualties. The situation was aggravated by the incoherent and divided chain of command in which Brigadier Ed Butler answered to President Karzai, the British ambassador (the senior British official in Afghanistan), the Permanent Joint Headquarters at Northwood and the International Security Assistance Force (ISAF) controlling the forces of the North Atlantic Treaty Organisation (NATO) based in Afghanistan.[40]

Thus, the campaign was undermined in a number of ways. The brutal and corrupt behaviour of the Afghan security forces together with the destruction wrought by the dispersed garrisons of 16 Air Assault Brigade reinforced the perception by the local population of the British as destructive invaders. The failure to provide security for the population created a security void in which most of the towns of Helmand were ruled by the Taliban, becoming no-go areas for the government. There was a lack of co-operation between civilians and the military who ran the British campaign in Helmand despite the rhetoric of a 'comprehensive approach'. The programme for development and reconstruction projects often lacked coherence and the lack of security meant that many remained unfulfilled. There was also criticism by the Army that funding for aid was spread across Afghanistan in support of development rather than being concentrated in Helmand in direct support of British operations to win hearts and minds. This was reflected in the failure to build and repair roads as the key to reconstruction and development in a country that had been suffering from war for some thirty years. Building roads allowed the security forces to regain the initiative. They not only allowed civilian agencies to provide rural development, bringing in economic and social benefits, but also forced the Taliban to attack to disrupt the road repair programme that they could not afford to ignore, fighting on ground chosen by the security forces.[41]

As in Iraq, the necessity of providing security for the local population had been fatally ignored by the British. Instead, the British concentrated on killing the Taliban, ignoring the real lessons of 'traditional' counterinsurgency during the eras of imperial occupation and decolonisation. There was no coherent strategy in Iraq and Afghanistan other than demonstrating to the Americans the British commitment to the 'special relationship'. These wars were waged partly to reaffirm Britain as a 'pivotal power' but instead the operational strain merely highlighted British limitations and the fragility of public support for such interventions. The difference between American and British approaches in Afghanistan is shown by the impact of the 1st Battalion, 5th Marines who replaced a besieged British garrison in the village of Nawa in June 2009. When the Americans arrived, no soldier left the base without being assaulted and there was no contact with or intelligence from the local population. However, after five months the town was once again thriving and IED attacks had fallen by 90 per cent as a result of a dynamic leader who established excellent relations with the local political elite and a new focus on understanding the local area.

By 'reinforcing failure' in Iraq and repeating these mistakes in Helmand, the British Army had by 2010 lost its reputation for being able to fight 'small wars' and win counterinsurgency campaigns, undermining the 'special relationship' with the Americans.

Ironically, by 2010 the British had begun to adapt and proof of their institutional learning was the strong emphasis on protecting the population and extending the Afghan government's authority shown by 6th Division in line with the American strategy. Previously, some brigade commanders, notably Brigadiers Andrew Mackay (52 Brigade) and James Chiswell (16 Air Assault Brigade) had placed greater importance on providing security for the local population than hunting down and engaging the Taliban, as 12 Mechanised Brigade attempted in 2007. But most other hard-driving and career-conscious brigade commanders who served before and after in Helmand did not as these methods were considered somewhat radical by their peers and not adopted widely. Nevertheless, from 2009, when the *Army Field Manual Volume 1, Part 10: Countering Insurgency*, which redefined its counterinsurgency doctrine, was published, British efforts to develop and implement appropriate tactics and cultural awareness meant that a repeat of the debacle in Basra was avoided. In April 2008, 16 Air Assault Brigade (Brigadier Mark Carleton-Smith) focused on protecting urban centres and developing governmental authority, deploying to the key towns. In Sangin, for example, great efforts were made from 2008 to win over the population and break the Taliban's link with the people by avoiding civilian casualties and working closely with the local government and Afghan troops. But, although improving, the British were unable to defeat the Taliban and it required the arrival of 20,000 US Marines in 2009 to turn the tide.[42]

Involvement in Afghanistan since the autumn of 2001 represented a huge chance that was squandered when resources and manpower were diverted for the invasion of Iraq in early 2003, leaving too few NATO forces to deal effectively with the return of al-Qaeda and the Taliban, who had re-emerged by 2006. In contrast to the Americans and British, who failed to develop a war-winning strategy, the Taliban adopted the tried-and-tested 'Fabian' strategy of outlasting their more powerful enemies. Abandoning human wave assaults and refusing direct engagements, they relied on hit-and-run tactics and IEDs to wear down their enemies. The insurgency spread to nearly every province of Afghanistan with assassinations, bombings and civilian deaths soaring. By 2007 the Taliban was once again a potent force in Afghanistan, using Pakistan as a safe haven. By 2008 the 33,000 Americans led by General David McKiernan in Afghanistan were still garrisoning heavily fortified bases in contrast to the tactics employed in Iraq where they had been moved from the bases into outposts to protect and secure the population. A counterinsurgency strategy had not been agreed with the Afghan government. Security was still deteriorating and the intervention force designed for conventional battle was stuck in a tribal quagmire. In 2009 the new ISAF Joint Command was established to bring Afghan, American and NATO forces under one unified command. Taking command in the autumn of 2009, General Stanley McChrystal began to implement the tactics developed in Iraq but lacked sufficient troops to take on the insurgents, close the border with Pakistan and secure any villages cleared during operations. When Petraeus was appointed to replace McChrystal in December 2009, President Obama announced two central tenets for the strategy, which he had to implement. An 'Afghan surge' of 30,000 additional American troops regained the initiative while a deadline, July 2011, was set for the reduction of the American presence in Afghanistan to begin.[43]

Arriving in Afghanistan in July 2010, Petraeus emphasised three key principles: the importance of civil-military co-operation; that the president had sent him to win the war, not to evacuate; and the transition of authority to the Afghans. Petraeus energised his command and his allies, setting a hectic schedule of meetings, briefings, appearances and visits to set out his strategy and to ensure that it was implemented by subordinate commanders whose actions would make the difference in ensuring either its success or failure. He sought to limit the use of kinetic firepower such as close air support, which had been relied on heavily during 2005–07 and produced unwanted civilian casualties. He also tried to establish a partnership with Afghan forces to reduce the damage caused by 'a misunderstanding or ignorance of local customs and behaviors'. In August 2010 Petraeus issued guidelines – the twenty-four commandments – to all forces in Afghanistan, complementing the *Counterinsurgency Manual* of 2006, known as 'King David's Bible', which was the first revision of the Army's counterinsurgency doctrine since Field Circular 100-20 of 1986. This document outlined the basics of Petraeus' counterinsurgency strategy. They were to secure and serve the local population; live among the people; pursue the enemy relentlessly; patrol on foot to learn about the local community; provide integrity and accurate information; treat the Afghans with respect and live up to the values that distinguished them from the enemy. These guidelines modified those issued by Petraeus in Iraq in 2007 to reflect conditions in Afghanistan. Whereas, according to Lieutenant Colonel John Nagl, most Army officers in 2003 knew more about the American Civil War than about counterinsurgency, many lessons had been learned by 2010 that had been disseminated and implemented between 2007 and 2008 during the Iraqi 'surge'.[44]

McChrystal had re-employed these methods in Afghanistan, identifying ninety districts (from 500) as key areas that had to be secured in order to defeat the Taliban. He focused on protecting the Afghan population before clearing the Taliban strongholds in the south-west, the south and east where the insurgency, funded by the illegal drugs industry, thrived in the mountainous terrain along the Pakistan border. With the insurgents largely removed from central Helmand province under McChrystal, Petraeus turned his attention to Kandahar province where the Taliban movement had first emerged and then to the mountains in the east along the border with Pakistan. For Petraeus the key to victory lay in providing security for the local population. He realised that their support was essential if the counterinsurgency campaign was to be successful. Some American units, notably the 10th Mountain Division in the Kunar Valley in 2005 and 2006, had already concentrated on providing security for the local population, rather than employing search-and-destroy operations in which security forces swept the countryside forlornly for the elusive enemy. This separated them from the insurgents, who were forced to attack. Similarly, from the summer of 2005, the 1st Brigade (82nd Airborne Division) in Kunar province attempted to build up partnerships with the Afghan and Pakistan security forces and focused its efforts on protecting the local population and implementing reconstruction of the infrastructure and roads.[45]

Afghanistan presented a more complex problem than Iraq, being larger in size and having harder terrain, although the level of violence was much lower. The rural insurgency in Afghanistan required a different approach to that employed against the urban insurgency in Iraq. Developing his 'clear, hold and build' strategy to drive the Taliban from their traditional strongholds in Helmand and Kandahar provinces, Petraeus employed night raids by Special Operations troops, which used good intelligence to target, capture or kill Taliban leaders. Taliban sanctuaries were cleared of insurgent networks by conventional forces and then held by local security forces whose task was to 'hold and build', building roads, medical facilities, schools, bazaars, houses and mosques, and supporting economic development. Petraeus sought to build up momentum by connecting areas that had been secured. 'Holding' operations to consolidate areas that had been cleared were the most challenging as they depended on effective local governance and security, which the Afghan security forces were unable to provide. Faced by a more effective opposition, many Taliban withdrew across the border into safe havens in Pakistan to fight another day. The ineffective Pakistani response prevented the Americans from destroying the defeated insurgents, who resorted to classic guerrilla tactics, launching hit-and-run attacks, laying IEDs and intimidating government 'loyalists'.[46]

The Americans made considerable efforts to build up the Afghan Army and police. The police, with a strength of 114,000, operated at the district and provincial level maintaining law and order. The Afghan Army grew from 97,000 in November 2009 when a NATO command was established under Lieutenant General William Caldwell, to more than 144,000 in October 2010 with the aim of reaching 164,000 by the autumn of 2011. It took time to train and equip both forces, which suffered from low literacy rates and high desertion and casualty rates. As well as being corrupt and ineffective, the Afghan police mistreated suspects and detainees, which was counterproductive, while the Army suffered from corruption, drugs, ethnic rivalry, shortages of management skills and poor leadership at all levels, which undermined its ability to shoulder the burden of fighting the Taliban. Criticisms of the number of US advisors provided and of their training suggest that, as in Vietnam, El Salvador and Iraq, the role of adviser was not one that attracted 'high-flyers' or was given a high priority in either the doctrine or organisational structure of the US Army. It relied on ad hoc improvisation rather than developing the formal institutional support that was required. There was also a chronic shortage of police trainers and mentors.[47]

In 2009, under McChrystal, Village Stability Operations had begun in which Special Force A-Teams of twelve soldiers lived among the local population in the villages. They learned their culture, provided medical care and clean water, and trained and equipped local villagers to defend themselves. It was hoped that the villagers' trust would eventually be won, securing areas that had been cleared by conventional Afghan, American and NATO forces and, when joined up with other areas, securing the countryside for the government. Petraeus extended this policy by creating a village police force, the Afghan Local Police (the latest and most developed version of ill-fated programmes to recruit indigenous auxiliary forces), which were trained by US Special Forces to undertake the critical role of protecting the villages. The village constabulary prevented the Taliban from re-entering villages that had been cleared and kept them from reasserting control while slowly expanding the influence of the security forces like inkblots. To do this Petraeus had to reassure Karzai, the Afghan President, who was concerned that these local police forces would quickly become militias out of his control. Moreover, the Taliban spring offensive of May 2011 showed that such gains were fragile. Attempts to provide community defence often floundered on poor training, lack of leadership and government support, the ineffectiveness of Afghan security forces and tribal feuds. Similarly, belated attempts to remove Taliban fighters from the battlefield and reintegrate them into local communities have struggled because of hasty implementation, an unclear amnesty policy and a lack of funding.[48]

There was a growing emphasis among Allied units of ISAF on 'waging influence', understanding Afghan culture and 'the dynamics of tribal and sub-tribes' in Afghanistan, which was 'one of the most challenging and complex COIN environments in history'. Such understanding helped troops to be more effective in working with the Afghan security forces, protecting the population, using local knowledge, building up and supporting the government and its development programmes, and neutralising the 'malign' influence of the insurgents.[49] Within the British Army several initiatives were employed in an attempt to generate 'Cultural Understanding', notably the use of military linguists as 'enablers' to establish good relationships with the leaders of the Afghan security forces. In 2007 an officer, trained to speak Pashtu, was appointed to act as liaison officer with the key Afghan leaders and as cultural adviser and influence officer to the newly created Battle Group (Centre South), providing information on the power dynamics in the district and liaising with local tribal leaders. Much of the cultural training and attempts to develop it were ad hoc, highlighting the importance (and previous neglect) of developing linguistic capability when training and educating the Army to conduct counterinsurgency campaigns.[50]

The 'surge' of 2011 in Afghanistan did not match the dramatic results achieved in Iraq during 2007 and 2008. The counterinsurgency campaign in Afghanistan was going to take much longer and require far more persistence. In regions such as Ghazni, Helmand and Kandahar, where the Taliban was deeply entrenched, there were not enough soldiers to protect the local population and develop the infrastructure, notably roads and schools. Incidents in which American air strikes caused civilian casualties continued to occur, leading the Karzai government to accuse the Americans of recklessly endangering civilian lives. Both McChrystal and Petraeus attempted to limit civilian casualties knowing that reckless use of kinetic firepower damaged the counterinsurgency campaign by alienating the local population. McChrystal emphasised 'courageous restraint' by the troops but this remained a challenging concept for an Army trained to be aggressive tactically and structured for conventional battles. Civilian casualties continued to bedevil relations with the Afghans. Similarly, the employment of Predator and Reaper *drones* over Afghanistan and Pakistan to attack insurgents was controversial. Although 'cost-effective' in reducing American casualties and providing an impressive display of sophisticated technology, they were an inferior substitute for a coherent strategy and counterproductive. Any militants who were killed became martyrs while heavy civilian casualties alienated not only the local population, providing recruits for the Taliban, but also causing international outrage. Nevertheless, the Taliban are almost exclusively Pashtun and the Afghan government was only able to retain power by harvesting support from Pashtun factions and non-Pashtun ethnic groups, such as the Tajiks, Uzbeks and Hazaras, who remained anti-Taliban.[51]

Conclusion

The campaigns in Iraq and Afghanistan have brought into question the validity of future counterinsurgency campaigns. The American and British Armies were unable to compensate for the lack of a central bureaucracy that the British in particular had taken for granted during previous counterinsurgencies. By the

1990s the colonial officials and their expertise in local languages, customs and bureaucratic structures, which had been the crucial, unsung factor in previous successes, were missing. Similarly, when confronted by state-building in developing countries, the Americans were crippled by the absence of a unified and co-ordinated civil and military bureaucracy to manage the counterinsurgency campaign, having never created a Colonial Office or service that could implement administrative control during interventions overseas. Instead of being the colonial masters with local credibility and natural authority, the Coalition forces were occupiers, lacking either legitimacy or local knowledge. The importance of 'cultural knowledge' in understanding the population of a non-Western country that is 'hosting' insurgents once again became clear. Moreover, most officers in Afghanistan and Iraq had little knowledge of counterinsurgency theory and practice, having been trained to fight a mechanised war with large units against the Soviet Army on the North German plain. In particular, the British had failed to build on their supposed 'counterinsurgency inheritance' of expertise from previous campaigns. The lessons of the British colonial experience had to be relearned. The most important were the necessity for pragmatic and realistic tactics, an understanding of the indigenous culture and political realities, and, above all, a coherent strategy based on clear policy. The interventions in Iraq and Afghanistan exposed serious flaws in doctrine, equipment and organisation. This led to a review and revitalisation of doctrine but, hindered by government policy and reluctance to commit resources, the British Army lacked the political backing and resources, notably manpower, to follow the Americans in changing strategy radically. The British Army could tinker with new techniques and re-evaluate doctrine in the light of its experiences in Afghanistan and Iraq but ultimately it lacked the political support to assign the resources that were necessary to implement a successful counterinsurgency strategy in either Basra or Helmand during 2003–14.[52] Ultimately, neither the Americans nor their allies were able to 'win' the wars in Afghanistan and Iraq by achieving a clear-cut victory without popular support for the governments that they were trying to maintain in power.

CONCLUSION

DIRTY WARS: THE FUTURE

There is a pressing need for a better understanding of counterinsurgency as it is arguably the future of warfare given the world trend towards the use of subversion and insurgency as part of a 'global insurgency' where the insurgents are no longer bound by narrow geographical boundaries. This represents one of the main challenges for political and military leadership during the twenty-first century. However, there is a growing questioning of whether much of the experience during the twentieth century of countries such as France, Portugal, Russia, the United Kingdom and the United States is still relevant, as many of these campaigns were waged during a colonial or post-colonial environment, often against Maoist insurgencies as part of the Cold War. While the main principles of conducting counterinsurgency remain the same, the challenges presented by new information technology and Islamic fundamentalism require new techniques in order to be able to achieve the same ends.[1] Recently there has been much more emphasis on studying 'the human terrain' of the area of operations. This includes the cultures, languages, ideologies, religions, economics, politics, group interests and social structures. This knowledge allows the security forces not only to understand and neutralise the insurgents but also to gain and maintain popular support, which is 'an essential objective for successful counterinsurgency'.[2]

The American and British governments and their military managers have remained attached to the concepts of propaganda and a centralised mass media that controlled information during the twentieth century. They have failed to understand and to adapt to the networked/information society connected by decentralised new media, such as the Internet, mobile phones, blogs, YouTube and social networking sites (notably Facebook and Twitter), which has emerged during the twenty-first century. This bypasses traditional media, challenging Western domination of the news media, and making it very difficult to control and manage information. The Americans and British have remained dominated by 'techno-war' and have fought conventional wars in the post–Cold War era beginning with the Gulf War in 1991 that led to conventional interventions in Kosovo, Afghanistan and Iraq. These interventions ignored the shift in War from 'techno-war' (operations relying on the technology of destruction and the managerial skills of the military) to 'image warfare' (where information and images disseminated by new media are more important than the operations on the ground). This shift followed al-Qaeda's attacks on the World Trade Center and the Pentagon on 11 September 2001, which announced the arrival of 'global terrorist warfare' or transnational terrorism. The significance of this swing from 'techno-war' to 'image war' was not fully grasped by the American and British military and rather than accepting the new challenge they attempted unsuccessfully to adapt 'techno-warfare', which retained its dominance over military thinking, to the new post-9/11 era. In the propaganda war or 'image war' any ammunition or 'image munitions', such as photographs or film, are employed to gain a military advantage against the superior conventional might of the Americans and British. These include 'unofficial' communications by political leaders, notably Osama bin Laden, suicide bombings, terror attacks, the taking and executions of hostages, and the exploitation of any government mistakes, such as the torture and mistreatment of Iraqi prisoners by the Americans at Abu Ghraib prison. Technological superiority, supported by attempts to control centrally the flow of information, proved to be no match for al-Qaeda's hijacking of the political agenda. In a media age control of information is no longer possible and the consequences increasingly unpredictable.[3]

This process has been made more complicated by the 'global' environment in which the security forces now have to operate while fighting the 'Global War against Terror'. As never before, counterinsurgency is 'a battle of ideas' or propaganda war, which a government and its security forces must win. Some campaigns, notably Palestine, Algeria, Aden, Vietnam and Northern Ireland, had an international dimension that gave a foretaste of the situation in the early twenty-first century. But the revolution in digital communications and the torrent of news and imagery has given insurgency a global dimension that campaigns in isolated colonies during the 1950s and 1960s simply did not have.[4] In the struggle between Western secularist governments and militant Islamists 'information and perception have become vitally important'. As a result, like security forces in the past, journalists have increasingly become targets for brutal terror campaigns by insurgent groups.[5] As one commentator has noted, the odds 'in modern counterinsurgency situations have shifted significantly in favour of the insurgent, rendering many historical lessons irrelevant'.[6]

The population affected by a modern insurgency is no longer just the local population in one well-defined area but multiple, global populations spread across the world, eroding the cohesion and increasing the complexity of the counterinsurgency campaign and making it difficult to define its centre of gravity. The development of global communications since the end of the Cold War has multiplied the central importance of the media in the vital propaganda and psychological battle to win the support of the population. The exploitation of imagery by insurgents as the primary instrument of their campaign has taken on a new significance as a result of the growing power of communications and the dispersal of populations across the globe. The use of highly visible and sensational terrorist attacks, 'the propaganda of the deed', which were first employed by organisations such as the IRA and PLO to goad security forces into overreacting with excessive force and as a means of catching the attention of the world via digital media, has increased dramatically. It has become central to the strategy employed by modern, global insurgents. In a war of ideas and images, the news value of an act of violence now outweighs its tactical value. The success of an insurgent campaign is currently measured in terms of recognition, notoriety and activism rather than the amount of territory it holds or the number of governments it has overthrown.[7] For example, the hunger strike and death of the IRA's Bobby Sands in 1981 resulted in anti-British demonstrations in Athens, Antwerp, Oslo, Brisbane and Chicago and a boycott of all British ships entering American ports.[8] As Anders Behring Breivik noted in his manifesto following his massacre of seventy-six young people in Norway, terrorists 'are in many ways sales representatives' and it is 'important' that they 'learn the basics of sales and marketing'.[9]

In the new Information Age, insurgencies have developed a new dynamism, exchanging information, deploying networks and establishing relationships of convenience with criminal gangs and local groups.[10] In particular, use of the Internet has greatly broadened the global capacity of insurgents to communicate, operate and organise securely. This enhances their ability to disseminate their message, recruit and seek external support, and gather information about the security forces. A combination of new technology, cheap and instantaneous communication, and increased mobility means that globalisation provides inherent advantages for pariah regimes, transnational terrorists and criminal groups over military forces who have less control over the movement of arms, information, people, resources and technology. This allows organisations such as al-Qaeda and Islamic State (IS) – formerly the Islamic State of Iraq and Syria (ISIS) – to react to circumstances more quickly and effectively. They can exploit ruthlessly any weaknesses and fill security voids in fragile states.[11] Operations in Kosovo had already highlighted that in the twenty-four-hour media age a well-prepared media set-up 'matters more than ever'. In particular this requires 'a more effective and speedier rebuttal system' to refute enemy propaganda, better use of Internet resources to disseminate 'a large quantity of background information' and better management of 'the twenty-four-hour media cycle in order to avoid gaps in the information flow'.[12] General Charles C. Krulak noted:

> The rapid diffusion of technology, the growth of a multitude of transnational factors, and the consequences of increasing globalization and economic interdependence, have coalesced to create national security challenges remarkable for their complexity ...[13]

The technological advances in media recording systems and near-instantaneous communications of the 'Information Age' has brought the reality of war into the viewers' homes and diminished significantly the ability to censor or limit coverage of military operations. In the face of instant, global news, soldiers have little control over the media. Moreover, marginalised by the mainstream Western media, Jihadist ideologues have turned to the Internet, using websites, forums and blogs to challenge this hegemony and reinforce the perception that Western media ignores Muslim suffering. Jihadist websites have grown from fourteen to more than 4,000 between 2000 and 2005 alone. Groups such as Al-Qaeda emphasise the importance of the Internet in reaching a global audience and providing near real-time coverage of news for Muslims. They use pirated software and hacking in order to circumvent security measures. By 2008 they were releasing a steady stream of videos worldwide through the new media. These show attacks on American troops, the last testaments of suicide bombers and the executions of hostages in order to encourage the faithful, radicalise and recruit followers. They also spread their militant message and provide online training on such varied subjects as building bombs, the collection of intelligence, the transfer of funds, recruitment and cyber security.[14] As General Petraeus concluded: 'The war is not only being fought on the ground in Iraq but also in cyberspace.'[15] The battle for public opinion remains of central importance but 'the sheer pervasiveness and responsiveness of new media' have changed 'the terms and content of the struggle', making it much more difficult for the security forces to win 'the information war' by controlling the press and the dissemination of news.[16]

As a consequence, 'every soldier is in a sense an ambassador' and 'needs to understand that improper actions aid the insurgent cause and that fighting an insurgency requires courage, patience, and, above all else, discipline'. Wars are now 'highly scrutinised' and, 'if the tactical battlefield is those countries' highways and roads, the strategic battlefield is the television screen, the Internet, and the covers of newspapers and magazines around the world'.[17]

The insurgent now targets the political decision-makers and public opinion rather than the armed forces or tactical formations, seeking to convince policy-makers that achieving their strategic goals will be too costly. Locked into a Cold War mindset of how to control and disseminate information, governments are in denial about the fragmenting media landscape of domestic and global audiences and fail to respond adequately.[18] One American general confessed that, 'We as a military are at risk of failing to understand the nature of the war we are fighting' – a war which has been characterised as 'a war of intelligence and a war of perceptions'.[19] Another admits that, 'We have consistently underestimated the importance the enemy places on the IO [Information Operations] campaign.' He believes that 'both our technological and organisational capability to disseminate IO and counter enemy propaganda' must be improved as 'currently, we do not respond well enough to deal effectively with enemies who can say whatever they want without retribution'. He also noted that 'our enemies in Iraq and Afghanistan' used 'the internet and associated technology to feed their sophisticated information campaign and to build better improvised explosive devices faster than we can field countermeasures or train service members to defeat them'.[20]

In contrast, owing to a tendency of the Blair, Brown and Bush governments to over-centralise the gathering and dissemination of information, there was a vacuum in Afghanistan where response times stretched 'into days and weeks' and it was 'far easier for the media to get information from the insurgents, who respond in minutes', than from the government and security forces, 'which respond well outside the window of the 24-hour news cycle', losing the 'campaign for influence over the Afghan people'.[21] A British officer, the SO1 Chief Public Affairs, HQ Allied Rapid Reaction Corps, agreed, noting that, 'The military do not always readily acknowledge information operations as a necessary factor in success; or, when they do, they are slow to really incorporate the principle into planning.'[22] Nevertheless, commanders were adapting and confronting the problem. For example, General David Richards as commander of ISAF in Afghanistan during 2006–07 had an information plan in response to the growing power of the media.[23]

Suppressing an insurgency is no longer a local, internal matter where domestic and international criticism can be controlled in a manner that favours the security forces. It is now one carried out by multiple, multinational agencies and forces against a global jihad under the glare of international media scrutiny.[24] The digital revolution of the new millennium means that the military are facing a radical transition in military affairs that 'calls for a complete rethink in the way modern states prepare for future war'.[25] Interventions to 'stabilise' Afghanistan and Iraq have done little to develop a strategy as a solution to this challenging situation, rather it has exacerbated the problem. The principles of how to fight a counterinsurgency remain the same but new techniques are required in order to implement them and to achieve success in this new, challenging environment. For example, thanks to the glare of the media, the imposition of draconian population control as employed in Algeria, Kenya and Malaya, is likely to be highly controversial and might extend the conflict by alienating public opinion both domestically and internationally.[26] It is much harder to cover up mistakes and the mistreatment of prisoners, notably in Iraq where the abuse of detainees became widely known very quickly.[27] Insurgents do have vulnerabilities. These include disrupting their organisation and networks, 'dividing and ruling' their loose coalitions by creating an alternative political agenda through reform and concessions, separating them from their support and cutting off their source of recruits, arms and finance, and monitoring and targeting their information technology infrastructure such as Internet sites.[28]

This has profound implications for any government (and its military advisers) that is contemplating involvement in a counterinsurgency campaign. Following interventions in Algeria and Vietnam respectively, the French and American armies were wary of further involvement in counterinsurgency campaigns. The French often propped up Francophile leaders in Africa, launching some twenty military interventions between 1963 and 1988, but they showed a marked reluctance to become too deeply involved in any protracted campaigns.[29] Similarly, American involvement in El Salvador was restricted to 'a strictly supporting role' during the 1980s. The operations in Afghanistan and Iraq have already led the Americans and their British and European allies to draw the same conclusion. It is unlikely that their troops will be committed to such interventions in the foreseeable future, especially as Parliament voted against the use of British forces in Syria. Lacking public support for intervention, governments may prefer to undertake limited, advisory roles rather than launching large-scale interventions. In the new Information Age these are much more under scrutiny by the international media and are much less easy to hide or to justify than hitherto. But it remains to be seen whether they can shun counterinsurgency operations totally in the future, if they wish to retain their global role as 'major players' in the world. It would be a major mistake for the military to attempt to forget the lessons of their involvement in Afghanistan and Iraq in a misguided belief that they can concentrate on conventional operations instead. It is doubtful that it is possible to ignore the multi-faceted problem of post-Maoist insurgency altogether, like the Americans, who eschew interventions and rely on drones and the CIA instead.[30] Some pundits have increasingly questioned the ideas, influence and importance of counterinsurgency, which is associated with 'dirty wars' and the misuse of history by its advocates (COINinistas).[31] But, as General Petraeus concludes, 'the Counterinsurgency Era is not over' and 'that is, quite simply, because the Insurgency Era is not over'.[32] Indeed, it has been calculated that only some 12–18 per cent of wars since 1945 have been conventional ones. Politicians and civil servants have been dangerously ignorant of the principles of counterinsurgency and in understanding its political demands when insurgency has always been and remains a highly political form of warfare.[33]

Events suggest that major powers, such as China, India, Russia and the United States, and their allies cannot ignore the likelihood of being involved in 'small wars'. They need to have a coherent and well thought out strategy to deal with insurgencies that threaten their interests. Simply ignoring them and hoping that they disappear will simply not suffice. The recent activities during 2014 of Tuareg separatists in Mali, al-Shabab militants in Kenya, Boko Haram in Nigeria, pro-Russian separatists in the Ukraine, the Taliban in Afghanistan and al-Sham and the Islamic State in Syria and Iraq all indicate that the problem will not go away. Both India (particularly the ongoing Naxalite-Maoist insurgency) and China (notably in Tibet and against the Uighurs in Xinjiang) have their own internal insurgencies by disaffected minorities and political groups to counter. The American, British and French Armies and other NATO forces should continue to search for answers to the multi-faceted problem of post-Maoist insurgency because their enemies are unlikely to fight on their terms, preferring unconventional to conventional warfare. They must train their soldiers for counterinsurgency among other missions and ensure that up-to-date doctrine and techniques are developed. But they need to have a well-developed strategy based on a clear sense of what they are trying to achieve and to fight limited wars less often and more wisely than the campaigns in Afghanistan and Iraq.

The real lesson from this book therefore is that the ability to conduct a successful counterinsurgency campaign is a useful tool or weapon for a nation to have alongside a more conventional capability. But politicians and military leaders have to understand the limitations of counterinsurgency and to be very careful when committing forces to such an operation to ensure that success is likely to be achieved. A classic counterinsurgency strategy can provide the means for defeating insurgents but often fails to offer a victory in which there are unambiguous winners and losers. It is not a panacea that will save failed states, establish Western-pattern democracies or alter the negative circumstances of a particular campaign where the support of the local population has already been lost by an unpopular regime. It also frequently takes a long time and demands large resources, placing great strain on the political will required to sustain this effort, especially if it is being sustained by a coalition force. Counterinsurgency is a necessary evil but one that requires significant commitment and resolve if it is to be successful.

NOTES

Preface
1   Robert Egnell, 'Winning "hearts and minds"?', p.282.
2   Frank Kitson, *Low Intensity Operations*, p.5; see also Lieutenant General Sir John Kiszely, 'Learning about Counterinsurgency', p.16.
3   Roger Trinquier, *Modern Warfare*, pp.3–9, 104.
4   General Sir Rupert Smith, *The Utility of Force*, p.372.
5   General Sir Rupert Smith, *The Utility of Force*, p.372.
6   David Kilcullen, *Out of the Mountains: The Coming Age of the Urban Guerrilla*.
7   Robert Thompson, *Revolutionary War in World Strategy 1945–1969*, pp.1–4.
8   Frank Kitson, *Bunch of Five*, p.282.
9   Lieutenant General K. Nagaraj (ed.), *Indian Army Doctrine*, pp.7–8, 69.
10 Che Guevara, *Guerrilla Warfare*, p.18.
11 Che Guevara, *Guerrilla Warfare*, p.23.
12 Mao Tse-tung (edited by Brigadier General Samuel B. Griffith), *Guerrilla Warfare*, p.43.
13 General Vo Nguyen Giap, *People's War, People's Army*, p.104.
14 Colonel C.E. Callwell, *Small Wars*, pp.125–6.
15 Mao Tse-tung, *Selected Writings of Mao Tse-tung*, p.161.
16 Mao Tse-tung, *Selected Writings of Mao Tse-tung*, p.165.
17 Mao Tse-tung (translation and introduction by Stuart R. Schram), *Basic Tactics*, pp.85–6.
18 Robert Taber, *The War of the Flea*, p.29.
19 Geoffrey Fairbairn, *Revolutionary Guerrilla Warfare*, pp.16–7, 39–43; General Vo Nguyen Giap, *Banner of People's War*, pp.7–13, 16–7.
20 Brigadier General Samuel B. Griffith, Introduction to Mao Tse-tung, *Guerrilla Warfare*, pp.4–5; Introduction by Stuart R. Schram to Mao Tse-tung, *Basic Tactics*, pp.19–20, 32–3; Dick Wilson, *The Long March, 1935*, pp.8, 26.
21 General Vo Nguyen Giap, *Banner of People's War*, p.83.
22 General Vo Nguyen Giap, *Banner of People's War*, pp.23–4.
23 Robert Taber, *The War of the Flea*, pp.16–7, 21.
24 Che Guevara, *Guerrilla Warfare*, p.13.
25 Brigadier General Samuel B. Griffith, Introduction to Mao Tse-tung, *Guerrilla Warfare*, pp.17–19, 26; see also General Vo Nguyen Giap, *Banner of People's War*, pp.83–4.

26 Mao Tse-tung, 'Strategic Problems in the Anti-Japanese Guerrilla War' (May 1938) in William J. Pomeroy (ed.), *Guerrilla Warfare and Marxism*, p.189.

27 General Vo Nguyen Giap, *Banner of People's War*, p.30.

28 Mao Tse-tung (edited by Brigadier General Samuel B. Griffith), *Guerrilla Warfare*, p.40.

29 General Vo Nguyen Giap, *People's War, People's Army*, p.48.

30 General Vo Nguyen Giap, *Banner of People's War*, p.18.

31 General Vo Nguyen Giap, *People's War, People's Army*, p.29.

32 General Vo Nguyen Giap, *Banner of People's War*, p.68.

33 C.P. Fitzgerald, *Revolution in China* (Cresset Press, 1952), pp.78–9, quoted in Geoffrey Fairbairn, *Revolutionary Guerrilla Warfare*, p.97.

34 Karl Marx, 'The Art of Insurrection' (September 1852) in William J. Pomeroy (ed.), *Guerrilla Warfare and Marxism*, p.53.

35 General Vo Nguyen Giap, *People's War, People's Army*, p.48.

36 Robert Taber, *The War of the Flea*, pp.53–4.

37 Mao Tse-Tung, *Selected Writings of Mao Tse-tung*, p.107.

38 General Vo Nguyen Giap, *People's War, People's Army*, p.29.

39 Mao Tse-Tung (translation and introduction by Stuart R. Schram), *Basic Tactics*, pp.56–7.

40 Elwin Verrier, *India's North-East Frontier in the Nineteenth Century*, pp.61–3, quoted in Geoffrey Fairbairn, *Revolutionary Guerrilla Warfare*, p.29.

41 Mao Tse-Tung (translation and introduction by Stuart R. Schram), *Basic Tactics*, pp.51–2.

42 Geoffrey Fairbairn, *Revolutionary Guerrilla Warfare*, pp.41–5.

43 Mao Tse-tung, 'Problems in the Guerrilla War of Resistance against Japan' (1939) in Gene Z. Hanrahan (ed.), *Chinese Communist Guerrilla Warfare Tactics*, p.6.

44 Mao Tse-tung (ed. Brigadier General Samuel B. Griffith), *Guerrilla Warfare*, p.41.

45 Mao Tse-tung (ed. Brigadier General Samuel B. Griffith), *Guerrilla Warfare*, p.83.

46 Mao Tse-tung (ed. Brigadier General Samuel B. Griffith), *Guerrilla Warfare*, p.91.

47 Robert Taber, *The War of the Flea*, p.43.

48 Mao Tse-tung, 'Problems in the Guerrilla War of Resistance against Japan' (1939) in Gene Z. Hanrahan (ed.), *Chinese Communist Guerrilla Warfare Tactics*, p.50.

49 General Vo Nguyen Giap, *Banner of People's War*, pp.42–4.

50 Mao Tse-tung, *Selected Writings of Mao Tse-tung*, p.173.

51 General Vo Nguyen Giap, *Banner of People's War*, p.85.

52 Colonel C.E. Callwell, *Small Wars*, p.130.

53 Colonel C.E. Callwell, *Small Wars*, pp.125–6.

Introduction

1 Thomas R. Mockaitis, 'The Origins of British Counterinsurgency', p.215; Rod Thornton, 'Minimum Force', p.224.

2 David Kilcullen, 'Twenty-Eight Articles', p.105.

3 Andrew Mumford, *The Counterinsurgency Myth*, p.10.

4 Major General Charles W. Gwynn, *Imperial Policing*, pp.11–3.

5 Frank Kitson, *Low Intensity Operations*, p.5.

6 Major General Charles W. Gwynn, *Imperial Policing*, pp.1–2.

7 Frank Kitson, *Low Intensity Operations*, p.50; see also Robert Thompson, *Defeating Communist Insurgency*, pp.50–2.

8 United States Field Manual No. 3–24, *Counterinsurgency*, Chapter 2, p.1.

9   Robin Evelegh, *Peace Keeping in a Democratic Society*, pp.48–59.
10 War Office, *Imperial Policing and Duties in Aid of the Civil Power*, p.12.
11 United States Marine Corps, *Small Wars Manual*, Chapter 1, p.18.
12 Andrew Mumford, *The Counterinsurgency Myth*, pp.10–11.
13 Frank Kitson, *Directing Operations*, p.50.
14 David Galula, *Counterinsurgency Warfare*, p.89.
15 Robert Thompson, *Defeating Communist Insurgency*, pp.55–7.
16 John Mackinlay, *The Insurgent Archipelago*, p.49.
17 D.M. Witty, 'A Regular Army in Counterinsurgency Operations', pp.401–2.
18 D.M. Witty, 'A Regular Army in Counterinsurgency Operations', pp.436–7.
19 Lieutenant General K. Nagaraj (ed.), *Indian Army Doctrine*, p.70.
20 David Galula, *Counterinsurgency Warfare*, pp.89–90; Thomas R. Mockaitis, 'The Origins of British Counterinsurgency', p.223; Mark Moyar, *A Question of Command*, pp.1–13, 83–6, 121–32; Andrew Mumford, *The Counterinsurgency Myth*, p.3; Kalev L. Sepp, 'Best Practices in Counterinsurgency', p.11.
21 Ian F.W. Beckett, *Modern Insurgencies and Counter-Insurgencies*, pp.102, 104; Andrew J. Birtle, *US Army Counterinsurgency and Contingency Operations Doctrine, 1942–1976*, pp.61–3, 329; Douglas S. Blaufarb, *The Counterinsurgency Era*, pp.22, 27–40; Major Daniel P. Bolger, *Scenes from an Unfinished War*, pp.39, 119, 121; Lawrence M. Greenberg, *The Hukbalahap Insurrection*, pp.82–9; Anthony James Joes, 'Counterinsurgency in the Philippines, 1898–1954', pp.50–1; Benedict J. Kerkvliet, *The Huk Rebellion*, pp.238 and 240–5; Andrew Mumford, *The Counterinsurgency Myth*, p.35; A.H. Peterson, G.C. Reinhardt and E.E. Conger (eds), *Symposium on the Role of Airpower in Counterinsurgency and Unconventional Warfare: The Philippine Huk Campaign*, pp.17–9; Colonel Napoleon D. Valeriano and Lieutenant Colonel Charles T.R. Bohannan, *Counter-Guerrilla Operations*, pp.29, 100–10, 139–41 and 207–8.
22 David A. Charters, 'From Palestine to Northern Ireland', pp.171–2; Andrew Mumford, *The Counterinsurgency Myth*, pp.9–10; United States Field Manual No. 3–24, *Counterinsurgency*, Chapter 7, pp.1–5.
23 Lieutenant General Peter W. Chiarelli and Major Patrick R. Michaelis, 'Winning the Peace', pp.18–9, 23–4.
24 Major Steven M. Miska, 'Growing the Iraqi Security Forces', p.65.
25 A. Deane-Drummond, *Riot Control* (London: Royal United Services Institute for Defence Studies, 1975), p.64.
26 Malaya Command, *The Conduct of Anti-Terrorist Operations in Malaya*, Chapters VIII–XV; East Africa Command, *A Handbook on Anti-Mau Mau Operations*, pp.11, 15–24, 34–55.
27 Brigadier Nigel R.F. Aylwin-Foster, 'Changing the Army for Counterinsurgency Operations', p.4.
28 Ian Gardiner, *In the Service of the Sultan*, p.174.
29 Robert D. Ramsey III, *Advising Indigenous Forces*, pp.52–7.
30 Ngo Quang Truong, *RVNAF and US Operational Co-operation and Co-ordinatiom*, p.173.
31 Lieutenant Colonel Lester W. Grau, 'Something Old, Something New', pp.46–7.
32 Brigadier Nigel R.F. Aylwin-Foster, 'Changing the Army for Counterinsurgency Operations', p.3.
33 David Galula, *Counterinsurgency Warfare*, pp.90–3; Lieutenant Colonel John J. McCuen, *The Art of Counter-Revolutionary War*, pp.182–92.
34 United States Field Manual No. 3–24, *Counterinsurgency*, Chapter 2, p.3.
35 Sir Robert Thompson, *Revolutionary War in World Strategy, 1945–1969*, p.96.

36   Andrew Mumford, *The Counterinsurgency Myth*, p.6; Major John S. Pustay, *Counterinsurgency Warfare*, p.91–3.

37   Roger Trinquier, *Modern Warfare*, pp.8.

38   British Army Field Manual, Volume 1, Part 10, *Countering Insurgency*, Chapter 1, pp.2–3.

39   David Galula, *Counterinsurgency Warfare*, pp.7–8, 74–5.

40   Andrew Mumford, *The Counterinsurgency Myth*, pp.12–13; Robert Thompson, *Defeating Communist Insurgency*, pp.148–9.

41   Douglas Porch, *The Portuguese Armed Forces and the Revolution*, pp.13 and 37–9; The Sunday Times Insight Team, *Insight on Portugal*, pp.11 and 22–3; W.S. van der Waals, *Portugal's War in Angola*, 1961–1974, p.xiv.

42   James S. Corum, *Bad Strategies*, p.22.

43   Major General Charles W. Gwynn, *Imperial Policing*, p.11.

44   David A Charters, 'From Palestine to Northern Ireland', pp.172–3 and 175; David Galula, *Counterinsurgency Warfare*, pp.10–11.

45   Ian F.W. Beckett, 'Robert Thompson and the British Advisory Mission to South Vietnam, 1961–1965', p.42; James Pritchard and M.L.R. Smith, 'Thompson in Helmand', p.66.

46   Robert Thompson, *Defeating Communist Insurgency*, p.171.

47   David Galula, *Counterinsurgency Warfare*, pp.103–4; Lieutenant Colonel John J. McCuen, *The Art of Counter-Revolutionary War*, pp.152–4, 225–31; Thomas R. Mockaitis, *Iraq and the Challenge of Counterinsurgency*, pp.23–4; Andrew Mumford, *The Counterinsurgency Myth*, pp.7–8.

48   Roger Trinquier, *Modern Warfare*, pp.3–9, 63–5.

49   Malaya Command, *The Conduct of Anti-Terrorist Operations in Malaya*, Chapter III, p.2.

50   David Galula, *Counterinsurgency Warfare*, pp.77–8.

51   David Galula, *Counterinsurgency Warfare*, pp.58, 83; David Kilcullen, *The Accidental Guerrilla*, pp.145–7; David Kilcullen, *Counterinsurgency*, p.10.

52   Eric Jardine, 'Population-Centric Counterinsurgency and the Movement of Peoples', p.271.

53   Andrew Mumford, *The Counterinsurgency Myth*, pp.6–7.

54   John M. Gates, 'Indians and Insurrectos', p.66.

55   Major General Charles W. Gwynn, *Imperial Policing*, pp.1–2.

56   Ian F.W. Beckett, *Modern Insurgencies and Counter-Insurgencies*, pp.46–7 and 91; Keith Jeffery, 'Intelligence and Counterinsurgency Operations', pp.131–2; Lieutenant Colonel John J. McCuen, *The Art of Counter-Revolutionary War*, pp.206–9; Major John S. Pustay, *Counterinsurgency Warfare*, pp.94–7, 110–3; William Sheehan, *A Hard Local War*, p.17; Martin Thomas, *Empires of Intelligence*, p.248; Robert Thompson, *Defeating Communist Insurgency*, p.117.

57   IWM, Papers of Field Marshal Lord Harding, AFH 10, Sir Hugh Foot to Sir John Martin, 22 April 1959.

58   Frank Kitson, *Low Intensity Operations*, p.143.

59   Major General Charles W Gwynn, *Imperial Policing*, pp.5–6.

60   Ministry of Defence, *Operation Banner*, p.8–15.

61   United States Marine Corps, *Small Wars Manual*, Chapter 1, p.18.

62   John A. Lynn, 'Patterns of Insurgency and Counterinsurgency', p.27.

63   Lieutenant General K. Nagaraj (ed.), *Indian Army Doctrine*, pp.70–1.

64   Lieutenant General K. Nagaraj (ed.), *Indian Army Doctrine*, p.68.

65   Major General Charles W. Gwynn, *Imperial Policing*, p.5.

66   Andrew Mumford, *The Counterinsurgency Myth*, p.14; David H. Ucko and Robert Egnell, *Counterinsurgency in Crisis*, pp.33–4.

67 United States Marine Corps, *Small Wars Manual*, Chapter 1, p.18.
68 British Army Field Manual, Volume 1, Part 10, *Countering Insurgency*, Chapter 3, pp.5–12.
69 Major General Charles W. Gwynn, *Imperial Policing*, pp.1–2.
70 United States Marine Corps, *Small Wars Manual*, Chapter 1, pp.24, 26, 31–2.
71 David French, *The British Way in Counterinsurgency, 1945–1967*, p.7; Wade Markel, 'Draining the Swamp', pp.35–43; Bing West, 'Counterinsurgency Lessons from Iraq', *Military Review* (March–April 2009), pp.5–6.
72 Robert Thompson, *Defeating Communist Insurgency*, pp.52–5.
73 David Galula, *Counterinsurgency Warfare*, pp.116–9; Roger Trinquier, *Modern Warfare*, pp.30–3, 73–6.
74 David Galula, *Counterinsurgency Warfare*, p.116.
75 Robert Thompson, *Defeating Communist Insurgency*, pp.143–7; Kalev L. Sepp, 'Best Practices in Counterinsurgency', p.10.
76 David Galula, *Counterinsurgency Warfare*, pp.65–6.
77 Calder Walton, *Empire of Secrets*, pp.200–9, 267–73.
78 *The Economist*, 'Kenya's new president: Will the new centre hold?', p.53.
79 Ian F.W. Beckett, 'The Soviet Experience', p.91; Thos G. Butson, *The Tsar's Lieutenant*, pp.137–40.
80 Ian F.W. Beckett, *Modern Insurgencies and Counter-Insurgencies*, pp.164–5; Anthony Clayton, *The Wars of French Decolonization*, p.121; David French, *The British Way in Counterinsurgency, 1945–1967*, pp.85–6; Peter Paret, *French Revolutionary Warfare from Indochina to Algeria*, pp.35–6; John Pimlott, 'The French Army', pp.65–6.
81 Robert Thompson, *Defeating Communist Insurgency*, p.123.
82 Andrew Mumford, *The Counterinsurgency Myth*, p.8.
83 David French, *The British Way in Counterinsurgency, 1945–1967*, pp.6–7; Wade Markel, 'Draining the Swamp', pp.35–48.
84 David Galula, *Counterinsurgency Warfare*, pp.111–3; see also Lieutenant Colonel John J. McCuen, *The Art of Counter-Revolutionary War*, pp.155–6, 231–4; Major John S. Pustay, *Counterinsurgency Warfare*, pp.100–3; Robert Thompson, *Defeating Communist Insurgency*, pp.121–42.
85 Tim Jones, 'The British Army, and Counter-Guerrilla Warfare in Transition, 1944–1952', pp.149–50.
86 Major John S. Pustay, *Counterinsurgency Warfare*, pp.100–1.
87 C.E. Callwell, *Small Wars: Their Principles and Practise*, p.143.
88 General Lord Bourne, 'The Direction of Anti-Guerrilla Operations', p.209.
89 Julian Paget, *Counterinsurgency Campaigning*, pp.163–4.
90 Frank Kitson, *Low Intensity Operations*, p.90; Robert Thompson, *Defeating Communist Insurgency*, pp.84–5; Roger Trinquier, *Modern Warfare*, pp.26–8.
91 War Office, *Imperial Policing and Duties in Aid of the Civil Power*, p.9.
92 Eliot Cohen, Conrad Crane, Jan Horvath and John Nagle, 'Principles, Imperatives and Paradoxes of Counterinsurgency', p.50.
93 David A. Charters, 'From Palestine to Northern Ireland', p.173; Lieutenant Colonel John J. McCuen, *The Art of Counter-Revolutionary War*, pp.113–9; Andrew Mumford, *The Counterinsurgency Myth*, pp.9 and 13–15; Robert Thompson, *Defeating Communist Insurgency*, pp.84–9.
94 David Galula, *Counterinsurgency Warfare*, p.72.
95 United States Field Manual No. 3–24, *Counterinsurgency*, Chapter 6, pp.19–22.
96 Robert Thompson, *Defeating Communist Insurgency*, p.85.
97 Major General Charles W. Gwynn, *Imperial Policing*, p.21.

98 Palestine Police Force, *Combined Military and Police Action for Platoon Commanders and Junior Police Ranks*, p.5.
99 Lieutenant Colonel John J. McCuen, *The Art of Counter-Revolutionary War*, pp.141–3; Andrew Mumford, *The Counterinsurgency Myth*, pp.9, 14; Georgina Sinclair (ed.), Introduction to *Globalising British Policing*, pp.xix–xxi.
100 Roger Trinquier, *Modern Warfare*, pp.43–51.
101 Tom Bowden, *The Breakdown of Public Security*, pp.1–3 and 12–3.
102 Robin Evelegh, *Peace Keeping in a Democratic Society*, p.64.
103 Major General Charles W. Gwynn, *Imperial Policing*, p.3.
104 Martin Thomas, *Empires of Intelligence*, p.6.
105 Anthony Short, *The Communist Insurrection in Malaya, 1948–1960*, p.143.
106 Charles Foley, *Island in Revolt*, p.68; John Newsinger, *British Counterinsurgency*, p.95.
107 Jim House and Neil MacMaster, *Paris 1961*, pp.48–87.
108 Robin Evelegh, *Peace Keeping in a Democratic Society*, pp.119–32.
109 Lieutenant Colonel Lester W. Grau, 'Something Old, Something New', p.45.
110 United States Field Manual No. 3–24, *Counterinsurgency*, Chapter 2, p.2.
111 United States Marine Corps, *Small Wars Manual*, Chapter 2, pp.19–20.
112 Quoted by Kate Utting, 'The Strategic Information Campaign', p.37.
113 Robert Thompson, *Defeating Communist Insurgency*, pp.90–102.
114 Colonel Tony Jeapes, *SAS: Operation Oman*, p.37.
115 Major John L. Clark, *Thinking Beyond Counterinsurgency*, pp.2, 21–7.
116 John Mackinlay, *The Insurgent Archipelago*, pp.54–9.
117 John Mackinlay, *The Insurgent Archipelago*, pp.92–8.
118 John Mackinlay, *The Insurgent Archipelago*, pp.81–92.
119 Robert Thompson, *Defeating Communist Insurgency*, pp.57–8; see also Lieutenant Colonel John J. McCuen, *The Art of Counter-Revolutionary War*, pp.119–24, 166–81.
120 Lieutenant Colonel John J. McCuen, *The Art of Counter-Revolutionary War*, pp.143–52; Thomas R. Mockaitis, *British Counterinsurgency in the post-imperial era*, p.24.
121 Lieutenant Colonel John J. McCuen, *The Art of Counter-Revolutionary War*, pp.218–25; Roger Trinquier, *Modern Warfare*, pp.66, 89–91.
122 War Office, *Imperial Policing and Duties in Aid of the Civil Power*, pp.1–2.
123 Ian F.W. Beckett, *Modern Insurgencies and Counter-Insurgencies*, pp.46–7 and 91; Keith Jeffery, 'Intelligence and Counterinsurgency Operations', pp.131–2; Lieutenant Colonel John J. McCuen, *The Art of Counter-Revolutionary War*, pp.206–9; Major John S. Pustay, *Counterinsurgency Warfare*, pp.94–7, 110–3; William Sheehan, *A Hard Local War*, p.17; Martin Thomas, *Empires of Intelligence*, p.248; Robert Thompson, *Defeating Communist Insurgency*, p.117.
124 Tim Jones, 'The British Army, and Counter-Guerrilla Warfare in Transition, 1944–1952', p.149; Lieutenant Colonel John J. McCuen, *The Art of Counter-Revolutionary War*, pp.214–8; Lieutenant General K. Nagaraj (ed.), *Indian Army Doctrine*, p.75; Major John S. Pustay, *Counterinsurgency Warfare*, pp.113–5; Robert Thompson, *Defeating Communist Insurgency*, pp.111–2.
125 Frank Kitson, *Low Intensity Operations*, p.133; see also Robert Thompson, *Defeating Communist Insurgency*, pp.111–2.
126 Lieutenant Colonel John J. McCuen, *The Art of Counter-Revolutionary War*, pp.107–13, 159–61, 210–14; Andrew Mumford, *The Counterinsurgency Myth*, pp.8–9 and 36.
127 Robert Thompson, *Defeating Communist Insurgency*, p.108.

128 British Army Field Manual, Volume 1, Part 10, *Countering Insurgency*, Chapter 4, pp.14–20; United States Field Manual No. 3–24, *Counterinsurgency*, Chapter 2, p.1 and Chapter 5, pp.18–23.

129 United States Marine Corps, *Small Wars Manual*, Chapter 2, p.44.

130 Tim Jones, 'The British Army, and Counter-Guerrilla Warfare in Transition, 1944–1952', p.149; Lieutenant Colonel John J. McCuen, *The Art of Counter-Revolutionary War*, pp.239–45; Lieutenant General K. Nagaraj (ed.), *Indian Army Doctrine*, p.75; Major John S. Pustay, *Counterinsurgency Warfare*, pp.113–5; Robert Thompson, *Defeating Communist Insurgency*, pp.117–9.

131 Alexander Alderson, 'Iraq and its Borders', pp.18–19; Andrew Mumford, *The Counterinsurgency Myth*, pp.20–2; Bard O'Neill, *Insurgency and Terrorism*, pp.114–17; Lieutenant Colonel John J. McCuen, *The Art of Counter-Revolutionary War*, pp.245–52; Major John S Pustay, *Counterinsurgency Warfare*, pp.103–4.

132 Captain Christopher Ford, 'Of Shoes and Sites', pp.87–91; Alice Hills, *Future War in Cities*, pp.17–9.

Chapter 1

1   Matthew Bennett, 'The German Experience', p.61; Philip W. Blood, *Hitler's Bandit Hunters*, p.7; Peter Lieb, *A Precursor of Modern Counterinsurgency Operations?*, pp.5, 7; Peter Lieb, 'Few Carrots and a Lot of Sticks', p.72; Martin Rink, 'The German wars of liberation 1807–1815', pp.828–40; Ben Shepherd, *War in the Wild East*, pp.1, 41.

2   Ben Shepherd, *War in the Wild East*, pp.41–2; Ben Shepherd, 'With the Devil in Titoland', pp.78–9.

3   Ben Shepherd, *War in the Wild East*, pp.41–2; Isabel V. Hull, *Absolute Destruction*, pp.117–9.

4   Matthew Bennett, 'The German Experience', p.61; Ben Shepherd, *War in the Wild East*, p.42; Geoffrey Wawro, *The Franco-Prussian War*, pp.237–8, 279.

5   Ben Shepherd, *War in the Wild East*, p.42.

6   Matthew Bennett, 'The German Experience', pp.61–2; Geoffrey Wawro, *The Franco-Prussian War*, pp.237–8, 257, 264–5, 279, 288–90; John Horne and Alan Kramer, *German Atrocities, 1914*, pp.141–2.

7   Matthew Bennett, 'The German Experience', p.61; Ben Shepherd, *War in the Wild East*, p.42.

8   Matthew Bennett, 'The German Experience', pp.61–2; Isabel V. Hull, *Absolute Destruction*, pp.117–30; Ben Shepherd, *War in the Wild East*, p.42; Geoffrey Wawro, *The Franco-Prussian War*, pp.309–10.

9   Jan-Bart Gewald, 'Learning to Wage and Win Wars in Africa', p.27.

10 Jan-Bart Gewald, 'Learning to Wage and Win Wars in Africa', p.26.

11 Philip W. Blood, *Hitler's Bandit Hunters*, pp.15–16; Sabine Dabringhaus, 'An Army on Vacation?', pp.459, 466–8; Jan-Bart Gewald, 'Learning to Wage and Win Wars in Africa', p.27; Isabel V. Hull, *Absolute Destruction*, pp.147–52; Eric Ouellet, 'Multinational Counterinsurgency', pp.515–8.

12 Jan-Bart Gewald, 'Learning to Wage and Win Wars in Africa', p.27.

13 Matthew Bennett, 'The German Experience', p.63; Sabine Dabringhaus, 'An Army on Vacation?', pp.459, 466, 470; Jan-Bart Gewald, 'Learning to Wage and Win Wars in Africa', p.27; Ben Shepherd, *War in the Wild East*, p.42; Woodruff D. Smith, *The German Colonial Empire*, pp.64, 224.

14 Jan-Bart Gewald, 'Learning to Wage and Win Wars in Africa', pp.23–5; Jan-Bart Gewald, 'Colonial Warfare', p.12; Woodruff D. Smith, *The German Colonial Empire*, p.106.

15 Jan-Bart Gewald, 'Learning to Wage and Win Wars in Africa', pp.24–5; Jan-Bart Gewald, 'Colonial Warfare', pp.11–2; Isabel V. Hull, *Absolute Destruction*, pp.155–7; Kirsten Zirkel, 'Military power in German colonial policy', p.97.

16 Matthew Bennett, 'The German Experience', p.65; L.H. Gann and Peter Duignan, *The Rulers of German Africa, 1884–1914*, pp.121–2; Heike Schmidt, 'Deadly Silence Predominates in the District', pp.205–10; Woodruff D. Smith, *The German Colonial Empire*, pp.105–7; Trutz von Trotha, 'The Fellows Can Just Starve', p.429; Bruce Vandervort, *Wars of Imperial Conquest in Africa, 1830–1914*, pp.202–3; Kirsten Zirkel, 'Military power in German colonial policy', pp.101–3; Andrew Zimmerman, 'What Do You Really Want in German East Africa, *Herr Professor?*', pp.419–20.

17 L.H. Gann and Peter Duignan, *The Rulers of German Africa, 1884–1914*, pp.121–2; Heike Schmidt, 'Deadly Silence Predominates in the District', pp.205–10; Woodruff D. Smith, *The German Colonial Empire*, pp.105–7; Bruce Vandervort, *Wars of Imperial Conquest in Africa, 1830–1914*, pp.202–3; Kirsten Zirkel, 'Military power in German colonial policy', pp.101–3; Andrew Zimmerman, 'What Do You Really Want in German East Africa, *Herr Professor?*', pp.419–20.

18 Matthew Bennett, 'The German Experience', p.63; David Olusoga and Casper W. Erichsen, *The Kaiser's Holocaust*, pp.124–33; Woodruff D. Smith, *The German Colonial Empire*, pp.53, 63–4.

19 Matthew Bennett, 'The German Experience', pp.63–4; David Olusoga and Casper W. Erichsen, *The Kaiser's Holocaust*, pp.133–7.

20 Isabel V. Hull, *Absolute Destruction*, p.27; David Olusoga and Casper W. Erichsen, *The Kaiser's Holocaust*, p.138; Woodruff D. Smith, *The German Colonial Empire*, p.64.

21 Ben Shepherd, *War in the Wild East*, pp.42–3.

22 Isabel V. Hull, *Absolute Destruction*, pp.22–5; David Olusoga and Casper W. Erichsen, *The Kaiser's Holocaust*, p.139; Woodruff D. Smith, *The German Colonial Empire*, p.140.

23 Matthew Bennett, 'The German Experience', pp.65–6; Helmut Bley, *South-West Africa under German Rule, 1894–1914*, p.155–69; Philip W. Blood, *Hitler's Bandit Hunters*, pp.17–8; Horst Drechsler, 'Let Us Die Fighting', pp.147, 150–67 and 207–14; Jan-Bart Gewald, *Herero Heroes*, pp.170–91; Isabel V. Hull, *Absolute Destruction*, pp.44–89; David Olusoga and Casper W. Erichsen, *The Kaiser's Holocaust*, pp.139–71 and 189; Ben Shepherd, *War in the Wild East*, pp.42–3; Union of South Africa, *Report on the Natives of South-West Africa and Their Treatment by Germany*, pp.34–5 and 58–67; Bruce Vandervort, *Wars of Imperial Conquest in Africa, 1830–1914*, pp.198–202; Woodruff D. Smith, *The German Colonial Empire*, pp.64–5; Kirsten Zirkel, 'Military power in German colonial policy', pp.100–1.

24 Matthew Bennett, 'The German Experience', pp.65–6; Horst Drechsler, 'Let Us Die Fighting', pp.179–99; L.H. Gann and Peter Duignan, *The Rulers of German Africa, 1884–1914*, pp.123–4; Werner Hillebrecht, 'The Nama and the war in the south', pp.151–4; David Olusoga and Casper W. Erichsen, *The Kaiser's Holocaust*, pp.174–88; Woodruff D. Smith, *The German Colonial Empire*, p.65; Kirsten Zirkel, 'Military power in German colonial policy', p.101.

25 Matthew Bennett, 'The German Experience', pp.65–6; Horst Drechsler, 'Let Us Die Fighting', pp.179–99; L.H. Gann and Peter Duignan, *The Rulers of German Africa, 1884–1914*, pp.123–5; Werner Hillebrecht, 'The Nama and the war in the south', pp.151–4; David Olusoga and Casper W. Erichsen, *The Kaiser's Holocaust*, pp.174–88, 192–206; Ben Shepherd, *War in the Wild East*, pp.43–5; Woodruff D. Smith, *The German Colonial Empire*, p.65; Kirsten Zirkel, 'Military power in German colonial policy', p.101.

26 L.H. Gann and Peter Duignan, *The Rulers of German Africa, 1884–1914*, pp.122–3.

27 Isabel V. Hull, *Absolute Destruction*, pp.44–58; Jeremy Sarkin, *Germany's Genocide of the Herero*, pp.244–5; Ben Shepherd, *War in the Wild East*, p.43; Woodruff D. Smith, *The German Colonial Empire*, p.65.

28 Matthew Bennett, 'The German Experience', p.62; John Horne and Alan Kramer, *German Atrocities 1914*, pp.13–23, 38–41, 142–6; Ben Shepherd, *War in the Wild East*, p.43.

29 Isabel V. Hull, *Absolute Destruction*, pp.103–58; Peter Lieb, 'Few Carrots and a Lot of Sticks', p.72; John Horne and Alan Kramer, *German Atrocities 1914*, pp.74–7, 94–139, 153–61, 419, 435–9; Ben Shepherd, *War in the Wild East*, p.44.

30 John Horne and Alan Kramer, *German Atrocities 1914*, p.166; Isabel V. Hull, *Absolute Destruction*, pp.230–42 and 248–57; Helen McPhail, *The Long Silence*, pp.158–85.

31 Peter Lieb, 'A Precursor of Modern Counterinsurgency Operations?', p.21.

32 Peter Lieb, 'A Precursor of Modern Counterinsurgency Operations?', pp.11–16, 21; Peter Lieb, 'Few Carrots and a Lot of Sticks', pp.72–3.

33 Peter Lieb, 'A Precursor of Modern Counterinsurgency Operations?', pp.15–7, 22.

34 Peter Lieb, 'A Precursor of Modern Counterinsurgency Operations?', pp.18, 24–5; Peter Lieb, 'Few Carrots and a Lot of Sticks', p.73.

35 Peter Lieb, 'A Precursor of Modern Counterinsurgency Operations?', pp.18–21.

36 Colin D. Heaton, *German Anti-Partisan Warfare in Europe, 1939–1945*, pp.114–5; John Horne and Alan Kramer, *German Atrocities 1914*, pp.79–84; Peter Lieb, 'A Precursor of Modern Counterinsurgency Operations?', p.25; Peter Lieb, 'Few Carrots and a Lot of Sticks', pp.73–4; Ben Shepherd, 'With the Devil in Titoland', p.94.

37 Peter Lieb, 'A Precursor of Modern Counterinsurgency Operations?', pp.22–3; Peter Lieb, 'Few Carrots and a Lot of Sticks', pp.79–81; Ben Shepherd, 'With the Devil in Titoland', pp.78–9.

38 Peter Lieb, 'Few Carrots and a Lot of Sticks', pp.80–1.

39 Ben Shepherd, 'With the Devil in Titoland', p.78.

40 Matthew Bennett, 'The German Experience', p.72; Colin D Heaton, *German Anti-Partisan Warfare in Europe, 1939–1945*, pp.104–5, 124, 127; Paul N. Hehn, *The German Struggle against Yugoslav Guerrillas in World War II*, p.143.

41 Colin D. Heaton, *German Anti-Partisan Warfare in Europe, 1939–1945*, p.128; see also Omer Bartov, *The Eastern Front, 1941–45*, pp.119–41 and Omer Bartov, *Hitler's Army*, pp.106–78.

42 Colin D. Heaton, *German Anti-Partisan Warfare in Europe, 1939–1945*, pp.15, 86, 104, 131; Paul N. Hehn, *The German Struggle against Yugoslav Guerrillas in World War II*, pp.143–4; Peter Lieb, 'Few Carrots and a Lot of Sticks', pp.74–5; Ben Shepherd, *War in the Wild East*, pp.32, 115–6.

43 Mark F. Cancian, 'The Wehrmacht in Yugoslavia', pp.77–9; Simon Corkery, *Anti-Partisan Warfare in the Balkans*, 1941–1945, p.33; Paul N. Hehn, *The German Struggle against Yugoslav Guerrillas in World War II*, p.4; Charles D. Melson, 'German Counterinsurgency in the Balkans', p.708; Ben Shepherd, 'With the Devil in Titoland', p.80.

44 Paul N. Hehn, *The German Struggle against Yugoslav Guerrillas in World War II*, pp.1–2.

45 Paul N. Hehn, *The German Struggle against Yugoslav Guerrillas in World War II*, p.6; Charles D Melson, 'German Counterinsurgency in the Balkans', p.707.

46 Paul N. Hehn, *The German Struggle against Yugoslav Guerrillas in World War II*, p.6.

47 Simon Corkery, *Anti-Partisan Warfare in the Balkans, 1941–1945*, p.45; Paul N. Hehn, *The German Struggle against Yugoslav Guerrillas in World War II*, pp.6–7.

48 Matthew Bennett, 'The German Experience', pp.74–5; Colin D. Heaton, *German Anti-Partisan Warfare in Europe, 1939–1945*, pp.100–1; Paul N. Hehn, *The German Struggle against Yugoslav Guerrillas in World War II*, p.1–2.

49 Matthew Bennett, 'The German Experience', pp.74–5; Simon Corkery, *Anti-Partisan Warfare in the Balkans, 1941–1945*, pp.iv, 8, 34, 38; Paul N. Hehn, *The German Struggle against Yugoslav Guerrillas in World War II*, p.142; Charles D. Melson, 'German Counterinsurgency in the Balkans', pp.711–2, 721–4; Ben Shepherd, 'With the Devil in Titoland', p.89.

50 Colin D. Heaton, *German Anti-Partisan Warfare in Europe, 1939–1945*, p.90.

51 Matthew Bennett, 'The German Experience', p.76; Colin D. Heaton, *German Anti-Partisan Warfare in Europe, 1939–1945*, p.100; Paul N. Hehn, *The German Struggle against Yugoslav Guerrillas in World War II*, pp.10–11, 28; Ben Shepherd, 'With the Devil in Titoland', pp.94–5.

52 Paul N. Hehn, *The German Struggle against Yugoslav Guerrillas in World War II*, p.29.

53 Paul N. Hehn, *The German Struggle against Yugoslav Guerrillas in World War II*, pp.9–10.

54 Mark F. Cancian, 'The Wehrmacht in Yugoslavia', pp.79–80, 82.

55 Colin D. Heaton, *German Anti-Partisan Warfare in Europe, 1939–1945*, p.97; Michael McConville, 'Knight's Move in Bosnia and the British Rescue of Tito: 1944', p.62; Ben Shepherd, 'With the Devil in Titoland', pp.81, 84–5, 87, 90.

56 Matthew Bennett, 'The German Experience', p.79; Colin D. Heaton, *German Anti-Partisan Warfare in Europe, 1939–1945*, pp.97–8; Michael McConville, 'Knight's Move in Bosnia and the British Rescue of Tito: 1944', p.62; Ben Shepherd, 'With the Devil in Titoland', p.92.

57 Matthew Bennett, 'The German Experience', pp.71–2, 79–80; Lieutenant Colonel Wayne D. Eyre, 'Operation RÖSSELSPRUNG and the Elimination of Tito, May 25, 1944', pp.344, 347–69; Michael McConville, 'Knight's Move in Bosnia and the British Rescue of Tito: 1944', pp.62, 66–8.

58 Matthew Cooper, *The Phantom War*, pp.20–1, 52–3; Peter Lieb, 'A Precursor of Modern Counterinsurgency Operations?', p.24; Edward B. Westermann, 'Partners in Genocide', pp.782–91.

59 Christopher R. Browning, *Ordinary Men*, pp.6–19, 24–5, 121–5; Daniel Jonah Goldhagen, *Hitler's Willing Executioners*, pp.181–282; Colin D. Heaton, *German Anti-Partisan Warfare in Europe, 1939–1945*, p.223; Peter Lieb, 'A Precursor of Modern Counterinsurgency Operations?', p.23; Peter Lieb, 'Few Carrots and a Lot of Sticks:', pp.75–6.

60 Omer Bartov, *The Eastern Front, 1941–45*, pp.3–5, 63–7, 83–7, 99, 106–26, 149–50; Omer Bartov, *Hitler's Army*, pp.4, 9–11, 60–2, 82–95, 127–35.

61 Colin D. Heaton, *German Anti-Partisan Warfare in Europe, 1939–1945*, pp.148, 207, 211; Peter Lieb, 'Few Carrots and a Lot of Sticks', pp.74–6; Ben Shepherd, *War in the Wild East:*, pp.163, 207, 211.

62 Colin D. Heaton, *German Anti-Partisan Warfare in Europe, 1939–1945*, pp.207 and 211; Peter Lieb, 'Few Carrots and a Lot of Sticks', pp.75–6.

63 Matthew Bennett, 'The German Experience', p.67; Colin D. Heaton, *German Anti-Partisan Warfare in Europe, 1939–1945*, p.207; Peter Lieb, 'Few Carrots and a Lot of Sticks', pp.78 and 82.

64 Peter Lieb, 'Few Carrots and a Lot of Sticks', pp.78 and 82–3.

65 Colin D. Heaton, *German Anti-Partisan Warfare in Europe, 1939–1945*, pp.146–7; Peter Lieb, 'Few Carrots and a Lot of Sticks', p.79.

66 Matthew Bennett, 'The German Experience', pp.60–1; Peter Lieb, 'Few Carrots and a Lot of Sticks', pp.80–1.

67 Matthew Bennett, 'The German Experience', p.71; Peter Lieb, 'Few Carrots and a Lot of Sticks', p.79.

68 Matthew Bennett, 'The German Experience', pp.71–2.

69 Matthew Bennett, 'The German Experience', p.80; Peter Lieb, 'Few Carrots and a Lot of Sticks', p.89.
70 Timo Noetzel and Benjamin Schreer, 'Counter-what? Germany and Counterinsurgency in Afghanistan', pp.42–4 and 'Missing Links', pp.17–21; Seibert, Bjoern, 'A Quiet Revolution', p.60.

Chapter 2

1   Ian F.W. Beckett, 'The Portuguese Army', p.136; John P. Cann, *Counterinsurgency in Africa*, p.93; Allen and Barbara Isaacman, *Mozambique*, pp.22–5; Malyn Newitt, *Portugal in Africa*, pp.49, 51–2; Douglas Porch, *The Portuguese Armed Forces and the Revolution*, p.11; Foreword by General Bernard E. Trainor to John P. Cann, *Counterinsurgency in Africa*, p.xi.
2   Douglas Porch, *The Portuguese Armed Forces and the Revolution*, pp.35–6; The Sunday Times Insight Team, *Insight on Portugal*, pp.13–5.
3   Douglas Porch, *The Portuguese Armed Forces and the Revolution*, p.37.
4   Patrick Chabal, 'Emergencies and Nationalist Wars in Portuguese Africa', pp.242–3; Norrie MacQueen, *The Decolonization of Portuguese Africa*, pp.26–7; Douglas Porch, *The Portuguese Armed Forces and the Revolution*, pp.37–9; The Sunday Times Insight Team, *Insight on Portugal*, pp.22–3; W.S. van der Waals, *Portugal's War in Angola, 1961–1974*, p.xiv.
5   Douglas Porch, *The Portuguese Armed Forces and the Revolution*, p.13.
6   The Sunday Times Insight Team, *Insight on Portugal*, p.11.
7   Peter Karibe Mendy, 'Portugal's Civilizing Mission in Colonial Guinea–Bissau', p.35.
8   Norrie MacQueen, *The Decolonization of Portuguese Africa*, p.23; Douglas Porch, *The Portuguese Armed Forces and the Revolution*, pp.75–6; The Sunday Times Insight Team, *Insight on Portugal*, p.15; W.S. van der Waals, *Portugal's War in Angola, 1961–1974*, pp.56, 109.
9   General Kaúlza de Arriaga, *The Portuguese Answer*, pp.17–26.
10 Douglas Porch, *The Portuguese Armed Forces and the Revolution*, p.84.
11 A.J. Venter, *The Zambesi Salient*, p.15.
12 Peter Karibe Mendy, 'Portugal's Civilizing Mission in Colonial Guinea–Bissau', pp.56–7; Douglas Porch, *The Portuguese Armed Forces and the Revolution*, p.76.
13 John P. Cann, *Counterinsurgency in Africa*, pp.37, 40–6 and 57, fn.28; Norrie MacQueen, *The Decolonization of Portuguese Africa*, pp.17–22.
14 John P. Cann, *Counterinsurgency in Africa*, pp.46–55; Patrick Chabal, 'Emergencies and Nationalist Wars in Portuguese Africa', pp.244–6; W.S. van der Waals, *Portugal's War in Angola, 1961–1974*, pp.114–6.
15 John P. Cann, *Counterinsurgency in Africa*, p.67.
16 John P. Cann, *Counterinsurgency in Africa*, pp.46–55; The Sunday Times Insight Team, *Insight on Portugal*, p.25; W.S. van der Waals, *Portugal's War in Angola, 1961–1974*, pp.114–6.
17 Gerald J. Bender, 'The Limits of Counterinsurgency', pp.332–3; John P. Cann, *Counterinsurgency in Africa*, pp.26–8; Norrie MacQueen, *The Decolonization of Portuguese Africa*, pp.23–5, 28–9; W.S. van der Waals, *Portugal's War in Angola, 1961–1974*, pp.56–8, 62–3, 71–5, 78–9, 88, 91.
18 Walter C. Opello, 'Guerrilla War in Portuguese Africa', p.34; The Sunday Times Insight Team, *Insight on Portugal*, p.15; W.S. van der Waals, *Portugal's War in Angola, 1961–1974*, pp.110–1; Douglas L. Wheeler, 'The Portuguese Army in Angola', p.431.
19 Douglas Porch, *The Portuguese Armed Forces and the Revolution*, p.53; W.S. van der Waals, *Portugal's War in Angola, 1961–1974*, p.112; Douglas L. Wheeler, 'The Portuguese Army in Angola', p.432.

20 Gerald J. Bender, 'The Limits of Counterinsurgency', pp.336–57; W.S. van der Waals, *Portugal's War in Angola, 1961–1974*, pp.119–20, 197–202; Douglas L. Wheeler, 'The Portuguese Army in Angola', pp.433–7.

21 W.S. van der Waals, *Portugal's War in Angola, 1961–1974*, pp.120–1.

22 Neil Bruce, *Portugal*, p.131; W.S. van der Waals, *Portugal's War in Angola, 1961–1974*, p.xi, 187–9, 194.

23 Neil Bruce, *Portugal*, pp.66–78; W.S. van der Waals, *Portugal's War in Angola, 1961–1974*, pp.136, 140–1, 183–9, 202 and 208; Douglas L. Wheeler, 'The Portuguese Army in Angola', pp.431–2.

24 W.S. van der Waals, *Portugal's War in Angola, 1961–1974*, pp.155–7, 184–7.

25 W.S. van der Waals, *Portugal's War in Angola, 1961–1974*, pp.195–6; A.J. Venter, *The Zambesi Salient*, p.57.

26 W.S. van der Waals, *Portugal's War in Angola, 1961–1974*, pp.xiv, 224, 231–2, 234 and 261.

27 Neil Bruce, *Portugal*, pp.20 and 81; Thomas H. Henriksen, *Revolution and Counterrevolution*, p.30; Norrie MacQueen, *The Decolonization of Portuguese Africa*, p.47; Walter C. Opello, 'Guerrilla War in Portuguese Africa', pp.29–33; Douglas Porch, *The Portuguese Armed Forces and the Revolution*, p.53; The Sunday Times Insight Team, *Insight on Portugal*, p.25; W.S. van der Waals, *Portugal's War in Angola, 1961–1974*, pp.84, 241.

28 General Kaúlza de Arriaga, *The Portuguese Answer*, p.70.

29 Ian F.W. Beckett, 'The Portuguese Army', pp.144–5; John P. Cann, *Counterinsurgency in Africa*, p.80; Thomas H. Henriksen, *Revolution and Counterrevolution*, p.30; Brendan F. Jundanian, 'Resettlement Programs', pp.519–40; Norrie MacQueen, *The Decolonization of Portuguese Africa*, pp.47–8; W.S. van der Waals, *Portugal's War in Angola, 1961–1974*, pp.241–2.

30 Thomas H. Henriksen, *Revolution and Counterrevolution*, pp.154–63.

31 John P. Cann, *Counterinsurgency in Africa*, p.120; Thomas H. Henriksen, *Revolution and Counterrevolution*, pp.54, 69, 120 and 128–33; Norrie MacQueen, *The Decolonization of Portuguese Africa*, pp.48–9; A.J. Venter, *The Zambesi Salient*, pp.35, 103–5, 266, 282.

32 Ian Beckett, 'The Portuguese Army', p.159; Thomas H. Henriksen, *Revolution and Counterrevolution*, pp.31–3, 40; Walter C. Opello, 'Guerrilla War in Portuguese Africa', pp.35–6; W.S. van der Waals, *Portugal's War in Angola, 1961–1974*, p.242; A.J. Venter, *The Zambesi Salient*, p.36.

33 Ian F.W. Beckett, 'The Portuguese Army', p.136; Neil Bruce, *Portugal*, pp.87–8; John P. Cann, *Brown Waters of Africa*, pp.143–7; Patrick Chabal, 'National Liberation in Portuguese Guinea, 1956–1974', pp.82–3; Mustafah Dhada, *Warriors at Work*, pp.14–37; Norrie MacQueen, *The Decolonization of Portuguese Africa*, pp.37–9; Norrie MacQueen, 'Portugal's First Domino', pp.212–3.

34 Mustafah Dhada, *Warriors at Work*, pp.xv, 23, 35–7; Norrie MacQueen, *The Decolonization of Portuguese Africa*, p.39; Douglas Porch, *The Portuguese Armed Forces and the Revolution*, p.53; The Sunday Times Insight Team, *Insight on Portugal*, pp.26–7; W.S. van der Waals, *Portugal's War in Angola, 1961–1974*, p.83.

35 Neil Bruce, *Portugal*, p.19; Norrie MacQueen, *The Decolonization of Portuguese Africa*, pp.39–40; Norrie MacQueen, 'Portugal's First Domino', pp.213–4; Douglas Porch, *The Portuguese Armed Forces and the Revolution*, p.53; The Sunday Times Insight Team, *Insight on Portugal*, pp.26–7; W.S. van der Waals, *Portugal's War in Angola, 1961–1974*, p.175; A.J. Venter, *Portugal's Guerrilla War*, p.33; A.J. Venter, *Portugal's War in Guinéa–Bissau*, p.25.

36 Neil Bruce, *Portugal*, pp.91–2; Mustafa Dhada, *Warriors at Work* , pp.xv, 37; Norrie
    MacQueen, 'Portugal's First Domino', pp.214–5; Douglas Porch, *The Portuguese
    Armed Forces and the Revolution*, p.56; The Sunday Times Insight Team, *Insight on
    Portugal*, pp.26–7; A.J. Venter, *The Zambesi Salient*, pp.280–1.
37 Ian F.W. Beckett, *Modern Insurgencies and Counter-Insurgencies*, p.137; Neil Bruce,
    Portugal, pp.88, 93–4; John P. Cann, *Counterinsurgency in Africa*, pp.76–7; Patrick
    Chabal, 'National Liberation in Portuguese Guinea, 1956–1974', pp.83–4; Mustafa
    Dhada, *Warriors at Work* , pp.37–9, 41–3, 248 (fn.164), 249 (fn.165); Norrie
    MacQueen, *The Decolonization of Portuguese Africa*, pp.39–40; Norrie MacQueen,
    'Portugal's First Domino', p.215; Douglas Porch, *The Portuguese Armed Forces and the
    Revolution*, pp.53–4; The Sunday Times Insight Team, *Insight on Portugal*, pp.26–7; A.J.
    Venter, *Portugal's Guerrilla War*, pp.33–4, 169–70.
38 John P. Cann, 'Operation *Mar Verde*, The Strike on Conakry, 1970', pp.71–9; Mustafa
    Dhada, *Warriors at Work*, pp.38–41; Norrie MacQueen, *The Decolonization of
    Portuguese Africa*, pp.38–40; Norrie MacQueen, 'Portugal's First Domino', pp.216–7;
    Douglas Porch, *The Portuguese Armed Forces and the Revolution*, pp.54–6; A.J. Venter,
    Portugal's Guerrilla War, pp.33–4, 61–3.
39 Norrie MacQueen, *The Decolonization of Portuguese Africa*, pp.40–1, 208–10; Norrie
    MacQueen, 'Portugal's First Domino', pp.217–23; Douglas Porch, *The Portuguese
    Armed Forces and the Revolution*, pp.56–8; W.S. van der Waals, *Portugal's War in Angola,
    1961–1974*, pp.238–9.
40 Patrick Chabal, 'National Liberation in Portuguese Guinea, 1956–1974', pp.83–4;
    Mustafah Dhada, *Warriors at Work*, pp.46–53; Norrie MacQueen, *The Decolonization
    of Portuguese Africa*, pp.40–2; Norrie MacQueen, 'Portugal's First Domino', pp.224–5;
    Douglas Porch, *The Portuguese Armed Forces and the Revolution*, p.58; A.J. Venter,
    Portugal's Guerrilla War, pp.172–84; W.S. van der Waals, *Portugal's War in Angola,
    1961–1974*, p.239.
41 Douglas Porch, *The Portuguese Armed Forces and the Revolution*, p.58.
42 Mustafah Dhada, *Warriors at Work*, pp.48–53; Norrie MacQueen, *The Decolonization
    of Portuguese Africa*, p.42; Norrie MacQueen, 'Portugal's First Domino', p.226; W.S.
    van der Waals, *Portugal's War in Angola, 1961–1974*, pp.239–41; A.J. Venter, *Portugal's
    Guerrilla War*, pp.50–2 and 97–8.
43 Douglas Porch, *The Portuguese Armed Forces and the Revolution*, pp.58, 62–3 and 121.
44 W.S. van der Waals, *Portugal's War in Angola, 1961–1974*, pp.112 and 136.
45 Douglas Porch, *The Portuguese Armed Forces and the Revolution*, pp.30, 63–4.
46 Douglas Porch, *The Portuguese Armed Forces and the Revolution*, pp.32, 64.
47 Ian F.W. Beckett, 'The Portuguese Army', p.153; Norrie MacQueen, *The
    Decolonization of Portuguese Africa*, pp.75–8; Douglas Porch, *The Portuguese Armed
    Forces and the Revolution*, pp.30, 65–6; The Sunday Times Insight Team, *Insight on
    Portugal*, pp.33–4.
48 Ian F,W, Beckett, 'The Portuguese Army', pp.136, 142; Gerald J. Bender, 'The Limits
    of Counterinsurgency', pp.331–2; Douglas Porch, *The Portuguese Armed Forces and the
    Revolution*, p.11; The Sunday Times Insight Team, *Insight on Portugal*, p.15.
49 Neil Bruce, *Portugal*, pp.21–2, 102; Mustafah Dhada, *Warriors at Work*, p.53; Lawrence
    S. Graham, 'The Military in Politics', p.231; Norrie MacQueen, *The Decolonization
    of Portuguese Africa*, pp.72–4; Douglas Porch, *The Portuguese Armed Forces and the
    Revolution*, pp.30, 75–6, 80, 83–5; The Sunday Times Insight Team, *Insight on Portugal*,
    pp.9–11, 41–3 and 46; Douglas L. Wheeler, 'The Military and the Portuguese
    Dictatorship, 1926–1974', p.207–8.
50 Douglas Porch, *The Portuguese Armed Forces and the Revolution*, pp.32–3.

51   Douglas Porch, *The Portuguese Armed Forces and the Revolution*, pp.33–4; Thomas H. Henriksen, *Revolution and Counterrevolution*, pp.37–8.

52   Ian F.W. Beckett, 'The Portuguese Army', pp.146–8.

53   Ian F.W. Beckett, 'The Portuguese Army', p.149; Neil Bruce, *Portugal*, pp.66–7.

54   Ian F.W. Beckett, 'The Portuguese Army', pp.151–2; Thomas H. Henriksen, *Revolution and Counterrevolution*, p.59; W.S. van der Waals, *Portugal's War in Angola, 1961–1974*, p.xiv; A.J. Venter, 'Why Portugal Lost its African Wars', pp.265–6.

55   Ian F.W. Beckett, 'The Portuguese Army', p.139–40; John P. Cann, *Counterinsurgency in Africa*, p.93; Douglas Porch, *The Portuguese Armed Forces and the Revolution*, p.30.

56   A.J. Venter, *The Zambesi Salient*, p.37.

57   Thomas H. Henriksen, 'People's War in Angola, Mozambique and Guinea–Bissau', pp.385–9.

58   Thomas H. Henriksen, 'Lessons from Portugal's Counterinsurgency Operations in Africa', p.33; A.J. Venter, 'Why Portugal Lost its African Wars', p.270.

59   Ian F.W. Beckett, 'The Portuguese Army', p.159; Thomas H. Henriksen, 'Some Notes on the National Liberation Wars in Angola, Mozambique and Guinea–Bissau', pp.31, 33, 35; Peter Karibe Mendy, 'Portugal's Civilizing Mission in Colonial Guinea–Bissau', p.57; W.S. van der Waals, *Portugal's War in Angola, 1961–1974*, p.xiv; A.J. Venter, 'Why Portugal Lost its African Wars', pp.226–7.

Chapter 3

1   Robert F. Baumann, *Russian–Soviet Unconventional Wars in the Caucasus, Central Asia, and Afghanistan*, pp.1–36; Karl Meyer, *The Dust of Empire*, pp.146–7, 153–4; Robert W. Schaefer, *The Insurgency in Chechnya and the North Caucasus*, pp.58–71.

2   Robert F. Baumann, *Russian–Soviet Unconventional Wars in the Caucasus, Central Asia, and Afghanistan*, pp.49–77; Anthony James Joes, *Urban Guerrilla Warfare*, p.132.

3   Ian F.W. Beckett, 'The Soviet Experience', pp.85–6.

4   Ian F.W. Beckett, 'The Soviet Experience', p.90; Thos G. Butson, *The Tsar's Lieutenant*, pp.137–9; Orlando Figes, *A People's Tragedy*, pp.599–600, 753–7.

5   Ian F.W. Beckett, 'The Soviet Experience', pp.91–2; Vadim J. Birstein, *Smersh*, p.195; Thos G. Butson, *The Tsar's Lieutenant*, pp.137–9; Orlando Figes, *A People's Tragedy*, pp.753–7, 768–9.

6   Ian F.W. Beckett, 'The Soviet Experience', pp.92–3; Vadim J. Birstein, *Smersh*, p.195; Thos G. Butson, *The Tsar's Lieutenant*, pp.122–8; Orlando Figes, *A People's Tragedy*, pp.768–9.

7   Ian F.W. Beckett, 'The Soviet Experience', pp.85, 100–1; Ian F.W. Beckett, *Modern Insurgencies and Counter-Insurgencies*, p.50; Scott R. McMichael, *Stumbling Bear*, p.40 and fn.2, pp.136–7; Alexander Marshall, 'Turkfront', pp.5, 18–20; Paul Robinson, 'Soviet Hearts-and-Minds Operations in Afghanistan', p.6; Alexander Statiev, *The Soviet Counterinsurgency in the Western Borderlands*, pp.17–19.

8   Alexander Statiev, *The Soviet Counterinsurgency in the Western Borderlands*, pp.17, 19.

9   Robert F. Baumann, *Russian–Soviet Unconventional Wars in the Caucasus, Central Asia, and Afghanistan*, pp.91–2; Alexander Marshall, 'Turkfront', p.5; Ian F.W. Beckett, 'The Soviet Experience', p.86.

10   Robert F. Baumann, *Russian–Soviet Unconventional Wars in the Caucasus, Central Asia, and Afghanistan*, pp.94–6, 104; Alexander Marshall, 'Turkfront', pp.7–9; Stephen Blank, 'Soviet Russia and Low-Intensity Conflict in Central Asia', pp.40–1.

11   Robert F. Baumann, *Russian–Soviet Unconventional Wars in the Caucasus, Central Asia, and Afghanistan*, pp.96–100; Alexander Marshall, 'Turkfront', p.10.

12 Alexander Marshall, 'Turkfront', pp.5, 18–20; Ian F.W. Beckett, 'The Soviet Experience', pp.96, 99, 101; Robert M. Cassidy, *Russia in Afghanistan and Chechnya*, p.9.

13 Robert F. Baumann, *Russian–Soviet Unconventional Wars in the Caucasus, Central Asia, and Afghanistan*, pp.101–4; Alexander Marshall, 'Turkfront', pp.10–12.

14 Robert F. Baumann, *Russian–Soviet Unconventional Wars in the Caucasus, Central Asia, and Afghanistan*, pp.104–7; Martha B. Olcott, 'The Basmachi or Freemen's Revolt in Turkestan 1918–24', p.360.

15 Robert F. Baumann, *Russian–Soviet Unconventional Wars in the Caucasus, Central Asia, and Afghanistan*, pp.107–13; Alexander Marshall, 'Turkfront', pp.12–13.

16 Robert F. Baumann, *Russian–Soviet Unconventional Wars in the Caucasus, Central Asia, and Afghanistan*, pp.113–15; Alexander Marshall, 'Turkfront', pp.15–7; Martha B. Olcott, 'The Basmachi or Freemen's Revolt in Turkestan 1918–24', pp.361–3; William S. Ritter, 'The Final Phase in the Liquidation of Anti-Soviet Resistance in Tadzhikistan', pp.486–8; Alexandre Bennigsen, *The Soviet Union and Muslim Guerrilla Wars, 1920–1981*, pp.5–6, 12–3.

17 Robert F. Baumann, *Russian–Soviet Unconventional Wars in the Caucasus, Central Asia, and Afghanistan*, pp.113–15; Alexandre Bennigsen, *The Soviet Union and Muslim Guerrilla Wars, 1920–1981*, pp.5–6; Andrei A. Doohovsky, 'Soviet Counterinsurgency in the Soviet Afghan War Revisited', pp.23–4; Alexander Marshall, 'Turkfront', pp.22–4; Stephen Blank, 'Soviet Russia and Low-Intensity Conflict in Central Asia', pp.41–3.

18 Alexandre Bennigsen, *The Soviet Union and Muslim Guerrilla Wars, 1920–1981*, pp.v–vi, 4.

19 Alexander Marshall, 'Turkfront', pp.24–5; Alexander Statiev, *The Soviet Counterinsurgency in the Western Borderlands*, pp.7–11.

20 Ian F.W. Beckett, 'The Soviet Experience', pp.86, 101; Alexander Statiev, *The Soviet Counterinsurgency in the Western Borderlands*, pp.1–4, 106–7.

21 *Robert F. Baumann, Russian–Soviet Unconventional Wars in the Caucasus, Central Asia, and Afghanistan*, p.129; Ian F.W. Beckett, 'The Soviet Experience', p.101; Stephen Blank, 'Soviet Russia and Low-Intensity Conflict in Central Asia: Three Case Studies', pp.70–1; Jeffrey Burds, 'The Early Cold War in Soviet West Ukraine, 1944–1948', pp.8, 17–24; Lester W. Grau and Michael A. Gress (ed.), *The Soviet–Afghan War: How a Superpower Fought and Lost*, p.xxiii; Lester W. Grau (ed.), *The Bear Went Over the Mountain*, p.13, 45–6, 52, 76 and 165; Lester W. Grau and Ali Ahmad Jalali (ed.), *The Other Side of the Mountain*, pp.xvi–xvii; Serhiy Kudelia, 'Choosing Violence in Irregular Wars', pp.158–60; Alexander Statiev, *The Soviet Counterinsurgency in the Western Borderlands*, pp.6–8, 108–14, 123–34; Larysa Zariczniak, 'Major Stepan Stebelski of the Ukrainian Insurgent Army', p.443; Yuri Zhukov, 'Examining the Authoritarian Model of Counterinsurgency', pp.447, 450.

22 Ian F.W. Beckett, 'The Soviet Experience', pp.86, 101; Jeffrey Burds, 'Gender and Policing in Soviet West Ukraine, 1944–1948', pp.294–6; Serhiy Kudelia, 'Choosing Violence in Irregular Wars', pp.161–4; Mark Kramer, 'Guerrilla Warfare, Counterinsurgency and Terrorism in the North Caucasus; The Military Dimension of the Russian–Chechen Conflict', p.255; George Reklaitis, 'Cold War Lithuania', pp.16–8; Dr Alexander Statiev, 'Motivations and Goals of Soviet Deportations in the Western Borderlands', pp.978, 988–90, 997–8; Alexander Statiev, *The Soviet Counterinsurgency in the Western Borderlands*, pp.139–40, 161–3, 172–9, 196–208, 253–71; Yuri Zhukov, 'Examining the Authoritarian Model of Counterinsurgency', pp.449, 452, 458.

23 Ian F.W. Beckett, 'The Soviet Experience', pp.86, 101; Mark Kramer, 'Guerrilla Warfare, Counterinsurgency and Terrorism in the North Caucasus; The Military

Dimension of the Russian–Chechen Conflict', p.255; George Reklaitis, 'Cold War Lithuania', pp.6, 18–21, 26; Alexander Statiev, *The Soviet Counterinsurgency in the Western Borderlands*, pp.209–29, 279–93; Yuri Zhukov, 'Examining the Authoritarian Model of Counterinsurgency', pp.447–8, 454–5.

24 Ian F.W. Beckett, 'The Soviet Experience', pp.86, 101; Jeffrey Burds, 'Gender and Policing in Soviet West Ukraine, 1944–1948', pp.279–85, 300–13; Mark Kramer, 'Guerrilla Warfare, Counterinsurgency and Terrorism in the North Caucasus; The Military Dimension of the Russian–Chechen Conflict', p.255; Serhiy Kudelia, 'Choosing Violence in Irregular Wars', pp.150–1, 163; George Reklaitis, 'Cold War Lithuania', pp.3–12, 21–9; Alexander Statiev, *The Soviet Counterinsurgency in the Western Borderlands*, pp.230–52; Yuri Zhukov, 'Examining the Authoritarian Model of Counterinsurgency', pp.448–9, 455–6.

25 Jeffrey Burds, 'Gender and Policing in Soviet West Ukraine, 1944–1948', pp.310–3, 316–8; Serhiy Kudelia, 'Choosing Violence in Irregular Wars', pp.160–1, 164–76; George Reklaitis, 'Cold War Lithuania', pp.34–5; Alexander Statiev, *The Soviet Counterinsurgency in the Western Borderlands*, pp.138, 161–3, 193–4, 207–8, 251–2, 270–1, 307–9, 331–8; Yuri Zhukov, 'Examining the Authoritarian Model of Counterinsurgency', pp.444–53, 458.

26 Stephen Blank, 'Soviet Russia and Low-Intensity Conflict in Central Asia', pp.70–1; Ian F.W. Beckett, 'The Soviet Experience', p.101; Robert F. Baumann, *Russian–Soviet Unconventional Wars in the Caucasus, Central Asia, and Afghanistan*, p.129; Rodric Braithwaite, *Afgantsy*, pp.127–34; Robert M. Cassidy, *War, Will, and Warlords*, pp.21–2; Lester W. Grau and Michael A. Gress, *The Soviet–Afghan War*, pp.xxiii, 19, 43; Lester W. Grau, *The Bear Went Over the Mountain*, pp.13, 45–6, 52, 76, 165; Lester W. Grau and Ali Ahmad Jalali, *The Other Side of the Mountain*, pp.88, 261, 355; Geraint Hughes, 'The Soviet–Afghan War, 1978–1989', pp.338–9; Seth G. Jones, *In the Graveyard of Empires*, pp.24–9, 34; Scott R. McMichael, 'The Soviet Army, Counterinsurgency, and the Afghan War', pp.23–8; Scott R. McMichael, *Stumbling Bear*, pp.1–12, 14–17.

27 Scott R. McMichael, *Stumbling Bear*, pp.40–4, 52–3.

28 Lester W. Grau and Michael A. Gress, *The Soviet–Afghan War*, p.43.

29 Stephen Blank, 'Afghanistan and Beyond', pp.223–4; Andrei A. Doohovsky, 'Soviet Counterinsurgency in the Soviet Afghan War Revisited', pp.63–9; Joseph J. Collins, 'The Soviet–Afghan War', p.54; Lester W. Grau and Michael A. Gress, *The Soviet–Afghan War*, pp.xxiii–xxiv; Scott R. McMichael, *Stumbling Bear*, pp.36–7, 99; Mark Urban, *War In Afghanistan*, p.65.

30 Alexander Alexiev, *Inside the Soviet Army in Afghanistan*, pp.35–60; Robert F. Baumann, *Russian–Soviet Unconventional Wars in the Caucasus, Central Asia, and Afghanistan*, p.139; Rodric Braithwaite, *Afgantsy*, pp.230–2; Gregory Feifer, *The Great Gamble*, pp.113, 123–4, 171–81, 214–9, 225–30; Gregory Feifer, *The Great Gamble*, pp.99–100, 125, 181–4; Scott R. McMichael, 'The Soviet Army, Counterinsurgency, and the Afghan War', pp.23–4; Scott R. McMichael, *Stumbling Bear*, pp.61–2; Major Stephen D. Pomper, 'Don't Follow the Bear', p.27; Brigadier Mohammad Yousaf and Mark Adkin, *The Battle for Afghanistan*, pp.51–6.

31 Alexander Alexiev, *Inside the Soviet Army in Afghanistan*, pp.15–18, 25–34; Robert F. Baumann, *Russian–Soviet Unconventional Wars in the Caucasus, Central Asia, and Afghanistan*, pp.140–3, 150–6; Stephen Blank, 'Soviet Russia and Low-Intensity Conflict in Central Asia', pp.70–1; Stephen Blank, 'Afghanistan and Beyond', pp.226–8; Rodric Braithwaite, *Afgantsy*, pp.164, 223–4; Andrei A. Doohovsky, 'Soviet Counterinsurgency in the Soviet Afghan War Revisited', pp.78–85; Matthew

J. Flynn, *Contesting History*, pp.65–6; Lester W. Grau and Michael A. Gress, *The Soviet–Afghan War*, pp.xx, 82–4, 208–9; Lester W. Grau and Ali Ahmad Jalali, *The Other Side of the Mountain*, pp.339–40; Geraint Hughes, 'The Soviet–Afghan War, 1978–1989', p.341; Scott R. McMichael, 'The Soviet Army, Counterinsurgency, and the Afghan War', pp.24–5, 27–31; Scott R. McMichael, *Stumbling Bear*, pp.16, 63–8, 80–92; Carl van Dyke, 'Kabul to Grozny', p.471.

32 Robert F. Baumann, *Russian–Soviet Unconventional Wars in the Caucasus, Central Asia, and Afghanistan*, pp.136–7, 142, 170–1; Robert M. Cassidy, *War, Will, and Warlords*, pp.20–1; Andrei A. Doohovsky, 'Soviet Counterinsurgency in the Soviet Afghan War Revisited', p.23; Gregory Feifer, *The Great Gamble*, pp.168–9, 175–6; Lester W. Grau and Michael A. Gress, *The Soviet–Afghan War*, p.xxv; Lester W. Grau, *The Bear Went Over the Mountain*, pp.xviii, 13, 45–6, 52, 75–6, 165; Lester W. Grau and Ali Ahmad Jalali, *The Other Side of the Mountain*, pp.xix–xx, 25, 57, 246, 261, 267, 339; Geraint Hughes, 'The Soviet–Afghan War, 1978–1989', p.339; Scott R. McMichael, *Stumbling Bear*, pp.36–7, 99; Mark Urban, *War In Afghanistan*, p.65; Brigadier Mohammad Yousaf and Mark Adkin, *The Battle for Afghanistan*, pp.4, 115–8.

33 Rodric Braithwaite, *Afgantsy*, pp.134–9; Geraint Hughes, 'The Soviet–Afghan War, 1978–1989', pp.339–40; Scott R. McMichael, *Stumbling Bear*, pp.107–8.

34 Major Joseph J. Collins, 'The Soviet Military Experience in Afghanistan', pp.16, 19–20; Andrei A. Doohovsky, 'Soviet Counterinsurgency in the Soviet Afghan War Revisited', pp.64–7, 74–7; Lester W. Grau and Michael A. Gress, *The Soviet–Afghan War*, pp.xix–xx, 92, 241–2; Lester W. Grau and Ali Ahmad Jalali, *The Other Side of the Mountain*, pp.95, 111; Scott R. McMichael, *Stumbling Bear*, pp.13, 59–61.

35 Robert F. Baumann, *Russian–Soviet Unconventional Wars in the Caucasus, Central Asia, and Afghanistan*, p.138; Alexandre Bennigsen, *The Soviet Union and Muslim Guerrilla Wars, 1920–1981*, pp.19–21; Stephen Blank, 'Soviet Russia and Low-Intensity Conflict in Central Asia: Three Case Studies', pp.60, 63, 65; Joseph J. Collins, 'Afghanistan', pp.32–3; Lester W. Grau and Michael A. Gress, *The Soviet–Afghan War*, pp.51–2.

36 Robert F. Baumann, *Russian–Soviet Unconventional Wars in the Caucasus, Central Asia, and Afghanistan*, pp.135–8; Alexandre Bennigsen, *The Soviet Union and Muslim Guerrilla Wars, 1920–1981*, pp.10–18; Rodric Braithwaite, *Afgantsy*, pp.134–9, 232; Andrei A. Doohovsky, 'Soviet Counterinsurgency in the Soviet Afghan War Revisited', pp.35–42; Paul Robinson, 'Soviet Hearts-and-Minds Operations in Afghanistan', pp.7–21; Mark Urban, *War In Afghanistan*, pp.117 and 167; Brigadier Mohammad Yousaf and Mark Adkin, *The Battle for Afghanistan*, pp.146–7.

37 Stephen Blank, 'Soviet Russia and Low-Intensity Conflict in Central Asia, pp.62–3; Robert F. Baumann, *Russian–Soviet Unconventional Wars in the Caucasus, Central Asia, and Afghanistan*, p.149; Joseph J. Collins, 'Afghanistan', pp.34–5; Lester W. Grau and Michael A. Gress, *The Soviet–Afghan War*, pp.64, 291, 305; Geraint Hughes, 'The Soviet–Afghan War, 1978–1989', p.340; Scott R. McMichael, *Stumbling Bear*, p.12; Major Stephen D. Pomper, 'Don't Follow the Bear', p.27; Mark Urban, *War In Afghanistan*, p.128.

38 Robert F. Baumann, *Russian–Soviet Unconventional Wars in the Caucasus, Central Asia, and Afghanistan*, pp.156–8, 169, 173–7; Stephen Blank, 'Afghanistan and Beyond', pp.224–5; Rodric Braithwaite, *Afgantsy*, pp.298–302; Major Joseph J. Collins, 'The Soviet Military Experience in Afghanistan', p.18; Gregory Feifer, *The Great Gamble*, p.161; Scott R. McMichael, *Stumbling Bear*, pp.43–4; Major Stephen D. Pomper, 'Don't Follow the Bear', pp.27–9; Brian Glyn Williams, 'Afghanistan after the Soviets', pp.925–8.

39 Rodric Braithwaite, *Afgantsy*, pp.243–6; Andrei A. Doohovsky, 'Soviet Counterinsurgency in the Soviet Afghan War Revisited', pp.22–3; Matthew J. Flynn,

Contesting History, pp.64–9; Geraint Hughes, 'The Soviet–Afghan War, 1978–1989', pp.342–3; Paul Robinson, 'Soviet Hearts-and-Minds Operations in Afghanistan', pp.21–2; Carl van Dyke, 'Kabul to Grozny', p.471.

40 Alexandre Bennigsen, *The Soviet Union and Muslim Guerrilla Wars, 1920–1981*, fn.3, p.23; Robert M. Cassidy, *Russia in Afghanistan and Chechnya*, p.31; Matthew Evangelista, *The Chechen Wars*, pp.13–5; Lester W. Grau and Michael A. Gress, *The Soviet–Afghan War*, p.xviii; Anthony James Joes, *Urban Guerrilla Warfare*, pp.133–4; Robert W. Schaefer, *The Insurgency in Chechnya and the North Caucasus*, pp.55–8, 92–107.

41 Robert M. Cassidy, *Russia in Afghanistan and Chechnya*, pp.11–7, 22–4, 29–32, 37–40, 44–9; Svante E. Cornell, 'International Reactions to Massive Human Rights Violations', pp.85–90; Anthony James Joes, *Urban Guerrilla Warfare*, pp.135–50; Matthew Evangelista, *The Chechen Wars*, pp.40–5; Matthew J. Flynn, *Contesting History*, p.78; Alice Hills, *Future War in Cities*, pp.151–8; Robert W. Schaefer, *The Insurgency in Chechnya and the North Caucasus*, pp.128–34; Lieutenant Colonel Timothy L. Thomas, 'Grozny 2000', p.52; Carl van Dyke, 'Kabul to Grozny', pp.474–81.

42 Robert M. Cassidy, *Russia in Afghanistan and Chechnya*, pp.11–7, 22–4, 29–32, 37–40, 44–9; Svante E. Cornell, 'International Reactions to Massive Human Rights Violations', pp.85–90; Matthew Evangelista, *The Chechen Wars*, pp.40–5; Major Dan Fayutkin, 'Russian–Chechen Information Warfare 1994–2006', pp.53–5; Matthew J. Flynn, *Contesting History*, pp.78–87; Matthew Janeczko, '"Faced with death, even a mouse bites": Social and religious motivations behind terrorism in Chechnya', p.429; Anthony James Joes, *Urban Guerrilla Warfare*, pp.135–50; Robert W. Schaefer, *The Insurgency in Chechnya and the North Caucasus*, pp.128–43; Carl van Dyke, 'Kabul to Grozny', pp.474–81.

43 Matthew Evangelista, *The Chechen Wars*, pp.63–85; Matthew J. Flynn, *Contesting History*, pp.87–9; Alice Hills, *Future War in Cities*, pp.151–8; Mark Kramer, 'Guerrilla Warfare, Counterinsurgency and Terrorism in the North Caucasus', pp.212–67; Emil Pain, 'The Second Chechen War', pp.60, 63–4; Robert W. Schaefer, *The Insurgency in Chechnya and the North Caucasus*, pp.4, 179–94, 210–3, 217–21; Lieutenant Colonel Timothy L. Thomas, 'Grozny 2000', pp.52–3.

44 Matthew Evangelista, *The Chechen Wars*, pp.63–85; Matthew J. Flynn, *Contesting History*, p.89; Mark Kramer, 'Guerrilla Warfare, Counterinsurgency and Terrorism in the North Caucasus', pp.212–67; Emil Pain, 'The Second Chechen War', pp.61–2; Robert W. Schaefer, *The Insurgency in Chechnya and the North Caucasus*, pp.1–5, 148–52, 163–70, 192–9, 209–10, 226–7, 239–41, 247, 251–3, 267–71, 276–81.

Chapter 4

1 Notably the Vendée (1793–96), Italy (1806–11), Spain (1808–14), Algeria (1830–48), Mexico (1862–66), Indochina (1875–1900), Madagascar (1883–97) and Morocco (1900–34).

2 Ian F.W. Beckett, Introduction, *The Roots of Counterinsurgency, 1900–1945*, p.14; Raymond F. Betts, *France and Decolonisation, 1900–1960*, p.6; John Chipman, *French Power in Africa*, pp.1–2, 4–9, 18–20; Jack Autrey Dabbs, *The French Army in Mexico, 1861–1867*, pp.84–105, 113–21, 160–82, 261–73; Flynn, *Contesting History*, pp.1–18; Michel L. Martin, 'From Algiers to N'Djamena', pp.79–81; Francis Toase, 'The French Experience', p.41.

3   Ian F.W. Beckett, *Modern Insurgencies and Counter-Insurgencies*, pp.26–7; Alan Forrest,
    'The insurgency of the Vendée', pp.800–12; Anthony James Joes, 'Insurgency and
    Genocide', pp.21–40; Anthony James Joes, *Resisting Rebellion*, pp.50–63; North,
    Jonathon, 'General Hoche and Counterinsurgency', pp.529–31; Peter Paret, *French
    Revolutionary Warfare from Indochina to Algeria*, p.43; Douglas Porch, *Counterinsurgency*,
    pp.5–9.

4   Ian F.W. Beckett, *Modern Insurgencies and Counter-Insurgencies*, p.27; Charles Esdaile,
    'Guerrillas and bandits in the Serranía de Ronda, 1810–1812', p.821; Milton Finley,
    The Most Montrous of Wars, pp.63–78, 97–111, 114–26, 132–47; John Lawrence Tone,
    The Fatal Knot, pp.85–7, 103, 108–22, 128–9, 144.

5   Ian F.W. Beckett, *Modern Insurgencies and Counter-Insurgencies*, pp.27–8; Ian F.W.
    Beckett, *The Roots of Counterinsurgency*, p.14; Douglas Porch, *Counterinsurgency*,
    pp.9–12; John Lawrence Tone, *The Fatal Knot*, pp.126, 128.

6   Ian F.W. Beckett, *Modern Insurgencies and Counter-Insurgencies*, p.29; Douglas Porch,
    'Bugeaud, Galliéni, Lyautey', pp.378, 380, 399–400; Barnett Singer and John
    Langdon, *Cultured Force*, p.67.

7   Ian F.W. Beckett, *Modern Insurgencies and Counter-Insurgencies*, pp.28, 40; Flynn,
    Contesting History, pp.6–11; Jean Gottman, 'Bugeaud, Galliéni, Lyautey', p.236;
    Michel L. Martin, 'From Algiers to N'Djamena', pp.81–4; Douglas Porch, 'Bugeaud,
    Galliéni, Lyautey', pp.378–9; Douglas Porch, *Counterinsurgency*, pp.19–21; Francis
    Toase, 'The French Experience', pp.43–4.

8   Christopher M. Andrew and A.S. Kanya-Forstner, *France Overseas*, p.10; Ian F.W.
    Beckett, Introduction, *The Roots of Counterinsurgency: Armies and Guerrilla Warfare,
    1900–1945*, pp.14–15; Ian F.W. Beckett, *Modern Insurgencies and Counter-Insurgencies*, p.41;
    William A. Hoisington, *Lyautey and the French Conquest of Morocco*, pp.16–7; Alistair
    Horne, *The French Army and Politics, 1870–1970*, pp.22, 28–9; A.S. Kanya-Forstner, *The
    Conquest of the Western Sudan*, pp.8–9; Douglas Porch, 'Bugeaud, Galliéni, Lyautey',
    pp.380–2, 386–7, 391, 405–6; Douglas Porch, *Counterinsurgency*, pp.21–5; Barnett
    Singer and John Langdon, *Cultured Force*, pp.64, 72–3, 88; A.T. Sullivan, *Thomas-Robert
    Bugeaud*, pp.127–32; Francis Toase, 'The French Experience', pp.42–3.

9   Douglas Porch, 'Bugeaud, Galliéni, Lyautey', p.377; Francis Toase, 'The French
    Experience', pp.44, 57.

10 Ian F.W. Beckett, *Modern Insurgencies and Counter-Insurgencies*, pp.40–1; Jean Gottman,
    'Bugeaud, Galliéni, Lyautey', p.237; Peter Paret, *French Revolutionary Warfare from
    Indochina to Algeria*, p.35.

11 Ian F.W. Beckett, *Modern Insurgencies and Counter-Insurgencies*, p.29; Michel L. Martin,
    'From Algiers to N'Djamena', p.86; Peter Paret, *French Revolutionary Warfare from
    Indochina to Algeria*, pp.105–6; Douglas Porch, *Counterinsurgency*, p.30; Francis Toase,
    'The French Experience', pp.42–3.

12 John Chipman, *French Power in Africa*, pp.37–44; Cohen, William B., *Rulers of Empire*,
    pp.21–36, 43–4, 72–9, 106–8, 124, 180; Moshe Gershovich, *French Military Rule
    in Morocco*, pp.24–8; Jean Gottman, 'Bugeaud, Galliéni, Lyautey', p.245; William
    A. Hoisington, *Lyautey and the French Conquest of Morocco*, pp.6–8, 14–6; Anthony
    Kirk-Greene, *Britain's Imperial Administrators, 1858–1966*, pp.151–63; Peter Paret, *French
    Revolutionary Warfare from Indochina to Algeria*, pp.105–6; Martin Thomas, *Empires of
    Intelligence*, pp.52–3, 56–7.

13 Ian F.W. Beckett, *Modern Insurgencies and Counter-Insurgencies*, pp.113–4; Raymond
    F. Betts, *France and Decolonisation, 1900–1960*, p.68; Alistair Horne, *The French
    Army and Politics, 1870–1970*, pp.2, 65; Robert Paxton, *Parades and Politics at Vichy*,
    pp.423, 426.

14 Ian F.W. Beckett, *Modern Insurgencies and Counter-Insurgencies*, pp.40–1; N.E. Bou-Nacklie, 'Tumult in Syria's Hama in 1925', pp.288–9; A. Horne, *A Savage War of Peace: Algeria 1954–1962*, pp.23–8; Joyce Laverty Miller, 'The Syrian Revolt of 1925', *International Journal of Middle East Studies*, Vol.8, No.4 (October 1977), pp.545–6; John Pimlott, 'The French Army', p.46; Douglas Porch, 'Bugeaud, Galliéni, Lyautey', pp.394, 396–400; Douglas Porch, *Conquest of Morocco*, pp.187–8; Douglas Porch, *Counterinsurgency*, pp.103–6; Barnett Singer and John Langdon, *Cultured Force*, pp.219–20, 229; Francis Toase, 'The French Experience', pp.56–7.

15 Anthony Clayton, *The Wars of French Decolonization*, pp.79, 83–5, 87; John Pimlott, 'The French Army', p.46.

16 Ian F.W. Beckett, *Modern Insurgencies and Counter-Insurgencies*, p.110; John Pimlott, 'The French Army', p.55.

17 Ian F.W. Beckett, *Modern Insurgencies and Counter-Insurgencies*, pp.114–5; Anthony Clayton, *The Wars of French Decolonization*, pp.55–7; John Pimlott, 'The French Army', pp.47, 49, 52, 55–7; Francis Toase, 'The French Experience', p.58; Lucien Bodard, *The Quicksand War: Prelude to Vietnam* (New York, 1967), p.3 quoted in Raymond F. Betts, *France and Decolonisation, 1900–1960*, p.81.

18 Raymond F. Betts, *France and Decolonisation, 1900–1960*, pp.80, 87; Anthony Clayton, *The Wars of French Decolonization*, fn.3, p.42–5; John Pimlott, 'The French Army', pp.51–2.

19 Anthony Clayton, *The Wars of French Decolonization*, p.48; John Pimlott, 'The French Army', pp.51–2.

20 Ian F.W. Beckett, *Modern Insurgencies and Counter-Insurgencies*, p.114; Raymond F. Betts, *France and Decolonisation, 1900–1960*, p.87; Anthony Clayton, *The Wars of French Decolonization*, p.52; John Pimlott, 'The French Army', pp.51–2; Xiaobing Li, *A History of the Modern Chinese Army*, pp.207–15; Alexander Zervoudakis, 'Nihil mirare, nihil contemptare, omnia intelligere', p.199.

21 Raymond F. Betts, *France and Decolonisation, 1900–1960*, p.88; Christopher C. Harmon, 'Illustrations of "Learning" in Counterinsurgency', p.354; Edward Geary Lansdale, *In the Midst of Wars*, p.111; John Pimlott, 'The French Army', p.52.

22 John Pimlott, 'The French Army', pp.52–3.

23 Anthony Clayton, *The Wars of French Decolonization*, pp.67, 75–6; Bernard Fall, *Hell in a Very Small Place*, p.8; John Pimlott, 'The French Army', pp.53–5; Alexander Zervoudakis, 'Nihil mirare, nihil contemptare, omnia intelligere', pp.209–25.

24 Ian F.W. Beckett, *Modern Insurgencies and Counter-Insurgencies*, p.115; Robert M. Cassidy, 'The Long Small War', pp.51–2; Anthony Clayton, *The Wars of French Decolonization*, fn.23, p.76; Bernard Fall, *Hell in a Very Small Place*, p.8; Peter Paret, *French Revolutionary Warfare from Indochina to Algeria*, p.43; John Pimlott, 'The French Army', pp.56–7; Douglas Porch, *The French Secret Service*, pp.326–34; Philippe Pottier, 'GCMA/GMI: A French Experience in Counterinsurgency during the French Indochina War', *Small Wars and Insurgencies*, Vol.16, No.2 (June 2005), pp.125–45; Captain André Souyris, 'An Effective Counterguerrilla Procedure', pp.87–9.

25 Ian F.W. Beckett, *Modern Insurgencies and Counter-Insurgencies*, pp.112–3; Raymond F. Betts, *France and Decolonisation, 1900–1960*, pp.81 and 87; Anthony Clayton, *The Wars of French Decolonization*, p.51; Edward Geary Lansdale, *In the Midst of Wars*, p.111; John Pimlott, 'The French Army', pp.55–6; Francis Toase, 'The French Experience', p.58.

26 Robert M. Cassidy, 'The Long Small War', pp.51–2; Anthony Clayton, *The Wars of French Decolonization*, pp.56, 62–4; Bernard Fall, *Hell in a Very Small Place*, p.8;

John Pimlott, 'The French Army', pp.56–7; Douglas Porch, *The French Secret Service*, pp.302–4; Alexander Zervoudakis, 'Nihil mirare, nihil contemptare, omnia intelligere', p.199.

27   Ian F.W. Beckett, *Modern Insurgencies and Counter-Insurgencies*, pp.113–4; Anthony Clayton, *The Wars of French Decolonization*, pp.55–6; John Pimlott, 'The French Army', pp.55–6.

28   J.S. Ambler, *The French Army in Politics, 1945–1962*, p.109; Ian F.W. Beckett, *Modern Insurgencies and Counter-Insurgencies*, p.160; Raymond F. Betts, *France and Decolonisation, 1900–1960*, p.92; Anthony Clayton, *The Wars of French Decolonization*, p.75; Alistair Horne, *The French Army and Politics, 1870–1970*, p.75; John Pimlott, 'The French Army', p.56.

29   Peter Paret, *French Revolutionary Warfare from Indochina to Algeria*, pp.6–8 and 100; John Pimlott, 'The French Army', pp.58–9, 66, 74; John Shy and Thomas W. Collier, 'Revolutionary War', pp.852–3; Ian F.W. Beckett, *Modern Insurgencies and Counter-Insurgencies*, p.159.

30   Ian F.W. Beckett, *Modern Insurgencies and Counter-Insurgencies*, p.159; Peter Paret, *French Revolutionary Warfare from Indochina to Algeria*, pp.12–15; John Pimlott, 'The French Army', pp.58–9.

31   John Shy and Thomas W. Collier, 'Revolutionary War', p.853.

32   Ian F.W. Beckett, *Modern Insurgencies and Counter-Insurgencies*, p.160; A.A. Cohen, *Galula*, pp.161–78; Peter Paret, *French Revolutionary Warfare from Indochina to Algeria*, pp.6–7 and 113–4.

33   Ian F.W. Beckett, *Modern Insurgencies and Counter-Insurgencies*, p.160; Anthony Clayton, *The Wars of French Decolonization*, p.131; A.A. Cohen, *Galula*, pp.135–7, 178–82, 241–50; David Galula, *Pacification in Algeria, 1956–1958*, pp.64–7; Alf Andrew Heggoy, *Insurgency and Counterinsurgency in Algeria*, pp.176–80; Peter Paret, *French Revolutionary Warfare from Indochina to Algeria*, p.8 and fn.4, p.143; John Pimlott, 'The French Army', p.60.

34   David Galula, *Pacification in Algeria, 1956–1958*, pp.64–7; see also A.A. Cohen, *Galula*, pp.161–71.

35   Alistair Horne, *A Savage War of Peace: Algeria 1954–1962*, p.166; Peter Paret, *French Revolutionary Warfare from Indochina to Algeria*, pp.6–8, 16–7, 101; John Pimlott, 'The French Army', pp.59–60.

36   Anthony Clayton, *The Wars of French Decolonization*, p.7; Alistair Horne, *The French Army and Politics, 1870–1970*, pp.74, 83–4; John Pimlott, 'The French Army', pp.6, 21–2, 58, 66.

37   Ian F.W. Beckett, *Modern Insurgencies and Counter-Insurgencies*, pp.159–60; Anthony Clayton, *The Wars of French Decolonization*, p p.76, 129–31; Peter Paret, *French Revolutionary Warfare from Indochina to Algeria*, pp.27–8, 111–2; John Pimlott, 'The French Army', p.66.

38   Peter Paret, *French Revolutionary Warfare from Indochina to Algeria*, p.111.

39   Anthony Clayton, 'The Sétif Uprising of May 1945', pp.17–8; Peter Paret, 'The French Army and La Guerre Révolutionnaire', pp.66–9; Peter Paret, *French Revolutionary Warfare from Indochina to Algeria*, pp.5, 28, 112–4; John Pimlott, 'The French Army', pp.60, 67; John Shy and Thomas W. Collier, 'Revolutionary War', pp.853–4.

40   Ariel C. Armony, *Argentina, the United States, and the Anti-Communist Crusade in Central America, 1977–1984*, pp.9–10, 193–4, fn.43; Hal Brands, *Latin America's Cold War*, pp.47–8, 79–80; John P. Cann, *Counterinsurgency in Africa*, pp.37, 40–6 and 57, fn.28; Lieutenant Colonel Abel Esterhuyse and Evert Jordaan, 'The South

African Defence Force and Counterinsurgency, 1966–1990', p.106; Philip H. Frankel, *Pretoria's Praetorians*, 46–70; Philip Frankel, *Soldiers in a Storm*, pp.38–40; Anita Grossman, 'The South African Military and Counterinsurgency', pp.89–90; Lewis, *Guerrillas and Generals*, pp.136–43; Norrie MacQueen, *The Decolonization of Portuguese Africa*, pp.17–22; Alex Marshall, 'Imperial Nostalgia, the Liberal Lie, and the Perils of Postmodern Counterinsurgency', pp.242–3; Robert L. Miller, 'Introduction to General Paul Aussaresses', *The Battle of the Casbah*, pp.xxvi–xxviii; Paul B. Rich, 'A historical overview of US counterinsurgency', pp.19–20; Annette Seegers, *The Military in the Making of Modern South Africa*, pp.133–4, 140–1.

41 Anthony Clayton, 'Algeria 1954', pp.65–8; Alistair Horne, *The French Army and Politics, 1870–1970*, p.78; Barnett Singer and John Langdon, *Cultured Force*, p.266; John Pimlott, 'The French Army', pp.60 and 62.

42 Anthony Clayton, *The Wars of French Decolonization*, pp.113–14 and 118.

43 Raymond F. Betts, *France and Decolonisation, 1900–1960*, p.105; John Pimlott, 'The French Army', p.62; Ian F.W. Beckett, *Modern Insurgencies and Counter-Insurgencies*, pp.161–2; Jim House and Neil MacMaster, *Paris 1961*, pp.2–3; Barnett Singer and John Langdon, *Cultured Force*, pp.302 and 305; Alistair Horne, *A Savage War of Peace*, p.26.

44 Ian F.W. Beckett, *Modern Insurgencies and Counter-Insurgencies*, p.162; Raymond F. Betts, *France and Decolonisation, 1900–1960*, pp.105–6; Alistair Horne, *The French Army and Politics, 1870–1970*, p.79; Alistair Horne, *A Savage War of Peace*, p.151; Jim House and Neil MacMaster, *Paris 1961*, pp.2–3; Douglas Porch, *The French Secret Service*, p.365.

45 Alexander Alderson, 'Iraq and its Borders', pp.19–20; Anthony Clayton, *The Wars of French Decolonization*, pp.134–6, 155; John Pimlott, 'The French Army', pp.64–5; Peter Paret, *French Revolutionary Warfare from Indochina to Algeria*, pp.33–5; Martin Thomas, 'Order before Reform', p.200.

46 Anthony Clayton, *The Wars of French Decolonization*, p.117; Lieutenant Colonel Frédéric Guelton, 'The French Army "Centre for Training and Preparation in Counter-Guerrilla Warfare" (CIPCG) at Arzew', pp.35–46; Alistair Horne, *A Savage War of Peace*, pp.26, 100–1, 112–4, 165–6.

47 Ian F.W. Beckett, *Modern Insurgencies and Counter-Insurgencies*, p.164; Anthony Clayton, *The Wars of French Decolonization*, pp.136–8, 160–1; Alf Andrew Heggoy, *Insurgency and Counterinsurgency in Algeria*, pp.183–4, 212–29; Jim House and Neil MacMaster, *Paris 1961*, pp.28–9, 54–5; Peter Paret, *French Revolutionary Warfare from Indochina to Algeria*, pp.42–5, 94.

48 Ian F.W. Beckett, *Modern Insurgencies and Counter-Insurgencies*, pp.164 and 166; Anthony Clayton, *The Wars of French Decolonization*, pp.117–8 and 137–9; Alf Andrew Heggoy, *Insurgency and Counterinsurgency in Algeria*, pp.147–8, 188–211; Alistair Horne, *A Savage War of Peace*, pp.83 and 108–9; Michel L. Martin, 'From Algiers to N'Djamena', p.90; Peter Paret, *French Revolutionary Warfare from Indochina to Algeria*, pp.46–52; Martin Thomas, 'Order before Reform', p.203.

49 Anthony Clayton, *The Wars of French Decolonization*, pp.115, 117; Lieutenant Colonel Frédéric Guelton, 'The French Army "Centre for Training and Preparation in Counter-Guerrilla Warfare" (CIPCG) at Arzew', pp.35–46; Yoav Gortzak, 'Using Indigenous Forces in Counterinsurgency Operations', pp.311–2; Alf Andrew Heggoy, *Insurgency and Counterinsurgency in Algeria*, pp.91–3, 146; Alistair Horne, *A Savage War of Peace*, pp.100–1, 112–4, 165–6; Jim House and Neil MacMaster, *Paris 1961*, p.54; Peter Paret, *French Revolutionary Warfare from Indochina to Algeria*, fn.7, p.146.

50 Ian F.W. Beckett, *Modern Insurgencies and Counter-Insurgencies*, pp.164–5; Anthony Clayton, *The Wars of French Decolonization*, p.121; Peter Paret, *French Revolutionary*

Warfare from Indochina to Algeria, pp.35–6; John Pimlott, 'The French Army', pp.65–6.

51 Anthony Clayton, *The Wars of French Decolonization*, pp.121, 136; Ian F.W. Beckett, *Modern Insurgencies and Counter-Insurgencies*, pp.164–5; Alf Andrew Heggoy, *Insurgency and Counterinsurgency in Algeria*, p.183; Peter Paret, *French Revolutionary Warfare from Indochina to Algeria*, p.35 and fn.7, p.146.

52 Ian F.W. Beckett, *Modern Insurgencies and Counter-Insurgencies*, pp.164–7; Anthony Clayton, *The Wars of French Decolonization*, p.141; Peter Paret, *French Revolutionary Warfare from Indochina to Algeria*, pp.36–7; John Pimlott, 'The French Army', pp.56–7.

53 Ian F.W. Beckett, *Modern Insurgencies and Counter-Insurgencies*, p.165; Anthony Clayton, *The Wars of French Decolonization*, p.139; Yoav Gortzak, 'Using Indigenous Forces in Counterinsurgency Operations', pp.316–28; Alistair Horne, *A Savage War of Peace*, pp.254–8, 537–8; Peter Paret, *French Revolutionary Warfare from Indochina to Algeria*, pp.40–1; Barnett Singer and John Langdon, *Cultured Force*, p.139; Martin Thomas, 'Order before Reform', p.200.

54 John Pimlott, 'The French Army', pp.65–6; Anthony Clayton, *The Wars of French Decolonization*, pp.159–60; Ian F.W. Beckett, *Modern Insurgencies and Counter-Insurgencies*, pp.164–5; Michel L. Martin, 'From Algiers to N'Djamena', p.86; Barnett Singer and John Langdon, *Cultured Force*, p.334; Martin Thomas, 'Order before Reform', p.200.

55 Ian F.W. Beckett, *Modern Insurgencies and Counter-Insurgencies*, pp.162 and 165; Raymond F. Betts, *France and Decolonisation, 1900–1960*, pp.104 and 109; Anthony Clayton, *The Wars of French Decolonization*, pp.158, 160 and 178; Lieutenant Colonel Philippe Francois, 'Waging Counterinsurgency in Algeria', pp.60–3; Yoav Gortzak, 'Using Indigenous Forces in Counterinsurgency Operations', pp.314–6; François-Marie Gougeon, 'The Challe Plan', pp.293–5, 298–314; Christopher C. Harmon, 'Illustrations of 'Learning' in Counterinsurgency', p.355; Alistair Horne, *A Savage War of Peace*, pp.317–9, 330–40; Alistair Horne, *The French Army and Politics, 1870–1970*, pp.85–6, 156 and 158; Jim House and Neil MacMaster, *Paris 1961*, pp.27–8, 63–6; Colonel Gilles Martin, 'War in Algeria', pp.53–5; John Pimlott, 'The French Army', p.66; Martin Thomas, 'Order before Reform', p.200; Sir Robert Thompson, *Revolutionary War in World Strategy, 1945–1969*, p.93.

56 Ian F.W. Beckett, *Modern Insurgencies and Counter-Insurgencies*, p.163; Raymond F. Betts, *France and Decolonisation, 1900–1960*, pp.106–7; Anthony Clayton, *The Wars of French Decolonization*, pp.131–2; Lieutenant Colonel Philippe Francois, 'Waging Counterinsurgency in Algeria', p.63; Alf Andrew Heggoy, *Insurgency and Counterinsurgency in Algeria*, pp.182–3, 233–44; Alistair Horne, *The French Army and Politics, 1870–1970*, pp.84–5 and 183–207; Jim House and Neil MacMaster, *Paris 1961*, pp.56–8; John Pimlott, 'The French Army', pp.62–4; Barnett Singer and John Langdon, *Cultured Force*, pp.312–4.

57 Ian F.W. Beckett, *Modern Insurgencies and Counter-Insurgencies*, pp.165–6; Raymond F. Betts, *France and Decolonisation, 1900–1960*, pp.24–5, 104–5; Anthony Clayton, *The Wars of French Decolonization*, pp.132–3 and fn.4, p.134; Lieutenant Colonel Philippe Francois, 'Waging Counterinsurgency in Algeria', pp.62, 65; Alf Andrew Heggoy, *Insurgency and Counterinsurgency in Algeria*, pp.235–44; Alistair Horne, *A Savage War of Peace*, pp.231–50; Alistair Horne, *The French Army and Politics, 1870–1970*, pp.85–6; John Pimlott, 'The French Army', pp.64, 66; Barnett Singer and John Langdon, *Cultured Force*, pp.316, 319–20.

58 Ian F.W. Beckett, *Modern Insurgencies and Counter-Insurgencies*, pp.162 and 165; Raymond F. Betts, *France and Decolonisation, 1900–1960*, pp.104 and 109; Anthony

Clayton, *The Wars of French Decolonization*, pp.158, 160 and 178; Lieutenant Colonel Philippe Francois, 'Waging Counterinsurgency in Algeria', pp.60–3; Yoav Gortzak, 'Using Indigenous Forces in Counterinsurgency Operations', pp.314–6; Christopher C. Harmon, 'Illustrations of "Learning" in Counterinsurgency', p.355; Alistair Horne, *A Savage War of Peace*, pp.317–9, 330–40; Alistair Horne, *The French Army and Politics, 1870–1970*, pp.85–6, 156 and 158; Jim House and Neil MacMaster, *Paris 1961*, pp.27–8, 63–6; Colonel Gilles Martin, 'War in Algeria', pp.53–5; John Pimlott, 'The French Army', p.66; Martin Thomas, 'Order before Reform', p.200; Sir Robert Thompson, *Revolutionary War in World Strategy, 1945–1969*, p.93.

59   Ian F.W. Beckett, *Modern Insurgencies and Counter-Insurgencies*, pp.162–3, 167–8; Raymond F. Betts, *France and Decolonisation, 1900–1960*, pp.108, 112–3; Anthony Clayton, *The Wars of French Decolonization*, pp.63–4, 162–3, 168–70; Alistair Horne, *A Savage War of Peace*, pp.14, 308–11, 349–72, 436–67, 480–504; Alistair Horne, *The French Army and Politics, 1870–1970*, pp.79–82; Colonel Gilles Martin, 'War in Algeria', pp.56–7; Barnett Singer and John Langdon, *Cultured Force*, p.337–9.

60   Raymond F. Betts, *France and Decolonisation, 1900–1960*, p.113; Dennis Chaplin, 'France – Military Involvement in Africa', pp.44–7; John Chipman, *French Power in Africa*, pp.114–67; Anthony Clayton, *The Wars of French Decolonization*, pp.12, 186; Michel L. Martin, 'From Algiers to N'Djamena', pp.100–7, 119 and 123–6; John Pimlott, 'The French Army', pp.46, 67–8, 72–3; Barnett Singer and John Langdon, *Cultured Force*, pp.348 and 350.

61   John Pimlott, 'The French Army', pp.46, 67–8 and 73.

62   John Chipman, *French Power in Africa*, pp.29–30; Alistair Horne, *A Savage War of Peace*, p.544; Alistair Horne, *The French Army and Politics, 1870–1970*, pp.87–8.

63   Colonel Gilles Martin, 'War in Algeria', pp.51, 57.

64   Lieutenant Colonel Henri Boré, 'Irregular Warfare in Africa', pp.38–40; Anthony Clayton, *The Wars of French Decolonization*, p.184; Alistair Horne, *The French Army and Politics, 1870–1970*, pp.88–9.

Chapter 5

1   Andrew J. Birtle, *US Army Counterinsurgency and Contingency Operations Doctrine, 1860–1941*, pp.4–15; Janine Davidson, *Lifting the Fog of Peace*, p.30; Daniel E. Fitz-Simons, 'Francis Marion the "Swamp Fox"', p.1; John Grenier, *The First Way of War*, pp.217–20; Armstrong Starkey, *European and Native American Warfare, 1675–1815*, pp.139, 141, 151–5; David H. Ucko, *The New Counterinsurgency Era*, pp.27–8; Robert Utley, 'Introduction', pp.1–8; Robert M. Utley and Wilcomb E. Washburn, *Indian Wars*, pp.108, 114–5; John D. Waghelstein, 'Regulars, Irregulars and Militia', pp.135–6, 141–3, 153.

2   Andrew J. Birtle, *US Army Counterinsurgency and Contingency Operations Doctrine, 1860–1941*, pp.16–18; John M. Gates, 'Indians and Insurrectos', pp.59–60, 64–5; John D. Waghelstein, 'The Mexican War and the American Civil War', pp.140–5; Russell F. Weigley, *History of the United States Army*, pp.187–9.

3   Ian F.W. Beckett, *Modern Insurgencies and Counter-Insurgencies*, p.30; Andrew J. Birtle, *US Army Counterinsurgency and Contingency Operations Doctrine, 1860–1941*, pp.23–30, 33–4; B. Franklin Cooling, 'A People's War', pp.121–2, 125; Janine Davidson, *Lifting the Fog of Peace*, pp.32–4; Larry Gordon, *The Last Confederate General*, pp.75–7; Colonel Carl E. Grant, 'Partisan Warfare, Model 1861–65', pp.44, 47; Mark Grimsley, *The Hard Hand of War*, p.112; Robert R. McKey, *The Uncivil War*, pp.72–94, 125–54; Mark Moyar, *A Question of Command*, pp.16–7; Daniel E. Sutherland, *A Savage Conflict*, pp.58–64, 76–83, 96, 123, 165–6.

4   Andrew J. Birtle, *US Army Counterinsurgency and Contingency Operations Doctrine, 1860–1941*, pp.36–7.

5   Michael Fellman, *Inside War*, p.113.

6   Ian F.W. Beckett, *Modern Insurgencies and Counter-Insurgencies*, pp.30–1; Andrew J. Birtle, *US Army Counterinsurgency and Contingency Operations Doctrine, 1860–1941*, pp.35–7; Don R. Bowen, 'Quantrill, James, Younger, *et al.*', p.42; B. Franklin Cooling, 'A People's War', p.116; Janine Davidson, *Lifting the Fog of Peace*, pp.32–4; Michael Fellman, *Inside War*, pp.126–7; Thomas Goodrich, *Black Flag*, pp.44–5; Mark Moyar, *A Question of Command*, pp.16–7; W. Wayne Smith, 'An Experiment in Counterinsurgency', pp.362–72, 375, 377, 380.

7   Andrew J. Birtle, *US Army Counterinsurgency and Contingency Operations Doctrine, 1860–1941*, p.37; Janine Davidson, *Lifting the Fog of Peace*, pp.32–4; Michael Fellman, *Inside War*, pp.94–7; Larry Gordon, *The Last Confederate General*, p.86; Mark Grimsley, *The Hard Hand of War*, pp.118–9; Mark Moyar, *A Question of Command*, pp.18–26.

8   Ian F.W. Beckett, *Modern Insurgencies and Counter-Insurgencies*, pp.30–1; Andrew J. Birtle, *US Army Counterinsurgency and Contingency Operations Doctrine, 1860–1941*, pp.38–40; Janine Davidson, *Lifting the Fog of Peace*, pp.32–4; John M. Gates, 'Indians and Insurrectos', p.65; Mark Grimsley, *The Hard Hand of War*, pp.174–8, 186–203; Robert R. McKey, *The Uncivil War*, pp.113–20; Daniel E. Sutherland, *A Savage Conflict*, pp.151–2, 242–5, 253–4.

9   Andrew J. Birtle, *US Army Counterinsurgency and Contingency Operations Doctrine, 1860–1941*, pp.30–6; B. Franklin Cooling, 'A People's War', p.128; Janine Davidson, *Lifting the Fog of Peace*, p.33; Michael Fellman, *Inside War*, pp.82–6; John M. Gates, 'Indians and Insurrectos', pp.64–5; Mark Grimsley, *The Hard Hand of War*, pp.38–9, 85–9; Daniel E. Sutherland, *A Savage Conflict*, pp.97–8, 123–9.

10 Brigadier Nigel R.F. Aylwin-Foster, 'Changing the Army for Counterinsurgency Operations', p.88; Andrew J. Birtle, *US Army Counterinsurgency and Contingency Operations Doctrine, 1860–1941*, pp.40–8; Robert M. Cassidy, 'Back to the Street without Joy', p.75; B. Franklin Cooling, 'A People's War', pp.124, 127; Michael Fellman, *Inside War*, pp.120–31; Donald S. Frazier, 'Out of Stinking Distance', pp.168–9; Brian McAllister Linn, 'The Philippines', p.54; Robert R. McKey, 'Bushwhackers, Provosts, and Tories', pp.172–7, 181–4; Robert R. McKey, *The Uncivil War*, pp.51–71, 99–113, 142–3, 168–70; Mark Moyar, *A Question of Command*, pp.26–8; Clyde R. Simmons, 'The Indian Wars and US Military Thought, 1865–1890', pp.60–1, 68–70; Daniel E. Sutherland, *A Savage Conflict*, pp.146, 209; David H. Ucko, *The New Counterinsurgency Era*, pp.26–8, Robert M. Utley, *The Indian Frontier of the American West 1846–1890*, pp.166–7.

11 Andrew J. Birtle, *US Army Counterinsurgency and Contingency Operations Doctrine, 1860–1941*, pp.55–67; Larry Cable, 'Reinventing the Round Wheel', pp.231–2; Brian W. Dippie, 'George A. Custer', p.104; Kendall D. Gott, *In Search of an Elusive Enemy*, p.9; Paul Andrew Hutton, *Phil Sheridan and his Army*, pp.67–9, 80–2, 98–100, 111–3, 128–9, 180–6; Paul Andrew Hutton, 'Philip H. Sheridan', pp.81–94; J'Nell L. Pate, 'Ranald S. Mackenzie', pp.178–87; Clyde R. Simmons, 'The Indian Wars and US Military Thought, 1865–1890', p.68; Iain R. Smith and Andreas Stucki, 'The Colonial Development of Concentration Camps (1868–1902)', pp.424–5; Robert M. Utley, *Frontier Regulars: The United States Army and the Indian, 1866–1890*, pp.44–56; Robert Utley, 'Introduction', p.7; Robert M. Utley, 'Total War on the American Indian Frontier', pp.401–8; Robert M. Utley and Wilcomb E. Washburn, *Indian Wars*, p.210; Bruce Vandervort, *Indian Wars of Mexico, Canada and the United States, 1812–1900*, pp.7–11, 72–4, 166–7, 169.

12 Andrew J. Birtle, *US Army Counterinsurgency and Contingency Operations Doctrine, 1860–1941*, pp.69–76, 90; Larry Cable, 'Reinventing the Round Wheel', pp.231–3; Robert Cassidy, 'Regular and Irregular Indigenous Forces for a Long War', p.45; Bruce J. Dingles, 'Benjamin H. Grierson', pp.165–6; Kendall D. Gott, *In Search of an Elusive Enemy*, pp.10–11, 32–40, 43–5; Jerome A. Greene, 'George Crook', pp.117–30; Paul Andrew Hutton, *Phil Sheridan and his Army*, pp.367–8; Robert M. Utley, *Frontier Regulars: The United States Army and the Indian, 1866–1890*, pp.53–5, 177–81, 192–8, 377–86; Robert Utley, 'Introduction', pp.5–6; Robert M. Utley, 'Nelson A. Miles', pp.222–3; Robert M. Utley, 'Total War on the American Indian Frontier', pp.408–9; Bruce Vandervort, *Indian Wars of Mexico, Canada and the United States, 1812–1900*, pp.202–3, 207–10; John D. Waghelstein, 'Preparing the US Army for the Wrong War, Educational and Doctrinal Failure 1865–91', pp.20–4; Robert N. Watt, 'Raiders of a Lost Art?', pp.3–4; Robert Wooster, *Nelson A Miles and the Twilight of the Frontier Army*, pp.144–59.

13 Andrew J. Birtle, *US Army Counterinsurgency and Contingency Operations Doctrine, 1860–1941*, pp.110–2; John Morgan Gates, *Schoolbooks and Krags*, pp.3–7, 22–42, 76–114; John M. Gates, 'The Pacification of the Philippines, 1898–1902', pp.79–81; Anthony James Joes, 'Counterinsurgency in the Philippines, 1898–1954', pp.38–45; Brian McAllister Linn, *The US Army and Counterinsurgency in the Philippine War, 1899–1902*, pp.1–17, 21–2; Brian McAllister Linn, *The Philippine War, 1899–1902*, pp.185–90; Glenn A. May, 'Was the Philippine–American War a "Total War"?', pp.439–40; Robert D. Ramsey III, *Savage Wars of Peace*, pp.1, 11–25.

14 Brian McAllister Linn, *The Philippine War, 1899–1902*, pp.198–200; Brian McAllister Linn, 'Intelligence and Low-Intensity Conflict in the Philippine War, 1899–1902', pp.90–6; Alfred W. McCoy, *Policing America's Empire*, pp.17–8.

15 John Morgan Gates, *Schoolbooks and Krags*, pp.156–78; Brian McAllister Linn, *The Philippine War, 1899–1902*, pp.189–91; Alfred W. McCoy, *Policing America's Empire*, pp.17–8; Mark Moyar, *A Question of Command*, p.76; Robert D. Ramsey III, *Savage Wars of Peace*, pp.39–50.

16 Andrew J. Birtle, *US Army Counterinsurgency and Contingency Operations Doctrine, 1860–1941*, pp.117–9; Robert M. Cassidy, 'The Long Small War', pp.49–50; Brian M. Linn, 'Provincial Pacification in the Philippines, 1900–1901', p.64; Brian McAllister Linn, 'Intelligence and Low-Intensity Conflict in the Philippine War, 1899–1902', pp.96–109; Brian McAllister Linn, 'The Philippines', p.53; Alfred W. McCoy, *Policing America's Empire*, pp.28–9, 33–6.

17 Andrew J. Birtle, *US Army Counterinsurgency and Contingency Operations Doctrine, 1860–1941*, pp.114–7; Timothy K. Deady, 'Lessons from a Successful Counterinsurgency', p.59; Brian McAllister Linn, 'Cerberus' dilemma', p.118; Alfred W. McCoy, *Policing America's Empire*, pp.82–92; Mark Moyar, *A Question of Command*, pp.75–6.

18 Edward M. Coffman, 'Batson of the Philippine Scouts', pp.68–72; Brian McAllister Linn, *The Philippine War, 1899–1902*, pp.203–4; Glenn A. May, *Battle for Batangas*, pp.259–60; Dr Richard L. Millett, *Searching for Stability*, pp.8–12; Mark Moyar, *A Question of Command*, pp.78–9; Robert D. Ramsey III, *Savage Wars of Peace*, pp.19–20, 57–8, 96; Lieutenant Colonel Richard W. Smith, 'Philippine Constabulary', pp.74–7.

19 Andrew J. Birtle, *US Army Counterinsurgency and Contingency Operations Doctrine, 1860–1941*, pp.123–32; Max Boot, *The Savage Wars of Peace*, p.116; John Morgan Gates, *Schoolbooks and Krags*, pp.187–200, 204–20; Brian McAllister Linn, *The US Army and Counterinsurgency in the Philippine War, 1899–1902*, pp.23–5; Brian McAllister Linn, *The Philippine War, 1899–1902*, pp.213–5; Robert D. Ramsey III,

Savage Wars of Peace, pp.51–64, 113–6; Robert D. Ramsey III, *A Masterpiece of Counterguerrilla Warfare*, p.4.

20 Andrew J. Birtle, *US Army Counterinsurgency and Contingency Operations Doctrine, 1860–1941*, pp.122–32; Max Boot, *The Savage Wars of Peace*, p.116; John Morgan Gates, *Schoolbooks and Krags*, pp.225–42; Brian McAllister Linn, *The US Army and Counterinsurgency in the Philippine War, 1899–1902*, pp.139, 145; Brian McAllister Linn, *The Philippine War, 1899–1902*, pp.223–4; Mark Moyar, *A Question of Command*, pp.76–7; Robert D. Ramsey III, *Savage Wars of Peace*, pp.51–64, 73, 77–90; Frank Schumacher, 'Marked Severities', pp.482–92.

21 Andrew J. Birtle, *US Army Counterinsurgency and Contingency Operations Doctrine, 1860–1941*, pp.133–5; Max Boot, *The Savage Wars of Peace*, pp.99–102, 120–2; John Morgan Gates, *Schoolbooks and Krags*, pp.252–6; Brian McAllister Linn, *The US Army and Counterinsurgency in the Philippine War, 1899–1902*, p.27; Brian McAllister Linn, *The Philippine War, 1899–1902*, pp.310–21; Glenn A. May, 'Was the Philippine–American War a "Total War"?', pp.443–6; Mark Moyar, *A Question of Command*, pp.80–3; Robert D. Ramsey III, *A Masterpiece of Counterguerrilla Warfare*, pp.5, 7.

22 Ian F.W. Beckett, 'The United States Experience', pp.105, 113–4; Andrew J. Birtle, *US Army Counterinsurgency and Contingency Operations Doctrine, 1860–1941*, p.129–31; Max Boot, *The Savage Wars of Peace*, pp.123–8; Ewing E. Booth, *My Observations and Experiences in the United States Army*, p.47, quoted in Robert D. Ramsey III, *A Masterpiece of Counterguerrilla Warfare*, p.3; John Morgan Gates, *Schoolbooks and Krags*, pp.256–64; Brian McAllister Linn, *The US Army and Counterinsurgency in the Philippine War, 1899–1902*, pp.27, 155, 159, 322–3; Alfred W. McCoy, *Policing America's Empire*, pp.80–1; Glenn A. May, *Battle for Batangas*, pp.242–69; Glenn A. May, 'Was the Philippine–American War a "Total War"?', pp.446–57; Mark Moyar, *A Question of Command*, pp.83–6; Robert D. Ramsey III, *Savage Wars of Peace*, pp.73, 90–103; Iain R. Smith and Andreas Stucki, 'The Colonial Development of Concentration Camps (1868–1902)', pp.424–5.

23 Stephen E. Ambrose, *Upton and the Army*, pp.87–111; Ian F.W. Beckett, 'The United States Experience', pp.116–7, 123; Andrew J. Birtle, *US Army Counterinsurgency and Contingency Operations Doctrine, 1860–1941*, pp.86–92, 136–9, 174, 181–2, 239, 279; Paul Andrew Hutton, *Phil Sheridan and his Army*, pp.159–62; Brian McAllister Linn, *The US Army and Counterinsurgency in the Philippine War, 1899–1902*, p.27; Brian McAllister Linn, *Guardians of Empire*, pp.47–8; Austin Long, *Doctrine of Eternal Recurrence*, p.4; Robert D. Ramsey III, *Savage Wars of Peace*, pp.113–6; Robert M. Utley, *Frontier Regulars: The United States Army and the Indian, 1866–1890*, pp.44–5; Russell F. Weigley, *History of the United States Army*, pp.273–81, 314–26; Russell F. Weigley, 'The Elihu Root Reforms and the Progressive Era', pp.18–22.

24 Ian F.W. Beckett, 'The United States Experience', p.117; Keith B. Bickel, *Mars Learning*, pp.18–9, 32–4, 38–9, 41–2, 69–98; Max Boot, *The Savage Wars of Peace*, pp.162–7, 171–80; Larry E. Cable, *Conflict of Myths*, p.97; Bruce Gudmundsson, 'The First of the Banana Wars'; pp.55, 61–9; Dr Richard L. Millett, *Searching for Stability*, pp.50–68.

25 Ian F.W. Beckett, 'The United States Experience', p.117; Keith B. Bickel, *Mars Learning*, pp.107–30; Max Boot, *The Savage Wars of Peace*, pp.167–71; Captain Stephen M. Fuller and Graham A. Cosmas, *Marines in the Dominican Republic, 1916–1924*, pp.22, 33–52; Victor H. Krulak, *First to Fight*, pp.190–1; Dr Richard L. Millett, *Searching for Stability*, pp.79–91.

26 Ian F.W. Beckett, 'The United States Experience', pp.117–23; Keith B. Bickel, *Mars Learning*, pp.155–79; Max Boot, *The Savage Wars of Peace*, pp.231–52; David C.

Brooks, 'US Marines, Miskitos and the Hunt for Sandino', pp.315–6, 320–1, 332–8; Larry E. Cable, *Conflict of Myths*, pp.97–108; Robert Cassidy, 'Regular and Irregular Indigenous Forces for a Long War', pp.44–5; First Lieutenant Neill Macaulay, 'Leading Native Troops', pp.32–5; First Lieutenant Neill Macaulay, 'Counterguerrilla Patrolling', pp.45–8; Dr Richard L. Millett, *Searching for Stability*, pp.98–109.

27 Andrew J. Birtle, *US Army Counterinsurgency and Contingency Operations Doctrine, 1942–1976*, pp.8–11; Michael McClintock, *Instruments of Statecraft*, pp.11–21, 59–73.

28 Andrew J. Birtle, *US Army Counterinsurgency and Contingency Operations Doctrine, 1942–1976*, pp.42–5; Paul F. Braim, *The Will to Win*, pp.158–61, 194–200, 204, 207; Larry E. Cable, *Conflict of Myths*, pp.9–11, 16–8; Robert M. Mages, 'Without the Need of a Single American Rifleman', pp.196; Edgar O'Balance, *The Greek Civil War, 1944–1949*, pp.141–8; Charles R. Shrader, *The Withered Vine*, pp.80–1, 218–21.

29 Andrew J. Birtle, *US Army Counterinsurgency and Contingency Operations Doctrine, 1942–1976*, pp.45–8, 51–2; Paul F. Braim, *The Will to Win*, pp.161–3, 170–80, 182–4; Larry E. Cable, *Conflict of Myths*, pp.11–7, 20; Michael McClintock, *Instruments of Statecraft*, p.12; Robert M. Mages, 'Without the Need of a Single American Rifleman', pp.196–8, 202–8; Edgar O'Ballance, *The Greek Civil War, 1944–1949*, pp.148, 156–7, 165–6.

30 Lieutenant Colonel Anastase Balcos, 'Guerrilla Warfare', p.54; Andrew J. Birtle, *US Army Counterinsurgency and Contingency Operations Doctrine, 1942–1976*, pp.48–51; Paul F. Braim, *The Will to Win*, pp.201–2; Tim Jones, 'The British Army, and Counter-Guerrilla Warfare in Transition, 1944–1952', p.149; Michael McClintock, *Instruments of Statecraft*, pp.13–7; Robert M. Mages, 'Without the Need of a Single American Rifleman', pp.209–10; Edgar O'Ballance, *The Greek Civil War, 1944–1949*, pp.166–9, 213–5; Lieutenant Colonel Edward R. Wainhouse, 'Guerrilla War in Greece, 1946–49', p.24; C.M. Woodhouse, *The Struggle for Greece, 1941–1949*, pp.233–4; Charles R. Shrader, *The Withered Vine*, pp.124–7.

31 Andrew J. Birtle, *US Army Counterinsurgency and Contingency Operations Doctrine, 1942–1976*, pp.52–5; Paul F. Braim, *The Will to Win*, pp.181, 187–92, 200–20; Larry E. Cable, *Conflict of Myths*, pp.18–29; Michael McClintock, *Instruments of Statecraft*, pp.13–7; Lieutenant Colonel John J. McCuen, *The Art of Counter-Revolutionary War*, pp.299–309; Robert M. Mages, 'Without the Need of a Single American Rifleman', pp.198–9, 205, 208; Edgar O'Ballance, *The Greek Civil War, 1944–1949*, pp.154, 167–202, 210–9; Charles R. Shrader, *The Withered Vine*, pp.69–72, 77–8, 81–4, 223–41, 256–67; John Shy and Thomas W. Collier, 'Revolutionary War', p.845; Sir Robert Thompson, *Revolutionary War in World Strategy, 1945–1969*, pp.47–51; Lieutenant Colonel Edward R. Wainhouse, 'Guerrilla War in Greece, 1946–49', pp.24–5; C.M. Woodhouse, *The Struggle for Greece, 1941–1949*, pp.233–8, 276.

32 Andrew J. Birtle, *US Army Counterinsurgency and Contingency Operations Doctrine, 1942–1976*, pp.52–5; Paul F. Braim, *The Will to Win*, pp.181, 187–92, 200–20; Larry E. Cable, *Conflict of Myths*, pp.18–29; Michael McClintock, *Instruments of Statecraft*, pp.13–7; Lieutenant Colonel John J. McCuen, *The Art of Counter-Revolutionary War*, pp.299–309; Robert M. Mages, 'Without the Need of a Single American Rifleman', pp.198–9, 205, 208; Edgar O'Ballance, *The Greek Civil War, 1944–1949*, pp.154, 167–202, 210–9; Charles R. Shrader, *The Withered Vine*, pp.69–72, 77–8, 81–4, 223–41, 256–67; John Shy and Thomas W. Collier, 'Revolutionary War', p.845; Sir Robert Thompson, *Revolutionary War in World Strategy, 1945–1969*, pp.47–51; C.M. Woodhouse, *The Struggle for Greece, 1941–1949*, pp.233–8, 276.

33 Andrew J. Birtle, *US Army Counterinsurgency and Contingency Operations Doctrine, 1942–1976*, pp.55–8; Larry E. Cable, *Conflict of Myths*, pp.44–52; Lawrence

M. Greenberg, *The Hukbalahap Insurrection*, pp.43–5, 56–78; Anthony James Joes, 'Counterinsurgency in the Philippines, 1898–1954', pp.48–9; Wray R. Johnson and Paul J. Dimech, 'Foreign Internal Defense and the Hukbalahap' pp.373–7; Benedict J. Kerkvliet, *The Huk Rebellion*, pp.210–8; Alfred W. McCoy, *Policing America's Empire*, pp.374–5; Mark Moyar, *A Question of Command*, pp.94–5.

34 Ian F.W. Beckett, *Modern Insurgencies and Counter-Insurgencies*, p.105; Andrew J. Birtle, *US Army Counterinsurgency and Contingency Operations Doctrine, 1942–1976*, pp.58–61; Larry E. Cable, *Conflict of Myths*, pp.52–7; Lawrence M. Greenberg, *The Hukbalahap Insurrection*, pp.100–2, 108–10; Benedict J. Kerkvliet, *The Huk Rebellion*, pp.160, 196–8, 210–1; Michael McClintock, *Instruments of Statecraft*, pp.101–6, 120–6; A.H. Peterson, G.C. Reinhardt and E.E. Conger, *Symposium on … The Philippine Huk Campaign*, pp.14–17; Robert Ross Smith, 'The Hukbalahap Insurgency', pp.36–9.

35 Ian F.W. Beckett, *Modern Insurgencies and Counter-Insurgencies*, p.104; Andrew J. Birtle, *US Army Counterinsurgency and Contingency Operations Doctrine, 1942–1976*, pp.61–3; Lawrence M. Greenberg, *The Hukbalahap Insurrection*, pp.82–9; Anthony James Joes, 'Counterinsurgency in the Philippines, 1898–1954', pp.50–1; Benedict J. Kerkvliet, *The Huk Rebellion*, pp.238, 240–5; Michael McClintock, *Instruments of Statecraft*, pp.106–15; Mark Moyar, *A Question of Command*, pp.97–101, 105–7; A.H. Peterson, G.C. Reinhardt and E.E. Conger, *Symposium on … The Philippine Huk Campaign*, pp.17–9; Colonel Napoleon D. Valeriano and Lieutenant Colonel Charles T.R. Bohannan, *Counter-Guerrilla Operations*, pp.29, 100–10, 139–41, 207–8.

36 Andrew J. Birtle, *US Army Counterinsurgency and Contingency Operations Doctrine, 1942–1976*, pp.63–4; Larry E. Cable, *Conflict of Myths*, pp.58–60, 66; Lawrence M. Greenberg, *The Hukbalahap Insurrection*, pp.89–92; Benedict J. Kerkvliet, *The Huk Rebellion*, pp.238–40; Mark Moyar, *A Question of Command*, pp.103–4; A.H. Peterson, G.C. Reinhardt and E.E. Conger, *Symposium on … The Philippine Huk Campaign*, pp.14–17, 29–33.

37 Andrew J. Birtle, *US Army Counterinsurgency and Contingency Operations Doctrine, 1942–1976*, pp.64–6; Larry E. Cable, *Conflict of Myths*, pp.60–7; Lawrence M. Greenberg, *The Hukbalahap Insurrection*, pp.110–11, 125–41; Anthony James Joes, 'Counterinsurgency in the Philippines, 1898–1954', pp.50–4; Benedict J. Kerkvliet, *The Huk Rebellion*, pp.233–48; Michael McClintock, *Instruments of Statecraft*, pp.83, 125–6; Mark Moyar, *A Question of Command*, pp.101–3; A.H. Peterson, G.C. Reinhardt and E.E. Conger, *Symposium on … The Philippine Huk Campaign*, pp.19–21; Colonel Napoleon D. Valeriano and Lieutenant Colonel Charles T.R. Bohannan, *Counter-Guerrilla Operations*, pp.117–23, 127–9, 131–9, 200.

38 Andrew J. Birtle, *US Army Counterinsurgency and Contingency Operations Doctrine, 1942–1976*, pp.86–8; Larry E. Cable, *Conflict of Myths*, pp.33–5; Peter Clemens, 'Captain James Hausman, US Army Military Advisor to Korea, 1946–48', pp.189–91; Bruce Cumings, *Korea's Place in the Sun*, pp.243–5; Mark J. Reardon, 'Chasing a Chameleon', pp.215–22; Major Robert K. Sawyer, *Military Advisors in Korea*, pp.12–33, 38–42.

39 Andrew J. Birtle, *US Army Counterinsurgency and Contingency Operations Doctrine, 1942–1976*, pp.89–97; Peter Clemens, 'Captain James Hausman, US Army Military Advisor to Korea, 1946–48', pp.190–1; Bruce Cumings, *Korea's Place in the Sun*, p.245; Soul Park, 'The Unnecessary Uprising', pp.359–61, 368–74; Mark J. Reardon, 'Chasing a Chameleon', pp.223, 226–7; Major Robert K. Sawyer, *Military Advisors in Korea*, pp.39–41, 57–66, 76–90, 186–7.

40 Andrew J. Birtle, *US Army Counterinsurgency and Contingency Operations Doctrine, 1942–1976*, pp.90–1, 93–4, 97–8; Larry E. Cable, *Conflict of Myths*, p.39; Bruce Cumings, *Korea's Place in the Sun*, p.246.

41 Andrew J. Birtle, *US Army Counterinsurgency and Contingency Operations Doctrine, 1942–1976*, pp.98–115; Larry E. Cable, *Conflict of Myths*, pp.35–41; Michael McClintock, *Instruments of Statecraft*, pp.19–20; Mark J. Reardon, 'Chasing a Chameleon', pp.223–8.

42 Andrew J. Birtle, *US Army Counterinsurgency and Contingency Operations Doctrine, 1942–1976*, pp.329–30; Major Daniel P. Bolger, *Scenes from an Unfinished War*, pp.8–11, 37–41, 119–24; Daniel P. Bolger, 'Unconventional Warrior', pp.67–71.

43 Andrew J. Birtle, *US Army Counterinsurgency and Contingency Operations Doctrine, 1942–1976*, pp.330–4; Major Daniel P. Bolger, *Scenes from an Unfinished War*, pp.46–59, 62–3, 79–87, 95–101, 111–15, 118–19; Daniel P. Bolger, 'Unconventional Warrior', pp.71–6.

44 Andrew J. Birtle, *US Army Counterinsurgency and Contingency Operations Doctrine, 1942–1976*, p.223.

45 Roger Hilsman, *To Move a Nation*, p.426.

46 Ian F.W. Beckett, *Modern Insurgencies and Counter-Insurgencies*, pp.184–5; Eric M. Bergerud, *The Dynamics of Defeat*, pp.85–90; Andrew J. Birtle, *US Army Counterinsurgency and Contingency Operations Doctrine, 1942–1976*, pp.223–7; Douglas S. Blaufarb, *The Counterinsurgency Era*, p.207; John M. Collins, 'Vietnam Postmortem: A Senseless Strategy', *Parameters*, Vol.VIII, No.1 (March 1978), pp.8–11; Lawrence Freedman, *Kennedy's Wars*, pp.287–92, 338–9; Michael T. Klare and Peter Kornbluh, 'The New Interventionism', pp.9–12; Andrew F. Krepinevich, *The Army and Vietnam*, pp.27–36, 64–5; Richard Lock-Pullan, *US Intervention Policy and Army Innovation*, pp.18–22; Michael McClintock, *Instruments of Statecraft*, pp.179–88; Charles Maechling, 'Counterinsurgecy', pp.21–35; John A. Nagl, 'Counterinsurgency in Vietnam', pp.134–5; John A. Nagl, *Counterinsurgency Lessons from Malaya and Vietnam*, pp.124–7, 131–3, 139–40; General Bruce Palmer Jr, *The 25-Year War*, pp.155–6 William Rosenau, *US Internal Security Assistance to South Vietnam*, pp.87–90; Lewis Sorley, *Honorable Warrior*, pp.152–7, 201–3.

47 Brian M. Jenkins, *The Unchangeable War*, p.3; Guenter Lewy, *America in Vietnam*, p.138.

48 Andrew J. Birtle, *US Army Counterinsurgency and Contingency Operations Doctrine, 1860–1941*, pp.136–9; Andrew J. Birtle, *US Army Counterinsurgency and Contingency Operations Doctrine, 1942–1976*, pp.229–76; Andrew F. Krepinevich, *The Army and Vietnam*, pp.36–55; Brian McAllister Linn, *The US Army and Counterinsurgency in the Philippine War, 1899–1902*, p.27; John A Nagl, *Counterinsurgency Lessons from Malaya and Vietnam*, pp.128–9, 137, 139; Robert D. Ramsey III, *Savage Wars of Peace*, pp.113–6; Sam C. Sarkesian, 'The American Response to Low-Intensity Conflict', pp.29–31, 34–9; John Shy and Thomas W. Collier, 'Revolutionary War', p.855; Lewis Sorley, *Thunderbolt*, pp.269–78; Robert L. Tonsetic, *Forsaken Warriors*, pp.8–9, 120–2; John D. Waghelstein, 'Post Vietnam Counterinsurgency Doctrine', pp.47–8.

49 Douglas Kinnard, *The War Managers*, pp.109–17; Major General Edward Lansdale, 'Contradictions in Military Culture', pp.40–2; Guenter Lewy, *America in Vietnam*, pp.118–9, 153–61; Thomas R. Mockaitis, *British Counterinsurgency, 1919–60*, pp.173–5; Ray, Captain James F., 'The District Advisor', pp.6–8; Lewis Sorley, *Westmoreland*, pp.86–8; Ronald H. Spector, *Advice and Support: The Early Years, 1941–1960*, pp.241–2, 286–91; Samuel Zaffiri, *Westmoreland*, pp.343–9.

50 Eric M. Bergerud, *The Dynamics of Defeat*, pp.24–7; Andrew J. Birtle, *US Army Counterinsurgency and Contingency Operations Doctrine, 1942–1976*, pp.307–12, 322–3; Jeffrey J. Clarke, *Advice and Support: The Final Years, 1965–1973*, p.7; Lawrence Freedman, *Kennedy's Wars*, p.290; Christopher K. Ives, *US Special Forces and*

Counterinsurgency in Vietnam, pp.72–3, 84–6; Andrew F. Krepinevich, *The Army and Vietnam*, pp.22–4; Michael McClintock, *Instruments of Statecraft*, pp.19–20; Mark Moyar, *A Question of Command*, pp.137–8; John A. Nagl, 'Counterinsurgency in Vietnam', pp.132–5; John A. Nagl, *Counterinsurgency Lessons from Malaya and Vietnam*, pp.120, 122–3; William Rosenau, *US Internal Security Assistance to South Vietnam*, p.67; Ronald H. Spector, 'The First Vietnamization', pp.111, 113–5; Ronald H. Spector, *Advice and Support: The Early Years, 1941–1960*, pp.221–2, 262–8, 278–302, 320–5, 343–57.

51 Andrew J. Birtle, *US Army Counterinsurgency and Contingency Operations Doctrine, 1942–1976*, pp.361, 373–5; Andrew F. Krepinevich, *The Army and Vietnam*, pp.3–4, 78–90, 99, 100, 134–61, 164–5; John A. Nagl, 'Counterinsurgency in Vietnam', p.137.

52 Ian F.W. Beckett, *Modern Insurgencies and Counter-Insurgencies*, pp.197–8; Andrew J. Birtle, *US Army Counterinsurgency and Contingency Operations Doctrine, 1942–1976*, pp.308–9, 312–13, 325, 327; Anne Blair, *There to the Bitter End*, pp.107–20; Peter M. Dunn, 'The American Army: The Vietnam War, 1965–1973', p.94; Christopher C. Harmon, 'Illustrations of "Learning" in Counterinsurgency', pp.364–5; Andrew F. Krepinevich, *The Army and Vietnam*, pp.24–6, 218–21, 227–30; Lieutenant General Victor H. Krulak, *First to Fight*, pp.187–8; James McAllister and Ian Schulte, 'The Limits of Influence in Vietnam', pp.32–5; Mark Moyar, *Phoenix and Birds of Prey*, pp.159–62; Mark Moyar, *A Question of Command*, pp.138–41; John A. Nagl, 'Counterinsurgency in Vietnam', pp.135–6, 145; John A. Nagl, *Counterinsurgency Lessons from Malaya and Vietnam*, pp.120–4; Jeffrey Race, *War Comes to Long An*, pp.61–2, 113–6; William Rosenau, *US Internal Security Assistance to South Vietnam*, pp.37–9, 53–5, 61–76, 101–2, 105–7, 117–9, 123–30; William Rosenau and Austin Lang, *The Phoenix Program and Contemporary Counterinsurgency*, p.7; Geoffrey D.T. Shaw, 'Policemen versus Soldiers', pp.58–69; Ronald H. Spector, 'The First Vietnamization', pp.111–3; Ronald H. Spector, *Advice and Support: The Early Years, 1941–1960*, pp.276–7, 320–5; General Lewis W. Walt, *Strange War, Strange Strategy*, pp.101–6.

53 Andrew J. Birtle, *US Army Counterinsurgency and Contingency Operations Doctrine, 1942–1976*, pp.309, 317–8; Christopher C. Harmon, 'Illustrations of "Learning" in Counterinsurgency', pp.364–6; R.W. Komer, *The Malayan Emergency in Retrospect*, pp.84–5; Michael McClintock, *Instruments of Statecraft*, pp.266–74; Dr Richard L Millett, *Searching for Stability*, pp.1–2.

54 Thomas L. Ahern, *Vietnam Declassified*, pp.76–90; Ian F.W. Beckett, 'Robert Thompson and the British Advisory Mission to South Vietnam, 1961–1965', pp.52–6, 58; Ian F.W. Beckett, *Modern Insurgencies and Counter-Insurgencies*, pp.198–9; Eric M. Bergerud, *The Dynamics of Defeat*, pp.33–8, 50–3; Andrew J. Birtle, *US Army Counterinsurgency and Contingency Operations Doctrine, 1942–1976*, pp.318–9; Douglas S. Blaufarb, *The Counterinsurgency Era*, pp.103–15, 122–7; Peter Busch, 'Killing the 'Vietcong', pp.137–56; Philip E. Catton, 'Counterinsurgency and Nation Building, pp.919–32, 938–9; Brigadier Richard L. Clutterbuck, *The Long Long War*, pp.66–73; Gregory A. Daddis, *No Sure Victory*, pp.57–8; Peter M. Dunn, 'The American Army: The Vietnam War, 1965–1973', pp.93–4; Geoffrey Fairbairn, *Revolutionary Guerrilla Warfare*, p.323–7; Lawrence Freedman, *Kennedy's Wars*, pp.336–7; Christopher C. Harmon, 'Illustrations of "Learning" in Counterinsurgency', pp.360–1; Christopher K. Ives, *US Special Forces and Counterinsurgency in Vietnam*, pp.82–7, 108–16; Andrew F. Krepinevich, *The Army and Vietnam*, pp.66–9, 88, 168, 170–1; Edward Miller, *Misalliance*, pp.177–84, 231–9; Mark Moyar, *Phoenix and Birds of Prey*, pp.36–7; Mark Moyar, *A Question of Command*, pp.141–4; John A. Nagl, 'Counterinsurgency in Vietnam', pp.135–6; John A. Nagl,

Counterinsurgency Lessons from Malaya and Vietnam, pp.130–1; William Rosenau, *US Internal Security Assistance to South Vietnam*, pp.101, 107–17.

55 Eric M. Bergerud, *The Dynamics of Defeat*, pp.90–3; Andrew J. Birtle, *US Army Counterinsurgency and Contingency Operations Doctrine, 1942–1976*, pp.373–4; Peter M. Dunn, 'The American Army: The Vietnam War, 1965–1973', pp.85, 89–92; Andrew F. Krepinevich, *The Army and Vietnam*, pp.167–8, 190–3; Guenter Lewy, *America in Vietnam*, pp.50–65; Richard Lock-Pullan, *US Intervention Policy and Army Innovation*, pp.31–5; John A. Nagl, 'Counterinsurgency in Vietnam', pp.137–9; John A. Nagl, *Counterinsurgency Lessons from Malaya and Vietnam*, pp.152–6; Jeffrey Race, *War Comes to Long An*, pp.224–36; Lewis Sorley, *A Better War*, p.4; Lewis Sorley, *Westmoreland*, pp.97–104.

56 Lieutenant Colonel George M. Shuffer Jr, 'Finish Them With Firepower', p.11.

57 Andrew J. Birtle, *US Army Counterinsurgency and Contingency Operations Doctrine, 1942–1976*, p.374.

58 Eric M. Bergerud, *The Dynamics of Defeat*, p.180; Andrew J. Birtle, *US Army Counterinsurgency and Contingency Operations Doctrine, 1942–1976*, pp.374–7, 381–2, 394–6; Robert A. Doughty, *The Evolution of US Army Tactical Doctrine, 1946–76*, pp.36–40; Peter M. Dunn, 'The American Army: The Vietnam War, 1965–1973', p.86; Major James F. Gebhardt, *Eyes Behind the Lines*, pp.49–59; James Kitfield, *Prodigal Soldiers*, pp.158–61; Andrew F. Krepinevich, *The Army and Vietnam*, pp.167–8, 191–2, 198–202; Guenter Lewy, *America in Vietnam*, pp.95–112; Richard Lock-Pullan, *US Intervention Policy and Army Innovation*, pp.35–6; John A. Nagl, 'Counterinsurgency in Vietnam', pp.137–9; Major General Frederick C. Weyand, 'Winning the People in Hau Nghia Province', p.53.

59 Andrew J. Birtle, *US Army Counterinsurgency and Contingency Operations Doctrine, 1942–1976*, pp.391, 399–401; Jeffrey J. Clarke, *Advice and Support: The Final Years, 1965–1973*, pp.180–1; William R. Corson, *The Betrayal*, pp.174–98; Peter M. Dunn, 'The American Army: The Vietnam War, 1965–1973', pp.98–9; Yoav Gortzak, 'The prospects of combined action', pp.138–53; Al Hemingway, *Our War Was Different*, pp.3–18, 39, 49–56, 105–9, 152–6, 177–8; Michael Hennessy, *Strategy in Vietnam*, pp.69, 76–8, 93–8, 111–3, 127; Captain Keith F. Kopets, 'The Combined Action Program', pp.78–80; Andrew F. Krepinevich, *The Army and Vietnam*, pp.172–7, 196–7; Victor H. Krulak, *First to Fight*, pp.180–92; Guenter Lewy, *America in Vietnam*, pp.116–7; Richard Lock-Pullan, *US Intervention Policy and Army Innovation*, pp.43–4; Mark Moyar, *A Question of Command*, pp.154–5; John A. Nagl, 'Counterinsurgency in Vietnam', p.138; John A. Nagl, *Counterinsurgency Lessons from Malaya and Vietnam*, pp.156–8; Michael Peterson, *The Combined Action Platoons*, pp.109–10, 123–4; Dr T.P. Schwartz, 'The Combined Action Program', pp.66, 69–71; Lewis Sorley, *Westmoreland*, pp.100–1; Lieutenant Colonel David H. Wagner, 'A Handful of Marines', pp.45–6; Lewis W. Walt, *Strange War, Strange Strategy*, pp.51–4, 105–12; Captain R.E. Williamson, 'A Briefing for Combined Action', pp.41–2; Lawrence Yates, 'A Feather in their Cap', pp.309–11, 315–21; Samuel Zaffiri, *Westmoreland*, pp.168–70.

60 Ian F.W. Beckett, *Modern Insurgencies and Counter-Insurgencies*, p.197; Eric M. Bergerud, *The Dynamics of Defeat*, pp.115–27; Gregory A. Daddis, *No Sure Victory*, pp.69–70, 93–4; Ronnie E. Ford, 'Intelligence and the Significance of Khe Sanh', p.145; Douglas Kinnard, *The War Managers*, pp.20–1, 39–46; James Kitfield, *Prodigal Soldiers*, pp.69–71, 73, 156–7; Andrew F. Krepinevich, *The Army and Vietnam*, pp.175–7; Austin Long, *Doctrine of Eternal Recurrence*, p.12; John A. Nagl, 'Counterinsurgency in Vietnam', pp.157–8; General Bruce Palmer Jr, *The 25-Year*

War, pp.58–64; Lewis Sorley, *Honorable Warrior*, pp.242–3; Lewis Sorley, *Westmoreland*, pp.66–8, 92–104, 114–20, 138–9, 188; Samuel Zaffiri, *Westmoreland*, pp.118–9, 162–3.

61 Gregory A. Daddis, *No Sure Victory*, p.94.

62 Xiaobing Li, *A History of the Modern Chinese Army*, pp.217–25.

63 Dale Andrade and Lieutenant Colonel James Willbanks, 'CORDS/Phoenix', pp.10–11, 20–2; Ian F.W. Beckett, *Modern Insurgencies and Counter-Insurgencies*, pp.201–2; Eric M. Bergerud, *The Dynamics of Defeat*, pp.81–3, 110–2, 216–26, 241–2, 262–72, 293–300; Andrew J. Birtle, *US Army Counterinsurgency and Contingency Operations Doctrine, 1942–1976*, pp.324–5, 366; Douglas S. Blaufarb, *The Counterinsurgency Era*, p.269; Jeffrey J. Clarke, *Advice and Support: The Final Years, 1965–1973*, pp.361–3; Major Ross Coffey, 'Revisiting CORDS', pp.30–2; Gregory A. Daddis, *No Sure Victory*, pp.152–3; Peter M. Dunn, 'The American Army: The Vietnam War, 1965–1973', pp.95–6; Douglas Kinnard, *The War Managers*, pp.81–99; Andrew F. Krepinevich, *The Army and Vietnam*, pp.239–57; Austin Long, *Doctrine of Eternal Recurrence*, pp.10–12, 17–9; Guenter Lewy, *America in Vietnam*, pp.85–90, 114, 133–53, 162–89; Mark Moyar, *Phoenix and Birds of Prey*, pp.242–78; Mark Moyar, *A Question of Command*, pp.160–4; John A. Nagl, 'Counterinsurgency in Vietnam', pp.141–3, 146; John A. Nagl, *Counterinsurgency Lessons from Malaya and Vietnam*, pp.159–60, 164–73; William Rosenau and Austin Long, *The Phoenix Program and Contemporary Counterinsurgency*, pp.7–14; Lewis Sorley, *Thunderbolt*, pp.192–3, 200–6, 228–41, 247, 254–6, 260–1; Lewis Sorley, 'To Change a War', pp.93–109; Lewis Sorley, 'To Change a War', pp.93–109; Lewis Sorley, *Honorable Warrior*, pp.227–41, 303; Lewis Sorley, A Better War, pp.xi–xiii, 6–7, 20–1, 28–30, 59–79, 138–43, 145–9, 224; Jeffrey Woods, *Counterinsurgency, the Interagency Process, and Vietnam*, p.109; Samuel Zaffiri, Westmoreland, pp.325–8.

64 Eric M. Bergerud, *The Dynamics of Defeat*, pp.315–21; Andrew J. Birtle, *US Army Counterinsurgency and Contingency Operations Doctrine, 1942–1976*, pp.325–7; Laurence E. Grinter, 'How They Lost: Doctrines, Strategies and Outcomes of the Vietnam War', pp.1114–8; Andrew F. Krepinevich, *The Army and Vietnam*, p.251; Guenter Lewy, *America in Vietnam*, pp.196–222, 272–99; Jeffrey H. Michaels, 'Helpless or Deliberate Bystander', pp.570–6; John A. Nagl, 'Counterinsurgency in Vietnam', pp.144–5; John A. Nagl, *Counterinsurgency Lessons from Malaya and Vietnam*, pp.173–4.

65 Cynthia J. Arnson, 'Window on the Past', pp.85–112; Neville Bolt, *The Violent Image: Insurgent Propaganda and the New Revolutionaries*, pp.234–5; Hal Brands, *Latin America's Cold War*, pp.6–7, 47–9,189–210; James S. Corum, 'Rethinking US Army Counterinsurgency Doctrine', pp.122–5; Conrad C. Crane, 'Avoiding Vietnam', p.v, 1–12; Janine Davidson, *Lifting the Fog of Peace*, pp.130–1, 138–9; Robert A. Doughty, *The Evolution of US Army Tactical Doctrine, 1946–76*, pp.40–2; Peter M. Dunn, 'The American Army', pp.99, 104–5; Ernest Evans, *Wars Without Splendor*, pp.89–109; Frederick H. Gareau, *State Terrorism and the United States*, pp.22–30; Todd R. Greentree, *Crossroads of Intervention*, pp.35–7, 97–101; Michael T. Klare and Peter Kornbluh, 'The New Interventionism', pp.3–7; Michael T. Klare, 'The Interventionalist Impulse', pp.53–5; Richard Lock-Pullan, *US Intervention Policy and Army Innovation*, pp.60–3, 74–6, 87–9, 104–5, 140–2; Carnes Lord, 'American Strategic Culture in Small Wars', pp.206–9; Carnes Lord, 'The Role of the United States in Small Wars', p.93; Michael McClintock, *Instruments of Statecraft*, pp.329–47, 421–3; Katherine E. McCoy, 'Trained to Torture?', pp.47–50; Charles Maechling Jr, 'Counterinsurgency', pp.45–8, 53–5; John A. Nagl, 'Counterinsurgency in Vietnam', pp.131–2, 146–8; John A. Nagl, *Counterinsurgency Lessons from Malaya and Vietnam*, pp.xiv–xv; John A. Nagl, 'An American View of Twenty-First Century

Counterinsurgency', p.13; Paul B. Rich, 'A historical overview of US counterinsurgency', p.21; Daniel Siegel and Joy Hackel, 'El Salvador', pp.112–35; Terry Terriff, 'Of Romans and Dragons', pp.137–49; John D. Waghelstein, 'Counterinsurgency Doctrine and Low-Intensity Conflict in the Post-Vietnam Era', pp.129–30; John D. Waghelstein, 'Ruminations of a Pachyderm or What I Learnt in the Counterinsurgency Business', pp.360–3, 368–9.

Chapter 6

1   Ian F.W. Beckett, 'The Study of Counterinsurgency', pp.47–9; Christopher C. Harmon, 'Illustrations of "Learning" in Counterinsurgency', pp.355–6; Alice Hills, *Future War in Cities*, pp.47–50; T.R. Moreman, 'Small Wars' and 'Imperial Policing', p.110; John A. Nagl, *Counterinsurgency Lessons from Malaya and Vietnam*, p.39; Hew Strachan, Introduction in Hew Strachan (ed.), *Big Wars and Small Wars*, p.8.

2   Gregor Davey, 'Conflicting worldviews, mutual incomprehension', pp.542–53; Caroline Kennedy-Pipe and Colin McInnes, 'The British Army in Northern Ireland 1969–1972', pp.204–5; Richard Popplewell, 'Lacking Intelligence', pp.320–1.

3   Douglas Porch, quoted by Benjamin Grob-Fitzgibbon, 'Securing the Colonies for the Commonwealth', p.15, fn.9.

4   Ian F.W. Beckett, 'Introduction', *The Roots of Counterinsurgency*, p.9; David French, *The British Way in Counterinsurgency, 1945–1967*, p.65; Matthew Hughes, 'The Banality of Brutality', p.354; Matthew Hughes, 'The practice and theory of British counterinsurgency', pp.538–9, 543; Matthew Hughes, 'Trouble in Palestine', pp.103–5; Nick Lloyd, 'The Armritsar Massacre and the Minimum Force Debate', p.384; Thomas R. Mockaitis, *British Counterinsurgency, 1919–60*, pp.17, 57, 67; Victoria Nolan, *Military Leadership and Counterinsurgency*, p.43–4; John Shy and Thomas W. Collier, 'Revolutionary War', p.830.

5   Ian F.W. Beckett, *The Roots of Counterinsurgency*, pp.9–10; Ian F.W. Beckett and John Pimlott, *Armed Forces & Modern Counterinsurgency*, p.3; Ian F.W. Beckett, *Modern Insurgencies and Counter-Insurgencies*, pp.38–40; Eversley Belfield, *The Boer War*, pp.102–5; Major Michael J. Lackman, 'The British Boer War and the French Algerian Conflict', pp.21–9, 54–9; Thomas R. Mockaitis, *British Counterinsurgency, 1919–60*, pp.18–9; Bill Nasson, *The South African War, 1899–1902*, pp.192–7; Thomas Pakenham, *The Boer War*, pp.493–5, 499, 536–7; John Pimlott, 'The British Army', pp.16–7; Major General E.K.G. Sixsmith, 'Kitchener and the Guerrillas in the Boer War', pp.208–12; S.B. Spies, *Methods of Barbarism?*, pp.108–13, 184–90, 201–6, 220–31, 273–5, 296; Keith Surridge, *Managing the South African War, 1899–1902*, pp.82–5, 113.

6   Ian F.W. Beckett, *Modern Insurgencies and Counter-Insurgencies*, p.40; Eversley Belfield, *The Boer War*, pp.106–8, 115–8, 125–6, 137; Byron Farwell, *The Great Anglo-Boer War*, pp.356–7, 361, 363–5; Bill Nasson, *The South African War, 1899–1902*, pp.200–2, 211–7; Thomas Pakenham, *The Boer War*, pp.540–2; John Pimlott, 'The British Army', pp.16–7; John Pimlott, 'The British Experience', p.21.

7   Ian F.W. Beckett, *The Roots of Counterinsurgency*, p.11; Ian F.W. Beckett, *Modern Insurgencies and Counter-Insurgencies*, pp.39–42; Byron Farwell, *The Great Anglo-Boer War*, pp.342–5; Albert Grunlingh, '"Protectors and friends of the people"?', pp.168–9; Bill Nasson, *The South African War, 1899–1902*, pp.212, 214, 224–6, 229; Thomas Pakenham, *The Boer War*, pp.488, 538–9; S.B. Spies, *Methods of Barbarism?*, p.276.

8   Ian F.W. Beckett, *The Roots of Counterinsurgency*, pp.9–10; Ian F.W. Beckett and John Pimlott, *Armed Forces & Modern Counterinsurgency*, p.3; Ian F.W. Beckett, *Modern*

Insurgencies and Counter-Insurgencies, pp.38–40; Alexander B. Downes, 'Draining the Sea by Filling the Graves', pp.423, 427–37; Eversley Belfield, *The Boer War*, pp.129–37, 141–8; Byron Farwell, *The Great Anglo-Boer War*, pp.348–56, 361–3, 392–4, 429–43; Major Michael J. Lackman, 'The British Boer War and the French Algerian Conflict', pp.21–9, 54–9; Marquess of Anglesey, *A History of the British Cavalry, 1816–1919*, Vol.4: 1899–1913, p.217; Thomas R. Mockaitis, *British Counterinsurgency, 1919–60*, pp.18–9; Bill Nasson, *The South African War, 1899–1902*, pp.208, 211–3, 217–21, 227–32; Victoria Nolan, *Military Leadership and Counterinsurgency*, pp.38–9; Victoria Nolan, *Military Leadership and Counterinsurgency*, p.38; Thomas Pakenham, *The Boer War*, pp.493–5, 499, 536–7; John Pimlott, 'The British Army', pp.16–7; Iain R. Smith and Andreas Stucki, 'The Colonial Development of Concentration Camps (1868–1902)', pp.425–31; S.B. Spies, *Methods of Barbarism?*, pp.108–13, 184–90, 201–6, 220–31, 273–5, 296; Keith Surridge, *Managing the South African War, 1899–1902*, pp.82–5, 113.

9 Ian F.W. Beckett, *Modern Insurgencies and Counter-Insurgencies*, p.39; Byron Farwell, *The Great Anglo-Boer War*, pp.396–420; Isabel V. Hull, *Absolute Destruction*, pp.186–7; Bill Nasson, *The South African War, 1899–1902*, pp.221–4, 232–3, 262–76; Victoria Nolan, *Military Leadership and Counterinsurgency*, pp.39–40; Thomas Pakenham, *The Boer War*, pp.438–41, 452–3 and 572; Iain R. Smith and Andreas Stucki, 'The Colonial Development of Concentration Camps (1868–1902)', pp.428–30.

10 Ian F.W. Beckett, *Modern Insurgencies and Counter-Insurgencies*, pp.16–7; David Benest, 'Aden to Northern Ireland, 1966–76', pp.124; Tom Bowden, *The Breakdown of Public Security*, pp.95–7 and 112–4; Major Pete Cottrell, 'Myth, The Military and Anglo-Irish Policing between 1913 and 1922', pp.15–7; Benjamin Grob-Fitzgibbon, 'Intelligence and Counterinsurgency', p.73; Peter Hart, *The I.R.A. at War, 1916–1923*, pp.3–4, 14–5, 63, 153–9, 165–75; J.B.E. Hittle, *Michael Collins and the Anglo-Irish War*, pp.42–54, 67–76, 87–92, 112, 118–9, 126, 232–3; W.H. Kautt, *Ambushes and Armour*, p.60; Brevet Major T.A. Lowe, 'Some Reflections of a Junior Commander upon "The Campaign" in Ireland, 1920 and 1921', pp.51–4; Thomas R. Mockaitis, *British Counterinsurgency, 1919–60*, pp.10–12, 20, 65 and 68; Victoria Nolan, *Military Leadership and Counterinsurgency*, pp.58–9; William Sheehan, *A Hard Local War*, pp.107–10, 112, 116–36; Charles Townshend, *The British Campaign in Ireland, 1919–21*, pp.1–4, 8–20, 30.

11 Michael T. Foy, *Michael Collins's Intelligence War*, pp.53–63; Peter Hart, *The I.R.A. at War, 1916–1923*, pp.223–58; J.B.E. Hittle, *Michael Collins and the Anglo-Irish War*, p.233; W.H. Kautt, *Ambushes and Armour*, p.79; D.M. Leeson, *The Black & Tans*, pp.8–12; Paul McMahon, *British Spies and Irish Rebels*, pp.26–9; Thomas R. Mockaitis, *British Counterinsurgency, 1919–60*, pp.66–7; Victoria Nolan, *Military Leadership and Counterinsurgency*, p.56; William Sheehan, *A Hard Local War*, pp.24–47, 102–7; Charles Townshend, *The British Campaign in Ireland, 1919–21*, pp.63–7.

12 Christopher Andrew, *The Defence of the Realm*, pp.106–9, 115–20; J.B.E. Hittle, *Michael Collins and the Anglo-Irish War*, pp.xvii–xix, 24–32, 81–7, 92–5; Thomas R. Mockaitis, *British Counterinsurgency, 1919–60*, pp.69–73 and 93; Victoria Nolan, *Military Leadership and Counterinsurgency*, pp.56–9; Charles Townshend, 'Policing Insurgency in Ireland, 1914–23', pp.34–5; Charles Townshend, *The British Campaign in Ireland, 1919–21*, pp.42–6, 55–7.

13 Tom Bowden, *The Breakdown of Public Security*, pp.18–20, 26–8, 45–7 and 121–3; Major Pete Cottrell, 'Myth, The Military and Anglo-Irish Policing between 1913 and 1922', pp.17–9; J.B.E. Hittle, *Michael Collins and the Anglo-Irish War*, pp.11–12, 114–7, 137–40, 142–3, 153, 178–80; W.H. Kautt, *Ambushes and Armour*, pp.78–80;

W.H. Kautt, *Ambushes and Armour*, p.79; D.M. Leeson, *The Black & Tans*, pp.24–38, 157–225; Thomas R. Mockaitis, *British Counterinsurgency, 1919–60*, pp.11–2, 18, 20, 65, 68; Victoria Nolan, *Military Leadership and Counterinsurgency*, pp.56–7; Richard Popplewell, 'Lacking Intelligence', pp.327–8; Charles Townshend, 'Policing Insurgency in Ireland, 1914–23', pp.33–47; Charles Townshend, *The British Campaign in Ireland, 1919–21*, pp.40–6, 92–7, 109–12, 130–1, 138–9.

14 David Benest, 'Aden to Northern Ireland, 1966–76', p.125; Tom Bowden, *The Breakdown of Public Security*, pp.88–9, 97–110 and 123–35; Michael T. Foy, *Michael Collins's Intelligence War*, pp.46–7, 93–6, 141–77; Benjamin Grob-Fitzgibbon, *Turning Points of the Irish Revolution*, pp.160–8; Benjamin Grob-Fitzgibbon, 'Intelligence and Counterinsurgency', p.74; Peter Hart, *The I.R.A. and its Enemies*, pp.93, 273–315; Peter Hart, *British Intelligence in Ireland, 1920–21*, pp.1–3 and 10–15; Peter Hart, *The I.R.A. at War, 1916–1923*, p.19; J.B.E. Hittle, *Michael Collins and the Anglo-Irish War*, pp.xxiii–xxiv, 117–37, 160–77, 231–2; Thomas R. Mockaitis, *British Counterinsurgency, 1919–60*, pp.73–6; William Sheehan, *A Hard Local War*, pp.71–90; Charles Townshend, *The British Campaign in Ireland, 1919–21*, pp.50–1, 123–30; Charles Townshend, 'The Irish Republican Army and the Development of Guerrilla Warfare, 1916–1921', pp.326–9.

15 Peter Hart, *The I.R.A. and its Enemies*, pp.93–6; J.B.E. Hittle, *Michael Collins and the Anglo-Irish War*, pp.113–4; Thomas R. Mockaitis, *British Counterinsurgency, 1919–60*, pp.11–12 and 149–52; Victoria Nolan, *Military Leadership and Counterinsurgency*, pp.57–8; William Sheehan, *A Hard Local War*, pp.18–9, 136–59, 164–7; Charles Townshend, *The British Campaign in Ireland, 1919–21*, pp.173–99.

16 Ian F.W. Beckett, *Modern Insurgencies and Counter-Insurgencies*, pp.16–8; J.B.E. Hittle, *Michael Collins and the Anglo-Irish War*, pp.125–7; Thomas R. Mockaitis, *British Counterinsurgency, 1919–60*, pp.74 and 76; Andrew Selth, 'Ireland and Insurgency', pp.303–4, 311–2; Calder Walton, *Empire of Secrets*, p.84.

17 'First Light', *Spectator*, Vol.197, No.6687, (24 August 1956), p.252, in Susan L. Carruthers, *Winning Hearts and Minds*, p.200.

18 Purnima Bose and Laura Lyons, 'Dyer Consequences: The Trope of Amritsar, Ireland, and the Lessons of the "Minimum" Force Debate', pp.203–7, 224; Gad Kroizer, 'From Dowbiggin to Tegart', p.80; Tim Jones, 'The British Army, and Counter-Guerrilla Warfare in Transition, 1944–1952', pp.149–50; Caroline Kennedy-Pipe and Colin McInnes, 'The British Army in Northern Ireland 1969–1972', p.206; Thomas R. Mockaitis, *British Counterinsurgency, 1919–60*, pp.12, 20, 24, 73–8, 83, 87, 118.

19 Tom Bowden, *The Breakdown of Public Security*, pp.145–51, 192–214; Edward Horne, *A Job Well Done*, pp.205–18, 228; Matthew Hughes, 'The practice and theory of British counterinsurgency', p.529; Thomas R. Mockaitis, *British Counterinsurgency, 1919–60*, pp.87–9; John Pimlott, 'The British Experience', pp.31–3; Andrew Selth, 'Ireland and Insurgency', pp.303–4.

20 Matthew Hughes, 'The Banality of Brutality ', p.313; Thomas R. Mockaitis, *British Counterinsurgency, 1919–60*, pp.33–5 and 157–8; John Pimlott, 'The British Experience', pp.33–4.

21 Edward Horne, *A Job Well Done*, pp.219–22; Matthew Hughes, 'The practice and theory of British counterinsurgency', pp.539–6 and 542; Matthew Hughes, 'The Banality of Brutality ', pp.313–4; Thomas R. Mockaitis, *British Counterinsurgency, 1919–60*, p.90; Jacob Norris, 'Repression and Rebellion', pp.27–38; John Pimlott, 'The British Experience', pp.34–5.

22 Tom Bowden, *The Breakdown of Public Security*, pp.1–3, 153–73, 217 (fn.65), 222–8; Edward Horne, *A Job Well Done*, pp.206; Gad Kroizer, 'From Dowbiggin to

Tegart', p.89; Thomas R. Mockaitis, *British Counterinsurgency, 1919–60*, pp.92–5; John Newsinger, *British Counterinsurgency*, p.4; John Pimlott, 'The British Experience', pp.36–7; Charles Townshend, 'The Defence of Palestine: Insurrection and Public Security, 1936–1939', pp.935–7.

23 Simon Anglim, 'Orde Wingate and the Special Night Squads', pp.26–32; Tom Bowden, *The Breakdown of Public Security*, pp.217 (fn.65), 245–8; Graham Ellison and Conor O'Reilly, 'From Empire to Iraq and the "War on Terror"', p.333; Hugh Foot, *A Start in Freedom*, pp.51–2; Edward Horne, *A Job Well Done*, pp.235–7; Matthew Hughes, 'From law and order to pacification', pp.16–17; Matthew Hughes, 'The Banality of Brutality', pp.331–5; Sir Charles Jeffries, *The Colonial Police*, pp.156–7; Gad Kroizer, 'From Dowbiggin to Tegart', pp.81 and 93; Thomas R. Mockaitis, *British Counterinsurgency, 1919–60*, pp.34, 102–3 and 159; John Newsinger, *British Counterinsurgency*, p.4; Jacob Norris, 'Repression and Rebellion', p.28; John Pimlott, 'The British Experience', pp.36–7; Victoria Nolan, *Military Leadership and Counterinsurgency*, pp.65–7; Richard Popplewell, 'Lacking Intelligence', pp.329–30; Charles Townshend, 'The Defence of Palestine: Insurrection and Public Security, 1936–1939', pp.937–8.

24 Ian F.W. Beckett, *Modern Insurgencies and Counter-Insurgencies*, p.47; Tom Bowden, 'The Politics of the Arab Rebellion in Palestine 1936–39', pp.166–9; Tom Bowden, *The Breakdown of Public Security*, pp.177–214, 217 (fn.65), 238–55; Hugh Foot, *A Start in Freedom*, pp.52–3; Matthew Hughes, 'From law and order to pacification', pp.10–8; Matthew Hughes, 'The Banality of Brutality ', pp.321–9; Thomas R. Mockaitis, *British Counterinsurgency, 1919–60*, p.159; Jacob Norris, 'Repression and Rebellion', pp.27–41; John Pimlott, 'The British Experience', pp.35–8.

25 Tom Bowden, *The Breakdown of Public Security*, pp.217 (fn.65), 248; Edward Horne, *A Job Well Done*, pp.237–9; Matthew Hughes, 'The practice and theory of British counterinsurgency', p.529; John Newsinger, *British Counterinsurgency*, p.4.

26 Ian F.W. Beckett, *The Roots of Counterinsurgency*, pp.7 and 12–3; Ian F.W. Beckett and John Pimlott, *Armed Forces & Modern Counterinsurgency*, pp.4–5; Ian F.W. Beckett, *Modern Insurgencies and Counter-Insurgencies*, pp.44–6; David A. Charters, 'From Palestine to Northern Ireland', pp.189–90 and 197; David A. Charters, *The British Army and Jewish Insurgency in Palestine, 1945–47*, pp.133–7; Tim Jones, *Postwar Counterinsurgency and the SAS, 1945–1952*, pp.10–12; Alastair MacKenzie, *Special Force*, p. 4; Thomas R. Mockaitis, *British Counterinsurgency, 1919–60*, pp.181–3; Victoria Nolan, *Military Leadership and Counterinsurgency*, pp.60–4.

27 Ian F.W. Beckett, *Modern Insurgencies and Counter-Insurgencies*, pp.89, 92–3; David Cesarani, *Major Farran's Hat*, pp.16–21; David A. Charters, *The British Army and Jewish Insurgency in Palestine, 1945–47*, pp.17–41, 131, 163–8; Bruce Hoffman, *The Failure of British Military Strategy within Palestine, 1939–1947*, pp.9–10; Matthew Hughes, 'Trouble in Palestine', p.106; Thomas R. Mockaitis, *British Counterinsurgency, 1919–60*, pp.100–4; John Newsinger, *British Counterinsurgency*, pp.4–11; Andrew Selth, 'Ireland and Insurgency', pp.303–4; Kate Utting, 'The Strategic Information Campaign', pp.39–51; Calder Walton, *Empire of Secrets*, pp.84, 101–3, 105–7, 112; Saul Zadka, *Blood in Zion*, pp.2–5, 10–12, 56–64, 141, 148, 172–83.

28 David A. Charters, *The British Army and Jewish Insurgency in Palestine, 1945–47*, pp.52–64; John Graham, *Ponder Anew*, pp.112–13; Bruce Hoffman, *The Failure of British Military Strategy within Palestine, 1939–1947*, pp.17–22; Matthew Hughes, 'Trouble in Palestine', p.107; Thomas R. Mockaitis, *British Counterinsurgency, 1919–60*, pp.42–3 and 104–7; John Newsinger, *British Counterinsurgency*, pp.11–6; Victoria Nolan, *Military Leadership and Counterinsurgency*, p.74; Major R.D. Wilson, *Cordon and Search: With the 6th Airborne Division in Palestine*, pp.24–40; Saul Zadka, *Blood in Zion*, pp.68–72.

29 David A. Charters, 'British Intelligence in the Palestine Campaign, 1945–47',
    pp.4, 40; David A. Charters, *The British Army and Jewish Insurgency in Palestine,
    1945–47*, pp.98–100, 117–20; Benjamin Grob-Fitzgibbon, *Imperial Endgame*, pp.46–8;
    Bruce Hoffman, *The Failure of British Military Strategy within Palestine, 1939–1947*,
    pp.22–4; Edward Horne, *A Job Well Done*, pp.297–307; Thomas R. Mockaitis,
    British Counterinsurgency, 1919–60, pp.42–3, 105 and 162–3; John Newsinger, *British
    Counterinsurgency*, pp.15–6, 20–2; Major R.D. Wilson, *Cordon and Search: With the 6th
    Airborne Division in Palestine*, pp.56–62, 66–77; Saul Zadka, *Blood in Zion*, pp.95–9,
    139–41.

30 Ian F.W. Beckett, *Modern Insurgencies and Counter-Insurgencies*, pp.94–5; David
    Cesarani, *Major Farran's Hat*, pp.59–63, 82–3, 87–178, 206–18; David A. Charters,
    'From Palestine to Northern Ireland', p.191; David A. Charters, 'Special Operations
    in Counterinsurgency: The Farran Case, Palestine 1947', pp.57–61; David A.
    Charters, *The British Army and Jewish Insurgency in Palestine, 1945–47*, pp.123, 135–8;
    John Graham, *Ponder Anew*, pp.105, 112–13; Bruce Hoffman, *The Failure of British
    Military Strategy within Palestine, 1939–1947*, pp.18–9, 33–4; Matthew Hughes,
    'Trouble in Palestine', pp.106–7; Keith Jeffery, 'Intelligence and Counterinsurgency
    Operations', p.128; Tim Jones, *SAS*, pp.72–84; Tim Jones, 'The British Army,
    and Counter-Guerrilla Warfare in Transition, 1944–1952', p.146; Thomas R.
    Mockaitis, *British Counterinsurgency, 1919–60*, pp.43–4, 109–11; John Newsinger,
    British Counterinsurgency, pp.27–8, 122; Victoria Nolan, *Military Leadership and
    Counterinsurgency*, pp.74–5; Calder Walton, *Empire of Secrets*, pp.109–11; Major R.D.
    Wilson, *Cordon and Search: With the 6th Airborne Division in Palestine*, pp.45–8; Saul
    Zadka, *Blood in Zion*, pp.171–2.

31 David Cesarani, *Major Farran's Hat*, pp.22–6, 37–43; David A. Charters, 'From
    Palestine to Northern Ireland', p.191; David A. Charters, *The British Army and
    Jewish Insurgency in Palestine, 1945–47*, pp.100–7; Bruce Hoffman, *The Failure
    of British Military Strategy within Palestine, 1939–1947*, pp.24–32; Thomas R.
    Mockaitis, *British Counterinsurgency, 1919–60*, pp.102 and 107–8; John Newsinger,
    British Counterinsurgency, pp.20, 22–6; Victoria Nolan, *Military Leadership and
    Counterinsurgency*, p.76; Saul Zadka, *Blood in Zion*, pp.7–8, 150–6.

32 David A. Charters, 'From Palestine to Northern Ireland', p.197; Thomas R.
    Mockaitis, *British Counterinsurgency, 1919–60*, p.108; John Newsinger, *British
    Counterinsurgency*, pp.14–5.

33 David A. Charters, 'British Intelligence in the Palestine Campaign, 1945–47',
    pp.11–21; David A. Charters, *The British Army and Jewish Insurgency in Palestine,
    1945–47*, pp.153–63; Ian Cobain, *Cruel Britannia*, pp.76–7; David French, *The
    British Way in Counterinsurgency, 1945–1967*, pp.24, 26; Benjamin Grob-Fitzgibbon,
    'Intelligence and Counterinsurgency', p.75; Edward Horne, *A Job Well Done*,
    pp.476–8; Matthew Hughes, 'Trouble in Palestine', p.106; Thomas R. Mockaitis,
    British Counterinsurgency, 1919–60, pp.108–9 and fn.24, p.140; John Newsinger, *British
    Counterinsurgency*, pp.16–17, 27; Calder Walton, *Empire of Secrets*, pp.108–9.

34 David A. Charters, 'British Intelligence in the Palestine Campaign, 1945–47',
    pp.3–4; Anthony Clayton, *The Wars of French Decolonization*, p.2; Matthew Hughes,
    'Trouble in Palestine', pp.106–8; Thomas R. Mockaitis, *British Counterinsurgency,
    1919–60*, p.111; Andrew Mumford, *The Counterinsurgency Myth*, p.3; John Newsinger,
    British Counterinsurgency, pp.1–2, 16–7, 29–30; John Shy and Thomas W. Collier,
    'Revolutionary War', p.845; Calder Walton, *Empire of Secrets*, pp.xxiii–xxxii.

35 Ian F.W. Beckett, *Modern Insurgencies and Counter-Insurgencies*, p.102; Huw Bennett,
    'A very salutary effect', pp.439–41; Douglas S. Blaufarb, *The Counterinsurgency Era*,

p.48; John Cloake, *Templer: Tiger of Malaya*, pp.236–40, 272–5; Brigadier Richard L. Clutterbuck, *The Long Long War*, pp.37–41, 83–5; Richard L. Clutterbuck, *Riot and Revolution in Singapore and Malaya, 1945–1963*, pp.180–3; Peter Dennis and Jeffrey Grey, *Emergency and Confrontation*, p.165; Donald Mackay, *The Domino That Stood*, pp.10–3; Thomas R. Mockaitis, *British Counterinsurgency, 1919–60*, pp.53–4, 111–3, 122; Andrew Mumford, *The Counterinsurgency Myth*, pp.26–8, 34–6, 47; John A. Nagl, *Counterinsurgency Lessons from Malaya and Vietnam*, pp.66–8; John Newsinger, *British Counterinsurgency*, pp.37–48; Kumar Ramakrishna, '"Transmogrifying" Malaya', pp.81–2; Anthony Short, *The Communist Insurrection in Malaya, 1948–1960*, pp.143, 151–8, 166–9, 379–86, 416–24; Brian Stewart, 'Winning in Malaya', p.276; A.J. Stockwell, 'Policing during the Malayan Emergency, 1948–60', p.113; Calder Walton, *Empire of Secrets*, pp.165–7, 184–5, 194–7.

36 Noel Barber, *The War of the Running Dogs*, pp.36, 62; Huw Bennett, 'A very salutary effect', pp.439–41; Brigadier Richard L. Clutterbuck, *The Long Long War*, pp.37–41, 49–53; John Coates, *Suppressing Insurgency*, pp.149–50, 159–62; Major F.A. Godfrey, *The History of the Suffolk Regiment, 1946–1959*, pp.49, 52; Donald Mackay, *The Domino That Stood* , pp.72–8, 101–2; Thomas R. Mockaitis, *British Counterinsurgency, 1919–60*, pp.53–4, 112–3, 162–7; Andrew Mumford, *The Counterinsurgency Myth*, pp.27–8, 30–1, 47; John A. Nagl, *Counterinsurgency Lessons from Malaya and Vietnam*, pp.66–8, 73–4, 96–7; John Newsinger, *British Counterinsurgency*, pp.41–8, 79–81; Victoria Nolan, *Military Leadership and Counterinsurgency*, pp.82, 88; Anthony Short, *The Communist Insurrection in Malaya, 1948–1960*, pp.95–112, 136–43, 152–6, 166–9; Calder Walton, *Empire of Secrets*, pp.165–7, 184–5, 194–7; David Young, *Four Five*, p.166.

37 David A. Charters, 'From Palestine to Northern Ireland', pp.203–5; Peter Dennis and Jeffrey Grey, *Emergency and Confrontation*, p.19; Major T.C. Edwards, '3d MarDiv Counterguerrilla Training', p.45; Raffi Gregorian, '"Jungle Bashing" in Malaya', pp.32–41, 43–4; Tim Jones, 'The British Army, and Counter-Guerrilla Warfare in Transition, 1944–1952', pp.168–9; Alastair MacKenzie, *Special Force*, p.54; Daniel Marston, 'Lost and Found in the Jungle', pp.98–106; Andrew Mumford, *The Counterinsurgency Myth*, p.38; John A. Nagl, *Counterinsurgency Lessons from Malaya and Vietnam*, pp.69–70, 97–8; John Newsinger, *British Counterinsurgency*, p.53.

38 David A. Charters, 'From Palestine to Northern Ireland', pp.197–8; John Cloake, *Templer: Tiger of Malaya*, pp.251–4; Brigadier Richard L. Clutterbuck, *The Long Long War*, pp.57–60; Peter Dennis and Jeffrey Grey, *Emergency and Confrontation*, p.16; David French, *The British Way in Counterinsurgency, 1945–1967*, pp.97–9; Tim Jones, *Postwar Counterinsurgency and the SAS, 1945–1952*, p.116; Thomas R. Mockaitis, *British Counterinsurgency, 1919–60*, pp.117–8 and 164–5; Andrew Mumford, *The Counterinsurgency Myth*, p.33; John A. Nagl, *Counterinsurgency Lessons from Malaya and Vietnam*, pp.71, 100–1; John Newsinger, *British Counterinsurgency*, p.50; Anthony Short, *The Communist Insurrection in Malaya, 1948–1960*, pp.239–40.

39 Ian F.W. Beckett, *Modern Insurgencies and Counter-Insurgencies*, p.102; Brigadier Richard L. Clutterbuck, *The Long Long War*, pp.60–4; David French, *The British Way in Counterinsurgency, 1945–1967*, pp.117–25, 180–1; Karl Hack, 'British Intelligence and Counterinsurgency in the Era of Decolonisation', pp.143–5; T.N. Harper, *The End of Empire and the Making of Malaya*, pp.175–82; Thomas R. Mockaitis, *British Counterinsurgency, 1919–60*, pp.115–7 and 119; Andrew Mumford, *The Counterinsurgency Myth*, pp.31–3 and 40–2; John A. Nagl, *Counterinsurgency Lessons from Malaya and Vietnam*, pp.74–5, 93–5, 98–9; John Newsinger, *British Counterinsurgency*, pp.49–51; Victoria Nolan, *Military Leadership and Counterinsurgency*,

pp.90–2; Kumar Ramakrisna, 'Content, Credibility and Context', pp.243–4, 248–50, 257–62; Anthony Short, *The Communist Insurrection in Malaya, 1948–1960*, pp.391–411; Brian Stewart, 'Winning in Malaya', pp.279–80; Calder Walton, *Empire of Secrets*, pp.183–6, 197–200.

40 Ian F.W. Beckett, *Modern Insurgencies and Counter-Insurgencies*, p.102; John Cloake, *Templer: Tiger of Malaya*, pp.248–50; David French, *The British Way in Counterinsurgency, 1945–1967*, pp.186–7; Raffi Gregorian, '"Jungle Bashing" in Malaya', pp.39–40; Benjamin Grob-FitzGibbon, *Imperial Endgame*, pp.143–6; T.N. Harper, *The End of Empire and the Making of Malaya*, pp.267–73; Sir Charles Jeffries, *The Colonial Police*, pp.78–81; John D. Leary, *Violence and the Dream People*, pp.108–17, 141–3, 148–52; Donald Mackay, *The Domino That Stood*, pp.74, 136–9; Alastair MacKenzie, *Special Force*, pp.63, 69–71 and 235; Lieutenant Colonel R.S.N. Mans, 'Counterinsurgency', p.49; Thomas R. Mockaitis, *British Counterinsurgency, 1919–60*, p.118; Andrew Mumford, *The Counterinsurgency Myth*, pp.33–4 and 36–8; John A Nagl, *Counterinsurgency Lessons from Malaya and Vietnam*, pp.76–7, 92, 100; John Newsinger, *British Counterinsurgency*, pp.56; Anthony Short, *The Communist Insurrection in Malaya, 1948–1960*, pp.124–32, 411–5, 439–56; Calder Walton, *Empire of Secrets*, pp.181–3.

41 John Cloake, *Templer: Tiger of Malaya*, pp.198, 227–35; Colonel Richard L. Clutterbuck, 'The SEP – Guerrilla Intelligence Source', pp.14–21; Brigadier Richard L. Clutterbuck, *The Long Long War*, pp.47–9; Richard L. Clutterbuck, *Riot and Revolution in Singapore and Malaya, 1945–1963*, p.178; John Coates, *Suppressing Insurgency*, pp.123–5; David French, *The British Way in Counterinsurgency, 1945–1967*, pp.28–33; Benjamin Grob-Fitzgibbon, 'Intelligence and Counterinsurgency', pp.75–7; Karl Hack, 'Corpses, Prisoners of War and Captured Documents', pp.211, 215–19; Karl Hack, 'British Intelligence and Counterinsurgency in the Era of Decolonisation, pp.127–34, 145–7; Sir Charles Jeffries, *The Colonial Police*, pp.80–1; Donald Mackay, *The Domino That Stood*, pp.80–2; Thomas R. Mockaitis, *British Counterinsurgency, 1919–60*, pp.119–20; Andrew Mumford, *The Counterinsurgency Myth*, pp.38–40; John A. Nagl, *Counterinsurgency Lessons from Malaya and Vietnam*, pp.92–3; John Newsinger, *British Counterinsurgency*, pp.52–7; Richard Popplewell, 'Lacking Intelligence', p.333; Anthony Short, *The Communist Insurrection in Malaya, 1948–1960*, pp.275–92; Brian Stewart, 'Winning in Malaya', pp.269–70; A.J. Stockwell, 'Policing during the Malayan Emergency, 1948–60', pp.110–9; Calder Walton, *Empire of Secrets*, pp.165–7, 171–4, 179–81, 186–94.

42 John Cloake, *Templer: Tiger of Malaya*, pp.265–72; Peter Dennis and Jeffrey Grey, *Emergency and Confrontation*, pp.16–9; Thomas R. Mockaitis, *British Counterinsurgency, 1919–60*, pp.119–23; Andrew Mumford, *The Counterinsurgency Myth*, pp.34–5; John A. Nagl, *Counterinsurgency Lessons from Malaya and Vietnam*, pp.87–90, 95–6; Victoria Nolan, *Military Leadership and Counterinsurgency*, pp.84–5, 117–20; Christopher Pugsley, *From Emergency to Confrontation*, pp.26–8; Kumar Ramakrishna, '"Transmogrifying" Malaya', pp.79–80, 83–92; A.J. Stockwell, 'Insurgency and Decolonisation during the Malayan Emergency', pp.340–2; Calder Walton, *Empire of Secrets*, p.179.

43 John Akehurst, *Generally Speaking*, pp.58–9; Douglas S. Blaufarb, *The Counterinsurgency Era*, pp.48–9; Walter C. Ludwig III, 'Managing Counterinsurgency', p.64; Thomas R. Mockaitis, *British Counterinsurgency, 1919–60*, pp.52 and 124; Andrew Mumford, *The Counterinsurgency Myth*, pp.29–30; John A. Nagl, *Counterinsurgency Lessons from Malaya and Vietnam*, pp.102–4; John Newsinger, *British Counterinsurgency*, pp.56–7; Simon C. Smith, 'General Templer and Counterinsurgency in Malaya',

pp.70–2; Professor Mary Turnbull, 'The Malayan Civil Service and the Transition to Independence', pp.274–85; Calder Walton, *Empire of Secrets*, pp.200–9, 334–9.

44 Ian F.W. Beckett, *Modern Insurgencies and Counter-Insurgencies*, p.98; John Erickson, Foreword to Donald Mackay, *The Domino That Stood*, p.xxii; David French, *The British Way in Counterinsurgency, 1945–1967*, p.179; Andrew Mumford, *The Counterinsurgency Myth*, pp.43–6; John Newsinger, *British Counterinsurgency*, pp.43–5, 56–9; Anthony Short, *The Communist Insurrection in Malaya, 1948–1960*, pp.244, 373–5; John Shy and Thomas W. Collier, 'Revolutionary War', p.854; Calder Walton, *Empire of Secrets*, p.197.

45 David Anderson, *Histories of the Hanged*, pp.47–51, 86–95, 125–35, 177–80; Ian F.W. Beckett, *Modern Insurgencies and Counter-Insurgencies*, pp.121–2; Huw Bennett, *Fighting the Mau Mau*, pp.17–8; Daniel Branch, *Defeating Mau Mau*, pp.6–8, 55–9; Robert S. Edgerton, *Mau Mau*, pp.69–72, 78–80, 239–41; Frank Furedi, 'Kenya: Decolonization through counterinsurgency', pp.147, 151, 159–61; Randall W. Heather, 'Intelligence and Counterinsurgency in Kenya, 1952–1956', pp.57–8; John Lonsdale, 'Mau Maus of the Mind', p.407; Thomas R. Mockaitis, *British Counterinsurgency, 1919–60*, pp.124–6; Andrew Mumford, *The Counterinsurgency Myth*, pp.49–51 and 65; John Newsinger, 'Minimum Force', p.228; John Newsinger, *British Counterinsurgency*, pp.60–9; David A. Percox, 'British Counterinsurgency in Kenya, 1952–56', pp.267–8; David Troup, 'Crime, politics and the policing in colonial Kenya, 1939–63', p.144.

46 David Anderson, *Histories of the Hanged*, pp.55–7, 61–3; Huw Bennett, *Fighting the Mau Mau*, pp.12–6; Daniel Branch, *Defeating Mau Mau*, pp.6–7, 47–9; Anthony Clayton, *Counterinsurgency in Kenya 1952–1960*, pp.13–5, 21–2; Robert S. Edgerton, *Mau Mau*, pp.65–8; Caroline Elkins, *Imperial Reckoning*, pp.31–8, 54–6; David French, *The British Way in Counterinsurgency, 1945–1967*, pp.115–6; Wunyabari O. Maloba, *Mau Mau and Kenya*, pp.76–8; Thomas R. Mockaitis, *British Counterinsurgency, 1919–60*, pp.125–6 Andrew Mumford, *The Counterinsurgency Myth*, pp.51–4; John Newsinger, 'Minimum Force', p.228; John Newsinger, *British Counterinsurgency*, pp.65–7; Victoria Nolan, *Military Leadership and Counterinsurgency*, pp.141–2; David A. Percox, 'British Counterinsurgency in Kenya, 1952–56', pp.277–82; David Troup, 'Crime, politics and the policing in colonial Kenya, 1939–63', p.140.

47 David Anderson, *Histories of the Hanged*, pp.51–2, 177–80; Huw Bennett, *Fighting the Mau Mau*, pp.18–9, 50–7, 234–40; David A. Charters, 'From Palestine to Northern Ireland', pp.198–9; Anthony Clayton, *Counterinsurgency in Kenya 1952–1960*, pp.7–11, 33–6; Caroline Elkins, *Imperial Reckoning*, pp.29–30, 52–3; Frank Furedi, 'Kenya: Decolonization through counterinsurgency', pp.149–50, 157–9; Benjamin Grob-Fitzgibbon, 'Intelligence and Counterinsurgency', p.77; Randall W. Heather, 'Intelligence and Counterinsurgency in Kenya, 1952–1956', pp.84–92; Thomas R. Mockaitis, *British Counterinsurgency, 1919–60*, pp.130–2; Andrew Mumford, *The Counterinsurgency Myth*, pp.49, 61–2; John Newsinger, 'Revolt and Repression in Kenya', p.171; Victoria Nolan, *Military Leadership and Counterinsurgency*, pp.155–62; David A. Percox, 'British Counterinsurgency in Kenya, 1952–56', pp.268–8, 287–92; David Troup, 'Crime, politics and the policing in colonial Kenya, 1939–63', pp.142–7; Calder Walton, *Empire of Secrets*, pp.244–6.

48 Ian F.W. Beckett, *Modern Insurgencies and Counter-Insurgencies*, pp.129 and 225; Huw Bennett, *Fighting the Mau Mau*, pp.29, 152–9, 243; Caroline Elkins, *Imperial Reckoning*, p.54; Randall W. Heather, 'Intelligence and Counterinsurgency in Kenya, 1952–1956', pp.96–9; Keith Jeffery, 'Intelligence and Counterinsurgency Operations', p.128; Thomas R. Mockaitis, *British Counterinsurgency, 1919–60*, p.132; Andrew Mumford,

The Counterinsurgency Myth, pp.58–9; John Newsinger, 'Minimum Force', p.229; John Newsinger, *British Counterinsurgency*, p.75; David A. Percox, 'British Counterinsurgency in Kenya, 1952–56', pp.305–6; Calder Walton, *Empire of Secrets*, pp.254–6.

49 David Anderson, *Histories of the Hanged*, pp.6–7, 136–9, 311–27; Huw Bennett, *Fighting the Mau Mau*, pp.160–93, 230–4; Anthony Clayton, *Counterinsurgency in Kenya 1952–1960*, pp.15–7, 47–52; Ian Cobain, *Cruel Britannia*, pp.78–90; Robert S. Edgerton, *Mau Mau*, pp.ix, 76–7, 173–201; Caroline Elkins, *Imperial Reckoning*, pp.56–72, 129–232; Randall W. Heather, 'Intelligence and Counterinsurgency in Kenya, 1952–1956', pp.92–4; Thomas R. Mockaitis, *British Counterinsurgency, 1919–60*, pp.126–9; Andrew Mumford, *The Counterinsurgency Myth*, pp.53–7; John Newsinger, 'Revolt and Repression in Kenya', p.180; John Newsinger, 'From Counterinsurgency to Internal Security', p.89; John Newsinger, 'Minimum Force', pp.234–5; John Newsinger, *British Counterinsurgency*, pp.69–70, 73–6; Victoria Nolan, *Military Leadership and Counterinsurgency*, pp.142–3, 149; David A. Percox, 'British Counterinsurgency in Kenya, 1952–56', pp.282–6 and 300–1; David Troup, 'Crime, politics and the policing in colonial Kenya, 1939–63', p.148; Calder Walton, *Empire of Secrets*, pp.237–8, 250–8.

50 Huw Bennett, *Fighting the Mau Mau*, pp.24–7, 220–5; Daniel Branch, *Defeating Mau Mau*, pp.107–18; Robert S. Edgerton, *Mau Mau*, pp.92–3; Caroline Elkins, *Imperial Reckoning*, pp.234–73; David French, *The British Way in Counterinsurgency, 1945–1967*, pp.118–24; Frank Furedi, 'Kenya: Decolonization through counter-insurgency', pp.155–7; Thomas R. Mockaitis, *British Counterinsurgency, 1919–60*, pp.129–30; Andrew Mumford, *The Counterinsurgency Myth*, p.54; John Newsinger, 'Minimum Force', p.229; John Newsinger, *British Counterinsurgency*, pp.73–4; Victoria Nolan, *Military Leadership and Counterinsurgency*, pp.148–9; David A. Percox, 'British Counterinsurgency in Kenya, 1952–56', pp.302–3; David Troup, 'Crime, politics and the policing in colonial Kenya, 1939–63', pp.147–8; Calder Walton, *Empire of Secrets*, p.243.

51 David Anderson, *Histories of the Hanged*, pp.239–43, 297–308; David Anderson, Huw Bennett and Daniel Branch, 'A Very British Massacre', *History Today*, pp.20–2; Ian F.W. Beckett, *Modern Insurgencies and Counter-Insurgencies*, pp.128–9; Huw Bennett, 'The Mau Mau Emergency as Part of the British Army's Post-War Counterinsurgency Experience', pp.150–6; Huw Bennett, 'The Other Side of the COIN', pp.647–51; Huw Bennett, *Fighting the Mau Mau*, pp.12, 14, 16, 40–2, 246–55; Daniel Branch, *Defeating Mau Mau*, pp.66–88; Anthony Clayton, *Counterinsurgency in Kenya 1952–1960*, pp.18–20, 28–30, 37–47, 57–9; Ian Cobain, *Cruel Britannia*, pp.78–90; Robert S. Edgerton, *Mau Mau*, pp.99–100, 150–62; Caroline Elkins, *Imperial Reckoning*, pp.42–4, 76–88, 276–80, 352; Benjamin Grob-Fitzgibbon, *Imperial Endgame*, pp.266–8; David French, *The British Way in Counterinsurgency, 1945–1967*, p.186; Randall W. Heather, 'Intelligence and Counterinsurgency in Kenya, 1952–1956', pp.58–67, 91–2; Dane Kennedy, 'Constructing the Colonial Myth of Mau Mau', p.246; Thomas R. Mockaitis, *British Counterinsurgency, 1919–60*, pp.44–8, 50–1, 125–7 and 130–1; Andrew Mumford, *The Counterinsurgency Myth*, pp.59–61; John Newsinger, 'Minimum Force', pp.229–35; John Newsinger, *British Counterinsurgency*, pp.76–81; David A. Percox, 'British Counterinsurgency in Kenya, 1952–56', pp.280 and 288; Rod Thornton, 'Minimum Force', pp.216–21; David Troup, 'Crime, politics and the policing in colonial Kenya, 1939–63', pp.140–2, 144–5, 149; Calder Walton, *Empire of Secrets*, pp.237, 243–4, 250–8.

52 Huw Bennett, 'The Mau Mau Emergency as Part of the British Army's Post-War Counterinsurgency Experience', p.150; Huw Bennett, *Fighting the Mau Mau*,

pp.129–46; Anthony Clayton, *Counterinsurgency in Kenya 1952–1960*, pp.13–5, 21–5, 31–2; General Sir George Erskine, 'Kenya – What is it all About', p.104; Frank Furedi, 'Kenya: Decolonization through counterinsurgency', pp.150–1; John Lonsdale, 'Mau Maus of the Mind', p.415; Thomas R. Mockaitis, *British Counterinsurgency, 1919–60*, pp.167–71.

53 David Anderson, *Histories of the Hanged*, pp.200–6, 268–72; Huw Bennett, *Fighting the Mau Mau*, pp.258–62; Anthony Clayton, *Counterinsurgency in Kenya 1952–1960*, pp.25–8, 30–1; Robert S. Edgerton, *Mau Mau*, pp.90–2; Caroline Elkins, *Imperial Reckoning*, pp.121–5; General Sir George Erskine, 'Kenya – What is it all About', pp.109–11; Frank Furedi, 'Kenya: Decolonization through counterinsurgency', pp.148–55; Fred Majdalany, *State of Emergency*, pp.216–7; Thomas R. Mockaitis, *British Counterinsurgency, 1919–60*, pp.130 and 168–71; Andrew Mumford, *The Counterinsurgency Myth*, pp.57–8 and 66–71; John Newsinger, 'Revolt and Repression in Kenya', pp.172–6; John Newsinger, 'Minimum Force', p.229; John Newsinger, *British Counterinsurgency*, pp.65–75; Victoria Nolan, *Military Leadership and Counterinsurgency*, pp.147–8; David A. Percox, 'British Counterinsurgency in Kenya, 1952–56', pp.292–308; Calder Walton, *Empire of Secrets*, p.249.

54 David French, *The British Way in Counterinsurgency, 1945–1967*, pp.204–18; John Lonsdale, 'Mau Maus of the Mind', p.394; Thomas R. Mockaitis, *British Counterinsurgency, 1919–60*, pp.138–9, 183–5, 188; Calder Walton, *Empire of Secrets*, pp.247–8, 267–73.

55 David French, *The British Way in Counterinsurgency, 1945–1967*, pp.48–9; John Newsinger, *British Counterinsurgency*, pp.84, 93; David Souter, 'An Island Apart', p.659; Stephen G. Xydis, 'The UN General Assembly as an instrument of Greek policy', pp.141–2.

56 David R. Devereux, *The Formulation of British Defence Policy towards the Middle East 1945–56*, pp.174–5; Thomas Ehrlich, 'Cyprus, the "Warlike Isle"', pp.1026–7; David Goldsworthy, 'Armed Struggle under late Colonialism', pp.538–9; Bruce Hoffman and Jennifer M. Taw, *Defense Policy and Low-Intensity Conflict*, pp.19–20; Robert Holland, *Britain and the Revolt in Cyprus 1954–1959*, p.32; John Newsinger, *British Counterinsurgency*, pp.84–5; Calder Walton, *Empire of Secrets*, pp.304–5.

57 531 House of Commons Debates (5th ser.), col.507–8 (1954) quoted in Thomas Ehrlich, 'Cyprus, the "Warlike Isle"', p.1029.

58 Thomas Ehrlich, 'Cyprus, the "Warlike Isle"', pp.1028–9; Benjamin Grob-Fitzgibbon, *Imperial Endgame*, pp.287–8; Bruce Hoffman and Jennifer M. Taw, *Defense Policy and Low-Intensity Conflict*, p.vii and 19; John Newsinger, *British Counterinsurgency*, p.96; Corran Purdon, *List the Bugle*, p.114.

59 C. Allen, *The Savage Wars of Peace*, p.139; Michael Carver, *War Since 1945*, pp.44–6; Michael Dewar, *Brush Fire Wars*, pp.70, 74; Bruce Hoffman and Jennifer M. Taw, *Defense Policy and Low-Intensity Conflict*, p.vii; Thomas R. Mockaitis, *British Counterinsurgency, 1919–60*, p.137; John Newsinger, *British Counterinsurgency*, pp.90–4; John Shy and Thomas W. Collier, 'Revolutionary War', p.845.

60 David Anderson, 'Policing and Communal Conflict: the Cyprus Emergency, 1954–60', pp.195, 208–9; Ian F.W. Beckett, *Modern Insurgencies and Counter-Insurgencies*, p.155; Panagiotis Dimitrakis, 'British Intelligence and the Cyprus Insurgency, 1955–1959', pp.379–87; Panagiotis Dimitrakis, *Military Intelligence in Cyprus*, pp.86–7; David French, *The British Way in Counterinsurgency, 1945–1967*, p.24; Major F.A. Godfrey, *The History of the Suffolk Regiment, 1946–1959*, pp.147–8, 152; George Grivas (edited by Charles Foley), *The Memoirs of General Grivas*, pp.76–7; Bruce Hoffman and Jennifer M. Taw, *Defense Policy and Low-Intensity Conflict*,

pp.19, 26; Kei Jeffery, 'Intelligence and Counterinsurgency Operations', p.125; Thomas R. Mockaitis, *British Counterinsurgency, 1919–60*, p.137; John Newsinger, *British Counterinsurgency*, pp.91–100, 106; Julian Paget, *Counterinsurgency Campaigning*, pp.122–4, 140, 184; Calder Walton, *Empire of Secrets*, pp.309–10.

61 David M. Anderson, 'Policing and communal conflict', pp.208–9; David A. Charters, 'From Palestine to Northern Ireland', p.199; Panagiotis Dimitrakis, 'British Intelligence and the Cyprus Insurgency, 1955–1959', pp.379–87; Panagiotis Dimitrakis, *Military Intelligence in Cyprus*, pp.77–9, 82–6; Charles Foley and W.I. Scobie, *The Struggle for Cyprus*, p.123; David French, *The British Way in Counterinsurgency, 1945–1967*, p.26; Bruce Hoffman and Jennifer M. Taw, *Defense Policy and Low-Intensity Conflict*, p.26; Thomas R. Mockaitis, *British Counterinsurgency, 1919–60*, p.137; Julian Paget, *Counterinsurgency Campaigning*, pp.122–4, 184; Calder Walton, *Empire of Secrets*, pp.307–10.

62 David M. Anderson, 'Policing and communal conflict', p.209; Ian F.W. Beckett, *Modern Insurgencies and Counter-Insurgencies*, pp.155–6; Michael Dewar, *Brush Fire Wars*, pp.75–7; Panagiotis Dimitrakis, *Military Intelligence in Cyprus*, pp.87–92, 99; Charles Foley and W.I. Scobie, *The Struggle for Cyprus*, pp.113–4, 123, 150–7; George Grivas, *Guerrilla Warfare and EOKA's Struggle*, p.52; Thomas Mockaitis, *British Counterinsurgency 1919–1960*, pp.171–3; John Newsinger, *British Counterinsurgency*, pp.98–102; Corran Purdon, *List the Bugle*, pp.114–5.

63 David M. Anderson, 'Policing and communal conflict', p.209; Ian F.W. Beckett, *Modern Insurgencies and Counter-Insurgencies*, p.156; Susan L. Carruthers, *Winning Hearts and Minds*, pp.200–1; Ian Cobain, *Cruel Britannia*, pp.92–9; Michael Dewar, *Brush Fire Wars*, pp.79–80; Lawrence Durrell, *Bitter Lemons*, pp.183, 242; Charles Foley, *Island in Revolt*, p.131; Charles Foley and W.I. Scobie, *The Struggle for Cyprus*, pp.123–4, 150; Hugh Foot, *A Start in Freedom*, pp.174–5; David French, *The British Way in Counterinsurgency, 1945–1967*, pp.113–4; Bruce Hoffman and Jennifer M. Taw, *Defense Policy and Low-Intensity Conflict*, pp.29–30; Thomas Mockaitis, *British Counterinsurgency 1919–1960*, pp.135–6, 188–9; John Newsinger, *British Counterinsurgency*, pp.97, 100–7; Julian Paget, *Counterinsurgency Campaigning*, p.146; Calder Walton, *Empire of Secrets*, pp.311–2.

64 David M. Anderson, 'Policing and communal conflict', p.211; Ian F.W. Beckett, *Modern Insurgencies and Counter-Insurgencies*, p.156; David A. Charters, 'From Palestine to Northern Ireland', pp.189–90; John Newsinger, *British Counterinsurgency*, pp.106–7; Calder Walton, *Empire of Secrets*, pp.314–5.

65 Peter Dennis and Jeffrey Grey, *Emergency and Confrontation*, pp.171–5; David Easter, 'British and Malaysian Covert Support for Rebel Movements in Indonesia during the "Confrontation", 1963–66', p.195; Raffi Gregorian, 'CLARET Operations and Confrontation, 1964–1966', pp.53–7; Major Peter J. Kramers, 'Konfrontsai in Borneo 1962–66', pp.64–8; Thomas R. Mockaitis, *British Counterinsurgency in the post-imperial era*, pp.14–8; Christopher Pugsley, *From Emergency to Confrontation*, pp.195–7; Justus M. van der Kroef, 'Communism and the Guerrilla War in Sarawak', pp.50–2, 56–9; Justus M. van der Kroef, 'The Sarawak–Indonesian Border Insurgency', pp.245–54, 258.

66 Ian F.W. Beckett, *Modern Insurgencies and Counter-Insurgencies*, p.127; Peter Dennis and Jeffrey Grey, *Emergency and Confrontation*, pp.239–43; Thomas R. Mockaitis, *British Counterinsurgency in the post-imperial era*, pp.17–8, 29–30; Christopher Tuck, 'Borneo 1963–66', pp.98–101; Jac Weller, 'British Weapons and Tactics in Malaysia', p.18.

67 Peter Dickens, *SAS: The Jungle Frontier*, pp.181–2; David French, *The British Way in Counterinsurgency, 1945–1967*, p.187; Raffi Gregorian, 'CLARET Operations and Confrontation, 1964–1966', pp.54–5, 70; Bob Hall and Andrew Ross, 'The Political

and Military Effectiveness of Commonwealth Forces in Confrontation 1963–66', p.245; Thomas R. Mockaitis, *British Counterinsurgency in the post-imperial era*, pp.19–21; Justus M. van der Kroef, 'The Sarawak–Indonesian Border Insurgency', pp.261–2.

68 Peter Dennis and Jeffrey Grey, *Emergency and Confrontation*, pp.253–62; Harold James and Denis Sheil-Small, *The Undeclared War*, pp.130–5; Thomas R. Mockaitis, *British Counterinsurgency in the post-imperial era*, pp.19–21, 25–7; Christopher Tuck, 'Borneo 1963–66', p.97; Christopher Tuck, 'Borneo, Counterinsurgency and War Termination', pp.115–6; General Sir Walter Walker, *Fighting On*, p.148; Jac Weller, 'British Weapons and Tactics in Malaysia', pp.23–4.

69 Ian F.W. Beckett, *Modern Insurgencies and Counter-Insurgencies*, p.127; David A. Charters, 'From Palestine to Northern Ireland', pp.219–20; Peter Dickens, *SAS: The Jungle Frontier*, pp.35–7, 54–9, 60, 71–3, 87–8; Raffi Gregorian, 'CLARET Operations and Confrontation, 1964–1966', pp.55–6; Alastair MacKenzie, *Special Force*, pp.112–4; Thomas R. Mockaitis, *British Counterinsurgency in the post-imperial era*, pp.27–32.

70 Ian F.W. Beckett, *Modern Insurgencies and Counter-Insurgencies*, pp.127–8; Raffi Gregorian, 'CLARET Operations and Confrontation, 1964–1966', pp.57–69; Peter Dennis and Jeffrey Grey, *Emergency and Confrontation*, pp.216–7, 246–53; Peter Dickens, *SAS: The Jungle Frontier*, pp.118–20, 125–32, 149, 155–7; David Easter, 'British Intelligence and Propaganda during the "Confrontation", 1963–1966', pp.84–90; David Easter, 'British and Malaysian Covert Support for Rebel Movements in Indonesia during the "Confrontation", 1963–66', pp.196–207; David Easter, *Britain and the Confrontation with Indonesia, 1960–66*, pp.193–7; Major Peter J. Kramers, 'Konfrontsai in Borneo 1962–66', p.70; Alastair MacKenzie, *Special Force*, pp.119–21; Thomas R. Mockaitis, *British Counterinsurgency in the post-imperial era*, pp.17, 31–9; Christopher Pugsley, *From Emergency to Confrontation*, pp.252–9; Christopher Tuck, 'Borneo 1963–66', pp.97–8.

71 Ian F.W. Beckett, *Modern Insurgencies and Counter-Insurgencies*, pp.127–8; David Easter, 'British Intelligence and Propaganda during the 'Confrontation', 1963–1966', pp.90–9; David Easter, *Britain and the Confrontation with Indonesia, 1960–66*, pp.167–70, 174–97; David Easter, 'Keep the Indonesian pot boiling', pp.57–8, 60–7; Raffi Gregorian, 'CLARET Operations and Confrontation, 1964–1966', pp.51–3, 57–69; Alastair MacKenzie, *Special Force*, pp.119–21; Thomas R. Mockaitis, *British Counterinsurgency in the post-imperial era*, pp.17, 20–4, 31–9; Christopher Tuck, 'Borneo 1963–66', pp.106–8.

72 David French, *The British Way in Counterinsurgency, 1945–1967*, pp.50–2, 127; Spencer Mawby, *British Policy in Aden and the Protectorates 1955–67*, pp.1–7, 29–34, 43–7, 100–2; Thomas R. Mockaitis, *British Counterinsurgency in the post-imperial era*, pp.44–7 and 50; Andrew Mumford, *The Counterinsurgency Myth*, pp.72–7 and 88–90; Jonathon Walker, *Aden Insurgency*, pp.22–36, 41–63.

73 David Benest, 'Aden to Northern Ireland, 1966–76', p.120; David A. Charters, 'From Palestine to Northern Ireland', pp.199–200; David French, *The British Way in Counterinsurgency, 1945–1967*, p.95; Spencer Mawby, *British Policy in Aden and the Protectorates 1955–67*, pp.81–5, 141–2; Thomas R. Mockaitis, *British Counterinsurgency in the post-imperial era*, pp.48–9, 57–8, 67; Andrew Mumford, *The Counterinsurgency Myth*, pp.80–1; John Newsinger, *British Counterinsurgency*, pp.126–9; Julian Paget, *Last Post*, pp.121–2, 125–32; Jonathon Walker, *Aden Insurgency*, pp.240–56.

74 David French, *The British Way in Counterinsurgency, 1945–1967*, pp.127–32; Spencer Mawby, *British Policy in Aden and the Protectorates 1955–67*, pp.102–5; Thomas R. Mockaitis, *British Counterinsurgency in the post-imperial era*, pp.49–54; Andrew

Mumford, *The Counterinsurgency Myth*, pp.79–80; John Newsinger, *British Counterinsurgency*, pp.114–8; Jonathon Walker, *Aden Insurgency*, pp.71–88, 93–111; David Young, *Four Five*, pp.326–7.

75 David French, *The British Way in Counterinsurgency, 1945–1967*, pp.127; David Benest, 'Aden to Northern Ireland, 1966–76', p.120; Thomas R. Mockaitis, *British Counterinsurgency in the post-imperial era*, pp.54–5 and 65–8; Andrew Mumford, *The Counterinsurgency Myth*, p.79; Jonathon Walker, *Aden Insurgency*, pp.120–3, 201–5.

76 David A. Charters, 'From Palestine to Northern Ireland', p.212; Tony Geraghty, *Who Dares Wins*, pp.79–83; Alastair MacKenzie, *Special Force*, pp.102–4 and 206; Spencer Mawby, *British Policy in Aden and the Protectorates 1955–67*, pp.16–18, 94–5, 134, 137–41; Thomas R. Mockaitis, *British Counterinsurgency in the post-imperial era*, pp.55–7; Andrew Mumford, *The Counterinsurgency Myth*, p.80; John Newsinger, *British Counterinsurgency*, pp.121–3; Julian Paget, *Last Post*, pp.119–20; Jonathon Walker, *Aden Insurgency*, pp.131–6, 181–97; John Willis, 'Colonial Police in Aden, 1937–1967', p.242.

77 David Benest, 'Aden to Northern Ireland, 1966–76', p.120; David A. Charters, 'From Palestine to Northern Ireland', p.219; Ian Cobain, *Cruel Britannia*, pp.99–109; David French, *The British Way in Counterinsurgency, 1945–1967*, pp.24–6; Clive Jones, 'Military intelligence and the war in Dhofar', p.639; Alfred W. McCoy, *A Question of Torture*, p.54; Spencer Mawby, *British Policy in Aden and the Protectorates 1955–67*, pp.137–8, 164–9; Thomas R. Mockaitis, *British Counterinsurgency in the post-imperial era*, pp.58–60, 64; Andrew Mumford, *The Counterinsurgency Myth*, pp.82–7; John Newsinger, *British Counterinsurgency*, pp.122–3, 128–9; Julian Paget, *Last Post*, pp.119, 150–2; Jonathon Walker, *Aden Insurgency*, pp.123–7, 141–7, 185–8, 259–71; Calder Walton, *Empire of Secrets*, pp.320–5; David Young, *Four Five*, pp.381–90.

78 Alexander Alderson, 'The British Approach to COIN and Stabilisation', p.66.

79 Clive Jones, 'Military intelligence and the war in Dhofar', pp.630–1; Spencer Mawby, *British Policy in Aden and the Protectorates 1955–67*, pp.152–64; Thomas R. Mockaitis, *British Counterinsurgency in the post-imperial era*, pp.60–6; Andrew Mumford, *The Counterinsurgency Myth*, pp.81–2 and 93–4; John Newsinger, *British Counterinsurgency*, pp.123–6, 130–1; Jonathon Walker, *Aden Insurgency*, pp.271–80, 285–93.

80 David French, *The British Way in Counterinsurgency, 1945–1967*, pp.52–3; Colonel Tony Jeapes, *SAS: Operation Oman*, pp.15–31; Spencer Mawby, *British Policy in Aden and the Protectorates 1955–67*, pp.170–6; Thomas R. Mockaitis, *British Counterinsurgency in the post-imperial era*, pp.72–4; John Newsinger, *British Counterinsurgency*, pp.132–42; John Pimlott, 'The British Army', pp.27–31.

81 John Akehurst, *We Won A War*, pp.33–7, 58–9; John Akehurst, *Generally Speaking*, pp.154–5; Marc R. DeVore, 'A More Complex and Conventional Victory', pp.161–3; Geraint Hughes, 'A "Model Campaign" Reappraised', pp.286–8, 291; Clive Jones, 'Military intelligence and the war in Dhofar', p.628; Alastair MacKenzie, *Special Force*, pp.238–9; Thomas R. Mockaitis, *British Counterinsurgency in the post-imperial era*, pp.74–6 and 92–3; John Newsinger, *British Counterinsurgency*, pp.142–5; John Pimlott, 'The British Army', pp.31–5.

82 Geraint Hughes, 'A "Model Campaign" Reappraised', pp.284–5; Thomas R. Mockaitis, *British Counterinsurgency in the post-imperial era*, pp.79–80 and 90–1.

83 Marc R. DeVore, 'A More Complex and Conventional Victory', pp.145, 153–5, 158–60, 162, 166; Geraint Hughes, 'A "Model Campaign" Reappraised', pp.283–4, 289–93; Colonel Tony Jeapes, *SAS: Operation Oman*, pp.36–7; Clive Jones, 'Military intelligence and the war in Dhofar', pp.631–2; Alastair MacKenzie, *Special Force*,

pp.141–2 and 150; Thomas R. Mockaitis, *British Counterinsurgency in the post-imperial era*, pp.74, 76–9, 81 and 90; John Newsinger, *British Counterinsurgency*, pp.143–5, 148–9; John Pimlott, 'The British Army', pp.33–5 and 42.

84 Ian F.W. Beckett, *Modern Insurgencies and Counter-Insurgencies*, p.228; Marc R. DeVore, 'A More Complex and Conventional Victory', pp.145, 159, 162–4; Geraint Hughes, 'A "Model Campaign" Reappraised', pp.294–6; Alastair MacKenzie, *Special Force*, pp.139, 143–5 and 147; Thomas R. Mockaitis, *British Counterinsurgency in the post-imperial era*, pp.80–8 and 91; John Newsinger, *British Counterinsurgency*, pp.148–9; John Pimlott, 'The British Army', pp.37–43.

85 Marc R. DeVore, 'A More Complex and Conventional Victory', pp.144–5, 161–2; Geraint Hughes, 'A "Model Campaign" Reappraised', pp.273, 286; Colonel Tony Jeapes, *SAS: Operation Oman*, p.11; Clive Jones, 'Military intelligence and the war in Dhofar', pp.632–9; Thomas R. Mockaitis, *British Counterinsurgency in the post-imperial era*, pp.88–4 and 91–2; John Newsinger, *British Counterinsurgency*, pp.143, 150; John Pimlott, 'The British Army', pp.25, 42–3.

86 J. Bowyer Bell, 'An Irish War', pp.245–7; Ian Cobain, *Cruel Britannia*, pp.135–203; Michael Dewar, *The British Army in Northern Ireland*, p.47; Aaron Edwards, 'Misapplying Lessons Learnt?', pp.307–15; Gaetano Joe Ilandi, 'Irish Republican Army Counterintelligence', pp.3–4; Caroline Kennedy-Pipe and Colin McInnes, 'The British Army in Northern Ireland 1969–1972', pp.203–7, 216–8, 222–3; Alfred W. McCoy, *A Question of Torture*, pp.54–8; Thomas R. Mockaitis, *British Counterinsurgency in the post-imperial era*, pp.96–104; Andrew Mumford, *The Counterinsurgency Myth*, pp.95–8, 104–5; Peter R. Neumann, *Britain's Long War*, pp.43–51, 56; John Newsinger, 'From Counterinsurgency to Internal Security', pp.90–5, 99; John Newsinger, *British Counterinsurgency*, pp.2, 151–64, 168–70; Rod Thornton, 'Getting it Wrong', pp.78–105; Christopher Tuck, 'Northern Ireland and the British Approach to Counterinsurgency', pp.172–3.

87 Ian F.W. Beckett, *Modern Insurgencies and Counter-Insurgencies*, p.226; David A. Charters, 'From Palestine to Northern Ireland', p.214; Alice Hills, *Future War in Cities*, p.102; Alastair MacKenzie, *Special Force*, pp.204–6 and 208–9; Ministry of Defence, *Operation Banner*, pp.2–7 and 8; Thomas R. Mockaitis, *British Counterinsurgency in the post-imperial era*, pp.104–7; Andrew Mumford, *The Counterinsurgency Myth*, pp.99–101, 106–8, and 122–3; Peter R. Neumann, *Britain's Long War*, pp.56–8, 106–7; John Newsinger, 'From Counterinsurgency to Internal Security', p.103; John Newsinger, *British Counterinsurgency*, pp.164–7, 187–90; Christopher Tuck, 'Northern Ireland and the British Approach to Counterinsurgency', pp.173–5.

88 Huw Bennett, 'From Direct Rule to Motorman', pp.513–4, 518–26; J. Bowyer Bell, 'An Irish War', pp.250–3; Robin Evelegh, *Peace Keeping in a Democratic Society*, pp.48–59; Gaetano Joe Ilandi, 'Irish Republican Army Counterintelligence', pp.6–9; Caroline Kennedy-Pipe and Colin McInnes, 'The British Army in Northern Ireland 1969–1972', p.216; Alastair MacKenzie, *Special Force*, p.196; Ministry of Defence, *Operation Banner*, pp.1.1, 2.9–2.11; Andrew Mumford, *The Counterinsurgency Myth*, pp.103–6; Peter R. Neumann, *Britain's Long War*, pp.79–80, 83–4; John Newsinger, *British Counterinsurgency*, pp.2, 171–2; M.L.R. Smith and Peter R. Neumann, 'Motorman's Long Journey', pp.426–30.

89 Gearóid Ó Faoleán, 'Ireland's Ho Chi Minh trail?', p.987; Gaetano Joe Ilandi, 'Irish Republican Army Counterintelligence', pp.4–9, 18–21; Thomas R. Mockaitis, *British Counterinsurgency in the post-imperial era*, pp.107–10 and 122–6; Andrew Mumford, *The Counterinsurgency Myth*, pp.108–13 and 123; John Newsinger, *British*

Counterinsurgency, pp.177–82; Christopher Tuck, 'Northern Ireland and the British Approach to Counterinsurgency', pp.175–7.

90 David A. Charters, 'From Palestine to Northern Ireland', pp.200–1, 205–7; Thomas R. Mockaitis, *British Counterinsurgency in the post-imperial era*, pp.116–9; Andrew Mumford, *The Counterinsurgency Myth*, pp.108–9.

91 Major John L. Clark, *Thinking Beyond Counterinsurgency*, pp.21–7; Tony Geraghty, *The Irish War*, pp.133–64; Alice Hills, *Future War in Cities*, pp.100–2; Jack Holland and Susan Phoenix: *Phoenix: Policing The Shadows*, pp.267–9; Ministry of Defence, *Operation Banner*, pp.2.15, 5.6–5.8; Andrew Mumford, *The Counterinsurgency Myth*, pp.12 and 118–21; Peter R. Neumann, *Britain's Long War*, pp.129–30, 137–9, 141–3; John Newsinger, *British Counterinsurgency*, pp.185–6, 191–4; Kevin Toolis, *Rebel Hearts*, pp.21–2, 50–1, 192–257.

92 Thomas R. Mockaitis, *British Counterinsurgency, 1919–60*, pp.173 and 188–9; Andrew Mumford, *The Counterinsurgency Myth*, pp.24, 70–1, and 153–4.

93 David French, *The British Way in Counterinsurgency, 1945–1967*, pp.204–18; Raffi Gregorian, '"Jungle Bashing" in Malaya', p.29; Thomas R. Mockaitis, *British Counterinsurgency in the post-imperial era*, pp.13–4, 24, 113; Thomas R. Mockaitis, 'The Origins of British Counterinsurgency', p.211; Andrew Mumford, *The Counterinsurgency Myth*, p.12; John A. Nagl, *Counterinsurgency Lessons from Malaya and Vietnam*, pp.xiv–xv; John Newsinger, 'From Counterinsurgency to Internal Security', p.89; David A. Percox, 'British Counterinsurgency in Kenya, 1952–56', p.267; Richard Popplewell, 'Lacking Intelligence', p.320; James Pritchard and M.L.R. Smith, 'Thompson in Helmand', p.67; Hew Strachan (ed.), *Big Wars and Small Wars*, p.6.

94 Alexander Alderson, 'The Army Brain', pp.11–4; Alderson, Alexander, 'Influence, the Indirect Approach and Manoeuvre', pp.36–7; Alexander Alderson, 'The British Approach to COIN and Stabilisation', p.66; General Sir Richard Dannatt, *Leading from the Front*, pp.94, 100–1; Alice Hills, *Future War in Cities*, pp.47–50; Richard Iron, 'The Charge of the Knights', p.54; Lieutenant General Sir John Kiszely, 'Thinking about the Operational Level', pp.40–2, and 'Learning about Counterinsurgency', pp.5–11; Daniel Marston, '"Smug and Complacent?" Operation TELIC', p.17; Colin McInnes, 'The Gulf War, 1990–1', p.163–4, 167–8; Colin McInnes, 'The British Army's New Way in Warfare', p.129; John A. Nagl, *Counterinsurgency Lessons from Malaya and Vietnam*, p.7; Hew Strachan (ed.), *Big Wars and Small Wars*, pp.6–7 and 17; James K. Wither, 'Basra's not Belfast', pp.616–8.

Chapter 7

1   Huw Bennett, 'Enmeshed in insurgency', p.505; Frank Ledwidge, *Losing Small Wars*, pp.144–46, 221; Andrew Mackay and Stephen Tatham, *Behavioural Conflict*, p.14; Ministry of Defence, *Operation Banner*, p.1–2.

2   Alexander Alderson, 'Too Busy to Learn', pp.281–4, 287; Jack Fairweather, *A War of Choice*, p.249; John Mackinlay, 'Is UK Doctrine Relevant to Global Insurgency?', p.35; John Mackinlay, *The Insurgent Archipelago*, pp.50–3, 67.

3   Frank Ledwidge, *Losing Small Wars*, pp.31, 192–209, 214–15; see also Jacob Kipp, Lester Grau, Karl Prinslow and Captain Don Smith, 'The Human Terrain System', pp.8–9 and Montgomery McFate, 'Anthropology and Counterinsurgency', pp.24–7.

4   Frank Ledwidge, *Losing Small Wars*, pp.217–23.

5   Robert M. Cassidy, *War, Will, and Warlords*, pp.40–1, 127–9, 132; Lieutenant General Peter W. Chiarelli with Major Stephen M. Smith, 'Learning from our Modern

Wars', p.4; Flynn, *Contesting History*, pp.vii–xi; Antonio Giustozzi, *Koran, Kalashnikov and Laptop*, pp.166–89; Lieutenant Colonel Lester W. Grau, 'Something Old, Something New', pp.46–7; Ahmed S. Hashim, *Insurgency and Counterinsurgency in Iraq*, pp.271–6, 303–41; David J. Kilcullen, *Counterinsurgency*, p.19; Mark Moyar, *A Question of Command*, p.198; John A. Nagl, 'An American View of Twenty-First Century Counterinsurgency', pp.13–4; Bruce R. Pirnie and Edward O'Connell, *Counterinsurgency in Iraq (2003–2006)*, pp.49–52; Seth G. Jones, *In the Graveyard of Empires*, pp.163–82; Frank Ledwidge, *Losing Small Wars*, pp.141–8, 192–209; David H. Ucko and Robert Egnell, *Counterinsurgency in Crisis*, pp.38–42.

6   British Army Field Manual, Volume 1, Part 10, *Countering Insurgency*, Chapter 3, p.15; Jack Fairweather, *A War of Choice*, pp.28–36; Alice Hills, 'Basra and the Referent Points of Twofold War', pp.39–42; Frank Ledwidge, *Losing Small Wars*, pp.16–7; Stephen C. Pelletière, *Losing Iraq*, pp.2–3; James K. Wither, 'Basra's not Belfast', p.620.

7   Colonel Alexander Alderson, 'The Validity of British Army Counterinsurgency Doctrine after the War in Iraq 2003–2009', pp.137, 140–1; Frank Ledwidge, *Losing Small Wars*, pp.18–29; Andrew Mumford, *The Counterinsurgency Myth*, pp.125–6, 132 and 137; Richard North, *The Ministry of Defeat*, pp.15–22; Hilary Synnott, 'Statebuilding in Southern Iraq', pp.18–9, 56, 131; David Ucko, 'Lessons from Basra', pp.136–8; David H. Ucko and Robert Egnell, *Counterinsurgency in Crisis*, pp.55–6; James K. Wither, 'Basra's not Belfast', pp.622–4.

8   Huw Bennett, 'Enmeshed in insurgency', pp.504–5, 508–9; Justin Maciejewski, 'Best effort', p.157; Peter R. Mansoor, 'The British Army and the Lessons of the Iraq War', pp.12–14; Daniel Marston, '"Smug and Complacent?" Operation TELIC', pp.16–21; David H. Ucko and Robert Egnell, *Counterinsurgency in Crisis*, pp.45–6.

9   Colonel Alexander Alderson, 'The Validity of British Army Counterinsurgency Doctrine after the War in Iraq 2003–2009', pp.137, 139, 147–56; Frank Ledwidge, *Losing Small Wars*, pp.23–4; Montgomery McFate, 'The Military Utility of Understanding Adversary Culture', pp.44–7; Peter R. Mansoor, 'The British Army and the Lessons of the Iraq War', p.12; Daniel Marston, '"Smug and Complacent?" Operation TELIC', pp.18–9; Mungo Melvin, 'Learning the Strategic Lessons from Afghanistan', p.59; Hilary Synnott, 'Statebuilding in Southern Iraq', pp.33–56, 138–9; David Ucko, 'Lessons from Basra', pp.135–6; David H. Ucko and Robert Egnell, *Counterinsurgency in Crisis*, pp.52–4; James K. Wither, 'Basra's not Belfast', p.613.

10 Paula Broadwell, *All In*, p.58; Patrick Cockburn, *The Occupation*, pp.1, 75, 82–6, 107–9, 124–6, 139; Flynn, *Contesting History*, pp.91–110; Ahmed S. Hashim, *Insurgency and Counterinsurgency in Iraq*, pp.xv–xviii, 17–47, 60, 125–38, 176–213, 340–1; Steven Metz, *Learning from Iraq*, pp.v–vi, 1–6, 24–31, 38–58; Mark Moyar, *A Question of Command*, pp.214–5, 224–6; Thomas R. Mockaitis, *The Iraq War*, pp.23–41; Stephen C. Pelletière, *Losing Iraq*, pp.67–74; Bruce R. Pirnie and Edward O'Connell, *Counterinsurgency in Iraq (2003–2006)*, pp.5–17, 21–48; Thomas E. Ricks, *Fiasco*, pp.195–200, 214–28, 290–7; James A. Russell, *Innovation, Transformation and War*, pp.1–2, 19–21.

11 Patrick Cockburn, *The Occupation*, pp.110–15; Jack Fairweather, *A War of Choice*, pp.1–2; Frank Ledwidge, *Losing Small Wars*, pp.32–6, 58–9; Peter R. Mansoor, 'The British Army and the Lessons of the Iraq War', pp.12–14; Daniel Marston, '"Smug and Complacent?" Operation TELIC', pp.16, 20; Andrew Mumford, *The Counterinsurgency Myth*, pp.12–13, 129–31; Richard North, *The Ministry of Defeat*, pp.26–8, 35, 53, 63, 72–3, 178–9, 185; Stephen C. Pelletière, *Losing Iraq*, pp.74–5;

David Ucko, 'Lessons from Basra', pp.142–51; David H. Ucko and Robert Egnell, *Counterinsurgency in Crisis*, pp.54–5; James K. Wither, 'Basra's not Belfast', pp.611–2.

12 Paula Broadwell, *All In*, pp.190–1; Jack Fairweather, *A War of Choice*, pp.39–40; Ahmed S. Hashim, *Insurgency and Counterinsurgency in Iraq*, pp.92–9; Mark Moyar, *A Question of Command*, pp.215–9; Thomas R. Mockaitis, *The Iraq War*, pp.33–6; Stephen C. Pelletière, *Losing Iraq*, pp.2, 56–9, 67–9; Thomas E. Ricks, *Fiasco*, pp.158–66; Linda Robinson, *Tell Me How This Ends*, pp.1–5; Neil Strachan, 'Abu Naji returns to Al-Kut', pp.111–2.

13 Jack Fairweather, *A War of Choice*, pp.138–45, 207–22, 252; David H. Ucko and Robert Egnell, *Counterinsurgency in Crisis*, pp.58–60.

14 Ian Cobain, *Cruel Britannia*, pp.279–80, 285–305; Jack Fairweather, *A War of Choice*, pp.2, 86–7, 252; Geraint Hughes, 'The Insurgencies in Iraq, 2003–2009', pp.161–2; Frank Ledwidge, *Losing Small Wars*, pp.29–30.

15 Alexander Alderson, 'Iraq and its Borders', pp.21–2; Colonel Duncan Barley, 'Training and Mentoring the Iraqi Army – the First Military Transition Teams', p.85; Lieutenant Colonel Simon Browne, 'Will One Size Ever MiTT All?', pp.39–41; Jack Fairweather, *A War of Choice*, pp.174–81, 198–9, 202; Colonel Charlie Herbert, 'Time for a Switch in Main Effort?', p.36; Richard Iron, 'The Charge of the Knights', p.57; Daniel Marston, '"Smug and Complacent?" Operation TELIC', pp.19; Lieutenant Colonel T.P. Robinson, 'Illuminating a Black Art', pp.35–8; Brigadier Sandy Storrie, 'First Do No Harm', pp.30–3.

16 Alexander Alderson, 'The British Approach to COIN and Stabilisation', p.67; Frank Ledwidge, *Losing Small Wars*, pp.34–5; Jack Fairweather, *A War of Choice*, pp.253 and 262; James Holland, 'The Way Ahead in Afghanistan', p.48; David H. Ucko and Robert Egnell, *Counterinsurgency in Crisis*, p.135; James K. Wither, 'Basra's not Belfast', p.618.

17 Jack Fairweather, *A War of Choice*, pp.274–5, 282–7, 301–3; Richard Iron, 'The Charge of the Knights', pp.61–2; Frank Ledwidge, *Losing Small Wars*, pp.38–41; Andrew Mumford, *The Counterinsurgency Myth*, pp.132–3; Richard North, *The Ministry of Defeat*, pp.54, 166–7.

18 Lieutenant General Peter W. Chiarelli and Major Patrick R. Michaelis, 'Winning the Peace', pp.19–23; Jack Fairweather, *A War of Choice*, pp.87–8; Frank Ledwidge, *Losing Small Wars*, pp.35–6.

19 Richard North, *The Ministry of Defeat*, pp.228–30, 240–2.

20 Richard North, *The Ministry of Defeat*, pp.85–7, 131.

21 Robert Egnell, 'Lessons from Helmand', pp.298–301; Jack Fairweather, *A War of Choice*, pp.248–51; Patrick Little, 'Lessons Unlearned', pp.11–6; David H. Ucko and Robert Egnell, *Counterinsurgency in Crisis*, pp.1–5.

22 Frank Ledwidge, *Losing Small Wars*, pp.3, 7 and fn.16, 212–3, 271; Patrick Little, 'Lessons Unlearned', pp.13–5.

23 Alexander Alderson, 'Counterinsurgency', pp.16–7; see also Patrick Little, 'Lessons Unlearned', pp.15–6.

24 Matt Cavanagh, 'Ministerial Decision-Making in the Run-up to the Helmand Deployment', p.51; Frank Ledwidge, *Losing Small Wars*, pp.3–5, 11, 151, 154; Richard North, *The Ministry of Defeat*, p.186; Thomas Ricks, *The Gamble*, p.277; James K. Wither, 'Basra's not Belfast', p.61–2.

25 Paula Broadwell, *All In*, p.242; Matt Cavanagh, 'Ministerial Decision-Making in the Run-up to the Helmand Deployment', p.51; Thomas Donnelly, 'The Cousins' Counterinsurgency Wars', pp.4–9; David Hastings Dunn and Andrew Futter,

'Short-Term Tactical Gains and Long-Term Strategic Problems', pp.196–200; Jack Fairweather, *A War of Choice*, pp.128–34, 248, 304–6, 313–18 and 332–9; Theo Farrell and Stuart Gordon, 'COIN machine', p.18; Matthew J. Flynn, *Contesting History*, pp.110–6; Richard Iron, 'The Charge of the Knights', pp.55–61; Frank Ledwidge, *Losing Small Wars*, pp.41–59; Andrew Mumford, *The Counterinsurgency Myth*, pp.126 and 134–5; Richard North, *The Ministry of Defeat*, pp.191, 196–9, and 203–5; Bruce R. Pirnie and Edward O'Connell, *Counterinsurgency in Iraq (2003–2006)*, p.19; David H. Ucko, *The New Counterinsurgency Era*, pp.112–6; David H. Ucko and Robert Egnell, *Counterinsurgency in Crisis*, pp.68–72.

26 Jack Fairweather, *A War of Choice*, pp.247–8; Geraint Hughes, 'The Insurgencies in Iraq, 2003–2009', p.160; David Kilcullen, *The Accidental Guerrilla*, pp.122–6; Linda Robinson, *Tell Me How This Ends*, pp.14–17.

27 Major Morgan Mann, 'The Power Equation', p.108.

28 Brigadier Nigel R.F. Aylwin-Foster, 'Changing the Army for Counterinsurgency Operations', pp.2–15.

29 Paula Broadwell, *All In*, pp.xxvii, 69, 82; Jack Fairweather, *A War of Choice*, pp.247–8; Richard North, *The Ministry of Defeat*, p.191; Thomas E. Ricks, *The Gamble*, pp.19–23, 27–8, 140–3; Linda Robinson, *Tell Me How This Ends*, pp.57, 99.

30 Alexander Alderson, 'Learning, Adapting, Applying', pp.13–7, 'Too Busy to Learn', pp.287–9, and 'The British Approach to COIN and Stabilisation', p.68; Paula Broadwell, *All In*, pp.203–4; Jack Fairweather, *A War of Choice*, p.248; Mark Moyar, *A Question of Command*, pp.242–7; John A. Nagl, 'An American View of Twenty-First Century Counterinsurgency', pp.14–6; Thomas E. Ricks, *The Gamble*, pp.24–32; Linda Robinson, *Tell Me How This Ends*, pp.77–81; David H. Ucko, *The New Counterinsurgency Era*, pp.103–12.

31 Jack Fairweather, *A War of Choice*, pp.248 and 289–90; Thomas E. Ricks, *Fiasco*, pp.195–200, 264–7; David H. Ucko, *The New Counterinsurgency Era*, p.76.

32 Colonel Alexander Alderson, 'The Validity of British Army Counterinsurgency Doctrine after the War in Iraq 2003–2009', pp.230–3; Major General John R.S. Batiste and Lieutenant Colonel Paul R. Daniels, 'The Fight for Samarra', pp.13–14, 19–21; Paula Broadwell, *All In*, pp.77, 189–94; Brian Burton and John Nagl, 'Learning as We Go', pp.309–10; Lieutenant General Peter W. Chiarelli and Major Patrick R. Michaelis, 'Winning the Peace', pp.16–19; Patrick Cockburn, *The Occupation* , pp.130–3; Dick Couch, *The Sheriff of Ramadi*, pp.101–3, 105–6, 130–4; Janine Davidson, *Lifting the Fog of Peace*, pp.94–5, 174; First Lieutenant Jason Goodale and First Lieutenant Jon Webre, 'The Combined Action Platoon in Iraq', pp.40–2; Bruce Hoffman, *Insurgency and Counterinsurgency in Iraq*, pp.8–9; Katie Ann Johnson, *A Reevaluation of the Combined Action Program as a Counterinsurgency Tool*, pp.24–6; Steven Metz, *Learning from Iraq*, pp.27–8; Major Steven M. Miska, 'Growing the Iraqi Security Forces', pp.66–9; Thomas R. Mockaitis, *The Iraq War*, pp.42–3; Mark Moyar, *A Question of Command*, pp.219–24, 228–31, 233–7, 239–42; Bruce R Pirnie and Edward O'Connell, *Counterinsurgency in Iraq (2003–2006)*, pp.39–41; Thomas E. Ricks, *Fiasco*, pp.228–32, 311–20, 367–70, 419–24; Thomas E. Ricks, *The Gamble*, pp.59–72, 95; Linda Robinson, *Tell Me How This Ends*, pp.67–78, 98; James A Russell, *Innovation, Transformation and War*, pp.62, 110–23, 137, 141–7; James A. Russell, 'Counterinsurgency American style', pp.78–9; Lieutenant Colonel Philip C. Skuta, 'Introduction to 2/7 CAP Platoon Actions in Iraq', p.35; Major Paul T. Stanton, 'Unit Immersion in Mosul', pp.63–9; David H. Ucko, *The New Counterinsurgency Era*, pp.75, 116, 131–2; Bob Woodward, *The War Within*, pp.36–8.

33 Colonel Mark F. Cancian, 'What Turned the Tide in Anbar?', pp.118–9; Colonel
    Anthony E. Deane, 'Providing Security Force Assistance in an Economy of Force
    Battle', pp.87–90; Jack Fairweather, *A War of Choice*, pp.290–9; David Kilcullen, *The
    Accidental Guerrilla*, pp.158–9; Major Andrew W. Koloski and Lieutenant Colonel
    John S. Kolasheski, 'Thickening the Lines', pp.42–53; Lieutenant Colonel Dale
    Kuchi, 'Testing Galula in Ameriyah', pp.72–80; Frank Ledwidge, *Losing Small Wars*,
    pp.248–9; Linda Robinson, *Tell Me How This Ends*, pp.96–8, 177–8, 251–4, 267–70;
    Major Niel Smith and Colonel Sean MacFarland, 'Anbar Awakens', pp.41–51; David
    H. Ucko, *The New Counterinsurgency Era*, pp.122–5; Bob Woodward, *The War Within*,
    p.381.
34 Matthew B. Arnold, 'The US "Surge" as a Collaborative Corrective for Iraq', p.27;
    Paula Broadwell, *All In*, pp.xxviii, 236–42; Lieutenant Colonel David G. Fivecoat
    and Captain Aaron T. Schwengler, 'Revisiting Modern Warfare', pp.80–2; David
    Kilcullen, *The Accidental Guerrilla*, pp.117, 127, 130, 135, 139, 148; Alfred W. McCoy,
    Policing America's Empire, pp.9–10; Lieutenant Colonel Jack Marr *et al*, 'Human
    Terrain Mapping', pp.20–4; Mark Moyar, *A Question of Command*, pp.225, 247–51;
    Richard North, *The Ministry of Defeat*, pp.191–3, 254; Paul B. Rich, 'A historical
    overview of US counterinsurgency', p.22; Linda Robinson, *Tell Me How This Ends*,
    pp.121–3, 163–7, 177–8; Lieutenant Colonel Thomas A. Seagrist, 'Combat Advising
    in Iraq', pp.65–72; David H. Ucko and Robert Egnell, *Counterinsurgency in Crisis*,
    pp.2–3, 61–8; Bob Woodward, *The War Within*, p.380.
35 Huw Bennett, 'Enmeshed in insurgency', p.507; Thomas Briggs, 'Naw Bahar
    District 2010–11', pp.129; Robert M. Cassidy, 'The Afghanistan Choice', pp.41–2,
    and *War, Will, and Warlords*, pp.31–2, 37–40; Major Christoff T. Gaub, 'Provincial
    Reconstruction Teams (PRTs)', pp.10–6; Antonio Giustozzi, *Koran, Kalashnikov
    and Laptop*, pp.2–6, 21–9, 123–9, 163–89; James Holland, 'The Way Ahead in
    Afghanistan', pp.47–8; Seth G. Jones, *In the Graveyard of Empires*, pp.88–119, 124–33,
    141–50, 163–82, 212, 248–55, 258–78; David J. Kilcullen, *The Accidental Guerrilla*,
    pp.60–2; Mark Moyar, *A Question of Command*, pp.203–4; Allan Orr, 'Recasting
    Afghan Strategy', pp.91–4; Emma Sky, 'The Lead Nation Approach', pp.22–6;
    Charles Style, 'Britain's Afghanistan Deployment in 2006', pp.40–1; David H. Ucko
    and Robert Egnell, *Counterinsurgency in Crisis*, pp.107–8; Syed Manzar Abbas Zaidi,
    'Pakistan's Anti-Taliban Counterinsurgency', pp.10–1.
36 Antonio Giustozzi, *Koran, Kalashnikov and Laptop*, pp.161–6; Seth G. Jones, *In the
    Graveyard of Empires*, pp.210–2; Frank Ledwidge, *Losing Small Wars*, pp.65–71, 130;
    Daniel Marston, 'British Operations in Helmand Afghanistan', p.2; Major (V)S.N.
    Miller, 'A Comprehensive Failure', pp.35–9; Mark Moyar, *A Question of Command*,
    pp.207–11.
37 Thomas Briggs, 'Naw Bahar District 2010–11', pp.124–7; Paula Broadwell, *All In*,
    pp.15, 23, 76–82; Captain James Cartwright, 'Operational Mentor and Liaison
    Team on Op Herrick 8', pp.13–5; Robert M. Cassidy, *War, Will, and Warlords*,
    pp.115–8; Rudra, Chaudhuri and Theo Farrell, 'Campaign Disconnect', pp.284–7;
    Warren Chin, 'British Counterinsurgency in Afghanistan', pp.210–12; James
    Fergusson, *A Million Bullets*, pp.233–5; Antonio Giustozzi, *Koran, Kalashnikov and
    Laptop*, pp.164; James Holland, 'The Way Ahead in Afghanistan', pp.47–8; Robert
    Johnson, 'Upstream engagement and downstream entanglements', pp.659–63; David
    Kilcullen, *The Accidental Guerrilla*, pp.46–7; Frank Ledwidge, *Losing Small Wars*,
    pp.2–3, 60–4, 71; Shivan Mahendrarajah, 'Conceptual failure, the Taliban's parallel
    hierarchies, and America's strategic defeat in Afghanistan', pp.92–112; Michael Rose,
    'Afghanistan', p.10.

38 Matt Cavanagh, 'Ministerial Decision-Making in the Run-up to the Helmand
    Deployment', pp.51–3; Warren Chin, 'British Counterinsurgency in Afghanistan',
    pp.220–1; Michael Clarke and Valentina Soria, 'Charging up the Valley', pp.81–7;
    Robert Egnell, 'Lessons from Helmand', pp.297, 302–8; Theo Farrell, 'Improving
    in War', pp.575–8; Theo Farrell and Stuart Gordon, 'COIN machine', pp.20–1;
    James Fergusson, *A Million Bullets*, pp.222–3; Anthony King, 'Understanding the
    Helmand campaign', pp.313–30; Frank Ledwidge, *Losing Small Wars*, pp.69–71
    and 74–9; James Pritchard and M.L.R. Smith, 'Thompson in Helmand', pp.70–3;
    Mike Ryan, *Battlefield Afghanistan*, pp.118–23; David H. Ucko and Robert Egnell,
    Counterinsurgency in Crisis, pp.77, 84–9.
39 Theo Farrell and Stuart Gordon, 'COIN machine', p.23; Frank Ledwidge, *Losing
    Small Wars*, pp.171–91.
40 Colonel Alexander Alderson, 'The Validity of British Army Counterinsurgency
    Doctrine after the War in Iraq 2003–2009', p.137; David Betz and Anthony
    Cormack, 'Hot War, Cold Comfort', p.27; Warren Chin, 'Colonial Warfare
    in a Post-Colonial State', pp.234–7; Robert Egnell, 'Lessons from Helmand',
    pp.308–10; Theo Farrell and Stuart Gordon, 'COIN machine', p.23; Anthony King,
    'Understanding the Helmand campaign', pp.317–8; Frank Ledwidge, *Losing Small
    Wars*, pp.84–94, 113, 133, 171–91; Michael Rose, 'Afghanistan', p.10; David H. Ucko
    and Robert Egnell, *Counterinsurgency in Crisis*, pp.90–2.
41 Warren Chin, 'Colonial Warfare in a Post-Colonial State', pp.237–40; Frank
    Ledwidge, *Losing Small Wars*, pp.2–3, 80–3; Major (V)S.N. Miller, 'A Comprehensive
    Failure, pp.36–7; Richard North, *The Ministry of Defeat*, p.219; James Pritchard and
    M.L.R. Smith, 'Thompson in Helmand', pp.73–80.
42 Alexander Alderson, 'Too Busy to Learn', p.295; Alexander Alderson, '"Learn
    from Experience" or "Never Again"', p.40; Alexander Alderson, 'The British
    Approach to COIN and Stabilisation', p.62; Huw Bennett, 'Enmeshed in
    insurgency', p.509; Nick Carter and Alexander Alderson, 'Partnering with Local
    Forces', pp.37–8; Robert Egnell, 'Lessons from Helmand', pp.309–13; Theo
    Farrell, 'Improving in War', pp.573, 578–83; Theo Farrell and Stuart Gordon,
    'COIN machine', pp.21–2; Major General Michael T. Flynn, Matt Pottinger
    and Paul Batchelor, 'Fixing Intel', p.34; James Holland, 'The Way Ahead in
    Afghanistan', pp.49–50; Frank Ledwidge, *Losing Small Wars*, pp.94–9 and 118;
    Patrick Porter, 'Why Britain Doesn't Do Grand Strategy', p.7; James Pritchard
    and M.L.R. Smith, 'Thompson in Helmand', p.75; David H. Ucko and Robert
    Egnell, *Counterinsurgency in Crisis*, pp.94–9, 102–4; Phil Weatherill, 'Notes from
    the Field', pp.92–6; Wilson, David and Gareth E. Conway, 'The Tactical Conflict
    Assessment Framework', p.11.
43 Duncan Barley, 'Rebuilding Afghanistan's Security Forces', p.57; Lieutenant General
    David W Barno, 'Fighting "The Other War"', pp.33–4; Paula Broadwell, *All In*,
    pp.11, 57–8; Jørgen W. Eriksen and Tormod Heier, 'Winter as the Number One
    Enemy', pp.64, 70; Frank Ledwidge, *Losing Small Wars*, pp.69–71 and 74–9; Allan
    Orr, 'Recasting Afghan Strategy', p.90.
44 Paula Broadwell, *All In*, pp.59–62; Robert M. Cassidy, *War, Will, and Warlords*,
    pp.49–50; Richard North, *The Ministry of Defeat*, pp.46–55.
45 Paula Broadwell, *All In*, p.26; Rudra, Chaudhuri and Theo Farrell, 'Campaign
    Disconnect', pp.272–5; Colonel (P.) Patrick Donahue and Lieutenant Colonel
    Michael Fenzel, 'Combating a Modern Insurgency', pp.27–37; David Kilcullen,
    The Accidental Guerrilla, pp.93–6; Richard North, *The Ministry of Defeat*, pp.31–3,
    113.

46 Warren Chin, 'British Counterinsurgency in Afghanistan', pp.210–12; Vikram
    Jagadish, 'Reconsidering American Strategy in South Asia', pp.37–43; Richard
    North, *The Ministry of Defeat*, pp.122–4, 131–5, 158–9; James A. Russell,
    'Counterinsurgency American style', pp.81–2.

47 Rudra, Chaudhuri and Theo Farrell, 'Campaign Disconnect', p.274; Antonio
    Giustozzi, 'The Afghan National Army', pp.36–41; Captain Daniel Helmer, 'Twelve
    Urgent Steps for the Adviser Mission in Afghanistan', pp.73–81; James Holland,
    'The Way Ahead in Afghanistan', pp.47–8; Major Michael D. Jason, 'Integrating the
    Advisory Effort in the Army', pp.27–32; David D. McKiernan, 'Recommitment and
    Shared Interests', pp.8–9; Lieutenant Colonel Dr John A. Nagl, 'Institutionalizing
    Adaptation', pp.22–32; Richard North, *The Ministry of Defeat*, pp.44, 161; Victor M.
    Rosello, 'Lessons from El Salvador', p.106.

48 John Alexander, '"Decomposing" an Insurgency', pp.48–53; Paula Broadwell, *All
    In*, pp.245–6 and 261–2; Robert M. Cassidy, *War, Will, and Warlords*, pp.119–21;
    Yoav Gortzak, 'The prospects of combined action', pp.137, 151; Richard North, *The
    Ministry of Defeat*, pp.42–4 and 157–9; Lieutenant Colonel Brian Petit, 'The Fight
    for the Village', pp.25–32; Michael Stevens, 'Community Defence in Afghanistan',
    pp.42–6, and 'Afghan Local Police in Helmand', pp.64–5; Jon Strandquist, 'Local
    defence forces and counterinsurgency in Afghanistan', pp.90–5.

49 Major Sean McKenna and Major Russell Hampsey, '"The COIN Warrior"'; Waging
    Influence', pp.3–15.

50 Captain Andrew Bell, 'Let's Get It Right', pp.39–45; Lieutenant Christopher Kent,
    'Speaking the Lingo', pp.67–9; Lieutenant Mike Martin, 'The Importance of
    Cultural Understanding to the Military', pp.44–9.

51 Paula Broadwell, *All In*, pp.166–8, 196, 282; Raffaello Pantucci, 'Deep Impact',
    pp.72–3; Jeffrey A. Sluka, 'Death from Above: UAVs and Losing Hearts and Minds',
    pp.70–6; Brian Glyn Williams, 'Afghanistan after the Soviets', pp.930–53.

52 Huw Bennett, 'Enmeshed in insurgency', pp.501–2, 506; Theo Farrell, 'Improving in
    War', pp.588–92; Stuart Griffin, 'Iraq, Afghanistan and the future of British military
    doctrine', pp.318–21; Trooper David Maddock, 'Donkeys v Mastiffs', p.50; Frank
    Ledwidge, *Losing Small Wars*, pp.3–5, 11, 151; Sergeant Bob Seely, 'Winning Through
    the People', p.17; James K. Wither, 'Basra's not Belfast', p.615.

Conclusion

1   John Mackinlay, 'Is UK Doctrine Relevant to Global Insurgency?', pp.35–40; John
    Mackinlay, *The Insurgent Archipelago*, pp.92–8.

2   British Army Field Manual, Volume 1, Part 10, *Countering Insurgency*, Chapter 3,
    pp.5–12.

3   Nathan Roger, *Image Warfare in the War on Terror*, pp.1–7, 12–9, 26–9, 45–9, 62–70,
    77–81, 102–3, 113–28, 140–52, 160–73.

4   John Mackinlay, 'Is UK Doctrine Relevant to Global Insurgency?', pp.35–40; John
    Mackinlay, *The Insurgent Archipelago*, pp.92–8.

5   Kenneth Payne, 'Media at War', pp.17–20.

6   Ashley Jackson, 'British Counterinsurgency in History: A Useful Precedent?' p.12.

7   Neville Bolt, *The Violent Image*, pp.1–8, 36, 240–55; John Mackinlay, 'Is UK Doctrine
    Relevant to Global Insurgency?', pp.36–8; John Mackinlay, *The Insurgent Archipelago*,
    pp.54–9.

8   Neville Bolt, 'Propaganda of the Deed and the Irish Republican Brotherhood',
    pp.52–3.

9   Quoted by Neville Bolt, *The Violent Image*, p.xxi.
10 Eliot Cohen, Crane; Lieutenant Colonel Conrad; Lieutenant Colonel Jan Horvath; and Lieutenant Colonel John Nagl, 'Principles, Imperatives, and Paradoxes of Counterinsurgency', p.49.
11 Captain Christopher Ford, 'Of Shoes and Sites', pp.87–91; Alice Hills, *Future War in Cities*, pp.17–9.
12 Alastair Campbell, 'Communications Lessons for NATO, the Military and Media', pp.32–4; Dr Jonathon Eyal, 'The Media and the Military', p.3.
13 General Charles C. Krulak, 'The Strategic Corporal: Leadership in the Three Block War', *Marines* (January 1999).
14 Akil N. Awan, 'Radicalization on the Internet?', pp.76–80; Akil N. Awan and Mina Al-Lami, 'Al-Qa'ida's Virtual Crisis', pp.54–62; Major P.W.D. Edwards, 'The Military–Media Relationship', pp.44–6; Philip Seib, 'The Al-Qaeda Media Machine', pp.74–80.
15 Philip Seib, 'The Al-Qaeda Media Machine', p.79.
16 Lieutenant General William B. Caldwell IV, Dennis M. Murphy and Anton Manning, 'Learning to Leverage New Media', pp.2–4.
17 Captain Jeffrey R. Sanderson and Captain Scott J. Akerley, 'Focusing Training', pp.77–8; see also Major P.W.D. Edwards, 'The Military–Media Relationship', p.45.
18 Neville Bolt, 'The Leak before the Storm', pp.47–51.
19 David W. Barno, 'Challenges in Fighting a Global Insurgency', pp.15, 17, 21.
20 Lieutenant General Peter W. Chiarelli with Major Stephen M. Smith, 'Learning from our Modern Wars', pp.10–1.
21 Captain Daniel Helmer, 'Twelve Urgent Steps for the Adviser Mission in Afghanistan', p.77.
22 Lieutenant Colonel M.H. Wenham, 'Information Operations – Main effort or Supporting Effort?', p.47.
23 John Mackinlay, 'Is UK Doctrine Relevant to Global Insurgency?', p.37.
24 Neville Bolt, *The Violent Image*, pp.47–50; Ashley Jackson, 'British Counterinsurgency in History: A Useful Precedent?' pp.18–20; John Mackinlay, *The Insurgent Archipelago*, pp.81–92.
25 Neville Bolt, *The Violent Image*, p.161.
26 Frank G. Hoffman, 'Neo-Classical Counterinsurgency?', pp.83–4.
27 Riddell, Peter, 'Armed Forces, Media and the Public', p.13.
28 Martin J. Muckian, 'Structural Vulnerabilities of Networked Insurgencies', pp.18–24.
29 Raymond F. Betts, *France and Decolonisation, 1900–1960*, p.113; Dennis Chaplin, 'France – Military Involvement in Africa', pp.44–7; John Chipman, *French Power in Africa*, pp.114–67; Anthony Clayton, *The Wars of French Decolonization*, pp.12, 186; Michel L. Martin, 'From Algiers to N'Djamena', pp.100–7, 119, and 123–6; John Pimlott, 'The French Army', pp.46, 67–8, 72–3; Victor M. Rosello, 'Lessons from El Salvador', p.102; Barnett Singer and John Langdon, *Cultured Force*, pp.348–50; David H. Ucko and Robert Egnell, *Counterinsurgency in Crisis*, pp.147–52.
30 Alexander Alderson, '"Learn from Experience" or "Never Again"' , pp.40–1, and 'Learning, Adapting, Applying', p.13; Raymond F. Betts, *France and Decolonisation, 1900–1960*, p.113; Dennis Chaplin, 'France – Military Involvement in Africa', pp.44–7; John Chipman, *French Power in Africa*, pp.114–67; Anthony Clayton, *The Wars of French Decolonization*, pp.12, 186; Matthew Ford, 'Influence without power?', pp.495–7; Michel L. Martin, 'From Algiers to N'Djamena', pp.100–7, 119 and 123–6; John Pimlott, 'The French Army', pp.46, 67–8, 72–3; Paul B. Rich, 'A historical overview of US counterinsurgency', pp.25–6; Victor M. Rosello, 'Lessons from El

Salvador', p.102; Barnett Singer and John Langdon, *Cultured Force*, pp.348–50; David H. Ucko and Robert Egnell, *Counterinsurgency in Crisis*, pp.147–52.

31 Paul B. Rich, 'A historical overview of US counterinsurgency', pp.5–7, 26–7; see, for example, Gian Gentile, *Wrong Turn*, Fred Kaplan, *The Insurgents*, and Douglas Porch, *Counterinsurgency*.

32 General David H. Petraeus, 'Reflections on the Counterinsurgency Era', p.82.

33 Alexander Alderson, 'Learning, Adapting, Applying', p.12; John Mackinlay, 'Counterinsurgency in Global Perspective', p.67; Hew Strachan, 'British counterinsurgency from Malaya to Iraq', p.11.

SELECT BIBLIOGRAPHY

Ahern, Thomas L., *Vietnam Declassified: The CIA and Counterinsurgency*, Lexington, Kentucky: University Press of Kentucky, 2010.

Akehurst, John, *We Won A War: The Campaign in Oman, 1965–1975*, Salisbury, Wiltshire: Michael Russell, 1982.

Akehurst, John, *Generally Speaking: 'Then Hurrah for the Life of a Soldier'*, Norwich: Michael Russell, 1999.

Alberts, Donald J., 'Insurgencies of Portuguese Africa: Angola, Mozambique, Guinea–Bissau' in Bard E. O'Neill, D.J. Alberts, and Stephen J. Rosetti (eds), *Political Violence: A Comparative Approach*, Arvada, Colorado: Phoenix Press, 1974.

Alberts, Donald J., 'Armed Struggle in Angola' in Bard E. O'Neill, William R. Heaton, and Donald J. Alberts (eds), *Insurgency in the Modern World*, Boulder, Colorado: Westview Press, 1980.

Alderson, Alexander, 'Counterinsurgency: Learn and Adapt: Can We Do Better?', *British Army Review*, Vol. 142 (Summer 2007), pp. 16–21.

Alderson, Alexander, 'Learning, Adapting, Applying: US Counterinsurgency Doctrine and Practice', *The RUSI Journal*, Vol. 152, No. 6 (December 2007), pp. 12–19.

Alderson, Alexander, 'US COIN Doctrine and Practice: An Ally's Perspective', *Parameters*, Vol. XXXVII (Winter 2007–08), pp. 33–45.

Alderson, Alexander, 'Influence, the Indirect Approach and Manoeuvre', *The RUSI Journal*, Vol. 157, No. 1 (February/March 2012), pp. 36–43.

Alderson, Alexander, 'Iraq and its Borders: The Role of Barriers in Counterinsurgency', *The RUSI Journal*, Vol. 153, No. 2 (April 2008), pp. 18–22.

Alderson, Colonel Alexander, 'The Validity of British Army Counterinsurgency Doctrine after the War in Iraq 2003–2009', Cranfield University: PhD Thesis, Defence Academy College of Management and Technology, November 2009.

Alderson, Alexander, 'The Army Brain: A Historical Perspective on Doctrine, Development and the Challenges of Future Conflict', *The RUSI Journal*, Vol. 155, No. 3 (June/July 2010), pp. 10–15.

Alderson, Alexander, 'The Historical Perspective on Doctrine, Development and the Challenges of Future Conflict', *The RUSI Journal*, Vol. 155, No. 3 (June/July 2010), pp. 10–15.

Alderson, Alexander, 'The British Approach to COIN and Stabilisation: A Retrospective on Developments since 2001', *The RUSI Journal*, Vol. 157, No. 4 (August/September 2012), pp. 62–71.

Alderson, Alexander, 'Too Busy to Learn: Personal Observations on British Campaigns in Iraq and Afghanistan' in Jonathan Bailey, Richard Iron and Hew Strachan (eds), *British Generals in Blair's Wars*, Farnham, Surrey: Ashgate Publishing, 2013.

Alderson, Alexander, '"Learn from Experience" or "Never Again"': What Next for UK Counterinsurgency?', *The RUSI Journal*, Vol. 159, No. 1 (February/March 2014), pp. 40–8

Aldrich, Richard J., Gary D. Rawnsley and Ming-Yeh T. Rawnsley (eds), *The Clandestine Cold War in Asia, 1945–65: Western Intelligence, Propaganda and Special Operations*, London: Frank Cass, 2000.

Alexander, Don W., *Rod of Iron: French Counterinsurgency Policy in Aragón during the Peninsular War*, Wilmington, Delaware: Scholarly Resources, 1985.

Alexander, John, '"Decomposing" an Insurgency: Reintegration in Afghanistan', *The RUSI Journal*, Vol.157, No.4 (August/September 2012), pp.48–54.

Alexiev, Alexander, *Inside the Soviet Army in Afghanistan*, Santa Monica, California: RAND, May 1988.

Allen, Charles, *The Savage Wars of Peace: Soldiers' Voices 1945–1989*, London: Futura, 1991.

Ambler, John Stewart, *The French Army in Politics, 1945–1962*, Columbus, Ohio: Ohio State University Press. 1966.

Ambrose, Stephen E., *Upton and the Army*, Baton Rouge, Louisiana: Louisiana State University Press, 1992.

Anderson, David M. and David Killingray (eds), *Policing the Empire: Government, Authority and Control, 1830–1940*, Manchester and New York: Manchester University Press, 1991.

Anderson, David and David Killingray (eds), *Policing and Decolonisation: Nationalism, Politics and the Police, 1917–65*, Manchester and New York: Manchester University Press, 1992.

Anderson, David, *Histories of the Hanged: The Dirty War in Kenya and the End of Empire*, London: Weidenfeld & Nicolson, 2005.

Anderson, David, Huw Bennett and Daniel Branch, 'A Very British Massacre', *History Today*, Vol.56, No.8 (August 2006), pp.20–2.

Anderson, David, 'Policing and Communal Conflict: the Cyprus Emergency, 1954–60' in David Anderson and David Killingray (eds), *Policing and Decolonisation: Nationalism, Politics and the Police, 1917–65*, Manchester and New York: Manchester University Press, 1992.

Andrade, Dale and Lieutenant Colonel James Willbanks, 'CORDS/Phoenix: counterinsurgency lessons from Vietnam for the future', *Military Review*, Vol.LXXXVI, No.2 (March–April 2006), pp.9–23.

Andrew, Christopher, *The Defence of the Realm: The Authorized History of MI5*, London: Penguin, 2010.

Andrew, Christopher M. and A.S. Kanya-Forstner, *France Overseas: The Great War and the Climax of French Imperial Expansion*, London: Thames & Hudson Ltd, 1981.

Anglesey, Marquess of, *A History of the British Cavalry, 1816–1919*, Vol.4: 1899–1913, London: Leo Cooper, 1986.

Anglim, Simon, 'Orde Wingate and the Special Night Squads: A Feasible Policy for Counter-terrorism?' in Tim Benbow and Rod Thornton (eds), *Dimensions of Counterinsurgency: Applying Experience to Practice*, London and New York: Routledge, 2008.

Armony, Ariel C., *Argentina, the United States, and the Anti-Communist Crusade in Central America, 1977–1984*, Athens, Ohio: Ohio University Press, 1997.

Arnold, Matthew B., 'The US "Surge" as a Collaborative Corrective for Iraq', *The RUSI Journal*, Vol.153, No.2 (April 2008), pp.24–9.

Arnson, Cynthia J., 'Window on the Past: A Declassified History of Death Squads in El Salvador' in Bruce B. Campbell and Arthur D. Brenner (eds), *Death Squads in Global Perspective: Murder with Deniability*, New York and Basingstoke, Hampshire: Palgrave Macmillan, 2002.

Aussaresses, General Paul, *The Battle of the Casbah: Terrorism and Counter-Terrorism in Algeria, 1955–1957*, New York: Enigma Books, 2006.

Awan, Akil N., 'Radicalization on the Internet?: The Virtual Propagation of Jihadist Media and its Effects', *The RUSI Journal*, Vol.152, No.3 (June 2007), pp.76–81.

Awan, Akil N. and Mina Al-Lami, 'Al-Qa'ida's Virtual Crisis', *The RUSI Journal*, Vol.154, No.1 (February 2009), pp.54–64.

Aylwin-Foster, Brigadier Nigel R.F., 'Changing the Army for Counterinsurgency Operations', *Military Review* (November–December 2005), pp.2–15.

Bailey, Jonathan, Richard Iron and Hew Strachan (eds), *British Generals in Blair's Wars*, Farnham, Surrey: Ashgate Publishing, 2013.

Baker, Colin, *Retreat from Empire: Sir Robert Armitage in Africa and Cyprus*, London and New York: I.B. Tauris, 1998.

Baker, Deane-Peter and Evert Jordaan (eds), *South Africa and Contemporary Counterinsurgency: Roots, Practices, Prospects*, Claremont, South Africa: UCT Press, 2010.

Balcos, Lieutenant Colonel Anastase, 'Guerrilla Warfare', *Military Review*, Vol.XXXVII, No.12 (March 1958), pp.49–54.

Bankwitz, Philip Charles Farwell, *Maxime Weygand and Civil Military Relations in Modern France*, Cambridge, Massachusetts: Harvard University Press, 1967.

Barber, Noel, *The War of the Running Dogs: The Malayan Emergency, 1948–1960*, London: Collins, 1971.

Barker, Dudley, *Grivas: Portrait of a Terrorist*, London: The Cresset Press, 1959.

Barley, Duncan, 'Rebuilding Afghanistan's Security Forces: Security Sector Reform in Contested State-Building', *The RUSI Journal*, Vol.153, No.3 (June 2008), pp.52–7.

Barley, Duncan, 'Training and Mentoring the Iraqi Army – the First Military Transition Teams', *British Army Review*, No.139 (Spring 2006), pp.85–91.

Barnes, Steven A., '"In a Manner Befitting Soviet Citizens": An Uprising in the Post-Stalin Gulag', *Slavic Review*, Vol.64, No.4 (Winter 2005), pp.823–50.

Barno, David W., 'Challenges in Fighting a Global Insurgency', *Parameters*, Vol.XXXVI (Summer 2006), pp.15–29.

Barno, Lieutenant General David W., 'Fighting "The Other War": Counterinsurgency Strategy in Afghanistan, 2003–2005', *Military Review* (September–October 2007), pp.32–42.

Bartov, Omer, *The Eastern Front, 1941–45, German Troops and the Barbarisation of Warfare*, Basingstoke, Hampshire: Macmillan, 1985.

Bartov, Omer, *Hitler's Army: Soldiers, Nazis, and War in the Third Reich*, New York and Oxford: Oxford University Press, 1992.

Batiste, Major General John R.S. and Lieutenant Colonel Paul R. Daniels, 'The Fight for Samarra: Full-Spectrum Operations in Modern Warfare', *Military Review* (May–June 2005), pp.13–21.

Baumann, Robert F., *Russian–Soviet Unconventional Wars in the Caucasus, Central Asia, and Afghanistan*, Fort Leavenworth, Kansas: Combat Studies Institute, US Army Command and General Staff College, Leavenworth Papers Number 20, 1993.

Beckett, Ian F.W., 'The Portuguese Army: The Campaign in Mozambique, 1964–1974' in Ian F.W. Beckett and John Pimlott (eds), *Armed Forces & Modern Counterinsurgency*, Beckenham, Kent: Croom Helm, 1985.

Beckett, Ian F.W. (ed.), *The Roots of Counterinsurgency: Armies and Guerrilla Warfare, 1900–1945*, London: Blandford Press, 1988.

Beckett, Ian F.W., 'The Soviet Experience' in Ian F.W. Beckett (ed.), *The Roots of Counterinsurgency: Armies and Guerrilla Warfare, 1900–1945*, London: Blandford Press, 1988.

Beckett, Ian F.W., 'The United States Experience' in Ian F.W. Beckett (ed.), *The Roots of Counterinsurgency: Armies and Guerrilla Warfare, 1900–1945*, London: Blandford Press, 1988.

Beckett, Ian F.W., 'The Study of Counterinsurgency: A British Perspective', *Small Wars and Insurgencies*, Vol.1, No.1 (April 1990), pp.47–53.

Beckett, Ian F.W., 'Robert Thompson and the British Advisory Mission to South Vietnam, 1961–1965', *Small Wars and Insurgencies*, Vol.8, No.3 (Winter 1997), pp.41–63.

Beckett, Ian F.W., *Modern Insurgencies and Counter-Insurgencies: Guerrillas and their Opponents*, London and New York: Routledge, 2001.

Beckett, Ian F.W., *Insurgency in Iraq: An Historical Perspective*, Carlisle, Pennsylvania: Strategic Studies Institute, US Army War College, 2005.

Beckett, Ian (ed.), *Modern Counterinsurgency*, Aldershot, Hampshire and Burlington, Vermont: Ashgate, 2007.

Beckett, Ian F.W. and John Pimlott (eds), *Armed Forces & Modern Counterinsurgency*, Beckenham, Kent: Croom Helm, 1985.

Belfield, Eversley, *The Boer War*, London: Leo Cooper, 1975.

Bell, Captain Andrew, 'Let's Get It Right – Culture and Language Training', *British Army Review*, No.154 (Spring/Summer 2012), pp.39–45.

Bell, J. Bowyer, 'Revolts Against the Crown: The British Response to Imperial Insurgency', *Parameters*, Vol.4, No.1 (1974), pp.31–46.

Bell, J. Bowyer, 'An Irish War: The IRA's Armed Struggle, 1969–90: Strategy as History Rules OK', *Small Wars and Insurgencies*, Vol.1, No.3 (December 1990), pp.239–65.

Benbow, Tim and Rod Thornton (eds), *Dimensions of Counterinsurgency: Applying Experience to Practice*, London and New York: Routledge, 2008.

Bender, Gerald J., 'The Limits of Counterinsurgency: An African Case', *Comparative Politics*, Vol.4, No.3 (April 1972), pp.331–60.

Benest, David, 'Aden to Northern Ireland 1966–76' in Hew Strachan (ed.), *Big Wars and Small Wars: The British Army and the Lessons of War in the 20th Century*, Abingdon, Oxon: Routledge, 2006.

Bennett, Huw, 'The Mau Mau Emergency as Part of the British Army's Post-War Counterinsurgency Experience', *Defense and Security Analysis*, Vol.23, No.2 (June 2007), pp.143–63.

Bennett, Huw, 'The Other Side of the COIN: Minimum Force and Exemplary Force in British Army Counterinsurgency in Kenya', *Small Wars and Insurgencies*, Vol.18, No.4 (2007), pp.638–64.

Bennett, Huw, '"A very salutary effect": The Counter-Terror Strategy in the Early Malayan Emergency, June 1948 to December 1949', *The Journal of Strategic Studies*, Vol.32, No.3 (June 2009), pp.415–44.

Bennett, Huw, 'From Direct Rule to Motorman: Adjusting British Military Strategy for Northern Ireland in 1972', *Studies in Conflict and Terrorism*, Vol.33, No.6 (2010), pp.511–12.

Bennett, Huw, *Fighting the Mau Mau: The British Army and Counterinsurgency in the Kenya Emergency*, Cambridge: Cambridge University Press, 2013.

Bennett, Huw, 'Enmeshed in insurgency: Britain's protracted retreats from Iraq and Afghanistan', *Small Wars and Insurgencies*, Vol.25, No.3 (2014), pp.501–21.

Bennett, Matthew, 'The German Experience' in Ian F.W. Beckett (ed.), *The Roots of Counterinsurgency: Armies and Guerrilla Warfare, 1900–1945*, London: Blandford Press, 1988.

Bennigsen, Alexandre, *The Soviet Union and Muslim Guerrilla Wars, 1920–1981: Lessons for Afghanistan*, Santa Monica, California: RAND, August 1981.

Bergerud, Eric M., *The Dynamics of Defeat: The Vietnam War in Hau Nghia Province*, Boulder, Colorado: Westview Press, 1991.

Betz, David and Anthony Cormack, 'Hot War, Cold Comfort: A Less Optimistic Take on the British Military in Afghanistan', *The RUSI Journal*, Vol.154, No.4 (August 2009), pp.26–9.

Betts, Raymond F., *France and Decolonisation, 1900–1960*, Basingstoke, Hampshire: Macmillan, 1991.

Bickel, Keith B., *Mars Learning: The Marine Corps Development of Small Wars Doctrine, 1915–1940*, Boulder, Colorado: Westview, 2001.

Birstein, Vadim J., *Smersh: Stalin's Secret Weapon*, London: Biteback Publishing, 2013.

Birtle, Andrew J., *US Army Counterinsurgency and Contingency Operations Doctrine, 1860–1941*, Washington, DC: US Army Center of Military History, 1998.

Birtle, Andrew J., *US Army Counterinsurgency and Contingency Operations Doctrine, 1942–1976*, Washington, DC: US Army Center of Military History, 2006.

Bjelajac, Slavko N., 'Unconventional Warfare: American and Soviet Approaches', *Annals of the American Academy of Political and Social Science*, Vol.341 (May 1962), pp.74–81.

Blair, Anne, *There to the Bitter End: Ted Serong in Vietnam*, Crows Nest, NSW, Australia: Allen & Unwin, 2001.

Blank, Stephen, 'Soviet Russia and Low-Intensity Conflict in Central Asia: Three Case Studies' in Lewis B. Ware (ed.), *Low-Intensity Conflict in the Third World*, Maxwell Air Force Base, Alabama: Air University Press, 1988.

Blank, Stephen, 'Afghanistan and Beyond: Reflections on the Future of Warfare', *Small Wars and Insurgencies*, Vol.3, No.3 (Winter 1992), pp.217–40.

Blaufarb, Douglas S., *The Counterinsurgency Era: US Doctrine and Performance, 1950 to the Present*, New York: Free Press, 1977.

Bley, Helmut, *South-West Africa under German Rule, 1894–1914*, London: Heinemann, 1971.

Blood, Philip W., *Hitler's Bandit Hunters: The SS and the Nazi Occupation of Europe*, Washington, DC: Potomac Books, 2008.

Boemeke, Manfred F., Roger Chickering and Stig Föster (eds), *Anticipating Total War: The German and American Experiences, 1871–1914*, Cambridge: Cambridge University Press, 1999.

Bolger, Major Daniel P., *Scenes from an Unfinished War: Low-Intensity Conflict in Korea, 1966–1969*, Fort Leavenworth, Kansas: Leavenworth Papers Number 19, Combat Studies Institute, US Army Command and General Staff College, 1991.

Bolger, Daniel P., 'Unconventional Warrior: General Charles H. Bonesteel III and the Second Korean Conflict 1966–69', *Small Wars and Insurgencies*, Vol.10, No.1 (Spring 1999), pp.65–77.

Bolt, Neville, 'Propaganda of the Deed and the Irish Republican Brotherhood: From the Politics of "Shock and Awe" to the "imagined Political Community"', *The RUSI Journal*, Vol.153, No.1 (February 2008), pp.48–54.

Bolt, Neville, 'The Leak before the Storm: What Wikileaks Tells us about Modern Communication', *The RUSI Journal*, Vol.155, No.4 (August/September 2010), pp.46–51.

Bolt, Neville, *The Violent Image: Insurgent Propaganda and the New Revolutionaries*, London: C Hurst & Company, 2012.

Boot, Max, *The Savage Wars of Peace: Small Wars and the Rise of American Power*, New York: Basic Books, 2003.

Boré, Lieutenant Colonel Henri, 'Irregular Warfare in Africa: A French Marine Experience', *Marine Corps Gazette*, Vol.90, No.7 (July 2006), pp.38–40.

Borovik, Artyom, *The Hidden War: A Russian Journalist's Account of the Soviet War in Afghanistan*, New York: Grove Press, 1990.

Bose, Purnima and Laura Lyons, 'Dyer Consequences: The Trope of Amritsar, Ireland, and the Lessons of the "Minimum" Force Debate', *Boundary* 2, Vol.26, No.2 (Summer 1999), pp.199–229.

Bou-Nacklie, N.E., 'Tumult in Syria's Hama in 1925: The Failure of a Revolt', *Journal of Contemporary History*, Vol.33, No.2 (April 1998), pp.273–89.

Bourne, General Lord, 'The Direction of Anti-Guerrilla Operations' in Major General J.L. Moulton, Brigadier C.N. Barclay and Air Vice-Marshal W.M. Yool (eds), *Brassey's Annual: The Armed Forces Year-Book, 1964*, London: William Clowes, 1964.

Bowden, Tom, 'The Politics of the Arab Rebellion in Palestine 1936–39', *Middle Eastern Studies*, Vol.11, No.2 (May 1975), pp.147–74.

Bowden, Tom, *The Breakdown of Public Security: The Case of Ireland, 1916–1921 and Palestine 1936–1939*, London and Beverly Hills, California: Sage Publications, 1977.

Bowen, Don R., 'Quantrill, James, Younger, *et al*: Leadership in a Guerrilla Movement, Missouri, 1861–1965', *Military Affairs*, Vol.XLI, No.1 (February 1977), pp.42–7.

Braim, Paul F., *The Will to Win: The Life of General James A Van Fleet*, Annapolis, Maryland: Naval Institute Press, 2001.

Braithwaite, Rodric, *Afgantsy: The Russians in Afghanistan 1979–89*, London: Profile Books, 2011.

Branch, Daniel, *Defeating Mau Mau, Creating Kenya: Counterinsurgency, Civil War, and Decolonization*, Cambridge: Cambridge University Press, 2009.

Brands, Hal, *Latin America's Cold War*, Cambridge, Massachusetts and London: Harvard University Press, 2010.

Briggs, Thomas, 'Naw Bahar District 2010–11: A case study of counterinsurgency Conducted by Naval Special Warfare in Afghanistan', *Small Wars and Insurgencies*, Vol.25, No.1 (2014), pp.122–36.

British Army Field Manual, Volume 1, Part 10, *Countering Insurgency*, London: Warfare Development, Ministry of Defence, October 2009.

Broadwell, Paula, with Vernon Loeb, *All In: The Education of David Petraeus*, New York: The Penguin Press, 2012.

Brooks, David C., 'US Marines, Miskitos and the Hunt for Sandino: The Rio Coco Patrol in 1928', *Journal of Latin American Studies*, Vol.21, No.2 (May 1989), pp.311–42.

Browne, Lieutenant Colonel Simon, 'Will One Size Ever MiTT All? – Reflections on Mentoring the Iraqi Army on Op TELIC 12', *British Army Review*, No.147 (Summer 2009), pp.39–41.

Browning, Christopher, *Ordinary Men: Reserve Battalion 101 and the Final Solution in Poland*, New York: HarperCollins, 1992.

Brownlee, Richard S, *Gray Ghosts of the Confederacy: Guerrilla Warfare in the West, 1861–1865*, Baton Rouge, Louisiana: Louisiana State University Press, 1986.

Bruce, Neil, *Portugal: The Last Empire*, Newton Abbot, Devon: David & Charles, 1975.

Bulloch, Gavin, 'The development of doctrine for counterinsurgency: the British experience', *British Army Review*, Vol.111 (1996), pp.21–4.

Bulloch, (Brigadier) Gavin, 'Military Doctrine and Counterinsurgency: A British Perspective', *Parameters*, Vol.XXVI, No.2 (Summer 1996), pp.4–16.

Bulloch, (Brigadier) Gavin, 'The Application of Military Doctrine to Counter Insurgency (COIN) Operations – A British Perspective', *Small Wars and Insurgencies*, Vol.7, No.2 (Autumn 1996), pp.165–77.

Burds, Jeffrey, 'AGENTURA: Soviet Informants' Networks and the Ukrainian Underground in Galicia, 1944–48', *East European Politics and Societies*, Vol.11, No.1 (Winter 1997), pp.89–130.

Burds, Jeffrey, 'The Early Cold War in Soviet West Ukraine, 1944–1948', *The Carl Beck Papers*, No.1505 (January 2001), University of Pittsburgh, Pennsylvania: Center for Russian and East European Studies, pp.1–69.

Burds, Jeffrey, 'Gender and Policing in Soviet West Ukraine, 1944–1948', *Cahiers du Monde russe*, Vol.42, No.2/4 (April–December 2001), pp.279–319.

Burds, Jeffrey, 'The Soviet War against "Fifth Columnists": The Case of Chechnya, 1942–4', *Journal of Contemporary History*, Vol.42, No.2 (April 2007), pp.267–314.

Burds, Jeffrey, 'Sexual Violence in Europe in World War II, 1939–1945', *Politics & Society*, Vol.37, No.1 (March 2009), pp.35–74.

Burleigh, Michael, *Germany Turns Eastward: A Study of Ostforschung in the Third Reich*, London: Pan Books, 2002.

Burton, Brian and John Nagl, 'Learning as We Go: The US Army Adapts to Counterinsurgency in Iraq, July 2004–December 2006', *Small Wars and Insurgencies*, Vol.19, No.3 (September 2008), pp.303–27.

Busch, Peter, 'Killing the "Vietcong": The British Advisory Mission and the Strategic Hamlet Programme', *The Journal of Strategic Studies*, Vol.25, No.1 (March 2002), pp.135–62.

Butson, Thos G., *The Tsar's Lieutenant: The Soviet Marshal*, New York: Praeger, 1984.

Byford-Jones, [Colonel] W., *Grivas and the Story of EOKA*, London: Robert Hale, 1959.

Cable, Larry E., *Conflict of Myths: The Development of American Counterinsurgency Doctrine and the Vietnam War*, New York: New York University Press, 1986.

Cable, Larry, 'Reinventing the Round Wheel: Insurgency, Counter-Insurgency, and Peacekeeping Post Cold War', *Small Wars and Insurgencies*, Vol.4, No.2 (Autumn 1993), pp.228–62.

Caldwell, Lieutenant General William B., IV, Mr Dennis M. Murphy and Mr Anton Manning, 'Learning to Leverage New Media: The Israeli Defense Forces in Recent Conflicts', *Military Review* (May–June 2009), pp.2–10.

Callwell, Major C.E., *Small Wars: Their Principles and Practice*, London: HMSO, 1899.

Callwell, Colonel C.E., *Small Wars: Their Principles and Practice*, London: HMSO, 1906.

Callwell, Colonel C.E., *Small Wars: Their Principles and Practice*, London: HMSO, 1906 (Reprinted, 1914).

Callwell, C.E., *Small Wars: A Tactical Textbook for Imperial Soldiers*, London: HMSO, 1906.

Callwell, Charles, *Small Wars: A Tactical Textbook for Imperial Soldiers*, London: Greenhill Books, 1990.

Campbell, Alastair, 'Communications Lessons for NATO, the Military and Media', *The RUSI Journal*, Vol.144, No.4 (1999), pp.31–6.

Campbell, Bruce B. and Arthur D. Brenner (eds), *Death Squads in Global Perspective: Murder with Deniability*, New York and Basingstoke, Hampshire: Palgrave Macmillan, 2002.

Cancian, Mark F., 'The Wehrmacht in Yugoslavia: Lessons of the Past?', *Parameters*, Vol. XXIII (Autumn 1993), pp.75–84.

Cancian, Colonel Mark F., 'What Turned the Tide in Anbar?', *Military Review* (September–October 2009), pp.118–21.

Cann, John P., 'Operation *Mar Verde*, The Strike on Conakry, 1970', *Small Wars and Insurgencies*, Vol.8, No.3 (Winter 1997), pp.64–81.

Cann, John P., *Counterinsurgency in Africa: The Portuguese Way of War, 1961–1974*, Westport, Connecticut: Greenwood Press, 1997.

Cann. John P, *Brown Waters of Africa: Portuguese Riverine Warfare, 1961–1974*, St Petersburg, Florida: Hailer Publishing, 2007.

Carruthers, Susan L., *Winning Hearts and Minds: British Governments, the Media and Colonial Counterinsurgency, 1944–1960*, London and New York: Leicester University Press, 1995.

Carter, Nick and Alexander Alderson, 'Partnering with Local Forces:Vintage Wine, New Bottles and Future Opportunities', *The RUSI Journal*, Vol.156, No.3 (June/July 2011), pp.34–40.

Cartwright, Captain James, 'Operational Mentor and Liaison Team on Op Herrick 8 – Experiences and the Way Forward', *British Army Review*, Vol.146 (Spring 2009), pp.13–18.

Carver, Michael, *War Since 1945*, London: Weidenfeld and Nicolson, 1980.

Cassidy, Robert M., *Russia in Afghanistan and Chechnya: Military Strategic Culture and the Paradoxes of Asymmetric Conflict*, Carlisle, Pennsylvania: Strategic Studies Institute, 2003.

Cassidy, Robert M., 'Back to the Street without Joy: Counterinsurgency Lessons from Vietnam and Other Small Wars', *Parameters*, Vol.XXXIV (Summer 2004), pp.73–83.

Cassidy, Robert M., 'The Long Small War: Indigenous Forces for Counterinsurgency', *Parameters* (Summer 2006), pp.47–62.

Cassidy, Robert, 'Regular and Irregular Indigenous Forces for a Long War', *The RUSI Journal*, Vol.152, No.1 (February 2007), pp.42–7.

Cassidy, Robert M., 'The Afghanistan Choice: Peace or Punishment in the Pashtun Belt', *The RUSI Journal*, Vol.155, No.4 (August/September 2010), pp.38–44.

Cassidy, Robert M., *War, Will, and Warlords: Counterinsurgency in Afghanistan and Pakistan, 2001–2011*, Quantico, Virginia: Marine Corps University Press, 2012.

Catton, Philip E., 'Counterinsurgency and Nation Building: The Strategic Hamlet Programme in South Vietnam, 1961–1963', *International History Review*, Vol.21, No.4 (December 1999), pp.918–40.

Cavanagh, Matt, 'Ministerial Decision-Making in the Run-up to the Helmand Deployment', *The RUSI Journal*, Vol.157, No.2 (April/May 2012), pp.48–54.

Cesarani, David, *Major Farran's Hat: Murder, Scandal and Britain's War against Jewish Terrorism, 1945–1948*, London: Vintage Books, 2010.

Chabal, Patrick, 'National Liberation in Portuguese Guinea, 1956–1974', *African Affairs*, Vol.80, No.318 (January 1981), pp.75–99.

Chabal, Patrick, 'Emergencies and Nationalist Wars in Portuguese Africa' in Robert Holland (ed.), *Emergencies and Disorder in the European Empire after 1945*, London: Frank Cass, 1994.

Chabal, Patrick, 'Angola and Mozambique: The Weight of History', *Portuguese Studies*, Vol.17 (2001), pp.216–32.

Chaplin, Dennis, 'France – Military Involvement in Africa', *Military Review*, Vol.LIX, No.1 (January 1979), pp.44–7.

Charters, David A., 'British intelligence in the Palestine Campaign, 1945–47', *Intelligence and National Security*, Vol. 6, No.1 (January 1991), pp.115–40.

Charters, David A., 'Special Operations in Counterinsurgency: The Farran Case, Palestine 1947, *The RUSI Journal*, Vol.124, No.2 (June 1979), pp.56–61.

Charters, David A., 'From Palestine to Northern Ireland: British Adaptation to Low-Intensity Operations' in David Charters and Maurice Tugwell (eds), *Armies in Low-Intensity Conflict*, London: Brassey's Defence, 1989.

Charters, David A., *The British Army and Jewish Insurgency in Palestine, 1945–47*, London: Macmillan, 1989.

Charters, David A., 'British Intelligence in the Palestine Campaign, 1945–47' in Ian Beckett (ed.), *Modern Counterinsurgency*, Aldershot, Hampshire and Burlington, Vermont: Ashgate, 2007.

Charters, David and Maurice Tugwell (eds), *Armies in Low-Intensity Conflict: A Comparative Analysis*, London: Brassey's Defence, 1989.

Chaudhuri, Rudra, and Theo Farrell, 'Campaign Disconnect: Operational Progress and Strategic Obstacles in Afghanistan, 2009–2011', *International Affairs*, Vol.87, No.2 (March 2011), pp.271–96.

Chiarelli, Lieutenant General Peter W. and Major Patrick R. Michaelis, 'Winning the Peace: The Requirement for Full-Spectrum Operations', *Military Review* (July–August 2005), pp.4–17.

Chiarelli, Lieutenant General Peter W. with Major Stephen M. Smith, 'Learning from our Modern Wars: The Imperatives of Preparing for a Dangerous Future', *Military Review* (September–October 2007), pp.2–15.

Chin, Warren, 'British Counterinsurgency in Afghanistan', *Defence and Security Analysis*, Vol.23, No.2 (June 2007), pp.201–25.

Chin, Warren, 'Colonial Warfare in a Post-Colonial State: British Military Operations in Helmand Province, Afghanistan', *Defence Studies*, Vol.10, No.1–2 (March–June 2010), pp.215–47.

Chipman, John, *French Power in Africa*, Oxford: Basil Blackwell, 1989.

Cincinnatus [C.B. Currey], *Self-Destruction: The Disintegration and Decay of the US Army during the Vietnam War*, New York: W.W. Norton, 1981.

Clark, Major John L., *Thinking Beyond Counterinsurgency: The Utility of a Balanced Approach to Amnesty, Reconciliation and Reintegration*, Fort Leavenworth, Kansas: School of Advanced Military Studies, US Army Command and General Staff College, 2008.

Clarke, Jeffrey J., *United States Army in Vietnam: Advice and Support: The Final Years, 1965–1973*, Washington, DC: US Army Center of Military History, US Government Printing Office, 1988.

Clarke, Michael and Valentina Soria, 'Charging up the Valley: British Decisions in Afghanistan', *The RUSI Journal*, Vol.156, No.4 (August/September 2011), pp.80–8.

Clayton, Anthony, *Counterinsurgency in Kenya 1952–1960*, Nairobi: Transafrica, 1976.

Clayton, Anthony, *France, Soldiers and Africa*, London: Brassey's, 1988.

Clayton, Anthony, *Three Marshals of France: Leadership after Trauma*, London: Brassey's, 1992.

Clayton, Anthony, 'The Sétif Uprising of May 1945', *Small Wars and Insurgencies*, Vol.3, No.1 (Spring 1992), pp.1–21.

Clayton, Anthony, *The Wars of French Decolonization*, London: Longman, 1994.

Clayton, Anthony, 'Algeria 1954: A Case Study', *The RUSI Journal*, Vol.144, No.5 (1999), pp.65–8.

Clemens, Peter, 'Captain James Hausman, US Army Military Advisor to Korea, 1946–48: The Intelligent Man on the Spot', *The Journal of Strategic Studies*, Vol.25, No.1 (March 2002), pp.163–98.

Cloake, John, *Templer: Tiger of Malaya: The Life of Field Marshal Sir Gerald Templer*, London: Harrap, 1985.

Clutterbuck, Colonel Richard L., 'The SEP – Guerrilla Intelligence Source', *Military Review*, Vol.XLII, No.10 (October 1962), pp.13–21.

Clutterbuck, Brigadier Richard L., *The Long Long War: Counterinsurgency in Malaya and Vietnam*, New York and Washington: Frederick A. Praeger, 1966.

Clutterbuck, Richard L., *Riot and Revolution in Singapore and Malaya, 1945–1963*, London: Faber & Faber, 1973.

Coates, John, *Suppressing Insurgency: An Analysis of the Malayan Emergency, 1948–1954*, Boulder, Colorado: Westview Press, 1992.

Cobain, Ian, *Cruel Britannia: A Secret History of Torture*, London: Portobello Books, 2012.

Cockburn, Patrick, *The Occupation: War and Resistance in Iraq*, London and New York: Verso Books, 2006.

Coffey, Major Ross, 'Revisiting CORDS: The Need for Unity of Effort to Secure Victory in Iraq', *Military Review*, Vol.LXXXVI, No.2 (March–April 2006), pp.24–34.

Coffman, Edward M., 'Batson of the Philippine Scouts', *Parameters*, Vol.VII, No.3 (1977), pp.68–72.

Cohen, A.A., *Galula: The Life and Writings of the French Officer Who Defined the Art of Counterinsurgency*, Westport, Connecticut: Praeger Publishers, 2012.

Cohen, Eliot, Lieutenant Colonel Conrad Crane, Lieutenant Colonel Jan Horvath; and Lieutenant Colonel John Nagl, 'Principles, Imperatives, and Paradoxes of Counterinsurgency', *Military Review*, Vol.LXXXVI, No.2 (March–April 2006), pp.49–53.

Cohen, William B., *Rulers of Empire: The French Colonial Service in Africa*, Stanford, California: Hoover Institution Press, Stanford University, 1971.

Collins, John M., 'Vietnam Postmortem: A Senseless Strategy', *Parameters*, Vol.VIII, No.1 (March 1978), pp.8–14.

Collins, Joseph J., 'Afghanistan: The Empire Strikes Out', *Parameters*, Vol.XII, No.1 (March 1982), pp.32–41.

Collins, Joseph J., 'The Soviet–Afghan War: The First Four Years', *Parameters*, Vol.XIV, No.2 (Summer 1984), pp.49–62.

Collins, Major Joseph J., 'The Soviet Military Experience in Afghanistan', *Military Review*, Vol.LXV, No.5 (May 1985), pp.16–28.

Colonial Office, *Historical Survey of the Origins and Growth of Mau Mau*, London: HMSO, 1960.

Cooke, James J., *New French Imperialism, 1880–1910: The Third Republic and Colonial Expansion*, Newton Abbot, Devon: David & Charles, 1973.

Cooling, B. Franklin, 'A People's War: Partisan Conflict in Tennessee and Kentucky' in Daniel E. Sutherland (ed.), *Guerrillas, Unionists and Violence on the Confederate Home Front*, Fayetteville, Arkansas: University of Arkansas Press, 1999.

Cooper, Matthew, *The Phantom War: The German Struggle against Soviet Partisans 1941–1944*, London: Macdonald & Jane's, 1979.

Cooper-Key, Major E.A., 'Some Reflections on Cyprus', *British Army Review*, No.5 (September 1957), pp.40–3.

Corkery, Simon (ed.), *Anti-Partisan Warfare in the Balkans, 1941–1945*, Milton Keynes: The Military Press, 1999.

Cornell, Svante E., 'International Reactions to Massive Human Rights Violations: The Case of Chechnya', *Europe-Asia Studies*, Vol.51, No.1 (January 1999).

Corson, William R., *The Betrayal*, New York: W.W. Norton, 1968.

Corum, James S., *Bad Strategies: How Major Powers Fail in Counterinsurgency*, Minneapolis, Minnesota: Zenith Press, 2008.

Corum, James S., 'Rethinking US Army Counterinsurgency Doctrine' in Tim Benbow and Rod Thornton (eds), *Dimensions of Counterinsurgency: Applying Experience to Practice*, London and New York: Routledge, 2008.

Cosmas, Graham A., '*Cacos* and *Caudillos*: Marines and Counterinsurgency in Hispaniola, 1915–1924' in William R. Roberts and Jack Sweetman (eds), *New Interpretations in Naval History*, Annapolis, Maryland: Naval Institute Press, 1991.

Cottrell, Major Pete, 'Myth, The Military and Anglo-Irish Policing between 1913 and 1922', *British Army Review*, No.133 (Winter 2003), pp.14–20.

Couch, Dick, *Down Range: Navy Seals in the War on Terrorism*, New York: Three Rivers Press, 2005.

Couch, Dick, *The Sheriff of Ramadi: Navy Seals and the Wining of Anbar*, Annapolis, Maryland: Naval Institute Press, 2008.

Crane, Conrad C., 'Avoiding Vietnam: The U.S. Army's Response to Defeat in Southeast Asia', *Strategic Studies Institute Monograph* (September 2002), pp.1–25.

Crawshaw, Nancy, *The Cyprus Revolt: An Account of the Struggle for Union with Greece*, London: George Allen & Unwin, 1978.

Cumings, Bruce, *Korea's Place in the Sun: A Modern History*, New York and London: W.W. Norton, 1997.

Dabbs, Jack Autrey, *The French Army in Mexico, 1861–1867: A Study in Military Government*, The Hague, The Netherlands: Mouton & Co, 1963.

Dabringhaus, Sabine, 'An Army on Vacation?: The German War in China, 1900–1901' in Manfred F. Boemeke, Roger Chickering and Stig Föster (eds), *Anticipating Total War: The German and American Experiences, 1871–1914*, Cambridge: Cambridge University Press, 1999.

Daddis, Gregory A., *No Sure Victory: Measuring U.S. Army Effectiveness and Progress in the Vietnam War*, New York: Oxford University Press, 2011.

Dannatt, General Sir Richard, *Leading from the Front*, Bantam, 2010.

Darling, Lieutenant General Sir Kenneth, 'British Counterinsurgency Experience: A Kermit Roosevelt Lecture', *Military Review*, Vol.XLV, No.1 (January 1965), pp.3–11.

Davey, Gregor, 'Conflicting worldviews, mutual incomprehension: The production of intelligence across Whitehall and the management of subversion during decolonisation, 1944–1966', *Small Wars and Insurgencies*, Vol.25, No.3 (2014), pp.539–59.

Davidson, Janine, *Lifting the Fog of Peace: How Americans Learned to Fight Modern War*, Ann Arbor, Michigan: University of Michigan Press, 2010.

Davis, Richard G. (ed.), *The US Army and Irregular Warfare, 1775–2007*, Washington, DC: Center of Military History, United States Army, 2008.

De Arriaga, General Kaúlza, *The Portuguese Answer*, London: Tom Stacey, 1973.

De La Gorce, Paul-Marie, *The French Army: A Military–Political History*, London: Weidenfeld and Nicolson, 1963.

Deady, Timothy K., 'Lessons from a Successful Counterinsurgency: The Philippines, 1899–1902', *Parameters*, Vol.XXXV (Spring 2005), pp.53–68.

Deane, Colonel Anthony E., 'Providing Security Force Assistance in an Economy of Force Battle', *Military Review* (January–February 2010), pp.80–90.

Dening, Major B.C., 'Modern Problems of Guerrilla Warfare', *Army Quarterly and Defence Journal*, Vol.13 (1927), pp.347–53.

Dennis, Peter, and Jeffrey Grey, *Emergency and Confrontation: Australian Military Operations in Malaya and Borneo, 1950–1966*, Sydney: Allen & Unwin, 1996.

De Treux, Lieutenant Colonel Kenneth M., *Contemporary Counterinsurgency (coin) Insights from the French-Algerian War (1954–1962)*, Carlisle Barracks, Pennsylvania: Thesis for Master of Strategic Studies, US Army War College, 2008.

Devereux, David R., *The Formulation of British Defence Policy towards the Middle East, 1948–56*, Basingstoke, Hampshire: Macmillan, 1990.

DeVore, Marc R., 'A More Complex and Conventional Victory: Revisiting the Dhofar Counterinsurgency, 1963–1975', *Small Wars and Insurgencies*, Vol.23, No.1 (March 2012), pp.144–73.

Dewar, Michael, *Brush Fire Wars: Minor Campaigns of the British Army since 1945*, London: Robert Hale, 1984.

Dewar, Michael, *The British Army in Northern Ireland*, London: Arms and Armour Press, 1985.

Dhada, Mustafah, *Warriors at Work: How Guinea Was Really Set Free*, Niwot, Colorado: University Press of Colorado, 1993.

Dhada, Mustafah, 'The Liberation War in Guinea–Bissau Reconsidered', *Journal of Military History*, Vol.62, No.3 (July 1998), pp.571–93.

Dickens, Peter, *SAS: The Jungle Frontier: 22 Special Air Service Regiment in the Borneo Campaign, 1963–1966*, London: Book Club Associates, 1983.

Dimitrakis, Panagiotis, 'British Intelligence and the Cyprus Insurgency, 1955–1959', *International Journal of Intelligence and Counterintelligence*, Vol.21, No.2 (Spring 2008), pp.375–94.

Dimitrakis, Panagiotis, *Military Intelligence in Cyprus: From the Great War to Middle East Crises*, London and New York: I.B. Tauris, 2010.

Dippie, Brian W., 'George A. Custer' in Paul Andrew Hutton (ed.), *Soldiers West: Biographies from the Military Frontier*, Lincoln, Nebraska and London: University of Nebraska Press, 1987.

Dingles, Bruce J., 'Benjamin H. Grierson' in Paul Andrew Hutton (ed.), *Soldiers West: Biographies from the Military Frontier*, Lincoln, Nebraska and London: University of Nebraska Press, 1987.

Dixon, Joe C. (ed.), *The American Military and the Far East: Proceedings of the Ninth Military History Symposium*, United States Air Force Academy, 1 October 1980, Washington, DC: US Government Printing Office, 1980.

Donahue, Colonel (P.) Patrick and Lieutenant Colonel Michael Fenzel, 'Combating a Modern Insurgency: Combined Task Force Devil in Afghanistan', *Military Review* (March–April 2008), pp.25–40.

Donnelly, Thomas, 'The Cousins' Counterinsurgency Wars', *The RUSI Journal*, Vol.154, No.3 (June 2009), pp.4–9.

Doohovsky, Andrei A., 'Soviet Counterinsurgency in the Soviet Afghan War Revisited: Analyzing the Effective Aspects of the Counterinsurgency Effort', Cambridge, Massachusetts: Harvard University MA Thesis, September 2009.

Dotolo, Frederick H., 'A long small war: Italian counterrevolutionary warfare in Libya, 1911 to 1932', *Small Wars and Insurgencies*, Vol.26, No.1 (2015), pp.158–80.

Doughty, Major Robert A., *The Evolution of US Army Tactical Doctrine, 1946–76*, Fort Leavenworth, Kansas: Leavenworth Papers No.1, Combat Studies Institute, US Army Command and General Staff College, 1979.

Downes, Alexander B., 'Draining the Sea by Filling the Graves: Investigating the Effectiveness of Indiscriminate Violence as a Counterinsurgency Strategy', *Civil Wars*, Vol.9, No.4 (December 2007), pp.420–44.

Drechsler, Horst, *'Let Us Die Fighting': The Struggle of the Herero and Nama against German Imperialism (1884–1915)*, London: Zed Press, 1980.

Dunn, David Hastings and Andrew Futter, 'Short-Term Tactical Gains and Long-Term Strategic Problems: The Paradox of the US Troop Surge in Iraq', *Defence Studies*, Vol.10, No.1–2 (March–June 2010), pp.195–214.

Dunn, Peter M., 'The American Army: The Vietnam War, 1965–1973' in Ian F.W. Beckett and John Pimlott (eds), *Armed Forces & Modern Counterinsurgency*, Beckenham, Kent: Croom Helm, 1985.

Durrell, Lawrence, *Bitter Lemons*, London: Faber & Faber, 1959.

Earle, Edward Mead (ed.), *Makers of Modern Strategy: Military Thought from Machiavelli to Hitler*, Princeton, New Jersey: Princeton University Press, 1971.

East Africa Command, *A Handbook on Anti-Mau Mau Operations*, Nairobi: The Government Printer, 1954.

Easter, David, 'British and Malaysian Covert Support for Rebel Movements in Indonesia during the "Confrontation", 1963–66', *Intelligence and National Security*, Vol.14, No.4 (Winter 1999), pp.195–208.

Easter, David, 'British Intelligence and Propaganda during the 'Confrontation', 1963–1966', *Intelligence and National Security*, Vol.16, No.2 (Summer 2001), pp.83–102.

Easter, David, *Britain and the Confrontation with Indonesia, 1960–1966*, London and New York: Tauris Academic Studies, 2004.

Easter, David, '"Keep the Indonesian pot boiling": Western covert intervention in Indonesia, October 1965–March 1966', *Cold War History*, Vol.5, No.1 (February 2005), pp.55–73.

Edgerton, Robert S., *Mau Mau: An African Crucible*, London: Tauris, 1990.

Edwards, Aaron, 'Misapplying Lessons Learnt? Analysing the Utility of British Counterinsurgency Strategy in Northern Ireland, 1971–76', *Small Wars and Insurgencies*, Vol.21, No.2 (June 2010), pp.303–30.

Edwards, Major P.W.D., 'The Military–Media Relationship – A Time to Redress the Balance?', *The RUSI Journal*, Vol.143, No.5 (October 1998), pp.43–9.

Edwards, Major T.C., '3d MarDiv Counterguerrilla Training: A Readiness Report', *Marine Corps Gazette*, Vol.47, No.51 (May 1963), pp.45–8.

Egnell, Robert, 'Winning hearts and minds? A critical analysis of counterinsurgency operations in Afghanistan', *Civil Wars*, Vol.12, No.3 (September 2010), pp.282–303.

Egnell, Robert, 'Lessons from Helmand, Afghanistan: what now for British Counterinsurgency?', *International Affairs*, Vol.87, No.2 (March 2011), pp.297–315.

Ehrlich, Thomas, 'Cyprus, the "Warlike Isle": Origins and Elements of the Current Crisis', *Stamford Law Review*, Vol.18, No.6 (May 1966), pp.1021–98.

Elkins, Caroline, *Imperial Reckoning: The Untold Story of Britain's Gulag in Kenya*, New York: Henry Holt, 2005.

Ellison, Graham and Conor O'Reilly, 'From Empire to Iraq and the "War on Terror"' in Georgina Sinclair (ed.), *Globalising British Policing*, Farnham, Surrey: Ashgate, 2011.

Epstein, Major David G., 'The Police Role in Counterinsurgency Efforts', *The Journal of Criminal Law, Criminology and Police Science*, Vol.59, No.1 (March 1968), pp.148–51.

Eriksen, Jørgen W. and Tormod Heier, 'Winter as the Number One Enemy: Lessons Learned from North Afghanistan', *The RUSI Journal*, Vol.154, No.5 (October 2009), pp.64–71.

Erskine, General Sir George, 'Kenya – What is it all About', *Journal of the Royal Artillery*, Vol.LXXXIII, No.2 (1956), pp.99–117.

Esdaile, Charles, 'Guerrillas and bandits in the Serranía de Ronda, 1810–1812', *Small Wars and Insurgencies*, Vol.25, No.4 (2014), pp.814–27.

Esterhuyse, Lieutenant Colonel Abel and Evert Jordaan, 'The South African Defence Force and Counterinsurgency, 1966–1990' in Deane-Peter Baker and Evert Jordaan (eds), *South Africa and Contemporary Counterinsurgency: Roots, Practices, Prospects*, Claremont, South Africa: UCT Press, 2010.

Etherington, Mark, *Revolt on the Tigris: The Al-Sadr Uprising and the Governing of Iraq*, Ithaca, New York: Cornell University Press, 2005.

Evangelista, Matthew, *The Chechen Wars: Will Russia go the Way of the Soviet Union?*, Washington, DC: Brooking Institution Press, 2002.

Evans, Ernest, *Wars Without Splendor: The U.S. Military and Low-Level Conflict*, New York, Westport, Connecticut and London: Greenwood Press, 1987.

Evans, Martin, *Algeria: France's Undeclared War*, Oxford: Oxford University Press, 2012.

Evelegh, Robin, *Peace Keeping in a Democratic Society: The Lessons of Northern Ireland*, London: C Hurst, 1978.

Eyal, Dr Jonathon, 'The Media and the Military: Continuing the Dialogue after Kosovo', *The RUSI Journal*, Vol.145, No.2 (April 2000), pp.37–43.

Eyre, Lieutenant Colonel Wayne D., 'Operation RÖSSELSPRUNG and the Elimination of Tito, May 25, 1944: A Failure in Planning and Intelligence Support', *Journal of Slavic Military Studies*, Vol.19, No.2 (2006), pp.343–76.

Fairbairn, Geoffrey, *Revolutionary Guerrilla Warfare: The Countryside Version*, Harmondsworth, Middlesex: Pelican Books, 1974.

Fairweather, Jack, *A War of Choice*, London: Jonathan Cape, 2011.

Fall, Bernard, *Hell in a Very Small Place: The Siege of Dien Bien Phu*, London: Pall Mall Press, 1966.

Faoleán, Gearóid Ó, 'Ireland's Ho Chi Minh trail? The Republic of Ireland's role in the Provisional IRA's bombing campaign, 1970–1976', *Small Wars and Insurgencies*, Vol.25, No.5–6 (2014), pp.976–91.

Farrell, Theo, 'Improving in War: Military Adaptation and the British in Helmand Province, Afghanistan, 2006–2009', *The Journal of Strategic Studies*, Vol.33, No.4 (August 2010), pp.567–94.

Farrell, Theo and Stuart Gordon, 'COIN machine: The British Military in Afghanistan', *The RUSI Journal*, Vol.154, No.3 (June 2009), pp.18–25.

Farwell, Byron, *The Great Anglo-Boer War*, New York, Hagerstown, San Francisco and London: Harper & Row, 1976.

Fayutkin, Major Dan, 'Russian–Chechen Information Warfare 1994–2006', *The RUSI Journal*, Vol.151, No.5 (October 2006), pp.52–5.

Feifer, Gregory, *The Great Gamble: The Soviet War in Afghanistan*, New York: Harper Perennial, 2010.

Fellman, Michael, *Inside War: The Guerrilla Conflict in Missouri during the American Civil War*, New York: Oxford University Press, 1989.

Fergusson, James, *A Million Bullets: The Real Story of the British Army in Afghanistan*, London: Corgi Books, 2009.

Figes, Orlando, *A People's Tragedy: The Russian Revolution, 1891–1924*, London: Jonathon Cape, 1996.

Finley, Milton, *The Most Montrous of Wars: The Napoleonic Guerrilla War in Southern Italy, 1806–1811*, Columbia, South Carolina: University of South Carolina Press, 1994.

Fitz-Simons, Daniel E., 'Francis Marion the "Swamp Fox": An Anatomy of a Low-Intensity Conflict', *Small Wars and Insurgencies*, Vol.6, No.1 (Spring 1995), pp.1–16.

Fivecoat, Lieutenant Colonel David G. and Captain Aaron T. Schwengler, 'Revisiting Modern Warfare: Counterinsurgency in the Mada'in Qada', *Military Review*, (November–December 2008), pp.77–87.

Flynn, Matthew J., *Contesting History: The Bush Counterinsurgency Legacy in Iraq*, Santa Barbara, California: Praeger, 2010.

Flynn, Major General Michael T., Captain Matt Pottinger; and Paul D. Batchelor, 'Fixing Intel: A Blueprint for Making Intelligence Relevant in Afghanistan', *Voices From the Field, Center for a New American Security* (January 2010), pp.1–26.

Foley, Charles, *Island in Revolt*, London: Longmans, Green, 1962.

Foley, Charles, *Legacy of Strife*, Harmondsworth, Middlesex: Penguin Books, 1964.

Foley, Charles and W.I. Scobie, *The Struggle for Cyprus*, Stanford, California: Hoover Institute Press, Stanford University, 1975.

Foot, Hugh, *A Start in Freedom*, London: Hodder and Stoughton, 1964.

Ford, Captain Christopher, 'Of Shoes and Sites: Globalization and Insurgency', *Military Review* (May–June 2007), pp.85–91.

Ford, Matthew, 'Influence without power? Reframing British concepts of military intervention after 10 years of counterinsurgency', *Small Wars and Insurgencies*, Vol.25, No.3 (2014), pp.495–500.

Ford, Ronnie E., 'Intelligence and the Significance of Khe Sanh', *Intelligence and National Security*, Vol.10, No.1 (January 1995), pp.144–69.

Forrest, Alan, 'The insurgency of the Vendée', *Small Wars and Insurgencies*, Vol.25, No.4 (2014), pp.800–13.

Foy, Michael T., *Michael Collins's Intelligence War: The Struggle Between the British and the IRA, 1919–1921*, Stroud, Gloucestershire: Sutton Publishing, 2006.

Francois, Lieutenant Colonel Philippe, 'Waging Counterinsurgency in Algeria: A French Point of View', *Military Review*, (September–October 2008), pp.56–67.

Frankel, Philip H., *Pretoria's Praetorians: Civil-Military Relations in South Africa*, Cambridge: Cambridge University Press, 1984.

Frankel, Philip, *Soldiers in a Storm: The Armed Forces in South Africa's Democratic Transition*, Boulder, Colorado: Westview Press, 2000.

Frazier, Donald S., '"Out of Stinking Distance": The Guerrilla War in Louisiana' in Daniel E. Sutherland (ed.), *Guerrillas, Unionists and Violence on the Confederate Home Front*, Fayetteville, Arkansas: University of Arkansas Press, 1999.

Freedman, Lawrence, *Kennedy's Wars: Berlin, Cuba, Laos, and Vietnam*, New York: Oxford University Press, 2002.

French, David, *The British Way in Counterinsurgency, 1945–1967*, Oxford and New York: Oxford University Press, 2011.

Fuller, Captain Stephen M. and Graham A. Cosmas, *Marines in the Dominican Republic, 1916–1924*, Washington, DC: US Marine Corps History and Museums Division, 1974.

Furedi, Frank, 'Kenya: Decolonization through counterinsurgency' in Anthony Gorst, Lewis Johnman, W. Scott Lucas (eds), *Contemporary British History 1931–61: Politics and the Limits of Policy*, London and New York: Pinter Publishers, 1991, pp.141–69

Galula, David, *Counterinsurgency Warfare: Theory and Practice*, London and Dunmow: Pall Mall Press, 1964.

Galula, David, *Pacification in Algeria, 1956–1958*, Santa Monica, California: RAND, 2006.

Gann, L.H. and Peter Duignan, *The Rulers of German Africa, 1884–1914*, Stanford, California: Stanford University Press, Hoover Institute Publications, 1977.

Gardiner, Ian, *In the Service of the Sultan: A First-Hand Account of the Dhofar Emergency*, Barnsley, South Yorkshire: Pen & Sword Military, 2006.

Gareau, Frederick H., *State Terrorism and the United States: From Counterinsurgency to the War on Terrorism*, London: Zed Books, 2004.

Gareev, Colonel General Makhmut Akhmetovich, *M. V. Frunze: Military Theorist*, London: Pergamon-Brassey's, 1988.

Gates, John Morgan, *Schoolbooks and Krags: The United States Army in the Philippines, 1898–1902*, Westport, Connecticut: Greenwood Press, 1973.

Gates, John M., 'The Pacification of the Philippines, 1898–1902' in Joe C. Dixon (ed.), *The American Military and the Far East: Proceedings of the Ninth Military History Symposium, United States Air Force Academy, 1–3 October 1980*, Washington, DC: US Government Printing Office, 1980.

Gates, John M., 'Indians and Insurrectos: The US Army's Experience with Insurgency', *Parameters*, Vol.XIII, No.1 (March 1983), pp.59–68.

Gaub, Major Christoff T., 'Provincial Reconstruction Teams (PRTs): Vietnam's Provincial Adviser Teams (PATs) Revisited?', Maxwell Air Force Base, Alabama, Air University, Air Command and Staff College, 2006.

Gebhardt, Major James F., *Eyes Behind the Lines: US Army Long-Range Reconnaissance and Surveillance Units*, Fort Leavenworth, Kansas: Combat Studies Institute Press, Global War on Terrorism, Occasional Paper 10, 2005.

Geffen, Lieutenant Colonel William (ed.), *Command and Commanders in Modern Warfare: Proceedings of the Second Military History Symposium, US Air Force Academy, 2–3 May 1968*, USAF Academy, Colorado: December 1969.

Gentile, Gian P., *Wrong Turn: America's Deadly Embrace of Counterinsurgency*, New York: New Press, 2013.

Geraghty, Tony, *Who Dares Wins: The Story of Special Air Service 1950–1980*, London: Arms and Armour Press, 1980.

Geraghty, Tony, *The Irish War: The Military History of a Domestic Conflict*, London: Harper/Collins, 1998.

Gerassi, John (ed.), *Venceremos: The speeches and writings of Ernesto Che Guevara*, London: Weidenfeld and Nicolson, 1968.

Gerolymatos, André, 'Greek Democracy on Trial: From Insurgency to Civil War, 1943–49', *Review of International Affairs*, Vol.2, No.31 (Spring 2003), pp.122–37.

Gerolymatos, André, 'Greek Democracy on Trial: From Insurgency to Civil War, 1943–49' in Efraim Inbar (ed.), *Democracies and Small Wars*, London: Routledge, 2003.

Gershovich, Moshe, *French Military Rule in Morocco: Colonialism and its Consequences*, London: Frank Cass, 2000.

Gewald, Jan-Bart, *Herero Heroes: A Socio-Political History of the Herero of Namibia, 1890–1923*, Oxford: James Currey, 1999.

Gewald, Jan-Bart, 'Learning to Wage and Win Wars in Africa: A Provisional History of German Military Activity in Congo, Tanzania, China and Namibia', Leiden, The Netherlands: African Studies Centre, Working Paper 60/2005, 2005.

Gewald, Jan-Bart, 'Colonial Warfare: Hehe and World War One, the wars besides Maji Maji in south-western Tanzania', Leiden, The Netherlands: African Studies Centre, Working Paper 63/2005, 2005.

Giap, General Vo Nguyen, *People's War, People's Army: The Viet Cong Insurrection Manual for Underdeveloped Countries*, New York: Frederick A. Praeger, 1962.

Giap, General Vo Nguyen, *Banner of People's War, The Party's Military Line*, New York, Washington and London: Praeger Publishers, 1970.

Giblin, James and Jamie Monson (eds), *Maji Maji: Lifting the Fog of War*, Leiden, The Netherlands: Brill, 2010.

Giustozzi, Antonio, *Koran, Kalashnikov and Laptop: The Neo-Taliban insurgency in Afghanistan*, London: Hurst, 2007.

Giustozzi, Antonio, *Empires of Mud: Wars and Warlords in Afghanistan*, London: Hurst, 2009.

Giustozzi, Antonio (ed.), *Decoding the New Taliban: Insights from the Afghan Field*, London: Hurst, 2009.

Giustozzi, Antonio, 'The Afghan National Army: Unwarranted Hope?', *The RUSI Journal*, Vol.154, No.6 (December 2009), pp.36–42.

Giustozzi, Antonio, 'The Taliban's "military courts"', *Small Wars and Insurgencies*, Vol.25, No.2 (2014), pp.284–96.

Giustozzi, Antonio and Mohammed Isaqzadeh, *Policing Afghanistan: The Politics of the Lame Leviathan*, London: Hurst, 2012.

Gleijeses, Piero, 'The First Ambassadors: Cuba's Contribution to Guinea–Bissau's War of Independence', *Journal of Latin American Studies*, Vol.29, No.1 (February 1997), pp.45–88.

Godfrey, Major F.A., *The History of the Suffolk Regiment, 1946–1959*, London: Leo Cooper, 1988.

Goldhagen, Daniel Josiah, *Hitler's Willing Executioners: Ordinary Germans and the Holocaust*, London: Abacus, 1997.

Goldworthy, David, 'Armed Struggle under late Colonialism', *The International History Review*, Vol.13, No.3 (August 1991), pp.538–47.

Goodale, First Lieutenant Jason and First Lieutenant Jon Webre, 'The Combined Action Platoon in Iraq', *Marine Corps Gazette*, Vol.89, No.4 (April 2005), pp.40–2.

Goodrich, Thomas, *Black Flag: Guerrilla Warfare on the Western Border, 1861–1865*, Bloomington, Indiana: Indiana University Press, 1995.

Gordon, Larry, *The Last Confederate General: John C. Vaughn and his East Tennessee Cavalry*, Minneapolis, Minnesota: Zenith Press, 2009.

Gorst, Anthony, Lewis Johnman and W. Scott Lucas (eds), *Contemporary British History 1931–61: Politics and the Limits of Policy*, London and New York: Pinter Publishers, 1991.

Gortzak, Yoav, 'Using Indigenous Forces in Counterinsurgency Operations: The French in Algeria, 1954–1962', *The Journal of Strategic Studies*, Vol.32, No.2 (April 2009), pp.307–33.

Gortzak, Yoav, 'The prospects of combined action: Lessons from Vietnam', *Small Wars and Insurgencies*, Vol.25, No.1 (2014), pp.137–60.

Gott, Kendall D., *In Search of an Elusive Enemy: The Victorio Campaign*, Fort Leavenworth, Kansas: Global War on Terrorism Occasional Paper 5, Combat Studies Institute Press, 2004.

Gott, Kendall D. and Michael G Brooks (eds), *The US Army and the Interagency Process: Historical Perspectives: The Proceedings of the Combat Institute 2008 Military History Symposium*, Fort Leavenworth, Kansas: Combat Studies Institute Press, 2008.

Gottman, Jean, 'Bugeaud, Galliéni, Lyautey: The Development of French Colonial Warfare' in Edward Mead Earle (ed.), *Makers of Modern Strategy: Military Thought from Machiavelli to Hitler*, Princeton, New Jersey: Princeton University Press, 1971.

Graham, John, *Ponder Anew: Reflections on the Twentieth Century*, Staplehurst, Kent: Spellmount, 1999.

Graham, Lawrence S., 'The Military in Politics: The Politicization of the Portuguese Armed Forces' in Lawrence S. Graham, and Harry M. Makler, *Contemporary Portugal: The Revolution and Its Antecedents*, Austin, Texas: University of Texas Press, 1979.

Graham, Lawrence S., and Harry M. Makler, *Contemporary Portugal: The Revolution and Its Antecedents*, Austin, Texas: University of Texas Press, 1979.

Grant, Colonel Carl E., 'Partisan Warfare, Model 1861–65', *Military Review*, Vol. XXXVIII, No.8 (November 1958), pp.42–56.

Grau, Lester W., (ed.), *The Bear Went Over the Mountain: Soviet Combat Tactics in Afghanistan*, Washington, DC: National Defense University Press, 1996.

Grau, Lieutenant Colonel Lester W., 'Something Old, Something New: Guerrillas, Terrorists, and Intelligence Analysis', *Military Review* (July–August 2004), pp.42–9.

Grau, Lester W., and Ali Ahmad Jalali (ed.), *The Other Side of the Mountain: Mujahideen Tactics in the Soviet–Afghan War*, Fort Leavenworth, Kansas: Foreign Military Studies Office, no date.

Grau, Lester W. and Michael A. Gress (eds), *The Soviet–Afghan War: How a Superpower Fought and Lost*, Lawrence, Kansas: University of Kansas Press, 2002.

Greenberg, Major Lawrence M., *The Hukbalahap Insurrection: A Case Study of a Successful Anti-Insurgency Operation in the Philippines – 1946–1955*, Washington, DC: United States Army Center of Military History, 1995.

Greene, Jerome A., 'George Crook' in Paul Andrew Hutton (ed.), *Soldiers West: Biographies from the Military Frontier*, Lincoln, Nebraska and London: University of Nebraska Press, 1987.

Greene, T.N. (ed.), *The Guerrilla and How to Fight Him*, New York: Frederick A Praeger, 1962.

Greentree, Todd R., *Crossroads of Intervention: Insurgency and Counterinsurgency Lessons from Central America*, Annapolis, Maryland: Naval Institute Press, 2009.

Gregorian, Raffi, 'CLARET Operations and Confrontation, 1964–1966' in Ian Beckett (ed.), *Modern Counterinsurgency*, Aldershot, Hampshire and Burlington, Vermont: Ashgate, 2007.

Gregorian, Raffi, 'Jungle Bashing' in Malaya': Towards a Formal Tactical Doctrine' in Ian Beckett (ed.), *Modern Counterinsurgency*, Aldershot, Hampshire and Burlington, Vermont: Ashgate, 2007.

Grenier, John, *The First Way of War: American War Making on the Frontier, 1607–1814*, New York: Cambridge University Press, 2005.

Grey, Stephen, *Operation Snakebite: The Explosive True Story of an Afghan Desert Siege*, London: Viking, 2009.

Griffin, Michael, *Reaping the Whirlwind: The Taliban Movement in Afghanistan*, London and Sterling, Virginia: Pluto Press, 2001.

Griffin, Stuart, 'Iraq, Afghanistan and the future of British military doctrine: from counterinsurgency to stabilization', *International Affairs*, Vol.87, No.2 (March 2011), pp.317–33.

Grimsley, Mark, *The Hard Hand of War: Union Military Policy toward Southern Civilians, 1861–1865*, Cambridge: Cambridge University Press, 1995.

Grinter, Laurence E., 'How They Lost: Doctrines, Strategies and Outcomes of the Vietnam War', *Asian Survey*, Vol.15, No.12 (1975), pp.1114–32.

Grinter, Lawrence E. and Peter M. Dunn (eds), *The American War in Vietnam: Lessons, Legacies, and Implications for Future Conflict*, New York, Westport, Connecticut and London: Greenwood Press, 1987.

Grivas, George (edited by Charles Foley), *The Memoirs of General Grivas*, London: Longmans, Green, 1964.

Grivas, George, *Guerrilla Warfare and EOKA's Struggle: A Politico–Military Study*, London: Longmans, Green, 1964.

Grob-Fitzgibbon, Benjamin, *Imperial Endgame: Britain's Dirty Wars and the End of Empire*, Basingstoke, Hampshire and New York: Palgrave Macmillan, 2011.

Grob-Fitzgibbon, Benjamin, 'Intelligence and Counterinsurgency: Case Studies from Ireland, Malaya and the Empire', *The RUSI Journal*, Vol.156, No.1 (February/March 2011), pp.72–9.

Grob-Fitzgibbon, Benjamin, 'Securing the Colonies for the Commonwealth: Counterinsurgency, Decolonization, and the Development of British Imperial Strategy in the Postwar Empire', *British Scholar*, Vol.II, Issue 1 (September 2009), pp.12–39.

Grob-Fitzgibbon, Benjamin, *Turning Points of the Irish Revolution: The Cost of Indifference, 1912–1921*, New York and Basingstoke: Palgrave Macmillan, 2007.

Grossman, Anita, 'The South African Military and Counterinsurgency: An Overview' in Deane-Peter Baker and Evert Jordaan, *South Africa and Contemporary Counterinsurgency: Roots, Practices, Prospects*, Claremont, South Africa: UCT Press, 2010.

Grunlingh, Albert, '"Protectors and Friends of the People"? The South African Constabulary in the Transvaal and Orange River Colony, 1900–08' in David M. Anderson and David Killingray (eds), *Policing the Empire: Government, Authority and Control, 1830–1940*, Manchester and New York: Manchester University Press, 1991.

Gudmundsson, Major Bruce, 'The First of the Banana Wars: US Marines in Nicaragua 1909–12' in Daniel Marston and Carter Malkasian (eds), *Counterinsurgency in Modern Warfare*, Botley, Oxford: Osprey Publishing, 2008.

Guelton, Lieutenant Colonel Frédéric, 'The French Army "Centre for Training" and Preparation in Counter-Guerrilla Warfare' (CIPCG) at Arzew', *Journal of Strategic Studies*, Vol.25, No.2 (June 2002), pp35–53.

Guevara, Che, *Reminiscences of the Cuban Revolutionary War*, London: George Allen and Unwin, 1968.

Guevara, Che, *Guerrilla Warfare*, Harmondsworth, Middlesex: Penguin, 1972.

Gventer, Celeste Ward, 'Keep the change: Counterinsurgency, Iraq, and historical understanding', *Small Wars and Insurgencies*, Vol.25, No.1 (2014), pp.242–53.

Gwynn, Major General Sir Charles W., *Imperial Policing*, London: Macmillan, 1936.

Hack, Karl, 'Corpses, Prisoners of War and Captured Documents: British and Communist Narratives of the Malayan Emergency, and the Dynamics of Intelligence Transformation' in Richard J. Aldrich, Gary D. Rawnsley and Ming-Yeh T. Rawnsley (eds), *The Clandestine Cold War in Asia, 1945–65: Western Intelligence, Propaganda and Special Operations*, London: Frank Cass, 2000.

Hack, Karl, 'British Intelligence and Counterinsurgency in the Era of Decolonisation: The Example of Malaya', *Intelligence and National Security*, Vol.14, No.2 (Summer 1999), pp.124–55.

Hack, Karl, 'The Malayan Emergency as Counterinsurgency Paradigm', *The Journal of Strategic Studies*, Vol.32, No.3 (June 2009), pp.383–414.

Hall, Bob and Andrew Ross, 'The Political and Military Effectiveness of Commonwealth Forces in Confrontation 1963–66', *Small Wars and Insurgencies*, Vol.19, No.2 (June 2008), pp.238–55.

Hamilton, Donald W., *The Art of Insurgency: American Military Policy and the Failure of Strategy in Southeast Asia*, Westport, Connecticut: Praeger, 1998.

Hannah, Norman B., *The Key to Failure: Laos & the Vietnam War*, Lanham, Maryland, New York and London: Madison Books, 1987.

Hanning, Hugh (ed.), *Lessons of the Vietnam War: Report of a Seminar held at the Royal United Service Institution on Wednesday, 12 February 1969*, Whitehall, London: Royal United Service Institution, 1969.

Hanrahan, Gene Z. (ed.), *Chinese Communist Guerrilla Warfare Tactics*, Boulder, Colorado: Paladin Press, 1974.

Harmon, Christopher C., 'Illustrations of "Learning" in Counterinsurgency' in Ian Beckett (ed.), *Modern Counterinsurgency*, Aldershot, Hampshire and Burlington, Vermont: Ashgate, 2007.

Harper, T.N., *The End of Empire and the Making of Malaya*, Cambridge: Cambridge University Press, 1999.

Harrison, Alexander, *Challenging de Gaulle: The OAS and the Counterrevolution in Algeria, 1954–1962*, Westport, Connecticut: Praeger, 1989.

Hart, Peter, *The I.R.A. and its Enemies: Violence and Community in Cork, 1916–1923*, Oxford: Clarendon Press, 1998.

Hart, Peter, *British Intelligence in Ireland, 1920–21*, Cork, Ireland: Cork University Press, 2002.

Hart, Peter, *The I.R.A. at War, 1916–1923*, Oxford: Oxford University Press, 2003.

Harvey, Claudia and Mark Wilkinson, 'The Value of Doctrine: Assessing British Officers' Perspectives', *The RUSI Journal*, Vol.154, No.6 (December 2009), pp.26–31.

Hashim, Ahmed S., *Insurgency and Counterinsurgency in Iraq*, London: C. Hurst & Company, 2006.

Haycock, Ronald (ed.), *Regular Armies and Insurgency*, London: Croom Helm, 1979.

Heather, Randall, 'Intelligence and Counterinsurgency in Kenya, 1952–56', *Intelligence and National Security*, Vol.5, Issue 3 (July 1990), pp.57–83.

Heather, Randall W., 'Intelligence and Counterinsurgency in Kenya, 1952–1956' in Ian Beckett (ed.), *Modern Counterinsurgency*, Aldershot, Hampshire and Burlington, Vermont: Ashgate, 2007.

Heaton, Colin D., *German Anti-Partisan Warfare in Europe, 1939–1945*, Atglen, Pennsylvania: Schiffler Publishing, 2001.

Heggoy, Alf Andrew, *Insurgency and Counterinsurgency in Algeria*, Bloomington, Indiana: Indiana University Press, 1972.

Hehn, Paul N., *The German Struggle against Yugoslav Guerrillas in World War II: German Counterinsurgency in Yugoslavia, 1941–1943*, Boulder, Colorado: East European Quarterly, East European Monographs No.LVII, 1979.

Helmer, Captain Daniel, 'Twelve Urgent Steps for the Adviser Mission in Afghanistan', *Military Review* (July–August 2008), pp.73–81.

Hemingway, Al, *Our War Was Different: Marine Combined Action Platoons in Vietnam*, London: Airlife Publishing, 1994.

Henderson, Lawrence W., *Angola: Five Centuries of Conflict*, Ithaca, New York: Cornell University Press, 1979.

Hennessy, Michael, *Strategy in Vietnam: The Marines and Revolutionary Warfare in I Corps, 1965–1972*, Westport, Connecticut: Frederick A Preager, 1997.

Henriksen, Thomas H., 'People's War in Angola, Mozambique and Guinea–Bissau', *Journal of Modern African Studies*, Vol.14, No.3 (September 1976), pp.377–99.

Henriksen, Thomas H., 'Some Notes on the National Liberation Wars in Angola, Mozambique and Guinea–Bissau', *Military Affairs*, Vol.XLI, No.1 (February 1977), pp.30–6.

Henriksen, Thomas H., 'Lessons from Portugal's Counterinsurgency Operations in Africa', *The RUSI Journal*, Vol.123, No.2 (June 1978), pp.31–5.

Henriksen, Thomas H., *Revolution and Counterrevolution: Mozambique's War of Independence, 1964–1974*, Westport, Connecticut: Greenwood Press, 1983.

Herbert, Colonel Charlie, 'Time for a Switch in Main Effort?: Metaphorically, counter-insurgency is about teaching people to fish, not about doing it for them', *British Army Review*, No.152 (Autumn 2011), pp.35–41.

Hillebrecht, Werner, 'The Nama and the war in the south' in Jürgen Zimmerer and Joachim Zeller (eds), *Genocide in German South-West Africa: The Colonial War (1904–1908) in Namibia and its Aftermath*, Pontypool, Wales: Merlin Press, 2010.

Hills, Alice, 'Basra and the Referent Points of Twofold War', *Small Wars and Insurgencies*, Vol.14, No.3 (Autumn 2003), pp.23–44.

Hills, Alice, *Future War in Cities: Rethinking a Liberal Dilemma*, London and New York: Frank Cass, 2004.

Hilsman, Roger, *To Move a Nation: The Politics of Foreign Policy in the Administration of John F. Kennedy*, Garden City, New York: Doubleday, 1967.

Hittle, J.B.E., *Michael Collins and the Anglo-Irish War: Britain's Counterinsurgency Failure*, Washington, DC: Potomac Books, 2011.

Hoffman, Bruce, *The Failure of British Military Strategy within Palestine, 1939–1947*, Jerusalem: Bar-Ilan University Press, 1983.

Hoffman, Bruce, *Insurgency and Counterinsurgency in Iraq*, Santa Monica, California: Occasional Paper, RAND, June 2004.

Hoffman, Bruce and Jennifer M. Taw, *Defense Policy and Low-Intensity Conflict: The Development of Britain's 'Small Wars' Doctrine during the 1950s*, Santa Monica, California: RAND, 1991.

Hoffman, Bruce and Jennifer Morrison Taw, *A Strategic Framework for Countering Terrorism and Insurgency*, Santa Monica, California: RAND, 1992.

Hoffman, Bruce, Jennifer Taw and David Arnold, *Lessons for Contemporary Counterinsurgencies: The Rhodesian Experience*, Santa Monica, California: RAND, 1991.

Hoffman, Frank G., 'Neo-Classical Counterinsurgency?', *Parameters*, Vol.XXXVII (Summer 2007), pp.71–87.

Hoffman, Jon T. (General Editor), *Tip of the Spear: U.S. Army Small Unit Action in Iraq, 2004–2007*, Washington, DC: Center of Military History, United States Army, 2009.

Hoisington, William A., *Lyautey and the French Conquest of Morocco*, Basingstoke, Hampshire: Macmillan, 1995.

Holland, Jack and Susan Phoenix, *Phoenix: Policing The Shadows: The Secret War Against Terrorism in Northern Ireland*, London: Hodder & Stoughton, 1996.

Holland, James, 'The Way Ahead in Afghanistan', *The RUSI Journal*, Vol. 153, No. 3 (June 2008), pp. 46–50.

Holland, Robert, *Britain and the Revolt in Cyprus, 1954–1959*, Oxford: Oxford University Press, 1998.

Horne, Alistair, *A Savage War of Peace: Algeria 1954–1962*, Harmondsworth, Middlesex: Peregrine Books, 1979.

Horne, Alistair, 'The French Army and the Algerian War, 1954–62' in Ronald Haycock (ed.), *Regular Armies and Insurgency*, London: Croom Helm, 1979.

Horne, Alistair, *The French Army and Politics, 1870–1970*, New York, Peter Bedrick Books, 1984.

Horne, Alistair, *The Fall of Paris: The Siege and the Commune, 1870–71*, Harmondsworth, Middlesex: Penguin Books, 1987.

Horne, Edward, *A Job Well Done: A History of the Palestine Police Force, 1920–1948*, Leigh-on-Sea, Essex: Palestine Police Old Comrades Benevolent Association, 1982.

Horne, John and Alan Kramer, *German Atrocities 1914: A History of Denial*, New Haven: Yale University Press, 2001.

Hosmer, Stephen T. and Sibylle O. Crane (eds), *Counterinsurgency: A Symposium, April 16–20, 1962*, Santa Monica, California: RAND, 1963.

Hosmer, Stephen, *The Army's Role in Counterinsurgency and Insurgency*, Santa Monica, California: RAND, November 1990.

House, Jim and Neil MacMaster, *Paris 1961: Algerians, State Terror, and Memory*, Oxford: Oxford University Press, 2006.

Huei, Pang Yang, 'Beginning of the End: ARVN and Vietnamisation (1969–72)', *Small Wars and Insurgencies*, Vol. 17, No. 3 (September 2006), pp. 287–310.

Hughes, Geraint, 'The Soviet–Afghan War, 1978–1989: An Overview', *Defence Studies*, Vol. 8, No. 3 (September 2008), pp. 326–50.

Hughes, Geraint, 'A "Model Campaign" Reappraised: The Counterinsurgency War in Dhofar, Oman 1965–75', *The Journal of Strategic Studies*, Vol. 32, No. 2 (April 2009), pp. 271–305.

Hughes, Geraint, 'The Insurgencies in Iraq, 2003–2009: Origins, Developments and Prospects', *Defence Studies*, Vol. 10, No. 1–2 (March–June 2010), pp. 152–76.

Hughes, Matthew, 'The Banality of Brutality: British Armed Forces and the Repression of the Arab Revolt in Palestine, 1936–39', *English Historical Review*, Vol. 124, No. 507 (April 2009), pp. 313–54.

Hughes, Matthew, 'The practice and theory of British counterinsurgency: the histories of the atrocities at the Palestinian villages of al-Bassa and Halhul, 1938–1939', *Small Wars and Insurgencies*, Vol. 20, Nos. 3–4 (September–December 2009), pp. 528–50.

Hughes, Matthew, 'From law and order to pacification: Britain's Suppression of the Arab Revolt in Palestine, 1936–39', *Journal of Palestine Studies*, Vol. XXXIX, No. 2 (Winter 2010), pp. 6–22.

Hughes, Dr Matthew, 'Trouble in Palestine', *British Army Review*, No. 154 (Spring/ Summer 2012), pp. 103–8.

Hull, Isabel V., *Absolute Destruction: Military Culture and the Practices of War in Imperial Germany*, Ithaca, New York: Cornell University Press, 2005.

Humbaraci, Arslan and Nicole Muchnik, *Portugal's Wars: Angola, Guinea–Bissao, Mozambique*, New York: Joseph Okpaku Publishing, 1974.

Hutton, Paul Andrew, *Phil Sheridan and his Army*, Lincoln, Nebraska: University of Nebraska Press, 1985.

Hutton, Paul Andrew (ed.), *Soldiers West: Biographies from the Military Frontier*, Lincoln, Nebraska and London: University of Nebraska Press, 1987.

Hutton, Paul Andrew, 'Philip H. Sheridan' in Paul Andrew Hutton (ed.), *Soldiers West: Biographies from the Military Frontier*, Lincoln, Nebraska and London: University of Nebraska Press, 1987.

Ilandi, Gaetano Joe, 'Irish Republican Army Counterintelligence', *International Journal of Intelligence and Counterintelligence*, Vol.23, No.1 (2010), pp.1–26.

Inbar, Efraim (ed.), *Democracies and Small Wars*, London: Routledge, 2003.

Iron, Richard, 'The Charge of the Knights: The British in Basra, 2008', *The RUSI Journal*, Vol.158, No.1 (February/March 2013), pp.54–62.

Isaacman, Allen and Barbara, *Mozambique: From Colonialism to Revolution, 1900–1982*, Boulder, Colorado: Westview Press, 1983.

Ives, Christopher K., *US Special Forces and Counterinsurgency in Vietnam: Military Innovation and Institutional Failure, 1961–63*, London and New York: Routledge, 2007.

Jackson, Ashley, 'British Counterinsurgency in History: A Useful Precedent?', *British Army Review*, No.139 (Spring 2006), pp.12–22.

Jacobs, Walter Darnell, *Frunze: The Soviet Clausewitz, 1885–1925*, The Hague, The Netherlands: Martinus Nijhoff, 1969.

Jagadish, Vikram, 'Reconsidering American Strategy in South Asia: Destroying Terrorist Sanctuaries in Pakistan's Tribal Areas', *Small Wars and Insurgencies*, Vol.20, No.1 (March 2009), pp.36–65.

James, Harold and Denis Sheil-Small, *The Undeclared War: The Story of the Indonesian Confrontation, 1963–1966*, London: Leo Cooper, 1971.

Janeczko, Matthew, '"Faced with death, even a mouse bites": Social and religious motivations behind terrorism in Chechnya', *Small Wars and Insurgencies*, Vol.25, No.2 (2014), pp.428–56.

Jardine, Eric, 'Population-Centric Counterinsurgency and the Movement of Peoples', *Small Wars and Insurgencies*, Vol.23, No.2 (May 2012), pp.264–94.

Jason, Major Michael D., 'Integrating the Advisory Effort in the Army: A Full-Spectrum Solution', *Military Review*, (September–October 2008), pp.27–32.

Jeapes, Colonel Tony, *SAS: Operation Oman*, London: William Kimber, 1985.

Jeffery, Keith, 'Intelligence and Counterinsurgency Operations: Some Reflections on the British Experience', *Intelligence and National Security*, Vol.2, No.1 (January 1987), pp.118–49.

Jeffries, Sir Charles, *The Colonial Police*, London: Max Parrish, 1952.

Jeffries, Sir Charles, *The Colonial Office*, London: George Allen & Unwin, 1956.

Jenkins, Brian M., *The Unchangeable War*, Santa Monica, California: RAND, November 1970.

Joes, Anthony James, 'Insurgency and Genocide: La Vendée', *Small Wars and Insurgencies*, Vol.9, No.3 (Winter 1998), pp.17–45.

Joes, Anthony James, *Resisting Rebellion: The History and Politics of Counterinsurgency*, Lexington, Kentucky: University Press of Kentucky, 2004.

Joes, Anthony James, *Urban Guerrilla Warfare*, Lexington, Kentucky: University Press of Kentucky, 2007.

Joes, Professor Anthony James, 'Counterinsurgency in the Philippines, 1898–1954' in
    Daniel Marston and Carter Malkasian (eds), *Counterinsurgency in Modern Warfare*,
    Botley, Oxford: Osprey Publishing, 2008.

Johnson, Katie Ann, *A Reevaluation of the Combined Action Program as a Counterinsurgency
    Tool*, Washington, DC: Georgetown University, 2008.

Johnson, Robert, 'Upstream engagement and downstream entanglements: The
    assumptions, opportunities, and threats of partnering', *Small Wars and Insurgencies*,
    Vol.25, No.3 (2014), pp.647–68.

Johnson, Wray R., *Vietnam and American Doctrine for Small Wars*, Bangkok: White Lotus, 2001.

Johnson, Wray R. and Paul J. Dimech, 'Foreign Internal Defense and the Hukbalahap: A
    Model Counterinsurgency' in Ian Beckett (ed.), *Modern Counterinsurgency*, Aldershot,
    Hampshire and Burlington, Vermont: Ashgate, 2007.

Jones, Clive, 'Military intelligence and the war in Dhofar: An appraisal', *Small Wars and
    Insurgencies*, Vol.25, No.3 (2014), pp.628–46.

Jones, Seth G., *Counterinsurgency in Afghanistan*, Santa Monica, California: RAND, 2008.

Jones, Seth G., *In the Graveyard of Empires: America's War in Afghanistan*, New York and
    London: W.W. Norton, 2009.

Jones, Tim, *Postwar Counterinsurgency and the SAS, 1945–1952: A Special Type of Warfare*,
    London and Portland, Oregon: Frank Cass, 2001.

Jones, Tim, *SAS: The First Secret Wars: The Unknown Years of Combat & Counterinsurgency*,
    London: I.B. Tauris, 2005.

Jones, Tim, 'The British Army, and Counter-Guerrilla Warfare in Transition, 1944–1952'
    in Ian Beckett (ed.), *Modern Counterinsurgency*, Aldershot, Hampshire and Burlington,
    Vermont: Ashgate, 2007.

Jones, Tim, 'The British Army, and Counter-Guerrilla Warfare in Greece, 1944–1949' in
    Ian Beckett (ed.), *Modern Counterinsurgency*, Aldershot, Hampshire and Burlington,
    Vermont: Ashgate, 2007.

Jundanian, Brendan F., 'Resettlement Programs: Counterinsurgency in Mozambique',
    Comparative Politics, Vol.6, No.4 (July 1974), pp.519–40.

Jureidini, Paul A., *Case Studies in Insurgency and Revolutionary Warfare: Algeria 1954–1962*,
    Washington DC: Special Operations Research Office, The American University,
    1963.

Kanya-Forstner, A.S., *The Conquest of the Western Sudan: A Study in French Military
    Imperialism*, Cambridge: Cambridge University Press, 1969.

Kautt, W.H., *Ambushes and Armour: The Irish Rebellion, 1919–1921*, Dublin and Portland,
    Oregon: Irish Academic Press, 2010.

Kelly, George Armstrong, *Lost Soldiers: The French Army and Empire in Crisis, 1947–1962*,
    Cambridge, Massachusetts: The MIT Press, 1965.

Kennedy, Dane, 'Constructing the Colonial Myth of Mau Mau', *The International Journal
    of African Historical Studies*, Vol.25, No.2 (1992), pp.241–60.

Kennedy-Pipe, Caroline and Colin McInnes, 'The British Army in Northern Ireland
    1969–1972: From Policing to Counter-Terror' in Ian Beckett (ed.), *Modern
    Counterinsurgency*, Aldershot, Hampshire and Burlington, Vermont: Ashgate, 2007.

Kent, Lieutenant Christopher, 'Speaking the Lingo', *British Army Review*, No.152
    (Autumn 2011), pp.67–9.

Kerkvliet, Benedict J., *The Huk Rebellion: A Study of Peasant Revolt in the Philippines*,
    Berkeley, California: University of California Press, 1982.

Kilcullen, Lieutenant Colonel David, '"Twenty-Eight Articles": Fundamentals of
    Company-level Counterinsurgency', *Military Review* (May–June 2006), pp.103–8.

Kilcullen, David, *The Accidental Guerrilla: Fighting Small Wars in the Midst of a Big One*, London: C. Hurst & Company, 2009.

Kilcullen, David, *Counterinsurgency*, London: C. Hurst & Company, 2010.

Kilcullen, David, *Out of the Mountains: The Coming Age of the Urban Guerrilla*, London: C. Hurst & Company, 2013.

Killingray, David and David Omissi (eds), *Guardians of Empire: The Armed Forces of the Colonial Powers, c.1700–1964*, Manchester and New York: Manchester University Press, 1999.

King, Anthony, 'Understanding the Helmand campaign: British operations in Afghanistan', *International Affairs*, Vol.86, No.2 (2010), pp.311–32.

Kinnard, Douglas, *The War Managers*, Wayne, New Jersey: Avery Publishing Group, 1985.

Kipp, Jacob, Lester Grau, Karl Prinslow and Captain Don Smith, 'The Human Terrain System: A CORDS for the 21st Century', *Military Review* (September–October 2006), pp.8–15.

Kirk-Greene, Anthony, *Britain's Imperial Administrators, 1858–1966*, Basingstoke, Hampshire: Macmillan, 2000.

Kiszely, Lieutenant General Sir John, 'Thinking about the Operational Level', *The RUSI Journal*, Vol.150, No.6 (December 2005), pp.38–43.

Kiszely, Lieutenant General Sir John, 'Learning about Counterinsurgency', *The RUSI Journal*, Vol.151, No.6 (December 2006), pp.16–21.

Kiszely, Lieutenant General Sir John, 'Learning about Counterinsurgency', *Military Review* (March–April 2007), pp.5–11.

Kitfield, James, *Prodigal Soldiers: How the Generation of Officers Born of Vietnam Revolutionized the American Style of War*, Washington and London: Brassey's, 1997.

Kitson, F.E., *Gangs and Counter-Gangs*, London: Barrie and Rockliffe, 1960.

Kitson, Frank, *Low Intensity Operations: Subversion, Insurgency, Peace-Keeping*, London: Faber & Faber, 1971.

Kitson, Frank, *Bunch of Five*, London: Faber & Faber, 1977.

Kitson, Frank, *Warfare as a Whole*, London: Faber & Faber, 1987.

Kitson, Frank, *Directing Operations*, London: Faber & Faber, 1989.

Klare, Michael T., 'The Interventionalist Impulse: U.S. Military Doctrine for Low-Intensity Warfare' in Michael T. Klare and Peter Kornbluh (eds), *Low-Intensity Warfare: Counterinsurgency, Proinsurgency, and Antiterrorism in the Eighties*, New York: Pantheon Books, 1988.

Klare, Michael T. and Peter Kornbluh (eds), *Low-Intensity Warfare: Counterinsurgency, Proinsurgency, and Antiterrorism in the Eighties*, New York: Pantheon Books, 1988.

Klare, Michael T. and Peter Kornbluh, 'The New Interventionism: Low-Intensity Warfare in the 1980s and Beyond' in Michael T. Klare and Peter Kornbluh (eds), *Low-Intensity Warfare: Counterinsurgency, Proinsurgency, and Antiterrorism in the Eighties*, New York: Pantheon Books, 1988.

Kogan, Boris, 'The Basmachi: Factors behind the Rise and Fall of an Islamic Insurgency in Central Asia', *Small Wars Journal* (March 2011).

Koloski, Major Andrew W. and Lieutenant Colonel John S. Kolasheski, 'Thickening the Lines: Sons of Iraq, A Combat Multiplier', *Military Review* (January–February 2009), pp.41–53.

Komer, R.W., *The Malayan Emergency in Retrospect: Organization of a Successful Counterinsurgency Effort*, Santa Monica, California: RAND, 1972.

Kopets, Captain Keith F., 'The Combined Action Program: Vietnam', *Military Review*, Vol.LXXXII, No.4 (July–August 2002), pp.78–81.

Kramer, Mark, 'Guerrilla Warfare, Counterinsurgency and Terrorism in the North Caucasus: The Military Dimension of the Russian–Chechen Conflict', *Europe–Asia Studies*, Vol. 57, No. 2 (March 2005).

Kramers, Major Peter J., 'Konfrontsai in Borneo 1962–66', *Military Review*, Vol. LXX, No. 11 (November 1990), pp. 64–74.

Krepinevich, Andrew F., Jr, *The Army in Vietnam*, Baltimore: Maryland: John Hopkins University Press, 1986.

Kroizer, Gad, 'From Dowbiggin to Tegart: Revolutionary Change in the Colonial Police in Palestine during the 1930s' in Georgina Sinclair (ed.), *Globalising British Policing*, Farnham, Surrey: Ashgate, 2011.

Krulak, Lieutenant General Victor H., *First to Fight: An Inside View of the US Marine Corps*, Annapolis, Maryland: Naval Institute Press, 1988.

Kuchi, Lieutenant Colonel Dale, 'Testing Galula in Ameriyah: The People are the Key', *Military Review* (March–April 2009), pp. 72–80.

Kudelia, Serhiy, 'Choosing Violence in Irregular Wars: The Case of Anti-Soviet Insurgency in Western Ukraine', *East European Politics and Societies*, Vol. 27, No. 1 (February 2013), pp. 149–81.

Lackman, Major Michael J., 'The British Boer War and the French Algerian Conflict: Counterinsurgency for Today', Fort Leavenworth, Kansas: Master of Military Art and Science Thesis, US Army Command and General Staff College, 2006.

Langley, Lester D., *The Banana Wars: United States Intervention in the Caribbean, 1898–1934*, Lexington, Kentucky: University Press of Kentucky, 1985.

Lansdale, Major General Edward, 'Contradictions in Military Culture' in W. Scott Thompson and Donaldson D. Frizzell (eds), *The Lessons of Vietnam*, London: Macdonald and Jane's, 1977.

Lansdale, Edward Geary, *In the Midst of Wars: An American's Mission to Southeast Asia*, New York: Fordham University Press, 1991.

Lazreg, Marnia, *Torture and the Twilight of Empire: From Algiers to Baghdad*, Princeton, New Jersey: Princeton University Press, 2008.

Leary, John D., *Violence and the Dream People: The Orang Asli in the Malayan Emergency, 1948–1960*, Athens, Ohio: Ohio University Press, 1995.

Ledwidge, Frank, *Losing Small Wars: British Military Failure in Iraq and Afghanistan*, New Haven, Connecticut and London: Yale University Press, 2011.

Leeson, D.M., *The Black & Tans: British Police and Auxiliaries in the Irish War of Independence, 1920–21*, Oxford: Oxford University Press, 2011.

Lewis, Paul H., *Guerrillas and Generals: The 'Dirty War' in Argentina*, Westport, Connecticut: Greenwood Press, 2001.

Lewy, Guenter, *America in Vietnam*, New York: Oxford University Press, 1978.

Lieb, Peter, *A Precursor of Modern Counterinsurgency Operations? The German Occupation of the Ukraine in 1918*, Salford: University of Salford, Working Papers in Military & International History Number 4, September 2007.

Lieb, Peter, 'Few Carrots and a Lot of Sticks: German Anti-Partisan Warfare in World War Two' in Daniel Marston and Carter Malkasian (eds), *Counterinsurgency in Modern Warfare*, Botley, Oxford: Osprey Publishing, 2008.

Lieven, Anatol, *Chechnya: Tombstone of Russian Power*, New Haven, Connecticut: Yale University Press, 1998.

Lieven, Anatol, 'Nightmare in the Caucasus', *The Washington Quarterly* (Winter 2000), pp. 145–59.

Linn, Brian M., 'Provincial Pacification in the Philippines, 1900–1901: The First District Department of North Luzon', *Military Affairs*, Vol. LI, No. 2 (April 1987), pp. 62–6.

Linn, Brian McAllister, *The US Army and Counterinsurgency in the Philippine War, 1899–1902*, Chapel Hill, North Carolina: University of North Carolina Press, 1989.

Linn, Brian McAllister, 'Intelligence and Low-Intensity Conflict in the Philippine War, 1899–1902', *Intelligence and National Security*, Vol.6, No.1 (January 1991), pp.90–114

Linn, Brian McAllister, '"We Will Go Heavily Armed": The Marines' Small War on Samar, 1901–1902' in William R. Roberts and Jack Sweetman (eds), *New Interpretations in Naval History*, Annapolis, Maryland: Naval Institute Press, 1991.

Linn, Brian McAllister, *Guardians of Empire: The U.S. Army and the Pacific, 1902–1940*, Chapel Hill, North Carolina and London: University of North Carolina Press, 1997.

Linn, Brian McAllister, 'Cerberus' dilemma: the US Army and internal security in the Pacific, 1902–1940' in David Killingray and David Omissi (eds), *Guardians of Empire: The Armed Forces of the Colonial Powers, c.1700–1964*, Manchester and New York: Manchester University Press, 1999.

Linn, Brian M., *The Philippine War, 1899–1902*, Lawrence, Kansas: University of Kansas Press, 2000.

Linn, Brian McAllister, 'The Philippines: Nationbuilding and Pacification', *Military Review* (March–April 2005), pp.46–54.

Little, Patrick, 'Lessons Unlearned: A Former Officer's Perspective on the British Army at War', *The RUSI Journal*, Vol.154, No.3 (June 2009), pp.10–16.

Lloyd, Nick, 'The Armritsar Massacre and the Minimum Force Debate', *Small Wars and Insurgencies*, Vol.21, No.2 (June 2010), pp.382–403.

Lock-Pullan, Richard, *US Intervention Policy and Army Innovation: From Vietnam to Iraq*, London and New York: Routledge, 2006.

Logevall, Fredrik, *Embers of War: The Fall of an Empire and the Making of America's Vietnam*, New York: Random House, 2012.

Long, Austin, *Doctrine of Eternal Recurrence: The US Military and Counterinsurgency Doctrine, 1960–1970 and 2003–2006*, Santa Monica, California: RAND, 2008.

Lonsdale, John, 'Mau Maus of the Mind: Making Mau Mau and Remaking Kenya', *Journal of African History*, Vol.31, No.3 (1990), pp.393–421.

Lord, Carnes, 'American Strategic Culture in Small Wars', *Small Wars and Insurgencies*, Vol.3, No.3 (Winter 1992), pp.205–16.

Lord, Carnes, 'The Role of the United States in Small Wars', *Annals of the American Academy of Political and Social Science*, Vol.541 (September 1995), pp.89–100.

Lowe, Brevet Major T.A., 'Some Reflections of a Junior Commander upon "The Campaign" in Ireland, 1920 and 1921', *Army Quarterly and Defence Journal*, Vol.5 (October 1922), pp.50–8.

Loyn, David, *Butcher and Bolt*, London: Hutchinson, 2008.

Ludwig, Walter C., III, 'Managing Counterinsurgency: Lessons from Malaya', *Military Review* (May–June 2007), pp.56–66.

Lynn, Dr John A., 'Patterns of Insurgency and Counterinsurgency', *Military Review* (July–August 2005), pp.22–7.

McAllister, James, 'The Lost Revolution: Edward Lansdale and the American Defeat in Vietnam 1964–1968', *Small Wars and Insurgencies*, Vol.14, No.2 (Summer 2003), pp.1–26.

McAllister, James, 'What Can One Man Do? Nuyen Doc Thang and the Limits of Reform in South Vietnam', *Journal of Vietnamese Studies*, Vol.4, No.2 (Summer 2009), pp.117–53.

McAllister, James and Schulte, Ian, 'The Limits of Influence in Vietnam: Britain, the United States and the Diem Regime, 1959–63', *Small Wars and Insurgencies*, Vol.17, No.1 (March 2006), pp.22–43.

Macaulay, First Lieutenant Neill, 'Leading Native Troops', *Marine Corps Gazette*, Vol.47, No.6 (June 1963), pp.32–5.

Macaulay, First Lieutenant Neill, 'Counterguerrilla Patrolling', *Marine Corps Gazette*, Vol.47, No.7 (July 1963), pp.45–8.

McClintock, Michael, *Instruments of Statecraft: U.S. Guerrilla Warfare, Counterinsurgency, and Counter-Terrorism, 1940–1990*, New York: Pantheon Books, 1992.

McConville, Michael, 'Knight's Move in Bosnia and the British Rescue of Tito: 1944', *The RUSI Journal*, Vol.142, No.6 (December 1997), pp.61–9.

McCoy, Alfred W., *A Question of Torture: CIA Interrogation, from the Cold War to the War on Terror*, New York: A Metropolitan/Owl Books, 2006.

McCoy, Alfred W., *Policing America's Empire: The United States, the Philippines, and the Rise of the Surveillance State*, Madison, Wisconsin: University of Wisconsin Press, 2009.

McCoy, Katherine E., 'Trained to Torture? The Human Rights Effects of Military Training at the School of the Americas', *Latin American Perspectives*, Vol.32, No.6 (November 2005), pp.47–64.

McCuen, Lieutenant Colonel John J., *The Art of Counter-Revolutionary War: Strategy of Counterinsurgency*, London: Faber & Faber, 1966.

McFate, Montgomery, 'The Military Utility of Understanding Adversary Culture', *JFQ: Joint Force Quarterly*, Issue 38, 3rd Quarter (2005), pp.42–8.

McFate, Montgomery, 'Anthropology and Counterinsurgency: The Strange Story of Their Curious Relationship', *Military Review* (March–April 2005), pp.24–38.

McInnes, Colin, 'The Gulf War, 1990–1' in Hew Strachan (ed.), *Big Wars and Small Wars: The British Army and the Lessons of War in the 20th Century*, Abingdon, Oxon: Routledge, 2006.

McInnes, Colin, 'The British Army's New Way in Warfare: A Doctrinal Misstep?', *Defense and Security Analysis*, Vol.23, No.2 (June 2007), pp.127–41.

Mackay, Andrew and Stephen Tatham, *Behavioural Conflict: From General to Strategic Corporal: Complexity, Adaptation and Influence*, Shrivenham, Oxfordshire: Defence Academy of the United Kingdom, Shrivenham Papers No.9, 2009.

Mackay, Donald, *The Domino That Stood: The Malayan Emergency, 1948–60*, London and Washington: Brassey's, 1997.

McKenna, Major Sean and Major Russell Hampsey, '"The COIN Warrior"; Waging Influence: Hints for the Counterinsurgency (COIN) Strategy in Afghanistan', *Special to Small Wars Journal* (2 June 2010), pp.1–16, www.smallwarsjournal.com.

MacKenzie, Alastair, *Special Force: The Untold Story of 22nd Special Air Service Regiment (SAS)*, London: I.B. Tauris, 2011.

McKey, Robert R., 'Bushwhackers, Provosts, and Tories: The Guerrilla War in Arkansas' in Daniel E. Sutherland (ed.), *Guerrillas, Unionists and Violence on the Confederate Home Front*, Fayetteville, Arkansas: University of Arkansas Press, 1999.

McKey, Robert R., *The Uncivil War: Irregular Warfare in the Upper South, 1861–1865*, Norman, Oklahoma: University of Oklahoma Press, 2004.

McKiernan, [General] David D., 'Recommitment and Shared Interests: Progress and the Future of Afghan National Security', *The RUSI Journal*, Vol.154, No.2 (April 2009), pp.6–11.

Mackinlay, John, 'War Lords', *The RUSI Journal*, Vol.143, No.2 (April 1998), pp.24–32.

Mackinlay, John, 'Is UK Doctrine Relevant to Global Insurgency?', *The RUSI Journal*, Vol.152, No.2 (April 2007), pp.35–8.

Mackinlay, John, 'Counterinsurgency in Global Perspective – An Introduction: Politicians need to Understand Insurgency', *The RUSI Journal*, Vol.152, No.6 (December 2007), pp.64–70.

Mackinlay, John, *The Insurgent Archipelago: From Mao to bin Laden*, London: C. Hurst & Company, 2009.

Mackinlay, John and Alison A-Baddawy, *Rethinking Counterinsurgency*, Santa Monica, California: RAND, 2008.

McMahon, Paul, *British Spies and Irish Rebels: British Intelligence and Ireland, 1916–1945*, Woodbridge, Suffolk: The Boydell Press, 2008.

McMichael, Scott R., 'The Soviet Army, Counterinsurgency, and the Afghan War', *Parameters*, Vol.19, No.4 (December 1989), pp.21–35.

McMichael, Scott R., *Stumbling Bear: Soviet Military Performance in Afghanistan*, London: Brassey's, 1991.

MacQueen, Norrie, *The Decolonization of Portuguese Africa: Metropolitan Revolution and the Dissolution of Empire*, London and New York: Longman, 1997.

MacQueen, Norrie, 'Portugal's First Domino: "Pluricontinentalism" and Colonial War in Guinéa–Bissau, 1963–1974', *Contemporary European History*, Vol.8, No.2 (July 1999), pp.209–30.

McPhail, Helen, *The Long Silence: Civilian Life under the German Occupation of Northern France, 1914–1918*, London: I.B. Tauris, 1999.

Maciejewski, Justin, '"Best effort": Operation Sinbad and the Iraq Campaign' in Jonathan Bailey, Richard Iron, and Hew Strachan (eds), *British Generals in Blair's Wars*, Farnham, Surrey and Burlington, Vermont: Ashgate Publishing, 2013.

Maddock, Trooper David, 'Donkeys v Mastiffs', *British Army Review*, No.147 (Summer 2009), pp.50–2.

Maechling, Charles, Jr, 'Counterinsurgency: The First Ordeal by Fire', in Michael T. Klare and Peter Kornbluh (eds), *Low-Intensity Warfare: Counterinsurgency, Proinsurgency, and Antiterrorism in the Eighties*, New York: Pantheon Books, 1988.

Mages, Robert M., 'Without the Need of a Single American Rifleman: James Van Fleet and his Lessons learned as Commander of the Joint United States Military Advisory and Planning Group during the Greek Civil War, 1948–1949' in Richard G. Davis (ed.), *The US Army and Irregular Warfare, 1775–2007*, Washington, DC: Center of Military History, United States Army, 2008.

Mahendrarajah, Shivan, 'Conceptual failure, the Taliban's parallel hierarchies, and America's strategic defeat in Afghanistan', *Small Wars and Insurgencies*, Vol.25, No.1 (2014), pp.91–121.

Mahnken, Thomas G., 'The American Way of War in the Twenty-first Century', *Review of International Affairs*, Vol.2, No.31 (Spring 2003), pp.73–84.

Mahnken, Thomas G., 'The American Way of War in the Twenty-first Century' in Efraim Inbar (ed.), *Democracies and Small Wars*, London: Routledge, 2003.

Majdalany, Fred, *State of Emergency: The Full Story of Mau Mau*, London: Longmans, 1962.

Malaya Command, *The Conduct of Anti-Terrorist Operations in Malaya* [ATOM], Singapore: Government Printing Office, 1954.

Maley, William, *Rescuing Afghanistan*, London: Hurst, 2006.

Maloba, Wunyabari O., *Mau Mau and Kenya: An Analysis of a Peasant Revolt*, Bloomington and Indianapolis, Indiana: Indiana University Press, 1993.

Mann, Major Morgan, 'The Power Equation: Using Tribal Politics in Counterinsurgency', *Military Review* (May–June 2007), pp.104–8.

Mans, Lieutenant Colonel R.S.N., 'Counterinsurgency: Victory in Malaya (Part One)', *Marine Corps Gazette*, Vol.47, No.1 (January 1963), pp.44–9.

Mansoor, Peter R., *Baghdad at Sunrise: A Brigade Commander's War in Iraq*, New Haven, Connecticut and London: Yale University Press, 2008.

Mansoor, Peter R., 'The British Army and the Lessons of the Iraq War', *British Army Review*, Vol. 147 (Summer 2009), pp. 11–14.

Manwaring, Max G. (ed.), *Uncomfortable Wars: Towards a New Paradigm of Low Intensity Conflict*, Boulder, Colorado: Westview Press, 1991.

Manwaring, Max G., *Shadows of Things Past and Images of the Future: Lessons for the Insurgencies in our Midst*, Carlisle, Pennsylvania: Strategic Studies Institute, US Army War College, November 2004.

Mao Tse-tung, *Selected Works of Mao Tse-Tung*, Volume Four, 1941–1945, New York: International Publishers, 1956.

Mao Tse-tung, *Selected Military Writings of Mao Tse-Tung*, Peking: Foreign Languages Press, 1963.

Mao Tse-tung (translation and introduction by Stuart R. Schram), *Basic Tactics*, London: Pall Mall Press, 1967.

Mao Tse-tung (edited by Brigadier General Samuel B. Griffith), *Guerrilla Warfare*, Garden City, New York: Anchor Press/Doubleday, 1978.

Markel, Wade, 'Draining the Swamp: The British Strategy of Population Control', *Parameters*, Vol. XXXVI (Spring 2006), pp. 35–48.

Marr, Lieutenant Colonel Jack, Major John Cushing, Major Brandon Garner and Captain Richard Thompson, 'Human Terrain Mapping: A Critical First Step to Winning the COIN Fight', *Military Review* (March–April 2008), pp. 18–24.

Marshall, Alexander, 'Turkfront: Frunze and the Development of Soviet Counterinsurgency in Central Asia' in Tom Everett-Heath (ed.), *Central Asia: Aspects of Transition*, London: Curzon Press, 2003.

Marshall, Alex, 'Imperial Nostalgia, the Liberal Lie, and the Perils of Postmodern Counterinsurgency', *Small Wars and Insurgencies*, Vol. 21, No. 2 (June 2010), pp. 233–58.

Marston, Daniel, 'Lost and Found in the Jungle: The Indian and British Army Jungle Warfare Doctrine for Burma, 1943–5, and the Malayan Emergency, 1948–60' in Hew Strachan (ed.), *Big Wars and Small Wars: The British Army and the Lessons of War in the 20th Century*, Abingdon, Oxon: Routledge, 2006.

Marston, Daniel, '"Smug and Complacent?" Operation TELIC: The Need for Critical Analysis', *British Army Review*, Vol. 147 (Summer 2009), pp. 16–23.

Marston, Daniel, 'British Operations in Helmand Afghanistan', *Small Wars Journal* (2009), www.smallwarsjournal.com.

Marston, Daniel and Carter Malkasian (eds), *Counterinsurgency in Modern Warfare*, Botley, Oxford: Osprey Publishing, 2008.

Martin, Colonel Gilles, 'War in Algeria: The French Experience', *Military Review* (July–August 2005), pp. 51–7.

Martin, Michel Louis, *Warriors to Managers: The French Military Establishment since 1945*, Chapel Hill, North Carolina: University of North Carolina Press, 1981.

Martin, Michel L., 'From Algiers to N'Djamena: France's Adaptation to Low-Intensity Wars, 1830–1987', in David Charters and Maurice Tugwell (eds), *Armies in Low-Intensity Conflict: A Comparative Analysis*, London: Brassey's Defence, 1989.

Martin, Lieutenant Mike, 'The Importance of Cultural Understanding to the Military', *British Army Review*, No. 147 (Summer 2009), pp. 44–9.

Mathias, Grégor, *Galula in Algeria: Counterinsurgency Practice versus Theory*, Santa Barbara, California: Praeger, 2011.

Mawby, Spencer, *British Policy in Aden and the Protectorates 1955–67: Last Outpost of a Middle East Empire*, London and New York: Routledge, 2005.

May, Glenn A., *Battle for Batangas: A Philippine Province at War*, New Haven, Connecticut: Yale University Press, 1991.

May, Glenn A., 'Was the Philippine-American War a "Total War"?' in Manfred F. Boemeke, Roger Chickering and Stig Föster (eds), *Anticipating Total War: The German and American Experiences, 1871–1914*, Cambridge: Cambridge University Press, 1999.

Mayes, Stanley, *Cyprus and Makarios*, London: Putnam, 1960.

Meehan, Shannon P. with Roger Thompson, *Beyond Duty: Life on the Frontline in Iraq*, Cambridge and Malden, Massachusetts: Polity Press, 2009.

Melnik, Constantin, *Insurgency and Counterinsurgency in Algeria*, Santa Monica, California: RAND, April 1964.

Melnik, Constantin, *The French Campaign against the FLN*, Santa Monica, California: RAND, September 1967.

Melson, Charles D., 'German Counterinsurgency in the Balkans: The Prinz Eugen Division Example, 1942–1944', *Journal of Slavic Military Studies*, Vol.20 (2007), pp.705–37.

Melvin, Mungo, 'Learning the Strategic Lessons from Afghanistan', *The RUSI Journal*, Vol.157, No.2 (April/May 2012), pp.56–61.

Menard, Orville D., *The Army and the Fifth Republic*, Lincoln, Nebraska: University of Nebraska Press, 1967.

Mendy, Peter Karibe, 'Portugal's Civilizing Mission in Colonial Guinea–Bissau: Rhetoric and Reality', *International Journal of African Historical Studies*, Vol.36, No.1 (2003), pp.35–58.

Metz, Steven, *Learning from Iraq: Counterinsurgency in American Strategy*, Carlisle, Pennsylvania: Strategic Studies Institute, US Army War College, January 2007.

Meyer, Karl, *The Dust of Empire: The Race for Supremacy in the Asian Heartland*, London: Abacus Books, 2004.

Michaels, Jeffrey H., 'Helpless or Deliberate Bystander: American Policy towards South Vietnam's Military Coups, 1954–1975', *Small Wars and Insurgencies*, Vol.25, No.3 (2014), pp.560–83.

Millett, Dr Richard L., *Searching for Stability: The U.S. Development of Constabulary Forces in Latin America and the Philippines*, Fort Leavenworth, Kansas: Occasional Paper 30, Combat Studies Institute Press, US Army Combined Arms Center, 2010.

Miller, Edward, *Misalliance: Ngo Dinh Diem, the United States, and the fate of South Vietnam*, Cambridge, Massachusetts and London: Harvard University Press, 2013.

Miller, Joyce Laverty, 'The Syrian Revolt of 1925', *International Journal of Middle East Studies*, Vol.8, No.4 (October 1977), pp.545–63.

Miller, Major (V)S.N., 'A Comprehensive Failure: British Civil-Military Strategy in Helmand Province', *British Army Review*, Vol.146 (Spring 2009), pp.35–40.

Ministry of Defence, *Operation Banner: An Analysis of Military Operations in Northern Ireland*, London: Prepared under the direction of the Chief of the General Staff, 2006.

Miska, Major Steven M., 'Growing the Iraqi Security Forces', *Military Review* (July–August 2005), pp.64–9.

Mockaitis, Thomas R., *British Counterinsurgency, 1919–60*, London: Macmillan, 1990.

Mockaitis, Thomas R., 'The Origins of British Counterinsurgency', *Small Wars and Insurgencies*, Vol.1, No.3 (December 1990), pp.209–25.

Mockaitis, Thomas R., *British Counterinsurgency in the post-imperial era*, Manchester: Manchester University Press, 1995.

Mockaitis, Thomas R., *The Iraq War: Learning from the Past, Adapting to the Present, and Planning for the Future*, Carlisle, Pennsylvania: Strategic Studies Institute, US Army War College, 2007.

Mockaitis, Thomas R., *Iraq and the Challenge of Counterinsurgency*, Westport, Connecticut and London: Praeger Security International, 2008.

Moreman, T.R., '"Small Wars" and "Imperial Policing": The British Army and the Theory and Practice of Colonial Warfare in the British Army 1919–1939', *Journal of Strategic Studies*, Vol. 19, No. 4 (1996), pp. 105–31.

Moreman, Tim: '"Watch and Ward": The Army in India and the North-West Frontier, 1920–1939' in Killingray, David, and David Omissi (eds), *Guardians of Empire: The Armed Forces of the Colonial Powers, c. 1700–1964*, Manchester and New York: Manchester University Press, 1999.

Moyar, Mark, *Triumph Forsaken: The Vietnam War, 1954–1965: Volume 1*, New York: Cambridge University Press, 2006.

Moyar, Mark, *Phoenix and Birds of Prey: Counterinsurgency and Counterterrorism in Vietnam*, Lincoln, Nebraska and London: University of Nebraska Press, 2007.

Moyar, Mark, *A Question of Command: Counterinsurgency from the Civil War to Iraq*, New Haven, Connecticut and London: Yale University Press, 2009.

Muckian, Martin J., 'Structural Vulnerabilities of Networked Insurgencies: Adapting to the New Adversary', *Parameters*, Vol. XXXVI (Winter 2006–2007), pp. 14–25.

Mumford, Andrew, *The Counterinsurgency Myth: The British Experience of Irregular Warfare*, London and New York: Routledge, 2012.

Murray, Colonel J.C., 'The Anti-Bandit War' in T.N. Greene (ed.), *The Guerrilla and How to Fight Him*, New York: Frederick A. Praeger, 1962.

Nagaraj, Lieutenant General K. (ed.), *Indian Army Doctrine*, Shimla, India: Headquarters Army Training Command, October 2004.

Nagl, John A., *Counterinsurgency Lessons from Malaya and Vietnam: Learning to Eat Soup with a Knife*, Westport, Connecticut: Praeger, 2002.

Nagl, John A., 'An American View of Twenty-First Century Counterinsurgency', *The RUSI Journal*, Vol. 152, No. 4 (August 2007), pp. 12–16.

Nagl, Lieutenant Colonel Dr John A., 'Institutionalizing Adaptation: It's Time for an Army Advisor Command', *Military Review*, (September–October 2008), pp. 22–32.

Nagl, Lieutenant Colonel John A., 'Counterinsurgency in Vietnam: American Organizational Culture and Learning' in Daniel Marston and Carter Malkasian (eds), *Counterinsurgency in Modern Warfare*, Botley, Oxford: Osprey Publishing, 2008.

Nasson, Bill, *The South African War, 1899–1902*, London, Sydney and Auckland: Arnold, 1999.

Neillands, Robin, *A Fighting Retreat: The British Empire 1947–1997*, London: Coronet, 1997.

Neumann, Peter R., *Britain's Long War: British Strategy in the Northern Ireland Conflict, 1969–98*, Basingstoke, Hampshire: Palgrave, 2003.

Newitt, Malyn, *Portugal in Africa: The Last Hundred Years*, London: C. Hurst, 1981.

Newsinger, John, 'Revolt and Repression in Kenya: The "Mau Mau" Rebellion, 1952–1960', *Science and Society*, Vol. 45 (Summer 1981), pp. 159–85.

Newsinger, John, 'From Counterinsurgency to Internal Security: Northern Ireland 1969–1992', *Small Wars and Insurgencies*, Vol. 6, No. 1 (Spring 1995), pp. 88–111.

Newsinger, John, *British Counterinsurgency: From Palestine to Northern Ireland*, Basingstoke, Hampshire: Palgrave, 2002.

Newsinger, John, 'Minimum Force, British Counterinsurgency and the Mau Mau Rebellion' in Ian Beckett (ed.), *Modern Counterinsurgency*, Aldershot, Hampshire and Burlington, Vermont: Ashgate, 2007.

Noetzel, Timo and Benjamin Schreer, 'Counter-what? Germany and Counterinsurgency in Afghanistan', *The RUSI Journal*, Vol. 153, No. 1 (February 2008), pp. 42–6.

Noetzel, Timo and Benjamin Schreer, 'Missing Links: The Evolution of German Counterinsurgency Thinking', *The RUSI Journal*, Vol. 154, No. 1 (February 2009), pp. 16–22.

Nolan, Victoria, *Military Leadership and Counterinsurgency: The British Army and Small War Strategy since World War II*, London and New York: I.B. Tauris, 2012.

Norris, Jacob, 'Repression and Rebellion: Britain's Response to the Arab Revolt in Palestine of 1936–39', *Journal of Imperial and Commonwealth History*, Vol. 36, No. 1 (March 2008), pp. 25–45.

North, Jonathon, 'General Hoche and Counterinsurgency', *The Journal of Military History*, Vol. 67, No. 2 (April 2003), pp. 529–40.

North, Richard, *The Ministry of Defeat: The British War in Iraq – 2003–2009*, London and New York: Continuum, 2009.

O'Ballance, Edgar, *The Algerian Insurrection, 1954–1962*, London: Faber & Faber, 1967.

O'Ballance, Edgar, *The Greek Civil War, 1944–1949*, London: Faber & Faber, 1966.

O'Ballance, Edgar, *The Indochina War, 1945–1954: A Study in Guerilla Warfare*, London: Faber & Faber, 1964.

O'Ballance, Edgar, *The Wars in Vietnam, 1954–1973*, London: Ian Allan, 1975.

Olcott, Martha B., 'The Basmachi or Freemen's Revolt in Turkestan 1918–24', *Soviet Studies*, Vol. 33, No. 3 (July 1981).

Olusoga, David and Casper W. Erichsen, *The Kaiser's Holocaust: Germany's Forgotten Genocide*, London: Faber & Faber, 2011.

O'Neill, Bard E., D.J. Alberts and Stephen J. Rosetti (eds), *Political Violence: A Comparative Approach*, Arvada, Colorado: Phoenix Press, 1974.

O'Neill, Bard E., *Insurgency and Terrorism: Inside Modern Revolutionary Warfare*, Dulles, Virginia: Brassey's, 1990.

O'Neill, Bard E., William R. Heaton and Donald J. Alberts (eds), *Insurgency in the Modern World*, Boulder, Colorado: Westview Press, 1980.

Opello, Walter C., 'Guerrilla War in Portuguese Africa: An Assessment of the Balance of Force in Mozambique', *Journal of Opinion*, Vol. 47, No. 23 (Summer 1974), pp. 29–37.

Orr, Allan, 'Recasting Afghan Strategy', *Small Wars and Insurgencies*, Vol. 20, No. 1 (March 2009), pp. 87–117.

Orsini, Dominique, 'Walking the Tightrope: Dealing with Warlords in Afghanistan's Destabilizing North', *The RUSI Journal*, Vol. 152, No. 5 (October 2007), pp. 46–50.

Osanka, Franklin Mark (ed.), *Modern Guerrilla Warfare: Fighting Communist Guerrilla Movements, 1941–1961*, New York: Free Press of Glencoe, 1962.

Ouellet, Eric, 'Multinational Counterinsurgency: The Western Intervention in the Boxer Rebellion 1900–1901', *Small Wars and Insurgencies*, Vol. 20, Nos. 3–4 (September–December 2009), pp. 507–27.

Paget, Julian, *Counterinsurgency Campaigning*, London: Faber & Faber, 1967.

Paget, Julian, *Last Post: Aden, 1964–67*, London: Faber & Faber, 1969.

Pain, Emil, 'The Second Chechen War: The Information Component', *Military Review*, Vol. LXXX, No. 4 (July–August 2000), pp. 59–69.

Pakenham, Thomas, *The Boer War*, London: Futura, 1982.

Palestine Police Force, *Combined Military and Police Action for Platoon Commanders and Junior Police Ranks*, Headquarters Palestine, June 1947.

Palmer, General Bruce, Jr, *The 25-Year War: America's Military Role in Vietnam*, New York: Touchstone, 1985.

Pantucci, Raffaello, 'Deep Impact: The Effect of Drone Attacks on British Counter-Terrorism', *The RUSI Journal*, Vol. 154, No. 5 (October 2009), pp. 72–6.

Papagos, Field Marshal Alexander, 'Guerrilla War' in Franklin Mark Osanka (ed.), *Modern Guerrilla Warfare: Fighting Communist Guerrilla Movements, 1941–1961*, New York: Free Press of Glencoe, 1962.

Paret, Peter, 'The French Army and La Guerre Révolutionnaire', *The RUSI Journal* (February 1959), pp.59–69.

Paret, Peter (ed.), *French Revolutionary Warfare from Indochina to Algeria: The Analysis of a Political and Military Doctrine*, London: Pall Mall Press for Princeton University, 1964.

Paret, Peter (ed.), *Makers of Modern Strategy: From Machiavelli to the Nuclear Age*, Oxford: Clarendon Press, 1986.

Park, Soul, 'The Unnecessary Uprising: Jeju Island Rebellion and South Korean Counterinsurgency Experience 1947–48', *Small Wars and Insurgencies*, Vol.21, No.2 (June 2010), pp.359–81.

Pate, J'Nell L., 'Ranald S. Mackenzie' in Paul Andrew Hutton (ed.), *Soldiers West: Biographies from the Military Frontier*, Lincoln, Nebraska and London: University of Nebraska Press, 1987.

Paxton, Robert O., *Parades and Politics at Vichy: The French Officer Corps under Marshal Pétain*, Princeton, New Jersey: Princeton University Press, 1966.

Payne, Kenneth, 'The Media at War: Ideology, Insurgency and Journalists in the Firing Line', *The RUSI Journal*, Vol.153, No.1 (February 2008), pp.16–21.

Pelletière, Stephen C., *Losing Iraq: Insurgency and Politics*, Westport, Connecticut and London: Praeger Security International, 2007.

Percox, David A., 'Internal Security and Decolonization in Kenya, 1956–63', *Journal of Imperial and Commonwealth History*, Vol.29, No.1 (2001), pp.92–116.

Percox, David A., 'British Counterinsurgency in Kenya, 1952–56: Extension of Internal Security Policy or prelude to Decolonisation?' in Ian Beckett (ed.), *Modern Counterinsurgency*, Aldershot, Hampshire and Burlington, Vermont: Ashgate, 2007.

Perkins, Ken, *A Fortunate Soldier*, London: Brassey's Defence Publishers, 1988.

Peterson, A.H., G.C. Reinhardt and E.E. Conger (eds), *Symposium on the Role of Airpower in Counterinsurgency and Unconventional Warfare: The Algerian War*, Santa Monica, California: RAND, July 1963.

Peterson, A.H., G.C. Reinhardt and E.E. Conger (eds), *Symposium on the Role of Airpower in Counterinsurgency and Unconventional Warfare: The Philippine Huk Campaign*, Santa Monica, California: RAND, July 1963.

Peterson, Michael, *The Combined Action Platoons*, New York: Frederick A. Praeger, 1989.

Petit, Lieutenant Colonel Brian, 'The Fight for the Village: Southern Afghanistan, 2010', *Military Review* (May–June 2011), pp.25–32.

Petraeus, General David H., 'Reflections on the Counterinsurgency Era', *The RUSI Journal*, Vol.158, No.4 (August/September 2013), pp.82–7.

Pimlott, John, 'The British Army: The Dhofar Campaign, 1970–1975' in Ian F.W. Beckett and John Pimlott (eds), *Armed Forces & Modern Counterinsurgency*, Beckenham, Kent: Croom Helm, 1985.

Pimlott, John, 'The French Army: From Indochina to Chad, 1946–1984' in Ian F.W. Beckett and John Pimlott (eds), *Armed Forces & Modern Counterinsurgency*, Beckenham, Kent: Croom Helm, 1985.

Pimlott, John, 'The British Experience' in Ian F.W. Beckett (ed.), *The Roots of Counterinsurgency: Armies and Guerrilla Warfare, 1900–1945*, London: Blandford Press, 1988.

Pirnie, Bruce R. and Edward O'Connell, *Counterinsurgency in Iraq (2003–2006)*, Santa Monica, California: RAND, 2008.

Pocock, Tom, *Fighting General: The Public and Private Campaigns of General Walter Walker*, London: Collins, 1973.

Pomeroy, William J. (ed.), *Guerrilla Warfare and Marxism: A Collection of Writings from Karl Marx to the present on Armed Struggles for Liberation and for Socialism*, London: Lawrence & Wishart, 1969.

Pomper, Major Stephen D., 'Don't Follow the Bear: The Soviet Attempt to Build Afghanistan's Military', *Military Review* (September–October 2005), pp.26–9.

Popplewell, Richard, '"Lacking Intelligence": Some Reflections on Recent Approaches in British Counterinsurgency, 1900–1960' in Ian Beckett (ed.), *Modern Counterinsurgency*, Aldershot, Hampshire and Burlington, Vermont: Ashgate, 2007.

Porch, Douglas, 'Bugeaud, Galliéni, Lyautey: The Development of French Colonial Warfare' in Peter Paret (ed.), *Makers of Modern Strategy: From Machiavelli to the Nuclear Age*, Oxford: Clarendon Press, 1986.

Porch, Douglas, *The Conquest of Morocco*, London: Jonathon Cape, 1986.

Porch, Douglas, *The Conquest of the Sahara*, Oxford: Oxford University Press, 1986.

Porch, Douglas, *Counterinsurgency: Exposing the Myths of the New Way of War*, Cambridge and New York: Cambridge University Press, 2013.

Porch, Douglas, 'Expendable soldiers', *Small Wars and Insurgencies*, Vol.25, No.3 (2014), pp.696–716.

Porch, Douglas, 'Reply to David Ucko', *Small Wars and Insurgencies*, Vol.25, No.1 (2014), pp.180–5.

Porch, Douglas, *The French Secret Service: From the Dreyfus Affair to the Gulf War*, London: Macmillan, 1996.

Porch, Douglas, *The March to the Marne: The French Army, 1871–1914*, Cambridge: Cambridge University Press, 2003.

Porch, Douglas, *The Portuguese Armed Forces and the Revolution*, London: Croom Helm, 1977.

Porch, Professor Douglas, 'French Imperial Warfare 1945–62' in Daniel Marston and Carter Malkasian (eds), *Counterinsurgency in Modern Warfare*, Botley, Oxford: Osprey Publishing, 2008.

Porter, Patrick, 'Why Britain Doesn't Do Grand Strategy', *The RUSI Journal*, Vol.155, No.4 (August/September 2010), pp.6–12.

Porter, Patrick, 'Goodbye to all that: On small wars and big choices', *Small Wars and Insurgencies*, Vol.25, No.3 (2014), pp.685–95.

Pottier, Philippe, 'GCMA/GMI: A French Experience in Counterinsurgency during the French Indochina War', *Small Wars and Insurgencies*, Vol.16, No.2 (June 2005), pp.125–46.

Pritchard, James and M.L.R. Smith, 'Thompson in Helmand: comparing theory to practice in British counterinsurgency operations in Afghanistan', *Civil Wars*, Vol.12, No.1–2 (March–June 2010), pp.65–90.

Pugsley, Christopher, *From Emergency to Confrontation: The New Zealand Forces in Malaya and Borneo, 1949–1966*, South Melbourne, Victoria: Oxford University Press, 2003.

Purdon, Corran, *List the Bugle: Reminiscences of an Irish Soldier*, Antrim, Northern Ireland: Greystone Books, 1993.

Pustay, Major John S., *Counterinsurgency Warfare*, New York: The Free Press, 1965.

Race, Jeffrey, *War Comes to Long An: Revolutionary Conflict in a Vietnamese Province*, Berkeley and Los Angeles, California and London: University of California Press, 1972.

Ramakrisna, Kumar, 'Content, Credibility and Context: Propaganda, Government Surrender Policy and the Malayan Communist Terrorist mass Surrenders of 1958' in Richard J. Aldrich, Gary D. Rawnsley and Ming-Yeh T. Rawnsley (eds), *The Clandestine Cold War in Asia, 1945–65: Western Intelligence, Propaganda and Special Operations*, London: Frank Cass, 2000.

Ramakrishna, Kumar, '"Transmogrifying" Malaya: The Impact of Sir Gerald Templer (1952–54)', *Journal of Southeast Asian Studies*, Vol.32, No.1 (February 2001), pp.79–92.

Ramsey, Robert D., III, *A Masterpiece of Counterguerrilla Warfare: BG J. Franklin Bell in the Philippines, 1901–1902*, Fort Leavenworth, Kansas: Combat Studies Institute Press, US Army Combined Arms Center, The Long War Series, Occasional Paper 25, 2007.

Ramsey, Robert D., III, *Advising Indigenous Forces: American Advisors in Korea, Vietnam, and El Salvador*, Fort Leavenworth, Kansas: Combat Studies Institute Press, Global War on Terrorism, Occasional Paper 18, 2006.

Ramsey, Robert D., III, *Savage Wars of Peace: Case Studies of Pacification in the Philippines, 1900–1902*, Fort Leavenworth, Kansas: Combat Studies Institute Press, The Long War Series, Occasional Paper 24, 2007.

Rashid, Ahmed, *Taliban: The Story of the Afghan Warlords*, London: Pan Books, 2001.

Ray, Captain James F., 'The District Advisor', *Military Review*, Vol.XLV, No.5 (May 1965), pp.3–8.

Reardon, Mark J., 'Chasing a Chameleon: The US Army Counterinsurgency Experience in Korea, 1945–1952' in Richard G. Davis (ed.), *The US Army and Irregular Warfare, 1775–2007*, Washington, DC: Center of Military History, United States Army, 2008.

Record, Jeffrey and W. Andrew Terrill, *Iraq and Vietnam: Differences, Similarities, and Insights*, Carlisle, Pennsylvania: Strategic Studies Institute, US Army War College, May 2004.

Reklaitis, George, 'Cold War Lithuania: National Armed Resistance and Soviet Counterinsurgency', *The Carl Beck Papers*, No.1806 (July 2007), University of Pittsburgh, Pennsylvania: Center for Russian and East European Studies, pp.1–43.

Renaud, Sean, 'A View from Chechnya: An Assessment of Russian Counterinsurgency during the two Chechen Wars and Future Implications', Palmerston North, New Zealand: Master of Arts in Defence and Strategic Studies, Massey University, 2010.

Rich, Paul B., 'A historical overview of US counterinsurgency', *Small Wars and Insurgencies*, Vol.25, No.1 (2014), pp.5–40.

Ricks, Thomas E., *Fiasco: The American Military Adventure in Iraq*, London: Allen Lane, 2006.

Ricks, Thomas E., *The Gamble: General Petraeus and the Untold Story of the American Surge in Iraq, 2005–2009*, London: Allen Lane, 2009.

Riddell, Peter, 'Armed Forces, Media and the Public: The Current "Crisis" in Historical and Political Perspective', *The RUSI Journal*, Vol.153, No.1 (February 2008), pp.12–15.

Rink, Martin, 'The German wars of liberation 1807–1815: The restrained insurgency', *Small Wars and Insurgencies*, Vol.25, No.4 (2014), pp.828–42.

Ritter, William S., 'The Final Phase in the Liquidation of Anti-Soviet Resistance in Tadzhikistan: Ibrahim Bek and the Basmachi, 1924–31', *Soviet Studies*, Vol.37, No.4 (October 1985).

Ritter, William S., 'Revolt in the Mountains: Fuzail Maksum and the Occupation of Garm, Spring 1929', *Journal of Contemporary History*, Vol.25, No.4 (October 1990).

Rittinger, Eric R., 'Exporting professionalism: US efforts to reform the armed forces in the Dominican Republic and Nicaragua, 1916–1933', *Small Wars and Insurgencies*, Vol.26, No.1 (2015), pp.136–57.

Robbins, Simon, 'The British Counterinsurgency in Cyprus', *Small Wars and Insurgencies*, Vol.23, Nos.4–5 (October–December 2012), pp.720–43.

Roberts, William R. and Jack Sweetman (eds), *New Interpretations in Naval History*, Annapolis, Maryland: Naval Institute Press, 1991.

Robinson, Adam, 'An Embarrassment of Riches? Britain's Lost Lessons from The Rhodesian Counterinsurgency War', London: MA Dissertation, Department of War Studies, King's College London, 2014.

Robinson, Linda, *Tell Me How This Ends: General David Petraeus and the Search for a Way Out of Iraq*, New York: Public Affairs, 2008.

Robinson, Paul, 'Soviet Hearts-and-Minds Operations in Afghanistan', *Historian*, Vol.72, No.1 (Spring 2010), pp.1–22.

Robinson, Lieutenant Colonel T.P., 'Illuminating a Black Art: Mentoring the Iraqi Army during Op TELIC 12', *British Army Review*, Vol.147 (Summer 2009), pp.35–8.

Roger, Nathan, *Image Warfare in the War on Terror*, Basingstoke, Hampshire and New York: Palgrave Macmillan, 2013.

Rose, Michael, 'Afghanistan: Some Recent Observations', *The RUSI Journal*, Vol.153, No.5 (October 2008), pp.8–13.

Rosello, Victor M., 'Lessons from El Salvador', *Parameters*, Vol.XXIII (Winter 1993–1994), pp.100–8.

Rosenau, William, *US Internal Security Assistance to South Vietnam: Insurgency, subversion and public order*, London and New York: Routledge, 2005.

Rosenau, William and Austin Long, *The Phoenix Program and Contemporary Counterinsurgency*, Santa Monica, California: RAND, 2009.

Rossolinski-Liebe, Grzegorz, 'The "Ukrainian National Revolution" of 1941', *Kritika: Explorations in Russian and Euroasian History*, Vol.12, No.1 (Winter 2011), pp.83–114.

Rubin, Barnett R., *The Fragmentation of Afghanistan*, New Haven, Connecticut: Yale University Press, 2002.

Russell, James A., *Innovation, Transformation and War: Counterinsurgency Operations in Anbar and Ninewa Provinces, Iraq, 2005–2007*, Stanford, California: Stanford University Press, 2011.

Russell, James A., 'Counterinsurgency American style: Considering David Petraeus and twenty-first century irregular war', *Small Wars and Insurgencies*, Vol.25, No.1 (2014), pp.69–90.

Ryan, Maria, '"Full spectrum dominance": Donald Rumsfeld, the Department of Defense, and US irregular warfare strategy, 2001–2008', *Small Wars and Insurgencies*, Vol.25, No.1 (2014), pp.41–68.

Ryan, Mike, *Battlefield Afghanistan*, Stroud, Gloucestershire: Spellmount, 2007.

Ryan, Mike, *Frontline Afghanistan: The Devil's Playground*, Stroud, Gloucestershire: Spellmount, 2010.

Sanderson, Captain Jeffrey R. and Captain Scott J. Akerley, 'Focusing Training: The Big Five for leaders', *Military Review* (July–August 2007), pp.71–8.

Sarkesian, Sam C., 'The American Response to Low-Intensity Conflict: The Formative Period', in David Charters and Maurice Tugwell (eds), *Armies in Low-Intensity Conflict: A Comparative Analysis*, London: Brassey's Defence, 1989.

Sarkesian, Sam C., *Unconventional Conflicts in a New Security Era: Lessons from Malaya and Vietnam*, Westport, Connecticut: Greenwood Press, 1993.

Sarkin, Jeremy, *Germany's Genocide of the Herero: Kaiser Wilhelm II, his General, his Settlers, his Soldiers*, Woodbridge, Suffolk: James Currey, 2011.

Sawyer, Major Robert K., *Military Advisors in Korea: KMAG in Peace and War*, Washington, DC: Office of the Chief of Military History, Department of the Army, 1962.

Schaefer, Robert W., *The Insurgency in Chechnya and the North Caucasus: From Gazavat to Jihad*, Santa Barbara, California: Praeger, 2011.

Scheipers, Sibylle, 'Counterinsurgency or irregular warfare? Historiography and the study of "small wars"', *Small Wars and Insurgencies*, Vol.25, Nos 5–6 (2014), pp.879–99.

Schmidt, Heike, '"Deadly Silence Predominates in the District": The Maji Maji War and its Aftermath in Ungoni' in James Giblin and Jamie Monson (eds), *Maji Maji: Lifting the Fog of War*, Leiden, The Netherlands: Brill, 2010.

Schumacher, Frank, '"Marked Severities": The Debate over Torture during America's Conquest of the Philippines, 1899–1902', *Amerikastudien / American Studies*, Vol.51, No.4 (2006), pp.475–98.

Schwartz, Dr T.P., 'The Combined Action Program: A Different Perspective', *Marine Corps Gazette*, Vol.83, No.2 (February 1999), pp.63–72.

Seagrist, Lieutenant Colonel Thomas A., 'Combat Advising in Iraq', *Military Review* (May–June 2010), pp.65–72.

Seegers, Annette, *The Military in the Making of Modern South Africa*, London and New York: I.B. Tauris, 1996.

Seely, Sergeant Bob, 'Winning Through the People: Dog Bark Patrolling, Illegal Vehicle Check Points – Why the Army needs to focus on conducting influence campaigns', *British Army Review*, No.152 (Autumn 2011), pp.17–26.

Seib, Philip, 'The Al-Qaeda Media Machine', *Military Review* (May–June 2008), pp.74–80.

Seibert, Bjoern, 'A Quiet Revolution: The Reform of the German Armed Forces', *The RUSI Journal*, Vol.157, No.1 (February/March 2012), pp.60–9.

Selth, Andrew, 'Ireland and Insurgency: The Lessons of History', *Small Wars and Insurgencies*, Vol.2, No.2 (August 1991), pp.299–322.

Sepp, Kalev L., 'Best Practices in Counterinsurgency', *Military Review* (May–June 2005), pp.8–12.

Shaw, Geoffrey D.T., 'Policemen versus Soldiers: The Debate Leading to MAAG Objections and Washington Rejections of the Core of British Counterinsurgency Advice', *Small Wars and Insurgencies*, Vol.12, No.2 (Summer 2001), pp.51–78.

Sheehan, William, *A Hard Local War: The British Army and the Guerrilla War in Cork, 1919–1921*, Stroud, Gloucestershire: The History Press, 2011.

Shepherd, Ben, *War in the Wild East: The German Army and Soviet Partisans*, Cambridge, Massachusetts: Harvard University Press, 2004.

Shepherd, Ben, 'With the Devil in Titoland: A Wehrmacht Anti-Partisan Division in Bosnia–Herzegovina, 1943', *War in History*, Vol.16, No.1, January 2009.

Short, Anthony, *The Communist Insurrection in Malaya, 1948–1960*, London: Frederick Muller, 1975.

Short, Anthony, 'The Malayan Emergency' in Ronald Haycock (ed.), *Regular Armies and Insurgency*, London: Croom Helm, 1979.

Shrader, Charles R., *The Withered Vine: Logistics and the Communist Insurgency in Greece, 1945–1949*, Westport, Connecticut: Praeger, 1999.

Shuffer, Lieutenant Colonel George M., Jr, 'Finish Them With Firepower', *Military Review*, Vol.XLVII, No.12 (December 1967), pp.11–15.

Shy, John and Thomas W. Collier, 'Revolutionary War' in Peter Paret (ed.), *Makers of Modern Strategy: From Machiavelli to the Nuclear Age*, Oxford: Clarendon Press, 1986.

Siegel, Daniel and Joy Hackel, 'El Salvador: Counterinsurgency Revisited' in Michael T. Klare and Peter Kornbluh (eds), *Low-Intensity Warfare: Counterinsurgency, Proinsurgency, and Antiterrorism in the Eighties*, New York: Pantheon Books, 1988.

Simmons, Clyde R., 'The Indian Wars and US Military Thought, 1865–1890', *Parameters*, Vol.XXII, No.1 (Spring 1992), pp.60–72.

Simpson, Captain Emile, 'Gaining the Influence Initiative: Why Kinetic Operations are Central to Influence in Southern Afghanistan', *British Army Review*, No.147 (Summer 2009), pp.67–71.

Simson, H.J., *British Rule, and Rebellion*, London: William Blackwood, 1937.

Sinclair, Georgina, *At the End of the Line: Colonial Policing and the Imperial Endgame, 1945–80*, Manchester and New York: Manchester University Press, 2006.

Sinclair, Georgina (ed.), *Globalising British Policing*, Farnham, Surrey: Ashgate, 2011.

Singer, Barnett and John Langdon, *Cultured Force: Makers and Defenders of the French Colonial Empire*, Madison, Wisconsin: University of Wisconsin Press, 2004.

Sixsmith, Major General E.K.G., 'Kitchener and the Guerrillas in the Boer War', *Army Quarterly*, Vol.104, No.2 (January 1974), pp.203–14.

Skuta, Lieutenant Colonel Philip C., 'Introduction to 2/7 CAP Platoon Actions in Iraq: The CAP philosophy show signs of success in Iraq', *Marine Corps Gazette*, Vol.89, No.4 (April 2005), p.35.

Sky, Emma, 'The Lead Nation Approach: The Case of Afghanistan', *The RUSI Journal*, Vol.151, No.6 (December 2006), pp.22–6.

Sluka, Jeffrey A., 'Death from Above: UAVs and Losing Hearts and Minds', *Military Review* (May–June 2011), pp.70–6.

Smith, Charles, 'Communal Conflict and Insurrection in Palestine, 1936–48' in David Anderson and David Killingray (eds), *Policing and Decolonisation: Nationalism, Politics and the Police, 1917–65*, Manchester and New York: Manchester University Press, 1992.

Smith, Iain R. and Andreas Stucki, 'The Colonial Development of Concentration Camps (1868–1902)', *Journal of Imperial and Commonwealth History*, Vol.39, No.3 (September 2011), pp.417–37.

Smith, John (ed.), *Administering Empire: The British Colonial Service in Retrospect*, London: University of London Press, 1999.

Smith, M.L.R. and Peter R. Neumann, 'Motorman's Long Journey: Changing the Strategic Setting in Northern Ireland', *Contemporary British History*, Vol.19, No.4 (2005), pp.413–35.

Smith, Lieutenant Colonel Richard W., 'Philippine Constabulary', *Military Review*, (May 1968), pp.73–80.

Smith, Major Niel and MacFarland, Colonel Sean, 'Anbar Awakens: The Tipping Point', *Military Review* (March-April 2008), pp.41–52.

Smith, Robert Ross, 'The Hukbalahap Insurgency', *Military Review*, Vol.XLV, No.6 (June 1965), pp.35–42.

Smith, General Sir Rupert, *The Utility of Force – The Art of War in the Modern World*, London: Allen Lane, 2005.

Smith, Simon C., 'General Templer and Counter-Insurgency in Malaya: Hearts and Minds, Intelligence, and Propaganda', *Intelligence and National Security*, Vol.16, No.3 (Autumn 2001), pp.60–78.

Smith, W. Wayne, 'An Experiment in Counterinsurgency: The Assessment of Confederate Sympathisers in Missouri', *Journal of Southern History*, Vol.35, No.3 (August 1969), pp.361–80.

Smith, Woodruff D., *The German Colonial Empire*, Chapel Hill, North Carolina: University of North Carolina Press, 1978.

Snyder, Timothy, 'The Causes of Ukrainian-Polish Ethnic Cleansing 1943', *Past & Present*, No.179 (May 2013), pp.197–234.

Sorley, Lewis, *Thunderbolt: Biography of General Creighton Abrams and the Army of His Times*, New York: Simon & Schuster, 1992.

Sorley, Lewis, 'To Change a War: General Harold K. Johnson and the PROVN Study', *Parameters*, Vol.XXVIII (Spring 1998), pp.93–109.

Sorley, Lewis, *Honorable Warrior: General Harold K. Johnson and the Ethics of Command*, Lawrence, Kansas: University Press of Kansas, 1998.

Sorley, Lewis, *A Better War: The Unexamined Victories and Final Tragedy of America's Last Years in Vietnam*, San Diego, New York and London: Harcourt, 2000.

Sorley, Lewis, *Westmoreland: The General Who Lost Vietnam*, Boston and New York: Houghton Mifflin Harcourt, 2011.

Souter, David, 'An Island Apart: A Review of the Cyprus Problem', *Third World Quarterly*, Vol.6, No.3 (July 1984), pp.657–74.

Southby-Tailyour, Ewen, *3 Commando Brigade: Helmand Assault*, London: Ebury Press, 2010.

Souyris, Captain André, 'An Effective Counterguerrilla Procedure', *Military Review*, Vol.XXXVI, No.12 (March 1957), pp.86–90.

Spector, Ronald H., 'The First Vietnamization: US Advisors in Vietnam, 1956–1960' in Joe C. Dixon (ed.), *The American Military and the Far East: Proceedings of the Ninth Military History Symposium, United States Air Force Academy, 1 October 1980*, Washington, DC: US Government Printing Office, 1980.

Spector, Ronald H., *United States Army in Vietnam: Advice and Support: The Early Years, 1941–1960*, Washington, DC: US Army Center of Military History, US Government Printing Office, 1985.

Spies, S.B., *Methods of Barbarism?: Roberts and Kitchener and Civilians in the Boer Republics, January 1900–May 1902*, Cape Town and Pretoria: Human & Rousseau, 1977.

Spiller, Roger J., 'In the Shadow of the Dragon: Doctrine and the US Army after Vietnam', *The RUSI Journal*, Vol.142, No.6 (December 1997), pp.41–54.

Stanton, Major Paul T., 'Unit Immersion in Mosul: Establishing Stability in Transition', *Military Review* (July–August 2006), pp.60–70.

Starkey, Armstrong, *European and Native American Warfare, 1675–1815*, Norman, Oklahoma: University of Oklahoma, 1998.

Statiev, Alexander, 'The Nature of Anti-Soviet Armed Resistance, 1942–44', *Kritika: Explorations in Russian and Eurasian History*, Vol.6, No.2 (Spring 2005), pp.281–314.

Statiev, Dr Alexander, 'Motivations and Goals of Soviet Deportations in the Western Borderlands', *The Journal of Strategic Studies*, Vol.28, No.6 (December 2005), pp.977–1003.

Statiev, Alexander, *The Soviet Counterinsurgency in the Western Borderlands*, Cambridge: Cambridge University Press, 2010.

Stewart, Brian, 'Winning in Malaya: An Intelligence Success Story' in Richard J. Aldrich, Gary D. Rawnsley and Ming-Yeh T. Rawnsley (eds), *The Clandestine Cold War in Asia, 1945–65: Western Intelligence, Propaganda and Special Operations*, London: Frank Cass, 2000.

Stevens, Michael, 'Community Defence in Afghanistan: A Model for Future Stabilisation?', *The RUSI Journal*, Vol.156, No.3 (June/July 2011), pp.42–7.

Stevens, Michael, 'Afghan Local Police in Helmand: Calculated Risk or Last Gamble', *The RUSI Journal*, Vol.158, No.1 (February/March 2013), pp.64–70.

Stewart, Rory, *Occupational Hazards: My Time Governing in Iraq*, London: Picador, 2006.

Stockwell, A.J., 'Policing during the Malayan Emergency, 1948–60: communism, communalism and decolonisation' in David Anderson and David Killingray (eds), *Policing and Decolonisation: Nationalism, Politics and the Police, 1917–65*, Manchester and New York: Manchester University Press, 1992.

Stockwell, A.J., 'Insurgency and Decolonisation during the Malayan Emergency' in Ian Beckett (ed.), *Modern Counterinsurgency*, Aldershot, Hampshire and Burlington, Vermont: Ashgate, 2007.

Storrie, Brigadier Sandy, '"First Do No Harm" – 7 Armoured Brigade in Southern Iraq', *British Army Review*, Vol.147 (Summer 2009), pp.29–34.

Strachan, Hew (ed.), *Big Wars and Small Wars: The British Army and the Lessons of War in the 20th Century*, Abingdon, Oxon and New York: Routledge, 2006.

Strachan, Hew, 'British counterinsurgency from Malaya to Iraq', *The RUSI Journal*, Vol.152, No.6 (December 2007), pp.8–11.

Strachan, Neil, 'Abu Naji returns to Al-Kut', *British Army Review*, No.137 (Summer 2005), pp.109–17.

Strandquist, Jon, 'Local defence forces and counterinsurgency in Afghanistan: Learning from the CIA's Village Defense Program in South Vietnam', *Small Wars and Insurgencies*, Vol.26, No.1 (2015), pp.90–113.

Style, Charles, 'Britain's Afghanistan Deployment in 2006', *The RUSI Journal*, Vol.157, No.2 (April/May 2012), pp.40–2.

Sullivan, Antony Thrall, *Thomas-Robert Bugeaud, France and Algeria, 1784–1849: Politics, Power and the Good Society*, Hamden, Connecticut: Archon Books, 1983.

Sun Tzu (translation and introduction by Samuel B. Griffith), *The Art of War*, London, Oxford and New York: Oxford University Press, 1979.

Surridge, Keith Terrance, *Managing the South African War, 1899–1902*, Woodbridge, Suffolk: Boydell Press for The Royal Historical Society, 1998.

Sutherland, Daniel E. (ed.), *Guerrillas, Unionists and Violence on the Confederate Home Front*, Fayetteville, Arkansas: University of Arkansas Press, 1999.

Sutherland, Daniel E., *A Savage Conflict: The Decisive Role of Guerrillas in the American Civil War*, Chapel Hill, North Carolina: University of North Carolina Press, 2009.

Synnott, Hilary, 'Statebuilding in Southern Iraq', *Survival*, Vol.47, No.2 (2005), pp.33–56.

Synnott, Hilary, *Bad Days in Basra: My Turbulent Times as Britain's Man in Southern Iraq*, London and New York: I.B. Tauris, 2008.

Taber, Robert, *The War of the Flea: A Study of Guerrilla Warfare, Theory and Practice*, London: Paladin, 1970.

Terriff, Terry, 'Of Romans and Dragons: Preparing the US Marine Corps for Future Warfare' in Tim Benbow and Rod Thornton (eds), *Dimensions of Counterinsurgency: Applying Experience to Practice*, London and New York: Routledge, 2008.

Thambipillay, R. (ed.), *The Malayan Police Force in the Emergency, 1948–1960*, Ipoh, Perak, Malaysia: Goodturn Hoover Publishing, 2003.

The Sunday Times Insight Team, *Insight on Portugal: The Year of the Captains*, London: Andre Deutsch, 1975.

Thomas, Martin, 'Order before reform: the spread of French military operations in Algeria, 1954–1958' in David Killingray and David Omissi (eds), *Guardians of Empire: The Armed Forces of the Colonial Powers, c.1700–1964*, Manchester and New York: Manchester University Press, 1999.

Thomas, Martin, *The French Empire between the Wars: Imperialism, Politics and Society*, Manchester: Manchester University Press, 2005.

Thomas, Martin, *Empires of Intelligence: Security Services and Colonial Disorder after 1914*, Berkeley, California: University of California Press, 2008.

Thomas, Martin, *Fight or Flight: Britain, France, and their Roads from Empire*, Oxford: Oxford University Press, 2014.

Thomas, Lieutenant Colonel Timothy L., 'Grozny 2000: Urban Combat Lessons Learned', *Military Review*, Vol.LXXX, No.4 (July–August 2000), pp.50–8.

Thompson, Major Michael A., USAF, *Lessons in Counterinsurgency: The French Campaign in Algeria*, Maxwell Air Force Base, Alabama: Air University, Air Command and Staff College, 2008.

Thompson, Sir Robert, *Defeating Communist Insurgency: The Lessons of Malaya and Vietnam*, London: Chatto & Windus, 1966.

Thompson, Sir Robert, *Revolutionary War in World Strategy, 1945–1969*, London: Secker & Warburg, 1970.

Thompson, Sir Robert G.K., 'Regular Armies and Insurgency' in Ronald Haycock (ed.), *Regular Armies and Insurgency*, London: Croom Helm, 1979.

Thompson, Sir Robert, *Make for the Hills: Memories of Far Eastern Wars*, London: Leo Cooper, 1989.

Thompson, W. Scott and Donaldson D. Frizzell (eds), *The Lessons of Vietnam*, London: Macdonald and Jane's, 1977.

Thornton, Rod, 'The British Army and the Origins of its Minimum Force Philosophy', *Small Wars and Insurgencies*, Vol.14, No.2 (2004), pp.83–106.

Thornton, Rod, 'Getting it Wrong: The Crucial Mistakes Made in the Early Stages of the British Army's Deployment in Northern Ireland (August 1969–March 1972)', *The Journal of Strategic Studies*, Vol.30, No.1 (February 2007), pp.73–107.

Thornton, Rod, 'Countering Arab Insurgencies: The British Experience' in Tim Benbow and Rod Thornton (eds), *Dimensions of Counterinsurgency: Applying Experience to Practice*, London and New York: Routledge, 2008.

Thornton, Rod, '"Minimum Force": A reply to Huw Bennett', *Small Wars and Insurgencies*, Vol.20, No.1 (2009), pp.215–26.

Toase, Francis, 'The French Experience', in Ian F.W. Beckett (ed.), *The Roots of Counterinsurgency: Armies and Guerrilla Warfare, 1900–1945*, London: Blandford Press, 1988.

Tone, John Lawrence, *The Fatal Knot: The Guerrilla War in Navarre and the Defeat of Napoleon in Spain*, Chapel, North Carolina: University of North Carolina Press, 1994.

Tonsetic, Robert L., *Forsaken Warriors: The Story of an American Advisor who Fought with the South Vietnamese Rangers and Airborne, 1970–71*, Philadelphia and Newbury, Berkshire: Casemate, 2009.

Toolis, Kevin, *Rebel Hearts: Journeys within the IRA's Soul*, London: Picador, 1995.

Townshend, Charles, *The British Campaign in Ireland, 1919–21: Development of Political and Military Policies*, London: Oxford University Press, 1975.

Townshend, Charles, 'The Irish Insurgency, 1918–21: The Military Problem' in Ronald Haycock (ed.), *Regular Armies and Insurgency*, London: Croom Helm, 1979.

Townshend, Charles, 'The Irish Republican Army and the Development of Guerrilla Warfare, 1916–1921', *The English Historical Review*, Vol.94, No.371 (April 1979), pp.318–45.

Townshend, Charles, *Britain's Civil Wars: Counterinsurgency in the Twentieth Century*, London: Faber & Faber, 1986.

Townshend, Charles, 'The Defence of Palestine: Insurrection and Public Security, 1936–1939', *The English Historical Review*, Vol.103, No.409 (October 1988), pp.917–49.

Townshend, Charles, 'Policing Insurgency in Ireland, 1914–23' in David Anderson and David Killingray (eds), *Policing and Decolonisation: Nationalism, Politics and the Police, 1917–65*, Manchester and New York: Manchester University Press, 1992.

Trinquier, Roger, *Modern Warfare: A French View of Counterinsurgency*, London and Dunmow: Pall Mall Press, 1964.

Trotha, Trutz von, '"The Fellows Can Just Starve": On Wars of "Pacification" in the African Colonies of Imperial Germany and the Concept of "Total War"' in Manfred F. Boemeke, Roger Chickering and Stig Föster (eds), *Anticipating Total War:*

The German and American Experiences, 1871–1914, Cambridge: Cambridge University Press, 1999.

Troup, David, 'Crime, politics and the policing in colonial Kenya, 1939–63' in David Anderson and David Killingray (eds), *Policing and Decolonisation: Nationalism, Politics and the Police, 1917–65*, Manchester and New York: Manchester University Press, 1992.

Truong, Ngo Quang, *RVNAF and US Operational Co-operation and Co-ordination*, Washington, DC: US Army Center of Military History, 1980.

Tuck, Christopher, 'Borneo 1963–66: Counterinsurgency Operations and War Termination', *Small Wars and Insurgencies*, Vol.15, No.3 (Winter 2004), pp.89–111.

Tuck, Christopher, 'Northern Ireland and the British Approach to Counterinsurgency', *Defence and Security Analysis*, Vol.23, No.2 (June 2007), pp.165–83.

Turnbull, Professor Mary, 'The Malayan Civil Service and the Transition to Independence' in John Smith (ed.), *Administering Empire: The British Colonial Service in Retrospect*, London: University of London Press, 1999.

Ucko, David H., *The New Counterinsurgency Era: Transforming the U.S. Military for Modern Wars*, Washington, DC: Georgetown University Press, 2009.

Ucko, David H., 'Critics gone wild: Counterinsurgency as the root of all evil', *Small Wars and Insurgencies*, Vol.25, No.1 (2014), pp.161–79.

Ucko, David H., 'Lessons from Basra: The Future of British Counter-insurgency', *Survival*, Vol.52, No.4 (August–September 2010), pp.131–57.

Ucko, David H. and Robert Egnell, *Counterinsurgency in Crisis: Britain and the Challenges of Modern Warfare*, New York: Columbia University Press, 2013.

Union of South Africa, *Report on the Natives of South-West Africa and Their Treatment by Germany*, London: His Majesty's Stationery Office, 1918.

United States Field Manual No. 3–24, *Counterinsurgency*, Washington, DC: Department of the Army and United States Marine Corps, December 2006.

United States Government, *U.S. Government Counterinsurgency Guide*, Washington DC: Bureau of Political-Military Affairs, January 2009.

United States Joint Publication 3–24, *Counterinsurgency Operations*, Washington DC: Joint Chiefs of Staff, Departments of Army, Navy and Air Force, October 2009.

United States Marine Corps, *Small Wars Manual*, Honolulu, Hawaii: University Press of the Pacific (Reprint from 1940), 2005.

Urban, Mark, *War In Afghanistan*, London: Macmillan, 1990.

Urban, Mark, *Task Force Black: The Explosive True Story of the SAS and the Secret War in Iraq*, London: Little, Brown, 2010.

Utley, Robert M., *Frontier Regulars: The United States Army and the Indian, 1866–1890*, New York: Macmillan, 1973.

Utley, Robert M., *The Indian Frontier of the American West 1846–1890*, Albuquerque, New Mexico: University of New Mexico Press, 1984.

Utley, Robert M., 'Introduction: The Frontier and the American Military Tradition' in Paul Andrew Hutton (ed.), *Soldiers West: Biographies from the Military Frontier*, Lincoln, Nebraska and London: University of Nebraska Press, 1987.

Utley, Robert M., 'Nelson A. Miles' in Paul Andrew Hutton (ed.), *Soldiers West: Biographies from the Military Frontier*, Lincoln, Nebraska and London: University of Nebraska Press, 1987.

Utley, Robert M., 'Total War on the American Indian Frontier' in Manfred F. Boemeke, Roger Chickering and Stig Föster (eds), *Anticipating Total War: The German and American Experiences, 1871–1914*, Cambridge: Cambridge University Press, 1999.

Utley, Robert M. and Wilcomb E. Washburn, *Indian Wars*, Boston and New York: Mariner Books, 2002.

Utting, Kate, 'The Strategic Information Campaign: Lessons from the British Experience in Palestine 1945–1948' in Tim Benbow and Rod Thornton (eds), *Dimensions of Counterinsurgency: Applying Experience to Practice*, London and New York: Routledge, 2008.

Valeriano, Colonel Napoleon D. and Lieutenant Colonel Charles T.R. Bohannan, *Counter-Guerrilla Operations: The Philippine Experience*, London: Pall Mall Press, 1962.

Van der Bijl, Nick, *Operation BANNER: The British Army in Northern Ireland 1969–2007*, Barnsley, South Yorkshire: Pen & Sword Military, 2009.

Van der Bijl, Nick, *The Cyprus Emergency: The Divided Island 1955–1974*, Barnsley, South Yorkshire: Pen & Sword Military, 2010.

Van der Kroef, Justus M., 'Communism and the Guerrilla War in Sarawak', *World Today*, Vol.20, No.2 (February 1964), pp.50–60.

Van der Kroef, Justus M., 'The Sarawak-Indonesian Border Insurgency', *Modern Asian Studies*, Vol.2, No.3 (1968), pp.245–65.

Van der Lijn, Jaïr, 'Comprehensive approaches, diverse coherences: the different levels of policy coherence in the Dutch 3D approach in Afghanistan', *Small Wars and Insurgencies*, Vol.26, No.1 (2015), pp.72–89.

Van der Waals, W.S., *Portugal's War in Angola, 1961–1974*, Rivonia, South Africa: Ashanti Publishing, 1993.

Van Dyke, Carl, 'Kabul to Grozny: A Critique of Soviet (Russian) Counterinsurgency Doctrine' in Ian Beckett (ed.), *Modern Counterinsurgency*, Aldershot, Hampshire and Burlington, Vermont: Ashgate, 2007.

Vandervort, Bruce, *Wars of Imperial Conquest in Africa, 1830–1914*, London: University College London Press, 1998.

Vandervort, Bruce, *Indian Wars of Mexico, Canada and the United States, 1912–1900*, New York: Routledge, 2006.

Venter, A.J., *The Terror Fighters: A Profile of Guerrilla Warfare in Southern Africa*, Cape Town, South Africa: Purnell, 1969.

Venter, A.J., *Portugal's War in Guinéa–Bissau*, Pasadena: California: California Institute of Technology, 1973.

Venter, A.J., *Portugal's Guerrilla War; The Campaign for Africa*, Cape Town, South Africa: John Malherbe, 1973.

Venter, A.J., *The Zambesi Salient: Conflict in Southern Africa*, Old Greenwich, Connecticut: Devlin-Adair, 1974.

Venter, A.J., 'Why Portugal Lost its African Wars' in A.J. Venter (ed.), *Challenge: Southern Africa within the African Revolutionary Context*, Rivonia, South Africa: Ashanti Publishing, 1989.

Venter, A.J. (ed.), *Challenge: Southern Africa within the African Revolutionary Context*, Rivonia, South Africa: Ashanti Publishing, 1989.

Venter, A.J., *Portugal's Guerrilla Wars in Africa: Lisbon's Three Wars in Angola, Mozambique and Portuguese Guinea, 1961–74*, Solihull, West Midlands: Helion and Company, 2013.

Waghelstein, John D., 'Counterinsurgency Doctrine and Low-Intensity Conflict in the Post-Vietnam Era' in Lawrence E. Grinter and Peter M. Dunn (eds), *The American War in Vietnam: Lessons, Legacies, and Implications for Future Conflict*, New York, Westport, Connecticut and London: Greenwood Press, 1987.

Waghelstein, John D., 'Post Vietnam Counterinsurgency Doctrine', *Military Review*, Vol.LXV, No.5 (May 1985), pp.42–9.

Waghelstein, John D., 'Preparing the US Army for the Wrong War, Educational and Doctrinal Failure 1865–91', *Small Wars and Insurgencies*, Vol.10, No.1 (Spring 1999), pp.1–33.

Waghelstein, John D, 'Regulars, Irregulars and Militia: The American Revolution', *Small Wars and Insurgencies*, Vol.6, No.2 (Autumn 1995), pp.133–58.

Waghelstein, John D, 'The Mexican War and the American Civil War: The American Army's Experience in Irregular as a Sub-set of a Major Conventional Conflict', *Small Wars and Insurgencies*, Vol.7, No.2 (Autumn 1996), pp.139–64.

Wagner, Lieutenant Colonel David H., 'A Handful of Marines', *Marine Corps Gazette*, Vol.52, No.3 (March 1968), pp.44–6.

Wagner, Steven, 'British Intelligence and the Jewish Resistance Movement in the Palestine Mandate, 1945–46', *Intelligence and National Security*, Vol.23, No.5 (October 2008), pp.629–57.

Wainhouse, Lieutenant Colonel Edward R., 'Guerrilla War in Greece, 1946–49: A Case Study', *Military Review*, Vol.XXXVII, No.3 (June 1957), pp.17–25.

Wainhouse, Colonel Edward R., 'Guerrilla War in Greece, 1946–49: A Case Study' in Franklin Mark Osanka (ed.), *Modern Guerrilla Warfare: Fighting Communist Guerrilla Movements, 1941–1961*, New York: Free Press of Glencoe, 1962.

Walker, Jonathon, *Aden Insurgency: The Savage War in Yemen, 1962–67*, Barnsley, South Yorkshire: Pen & Sword Military, 2011.

Walker, General Sir Walter, *Fighting On*, London: New Millennium, 1997.

Walt, General Lewis W., *Strange War, Strange Strategy: A General's Report on Vietnam*, New York: Funk & Wagnalls, 1970.

Walton, Calder, *Empire of Secrets: British Intelligence, the Cold War and the Twilight of Empire*, London: HarperPress, 2013.

War Office, *Imperial Policing and Duties in Aid of the Civil Power*, London: June 1949.

Ware, Lewis B. (ed.), *Low-Intensity Conflict in the Third World*, Maxwell Air Force Base, Alabama: Air University Press, 1988.

Watt, Robert N., 'Raiders of a Lost Art? Apache War and Society', *Small Wars and Insurgencies*, Vol.13, No.3 (Autumn 2002), pp.1–28.

Wawro, Geoffrey, *The Franco-Prussian War: The German Conquest of France in 1870–1871*, Cambridge: Cambridge University Press, 2003.

Weatherill, Phil, 'Notes from the Field: Targeting the Centre of Gravity: Adapting Stabilisation in Sangin', *The RUSI Journal*, Vol.156, No.4 (August/September 2011), pp.90–8.

Weigley, Russell F., *History of the United States Army*, New York: Macmillan, 1967.

Weigley, Russell F., 'The Elihu Root Reforms and the Progressive Era' in Lieutenant Colonel William Geffen (ed.), *Command and Commanders in Modern Warfare: Proceedings of the Second Military History Symposium, US Air Force Academy, 2–3 May 1968*, USAF Academy, Colorado: December 1969.

Weller, Jac, 'British Weapons and Tactics in Malaysia', *Military Review*, (November 1966), pp.17–24.

Wenham, Lieutenant Colonel M.H., 'Information Operations – Main effort or Supporting Effort?', *British Army Review*, Vol.146 (Spring 2009), pp.46–52.

West, Bing, 'Counterinsurgency Lessons from Iraq', *Military Review* (March–April 2009), pp.2–12.

Westermann, Edward B., 'Partners in Genocide: The German Police and the Wehrmacht in the Soviet Union', *The Journal of Strategic Studies*, Vol.31, No.5 (October 2008), pp.771–96.

Weyand, Major General Frederick C., 'Winning the People in Hau Nghia Province', *Army*, Vol.17, No.1 (January 1967), pp.52–5.

Whalen, Major Autum C., USAF, *Intelligence Successes and Failures of the Rif Rebellion*, Maxwell Air Force Base, Alabama: Air University, Air Command and Staff College, 2006.

Wheeler, Douglas L., 'The Portuguese Army in Angola', *Journal of Modern African Studies*, Vol.7, No.3 (October 1969), pp.425–39.

Wheeler, Douglas L., 'The Military and the Portuguese Dictatorship, 1926–1974: "The Honor of the Army"' in Lawrence S. Graham, and Harry M. Makler, *Contemporary Portugal: The Revolution and Its Antecedents*, Austin, Texas: University of Texas Press, 1979.

Williams, Brian Glyn, 'Afghanistan after the Soviets: From jihad to tribalism', *Small Wars and Insurgencies*, Vol.25, Nos 5–6 (2014), pp.924–56.

Williamson, Captain R.E., 'A Briefing for Combined Action', *Marine Corps Gazette*, Vol.52, No.3 (March 1968), pp.41–3.

Willis, John, 'Colonial Police in Aden, 1937–1967' in Georgina Sinclair (ed.), *Globalising British Policing*, Farnham, Surrey: Ashgate, 2011.

Wilson, David and Gareth E. Conway, 'The Tactical Conflict Assessment Framework: A Short-lived Panacea', *The RUSI Journal*, Vol.154, No.1 (February 2009), pp.10–15.

Wilson, Dick, *The Long March, 1935: The Epic of Chinese Communism's Survival*, Harmondsworth, Middlesex: Penguin Books, 1977.

Wilson, Major R.D., *Cordon and Search: With the 6th Airborne Division in Palestine*, Aldershot, Surrey: Gale & Polden, 1949.

Windrow, Martin, *The Last Valley: Dien Bien Phu and the French Defeat in Vietnam*, London: Weidenfeld & Nicolson, 2004.

Windrow, Martin, *Our Friends Beneath the Sands: The Foreign Legion in France's Colonial Conquests, 1870–1935*, London: Phoenix, 2011.

Winslow, Donna, René Moelker and Françoise Companjen, 'Glocal Chechnya from Russian sovereignty to pan-Islam autonomy', *Small Wars and Insurgencies*, Vol.24, No.1 (2013), pp.129–51.

Wither, James K., 'Basra's not Belfast: The British Army, "Small Wars" and Iraq', *Small Wars and Insurgencies*, Vol.20, Nos 3–4 (September–December 2009), pp.611–35.

Witty, D.M., 'A Regular Army in Counterinsurgency Operations: Egypt in North Yemen, 1962–1967', *Journal of Military History*, Vol. 65, No.2 (April 2002), pp.401–39.

Wolf, Charles, Jr, *Insurgency and Counterinsurgency: New Myths and Old Realities*, Santa Monica, California: RAND, 1965.

Wolf, Charles, Jr, *Controlling Small Wars*, Santa Monica, California: RAND, 1968.

Woodhouse, C.M., *The Struggle for Greece, 1941–1949*, London: Hart-Davis, MacGibbon, 1976.

Woods, Jeffrey, 'Counterinsurgency, the Interagency Process, and Vietnam: The American Experience' in Kendall D. Gott and Michael G. Brooks (eds), *The US Army and the Interagency Process: Historical Perspectives: The Proceedings of the Combat Institute 2008 Military History Symposium*, Fort Leavenworth, Kansas: Combat Studies Institute Press, 2008.

Woodward, Bob, *The War Within: A Secret White House History, 2006–2008*, London, New York, Sydney and Toronto: Simon & Schuster, 2008.

Wooster, Robert, *Nelson A. Miles and the Twilight of the Frontier Army*, Lincoln, Nebraska and London: Bison Books, University of Nebraska Press, 1996.

Xiaobing Li, *A History of the Modern Chinese Army*, Lexington, Kentucky: The University Press of Kentucky, 2007.

Xydis, Stephen G., 'The UN General Assembly as an Instrument of Greek Policy: Cyprus, 1954–58', *The Journal of Conflict Resolution*, Vol.12, No.2 (June 1968), pp.141–58.

Yarborough, William P., 'Unconventional Warfare: One Military View', *Annals of the American Academy of Political and Social Science*, Vol.341 (May 1962), pp.1–7.
Yates, Lawrence, 'A Feather in their Cap? The Marines' Combined Action Program in Vietnam' in William R. Roberts and Jack Sweetman (eds), *New Interpretations in Naval History*, Annapolis, Maryland: Naval Institute Press, 1991.
Young, David, *Four Five: The Story of 45 Commando, Royal Marines, 1943–1971*, London: Cooper, 1972.
Yousaf, Brigadier Mohammad and Mark Adkin, *The Battle for Afghanistan: The Soviets versus the Mujahideen during the 1980s*, Barnsley, South Yorkshire: Pen & Sword, 2007.

Zadka, Saul, *Blood in Zion: How the Jewish Guerrillas drove the British out of Palestine*, London and Washington: Brassey's, 1995.
Zaffiri, Samuel, *Westmoreland: A Biography of General William C. Westmoreland*, New York: William Morrow, 1994.
Zaidi, Syed Manzar Abbas, 'Pakistan's Anti-Taliban Counterinsurgency', *The RUSI Journal*, Vol.155, No.1 (February/March 2010), pp.10–19.
Zariczniak, Larysa, 'Major Stepan Stebelski of the Ukrainian Insurgent Army: Examples of Insurgency leadership and Tactics', *Small Wars and Insurgencies*, Vol.22, No.3 (July 2011), pp.435–47.
Zervoudakis, Alexander, 'Nihil mirare, nihil contemptare, omnia intelligere: Franco-Vietnamese Intelligence in Indochina, 1950–1954', *Intelligence and National Security*, Vol.13, No.1 (Spring 1998), pp.195–231.
Zervoudakis, Alexander, 'A Case of Successful Pacification: the 584th Bataillon du Train at Bordj de l'Agha (1956–57)', *Journal of Strategic Studies*, Vol.25, No.2 (June 2002), pp.54–64.
Zhukov, Yuri, 'Examining the Authoritarian Model of Counterinsurgency: The Soviet Campaign Against the Ukrainian Insurgent Army', *Small Wars and Insurgencies*, Vol.18, No.3 (September 2007), pp.439–66.
Zimmerer, Jürgen and Joachim Zeller (eds), *Genocide in German South-West Africa: The Colonial War (1904–1908) in Namibia and its Aftermath*, Pontypool, Wales: Merlin Press, 2010.
Zimmerman, Andrew, '"What Do You Really Want in German East Africa, *Herr Professor?*" Counterinsurgency and the Science Effect in Colonial Tanzania', *Comparative Studies in Society and History*, Vol.48, No.2 (April 2006), pp.419–61.
Zirkel, Kirsten, 'Military Power in German Colonial Policy: The *Schutztruppen* and their Leaders in East and South-West Africa, 1888–1918' in David Killingray and David Omissi (eds), *Guardians of Empire: The Armed Forces of the Colonial Powers, c.1700–1964*, Manchester and New York: Manchester University Press, 1999.

INDEX

You may also be interested in …

978 0 7509 5584 3

In 2007, journalist Nick Allen quit a secure job in Pakistan as a news agency writer to experience the life of foreign troops fighting the Taliban in Afghanistan. Over several years he journeyed as an embedded reporter with a dozen armies, working his way through placid backwaters to remote, savage hotspots where daily clashes with insurgent forces were the norm.